de Gruyter Expositions in Mathematics 30

de Gruyter Expositions in Mathematics

1 The Analytical and Topological Theory of Semigroups, *K. H. Hofmann, J. D. Lawson, J. S. Pym* (Eds.)

2 Combinatorial Homotopy and 4-Dimensional Complexes, *H. J. Baues*

3 The Stefan Problem, *A. M. Meirmanov*

4 Finite Soluble Groups, *K. Doerk, T. O. Hawkes*

5 The Riemann Zeta-Function, *A. A. Karatsuba, S. M. Voronin*

6 Contact Geometry and Linear Differential Equations, *V. E. Nazaikinskii, V. E. Shatalov, B. Yu. Sternin*

7 Infinite Dimensional Lie Superalgebras, *Yu. A. Bahturin, A. A. Mikhalev, V. M. Petrogradsky, M. V. Zaicev*

8 Nilpotent Groups and their Automorphisms, *E. I. Khukhro*

9 Invariant Distances and Metrics in Complex Analysis, *M. Jarnicki, P. Pflug*

10 The Link Invariants of the Chern-Simons Field Theory, *E. Guadagnini*

11 Global Affine Differential Geometry of Hypersurfaces, *A.-M. Li, U. Simon, G. Zhao*

12 Moduli Spaces of Abelian Surfaces: Compactification, Degenerations, and Theta Functions, *K. Hulek, C. Kahn, S. H. Weintraub*

13 Elliptic Problems in Domains with Piecewise Smooth Boundaries, *S. A. Nazarov, B. A. Plamenevsky*

14 Subgroup Lattices of Groups, *R. Schmidt*

15 Orthogonal Decompositions and Integral Lattices, *A. I. Kostrikin, P. H. Tiep*

16 The Adjunction Theory of Complex Projective Varieties, *M. C. Beltrametti, A. J. Sommese*

17 The Restricted 3-Body Problem: Plane Periodic Orbits, *A. D. Bruno*

18 Unitary Representation Theory of Exponential Lie Groups, *H. Leptin, J. Ludwig*

19 Blow-up in Quasilinear Parabolic Equations, *A. A. Samarskii, V. A. Galaktionov, S. P. Kurdyumov, A. P. Mikhailov*

20 Semigroups in Algebra, Geometry and Analysis, *K. H. Hofmann, J. D. Lawson, E. B. Vinberg* (Eds.)

21 Compact Projective Planes, *H. Salzmann, D. Betten, T. Grundhöfer, H. Hähl, R. Löwen, M. Stroppel*

22 An Introduction to Lorentz Surfaces, *T. Weinstein*

23 Lectures in Real Geometry, *F. Broglia* (Ed.)

24 Evolution Equations and Lagrangian Coordinates, *A. M. Meirmanov, V. V. Pukhnachov, S. I. Shmarev*

25 Character Theory of Finite Groups, *B. Huppert*

26 Positivity in Lie Theory: Open Problems, *J. Hilgert, J. D. Lawson, K.-H. Neeb, E. B. Vinberg* (Eds.)

27 Algebra in the Stone-Čech Compactification, *N. Hindman, D. Strauss*

28 Holomorphy and Convexity in Lie Theory, *K.-H. Neeb*

29 Monoids, Acts and Categories, *M. Kilp, U. Knauer, A. V. Mikhalev*

Relative Homological Algebra

by

Edgar E. Enochs
Overtoun M. G. Jenda

Walter de Gruyter · Berlin · New York 2000

Authors

Edgar E. Enochs
Department of Mathematics
University of Kentucky
Lexington
KY 40506−0027
USA
enochs@ms.uky.edu

Overtoun M. G. Jenda
Department of Mathematics
Auburn University
Auburn
AL 36849−5307
USA
jendaov@mail.auburn.edu

1991 Mathematics Subject Classification:
13-02, 16-02; 13C14, 13C15, 13H10, 16D40, 16D50, 16E10, 16P40, 18G10, 18G15, 18G25

Keywords:
Module, injecture, flat, cover, envelope

♾ Printed on acid-free paper which falls within the guidelines of the ANSI to ensure permanence and durability.

Library of Congress − Cataloging-in-Publication Data

Enochs, Edgar E.
  Relative homological algebra / Edgar E. Enochs and Overtoun
M. G. Jenda.
    p. cm. − (De Gruyter expositions in mathematics,
ISSN 0938-6572)
Includes bibliogaphical references and index.
  ISBN 3 11 016633 X (alk. paper)
    1. Algebra, Homological.   I. Jenda, Overtoun M. G.   II. Title.
III. Series.

QA169.E6 2000
512′.55−dc21   99-086236

Die Deutsche Bibliothek − Cataloging-in-Publication Data

Enochs, Edgar E.:
Relative homological algebra / by Edgar E. Enochs ; Overtoun M. G.
Jenda. − Berlin ; New York : de Gruyter, 2000
  (De Gruyter expositions in mathematics ; 30)
  ISBN 3-11-016633-X

Typesetting using the author's TeX files: I. Zimmermann, Freiburg.
Printing: WB-Druck GmbH & Co., Rieden/Allgäu. Binding: Lüderitz & Bauer-GmbH, Berlin.
Cover design: Thomas Bonnie, Hamburg.

For our wives
Louise and Claudine

and children
Corinne, Mary Jane, Kathryn, Maureen, Madelaine, Anne, and John
and
Emily and Overtoun, Jr

Preface

The subject of relative homological algebra was introduced by S. Eilenberg and J. C. Moore in their 1965 AMS Memoir 'Foundations of Relative Homological Algebra'. We now have in hand more theorems guaranteeing the existence of precovers, covers, preenvelopes and envelopes. These are basic objects of the subject and are used to construct resolutions and then left and right derived functors. Also, several new useful ideas have come into play since the appearance of Eilenberg and Moore's work. Among others these include the various versions of what is now known as Wakamatsu lemma, the notions of special precovers and preenvelopes and the orthogonality of classes of objects of an abelian category with respect to the extension functor. Hence it seems opportune to now give a systematic treatment of this subject along with the new developments and applications.

This book is aimed at graduate students. For that reason, we have attempted to make the book a reasonably self-contained treatment of the subject requiring only familiarity with basic notions in module and ring theory at the level of Basic Algebra I by N. Jacobson [117].

The first three chapters give the basic tools and notation that will be used throughout the book. This material constitutes notes from our lectures at our respective universities and is suitable for an introductory course in module and ring theory.

The material in chapter four which deals with torsion free covers over integral domains is not essential to what follows in the book, but the ideas and proofs in this chapter give the flavor of what is to come. Chapter five gives information about precovers and covers and chapter six deals with preenvelopes and envelopes. Chapter seven introduces the notion of cotorsion theory which is used to prove the existence of special covers and envelopes. Chapter eight introduces balance (on the left or the right) of a functor of two variables. Balance means that we have two specific kinds of resolutions of each of the two variables each of which can be used to compute the relative derived functors. We show that the basic functors Hom and Tensor are balanced using resolutions different from the usual projective, injective and flat resolutions. This allows us to compute useful versions of the Extension and Torsion functors with negative indices. We consider chapters five, six, seven and eight as the heart of the book. This material together with chapters four and nine is suitable for a course in relative homological algebra and its applications to commutative and noncommutative algebra.

The remainder of the book gives applications to ring theory and is more specialized. The commutative rings that we consider include local Cohen–Macaulay rings admitting a dualizing module with the Gorenstein local rings as a special case. For example, we prove Auslander's announced (but unpublished) result concerning the existence of maximal Cohen–Macaulay approximations over Gorenstein local rings. We also consider a noncommutative version of Gorenstein rings which we call Iwanaga–Gorenstein rings. Over these rings there is an especially pleasant application of relative homological algebra. We define relative versions of projective, injective and flat modules which we label Gorenstein. We prove that over an Iwanaga–Gorenstein ring there are enough Gorenstein projectives, injectives and flats (that is, the appropriate precovers and preenvelopes exist). We then show that Hom and Tensor are balanced when we use these Gorenstein versions of the projective, injective and flat modules to compute the resolutions over Iwanaga–Gorenstein rings. Then we prove that these rings have finite global dimension in this situation.

This book was planned while we were visiting the Department of Mathematics and Computer Science at University of Antwerp, UIA, Belgium. We would like to thank the department for its hospitality.

We would like to thank the many colleagues who have discussed and studied these topics with us through the years. These include (but are not limited to) Tom Cheatham, Richard Belshoff, Jinzhong Xu, Mark Teply, and Freddy Van Oystaeyen. We also appreciate the interest our colleagues in Spain have shown in this subject. These include Juan Martinez, Alberto del Valle, Manolo Saorin and Pepe Asensio in Murcia and Blas Torrecillas, Juan Ramon Garcia Rozas, Luis Oyonarte, Juan Antonio Lopez Ramos, and Maria Jesus Asensio in Almeria.

Finally, we would like to thank Mrs. Rosie Torbert for the excellent job she has done typing the manuscript.

Edgar E. Enochs
Overtoun M. G. Jenda

Contents

Preface   vii

1   Basic Concepts   1
    1.1   Zorn's lemma, ordinal and cardinal numbers 1
    1.2   Modules 7
    1.3   Categories and functors 17
    1.4   Complexes of modules and homology 25
    1.5   Direct and inverse limits 31
    1.6 I-adic topology and completions 36

2   Flat Modules, Chain Conditions and Prime Ideals   40
    2.1   Flat modules 40
    2.2   Localization 44
    2.3   Chain conditions 46
    2.4   Prime ideals and primary decomposition 51
    2.5   Artin–Rees lemma and Zariski rings 61

3   Injective and Flat Modules   68
    3.1   Injective modules 68
    3.2   Natural identities, flat modules, and injective modules 75
    3.3   Injective modules over commutative noetherian rings 84
    3.4   Matlis duality 88

4   Torsion Free Covering Modules   93
    4.1   Existence of torsion free precovers 93
    4.2   Existence of torsion free covers 95
    4.3   Examples 97
    4.4   Direct sums and products 101

5   Covers   105
    5.1 $\mathcal{F}$-precovers and covers 105
    5.2   Existence of precovers and covers 107
    5.3   Projective and flat covers 110
    5.4   Injective covers 120
    5.5   Direct sums and T-nilpotency 125

6   Envelopes   129
    6.1 $\mathcal{F}$-preenvelopes and envelopes 129
    6.2   Existence of preenvelopes 130
    6.3   Existence of envelopes 132
    6.4   Direct sums of envelopes 134
    6.5   Flat envelopes 136
    6.6   Existence of envelopes for injective structures 139
    6.7   Pure injective envelopes 144

7   Covers, Envelopes, and Cotorsion Theories   152
    7.1   Definitions and basic results 152
    7.2   Fibrations, cofibrations and Wakamatsu lemmas 154
    7.3   Set theoretic homological algebra 160
    7.4   Cotorsion theories with enough injectives and projectives 162

8   Relative Homological Algebra and Balance   167
    8.1   Left and right $\mathcal{F}$-resolutions 167
    8.2   Derived functors and balance 169
    8.3   Applications to modules 177
    8.4 $\mathcal{F}$-dimensions 180
    8.5   Minimal pure injective resolutions of flat modules 194
    8.6 λ and μ-dimensions 203

9   Iwanaga–Gorenstein and Cohen–Macaulay Rings and Their Modules   211
    9.1   Iwanaga–Gorenstein rings 211
    9.2   The minimal injective resolution of R 215
    9.3   More on flat and injective modules 223
    9.4   Torsion products of injective modules 226
    9.5   Local cohomology and the dualizing module 229

10 Gorenstein Modules   239
    10.1 Gorenstein injective modules 239
    10.2 Gorenstein projective modules 246
    10.3 Gorenstein flat modules 253
    10.4 Foxby classes 258

11 Gorenstein Covers and Envelopes   269
    11.1 Gorenstein injective precovers and covers 269
    11.2 Gorenstein injective preenvelopes 270
    11.3 Gorenstein injective envelopes 274
    11.4 Gorenstein essential extensions 277
    11.5 Gorenstein projective precovers and covers 279
    11.6 Auslander's last theorem (Gorenstein projective covers) 284
    11.7 Gorenstein flat covers 288

11.8 Gorenstein flat and projective preenvelopes 292

12 Balance over Gorenstein and Cohen–Macaulay Rings   294
   12.1 Balance of $\mathrm{Hom}(-,-)$ 294
   12.2 Balance of $-\otimes-$ 298
   12.3 Dimensions over n-Gorenstein rings 300
   12.4 Dimensions over Cohen–Macaulay rings 305
   12.5 Ω-Gorenstein modules 307

Bibliographical Notes   319

Bibliography   321

Index   331

Chapter 1

Basic Concepts

In this chapter we introduce basic terminology, notation, and results concerning set theory, modules, categories, and complexes.

1.1 Zorn's lemma, ordinal and cardinal numbers

We start by introducing an informal and naive set theory. To avoid the usual contradictions of set theory we will use the term *class* for a collection which may be too large to be called a set. Some definitions which we give concerning sets can obviously be applied to classes.

Definition 1.1.1. A *partially ordered set* is a set X with a relation $\leq$ such that (1) $x \leq x$, (2) $x \leq y$ and $y \leq z$ implies $x \leq z$, and (3) $x \leq y$ and $y \leq x$ implies $x = y$, for all $x, y, z \in X$. A partially ordered set X is said to be *totally ordered* if for all $x, y \in X$ either $x \leq y$ or $y \leq x$. If $S \subset X$ for the partially ordered set X, then S has an *induced order* with $x \leq y$ in S exactly when this relation holds in X.

Definition 1.1.2. If X is a partially ordered set, an element $x \in X$ is said to be an *upper bound* of a subset $S \subset X$ if $y \leq x$ for all $y \in S$. An $x \in X$ is said to be a *maximal element* of X if $x \leq y$ for $y \in X$ implies $x = y$. The partially ordered set X is said to be *inductively ordered* if every subset $S \subset X$ which is totally ordered with the induced order has an upper bound in X.

Theorem 1.1.3 (Zorn's lemma). *Every inductively ordered set has a maximal element.*

Zorn's lemma is implied by (and in fact equivalent to) the *axiom of choice* which states that the cartesian product of a nonempty family of nonempty sets is nonempty. There are other versions of the result which can be found in standard books on set theory and logic.

Definition 1.1.4. An element x of a partially ordered set X is said to be a *least element* of X if $x \leq y$ for all $y \in X$. A totally ordered set X is said to be *well ordered* if every nonempty subset S has a least element (when S is given the induced order).

We note that a totally ordered set X is well ordered if and only if whenever we have $x_0 \geq x_1 \geq x_2 \geq x_3 \geq \cdots$ for elements x_n of X there is an n_0 such that $x_n = x_{n_0}$ for $n \geq n_0$.

Definition 1.1.5. Two well ordered sets X and Y are said to be *isomorphic* if there is a bijection $f : X \to Y$ which preserves order, that is, $x_1 \leq x_2$ in X implies $f(x_1) \leq f(x_2)$ in Y. Such an $f : X \to Y$ is called an *isomorphism*. We note that if f is an isomorphism, then so is f^{-1}.

Definition 1.1.6. If X is a well ordered set, then a subset $S \subset X$ is said to be a *segment* of X if $y \leq x$ and $x \in S$ implies $y \in S$.

The union and intersection of any collection of segments of the well ordered set X are segments of X. If S and T are segments of X, then either $S \subset T$ or $T \subset S$.

X is a segment of X. If $S \neq X$ is a segment of X and $x \in X$ is the least element of X not in S, then $S = \{y : y \in X, y < x\}$. Conversely, any such S (for any $x \in X$) is a segment of X. Thus the set of proper segments of X ordered by inclusion is a well ordered set which is isomorphic to X. So we see that if $S_0 \supset S_1 \supset S_2 \supset \cdots$ are segments of X, then for some $n_0 \geq 0$, $S_n = S_{n_0}$ for all $n \geq n_0$.

Proposition 1.1.7. *Every set can be well ordered.*

Proof. Let X be a set. We consider the set $\mathcal{X}$ of well ordered sets S such that as a set $S \subset X$. We order these sets so that $S \leq T$ if and only if S is a segment of T and if the order on S is induced by that on T. $S = \emptyset$ is an example of such an S. If $\mathcal{C} \subset \mathcal{X}$ is a nonempty totally ordered subset of $\mathcal{X}$, there is a unique way to order $Y = \bigcup_{S \in \mathcal{C}} S$ so that the order induced on each $S \in \mathcal{C}$ is the original order on S. Then in fact Y is well ordered (and each $S \in \mathcal{C}$ is a segment of Y), that is, $\mathcal{C}$ has an upper bound in $\mathcal{X}$. So by Zorn's lemma, there is a maximal element S of $\mathcal{X}$. If $S \neq X$ and $y \in X$, $y \notin S$, we order $T = S \cup \{y\}$ with $x < y$ for all $x \in S$ and with the induced order on S the original order on S. Then T is well ordered and $S < T$, contradicting the choice of S. Hence $S = X$ and so X can be well ordered. $\square$

Theorem 1.1.8. *Suppose S is a segment of the well ordered set X and that $f : X \to S$ is an isomorphism. Then $S = X$ and $f = \mathrm{id}_X$.*

Proof. We only need to argue that $f(x) = x$ for all $x \in X$. If this is not the case, let x be the least element of X such that $f(x) \neq x$. Then $f(x) > x$, for if $f(x) < x$ we have $f(f(x)) = f(x)$ by the choice of x. But this means f is not an injection. So $f(x) > x$ and hence $x \in S$. Now let $x = f(y)$. Then $f(x) > f(y)$ and so $x > y$. So again by the choice of x, $f(y) = y$. But then $x = y$ contradicting the fact that $f(x) \neq x$. $\square$

Corollary 1.1.9. *If X and Y are well ordered sets, then there is at most one isomorphism $f : X \to Y$.*

Proof. If $f, g : X \to Y$ are isomorphisms, then by the preceding result, $g^{-1} \circ f = \mathrm{id}_X$. Hence $f = g$. $\square$

Theorem 1.1.10. *If X and Y are well ordered sets, then exactly one of the following holds:*

(a) *X is isomorphic to Y,*

(b) *X is isomorphic to a proper segment of Y,*

(c) *Y is isomorphic to a proper segment of X.*

Proof. We first argue that one of (a), (b) or (c) holds. We consider the set of pairs (S, T) where S and T are segments of X and Y respectively and where S is isomorphic to T. Given two such pairs $(S, T), (S', T')$, we write $(S, T) \le (S', T')$ if $S \subset S'$ and $T \subset T'$. Note that the isomorphisms (necessarily unique) $f : S \to T$ and $f' : S' \to T'$ agree on S. For T and $f'(S)$ are isomorphic segments of Y with $y \mapsto f'(f^{-1}(y))$ the isomorphism. By Theorem 1.1.8, $f'(S) = T$ and $f'(f^{-1}(y)) = y$ for $y \in T$. So with $y = f(x)$ we get $f'(x) = f(x)$ for $x \in S$.

By taking unions, we see that any chain of such pairs (S, T) has an upper bound (note that $(\emptyset, \emptyset)$ is such a pair and so by Zorn's lemma there is a maximal pair (S, T)). If $S \ne X$ and $T \ne Y$, let x_0 be the least element of X not in S and let y_0 be the least element of Y not in T. Then $S' = S \cup \{x_0\}$, $T' = T \cup \{y_0\}$ are isomorphic segments of X and Y respectively. Since $(S, T) < (S', T')$, this contradicts our choice of (S, T).

It is not hard then to see by Theorem 1.1.8 that no two of (a), (b) and (c) can simultaneously be true. $\square$

If X and Y are well ordered sets, we write $\mathrm{Ord}\, X = \mathrm{Ord}\, Y$ if X and Y are isomorphic. We write $\mathrm{Ord}\, X < \mathrm{Ord}\, Y$ if X is isomorphic to a proper segment of Y and $\mathrm{Ord}\, X > \mathrm{Ord}\, Y$ if Y is isomorphic to a proper segment of X. Then by Theorem 1.1.10 we see (without actually specifying what $\mathrm{Ord}\, X$ is) that $\{\mathrm{Ord}\, X : X \text{ is a set}\}$ is a totally ordered class. Any two finite well ordered sets X, Y with n elements are isomorphic. So we write $\mathrm{Ord}\, X = n$ for such an X. If $\mathbb{N}$ is the set of natural numbers with the usual order, we write $\mathrm{Ord}(\mathbb{N}) = \omega$. If $\alpha = \mathrm{Ord}\, X$ and $\beta = \mathrm{Ord}\, Y$ with X and Y well ordered sets with $X \cap Y = \emptyset$, we let $\alpha + \beta = \mathrm{Ord}(X \cup Y)$ where $X \cup Y$ is ordered so that $x < y$ for all $x \in X$, $y \in Y$ and such that the induced orders on X and Y are the original orders. We note that $X \cup Y$ is well ordered with this order.

An *ordinal number* $\alpha > 0$ with $\alpha = \mathrm{Ord}\, X$ is said to be a *limit ordinal* if X has no largest element. By the definition of addition of ordinals above, we see that $\alpha > 0$ is a limit ordinal if and only if $\alpha \ne \beta + 1$ for any ordinal β.

Theorem 1.1.11. *The class $\{\mathrm{Ord}\, X : X \text{ is a set}\}$ is well ordered.*

Proof. The statement means that any nonempty set of ordinals has a least element. So it suffices to argue that if $(X_i)_{i \in I}$ is a nonempty family of well ordered sets, then for some $j \in I$, X_j is isomorphic to a segment of each $X_i, i \in I$. Suppose this is not the case. Let $i_0 \in I$. Then by Theorem 1.1.10 we see that there is an $i_1 \in I$ so that X_{i_1} is isomorphic to a proper segment S_1 of X_{i_0}. Then repeating the argument with i_1 replacing i_0 we see that there is an $i_2 \in I$ with X_{i_2} isomorphic to a proper segment of X_{i_1}. But this implies X_{i_2} is isomorphic with a segment S_2 of X_{i_0} with $S_2 \subsetneqq S_1$. Repeating the argument we see that we can find segments $S_1 \supsetneqq S_2 \supsetneqq S_3 \supsetneqq S_4 \supsetneqq \cdots$ of X_{i_0}. But this is impossible. $\square$

Definition 1.1.12. If X and Y are sets, we say that X and Y have the same *cardinality* if there is a bijection $f : X \to Y$, and we write Card $X =$ Card Y. We say Card $X \le$ Card Y if there is an injection $f : X \to Y$.

Theorem 1.1.13. *If X and Y are sets, then either* Card $X \le$ Card Y *or* Card $Y \le$ Card X.

Proof. We consider subsets $S \subset X \times Y$ with the property that if $(x, y), (x', y') \in S$ and $(x, y) \ne (x', y')$ then $x \ne x'$, $y \ne y'$. Among such subsets one can clearly pick a maximal one, say T, by Zorn's lemma. We claim that T is the graph of a function from X to Y or $T' = \{(y, x) : (x, y) \in T\}$ is the graph of a function from Y to X. For if neither holds, then for some $x_0 \in X$, $(x_0, y) \notin T$ for any $y \in Y$ and for some $y_0 \in Y$, $(x, y_0) \notin T$ for any $x \in X$. Then we note that $T \cup \{(x_0, y_0)\} \supsetneqq T$ and has the property above, contradicting the maximality of T. Hence either T is a graph of a function from X to Y so that there is an injection $X \to Y$ or T' is a graph of a function from Y to X and so there is an injection $Y \to X$. $\square$

Theorem 1.1.14 (Cantor, Schröder, Bernstein). *Let X and Y be sets. If* Card $X \le$ Card Y *and* Card $Y \le$ Card X, *then* Card $X =$ Card Y.

Proof. Let $f : X \to Y$ and $g : Y \to X$ be injections and let $h = gf$, $R = X - g(Y)$, and $A = R \cup h(R) \cup h^2(R) \cup \cdots$ so that $h(A) \subset A \subset X$. Hence if we let $A' = f(A)$, then $g(A') = h(A) \subset A$. Now let $B = X - A$, $B' = Y - A'$. Then $A \cap B = \emptyset$, $A' \cap B' = \emptyset$, $A \cup B = X$, $A' \cup B' = Y$, and Card $A =$ Card A'. So if Card $B =$ Card B', then Card $X =$ Card$(A \cup B) =$ Card $A +$ Card $B =$ Card $A' +$ Card $B' =$ Card$(A' \cup B') =$ Card Y and so we are done. Thus it suffices to show that Card $B =$ Card B'. But then we only need to argue that $g(B') = B$. So let $x \in B$. Then $x \notin A$ and so $x \notin R$ since $R \subset A$. Hence $x \in g(Y)$ and so $x = g(y)$ for some $y \in Y$. If $y \in A'$, then $g(y) = x \in A$, a contradiction. So $y \in B'$ and thus $x \in g(B')$. Hence $B \subset g(B')$. To show $g(B') \subset B$ we let $y \in B'$ and argue $g(y) \in B$. Suppose $g(y) \in A$. Then $g(y) \notin R$ since $R = X - g(Y)$ and so $g(y) \in h^n(R)$ for some $n \ge 1$. Let $g(y) = h^n(z)$ for some $z \in R \subset A$. Then $g(y) = h(h^{n-1}(z)) = g(f(h^{n-1}(z)))$. So $y = f(h^{n-1}(z))$. But $h^{n-1}(z) \in A$. Hence $y \in A'$, a contradiction since $y \in B'$. So $g(y) \in B$ and we are done. $\square$

Theorem 1.1.15. *The class* $\{$Card $X : X$ *a set*$\}$ *is well ordered.*

Proof. We need to argue that if $(X_i)_{i \in I}$ is any nonempty family of sets then there is an $i_0 \in I$ such that for each $i \in I$ there is an injection $X_{i_0} \to X_i$. We consider subsets $S \subset \prod_{i \in I} X_i$ with the property that if $(x_i)_{i \in I}$, $(y_i)_{i \in I} \in S$ and if $(x_i)_{i \in I} \neq (y_i)_{i \in I}$ then $x_i \neq y_i$ for all $i \in I$. Partially ordering these S by inclusion we see that an application of Zorn's lemma gives a maximal such S. If $\pi_i(S) \neq X_i$ for each $i \in I$, then choosing $y_i \in X_i$, $y_i \notin \pi_i(S)$ for each i, we see that the set $S \cup \{(y_i)_{i \in I}\}$ contradicts the maximality of S. Hence for some $i_0 \in I$, $\pi_{i_0}(S) = X_{i_0}$. Then given $x \in X_{i_0}$ there is a unique $(x_i)_{i \in I} \in S$ with $x_{i_0} = x$. Hence for $i \in I$ we can define a function $X_{i_0} \to X_i$ which, with this notation, maps x to x_i. By the property imposed on S, this function is an injection. $\qquad \square$

As usual, we use the symbols $0, 1, 2, 3, \ldots, n, \ldots$ to denote the finite cardinals. The infinite cardinals are written $\aleph_\alpha$ where α is an ordinal number. So $\aleph_0$ is the smallest infinite cardinal. Hence $\aleph_0 = \operatorname{Card} \mathbb{N}$. Then for any $\aleph_\alpha$, $\aleph_{\alpha+1}$ is the least cardinal number larger than $\aleph_\alpha$. If β is a limit ordinal, $\aleph_\beta$ is the least cardinal number greater that $\aleph_\alpha$ for all ordinals $\alpha < \beta$.

Definition 1.1.16. Given cardinal numbers m_1 and m_2 with $m_1 = \operatorname{Card} X_1$, $m_2 = \operatorname{Card} X_2$, we define $m_1 m_2$ to be $\operatorname{Card}(X_1 \times X_2)$ and define $m_1 + m_2$ to be $\operatorname{Card}(X_1 \cup X_2)$ if $X_1 \cap X_2 = \emptyset$.

We see that $n + \aleph_0 = \aleph_0$ for any finite $n \geq 0$. Also, the usual arguments show that $\aleph_0 + \aleph_0 = \aleph_0$, $n \cdot \aleph_0 = \aleph_0$ if $n \geq 1$ and that $\aleph_0 \cdot \aleph_0 = \aleph_0$ (that is, $\aleph_0^2 = \aleph_0$).

Proposition 1.1.17. *For any infinite cardinal $\aleph_\alpha$, $\aleph_\alpha^2 = \aleph_\alpha$.*

Proof. Clearly $\aleph_\alpha \leq \aleph_\alpha^2$ since there is an injection $X \to X \times X$ for any set X. So if $\aleph_\alpha^2 = \aleph_\alpha$ fails then $\aleph_\alpha^2 > \aleph_\alpha$. So assume $\aleph_\alpha^2 > \aleph_\alpha$ for some α. We can then assume $\aleph_\alpha$ is the least infinite cardinal with $\aleph_\alpha^2 > \aleph_\alpha$. Let $\aleph_\alpha = \operatorname{Card} X$ for some set X. By Proposition 1.1.7, X can be well ordered. We consider the set of segments $S \subset X$ such that $\operatorname{Card} S = \aleph_\alpha$ (for example, $S = X$). Since the segments are well ordered by inclusion, there is a least such S. So we can suppose that $S = X$. This then means that for every proper segment T of X, $\operatorname{Card} T < \operatorname{Card} X$. Hence by our assumption on $\aleph_\alpha = \operatorname{Card} X$, $\operatorname{Card} T^2 = \operatorname{Card} T$ for every infinite proper segment T.

We now order $X \times X$ so that $(x_1, y_1) \leq (x_2, y_2)$ if $\sup\{x_1, y_1\} < \sup\{x_2, y_2\}$, so that $(x_1, y_1) \leq (x_2, y_2)$ if $\sup\{x_1, y_1\} = \sup\{x_2, y_2\}$ and if $x_1 < x_2$, and so that $(x_1, y_1) \leq (x_2, y_2)$ if $\sup\{x_1, y_1\} = \sup\{x_2, y_2\}$ and $x_1 = x_2$ and $y_1 \leq y_2$. Then it is easy to see that $Y = X \times X$ is well ordered. Now we apply Theorem 1.1.10. If Y is isomorphic to a segment of X (possibly X itself) then $\aleph_\alpha^2 = \operatorname{Card} Y \leq \operatorname{Card} X = \aleph_\alpha$, contradicting our choice of $\aleph_\alpha$. So suppose X is isomorphic to a proper segment U of $Y = X \times X$. Then there is a $z \in X$ so that $(x, y) < (z, z)$ for all $(x, y) \in U$. Let T be the segment of X determined by z, that is $T = \{x : x \in X, x < z\}$. Then noting that T is infinite and that $\operatorname{Card} T < \operatorname{Card} X$ we have $\operatorname{Card} T^2 = \operatorname{Card} T$. But $U \subset T \times T$. So $\operatorname{Card} U \leq \operatorname{Card} T$. But then $\operatorname{Card} X = \operatorname{Card} U \leq \operatorname{Card} T < \operatorname{Card} X$. This gives a contradiction and so proves the proposition. $\qquad \square$

Proposition 1.1.18 (The principle of transfinite induction). *If $\beta \geq 0$ is an ordinal, let $X = \{\alpha : \alpha$ is an ordinal number, $\alpha < \beta\}$. Let $S \subset X$. If (1) $0 \in S$, (2) $\alpha + 1 < \beta$ and $\alpha \in S$ implies $\alpha + 1 \in S$, and (3) $\gamma < \beta$ is a limit ordinal and $\alpha \in S$ for all $\alpha < \gamma$ implies $\gamma \in S$, then $S = X$.*

Proof. If $S \neq X$, let γ be the least ordinal $\gamma < \beta$ which is not in S. By (1), $\gamma > 0$. If γ is not a limit ordinal then $\gamma = \alpha + 1$. But then $\alpha \in S$ and so $\alpha + 1 = \gamma \in S$ by (2). If γ is a limit ordinal we get $\gamma \in S$ by (3) and so we get a contradiction in both cases. Hence $S = X$. $\square$

When $\beta = \omega$, we have $X = \mathbb{N}$ and we get the usual induction ((3) does not apply).

If we are given a statement P_α for each $\alpha < \beta$ and we let S be those α for which P_α is true. Then to argue all P_α are true we only need check (1), (2) and (3) for S. For example, if we can argue P_α implies $P_{\alpha+1}$ for $\alpha + 1 < \beta$, then we get (2) for S.

There is an analogous *principle of transfinite construction*. This principle says, for example, that in order to construct a set M_α for all $\alpha < \beta$, it suffices to give M_0, to show how to get $M_{\alpha+1}$ from M_α when $\alpha + 1 < \beta$, and how to get M_γ from all the $M_\alpha, \alpha < \gamma$ when $\gamma < \beta$ is a limit ordinal.

Exercises

1. Argue that if $f : Y \to Z$ is a surjective function, then Card $Z \leq$ Card Y.

2. If X is an infinite set, argue that X admits a partition $\mathcal{P}$ into countable subsets $S \subset X$. (That is, (1) if $S \in \mathcal{P}$ then $S \subset X$ and Card $S = \aleph_0$, (2) if $S, T \in \mathcal{P}$ then $S = T$ or $S \cap T = \emptyset$, (3) $X = \bigcup_{S \in \mathcal{P}} S$).
   Hint: Use Zorn's lemma on sets $\mathcal{P}$ satisfying (1) and (2). If $\mathcal{P}$ is a maximal such set, argue that $X - \bigcup_{S \in \mathcal{P}}$ is finite.

3. If $\aleph_\alpha$ and $\aleph_\beta$ are infinite cardinals, argue that $\aleph_\alpha + \aleph_\beta = \aleph_\alpha \cdot \aleph_\beta = \aleph_\gamma$ where $\gamma = \sup(\alpha, \beta)$.

4. a) Let X be an infinite set and let $X^1 = X$, $X^2 = X \times X$ and in general $X^{n+1} = X^n \times X$ for $n \geq 1$. Let $Y = \bigcup_{n=1}^\infty X^n$. Argue that Card $Y \leq \aleph_0 \cdot$ Card $X =$ Card X and so deduce that Card $Y =$ Card X.

   b) Use (a) and Problem 1 to argue that if $\mathcal{F}(X)$ is the set of finite subsets of X, then Card $\mathcal{F}(X) =$ Card X.

5. a) Let X be any set and let $\mathcal{P}(X)$ be the set of all subsets of X. Argue that there is no surjection $\sigma : X \to \mathcal{P}(X)$ by arguing that for any function $\sigma : X \to \mathcal{P}(X)$ the set $Y = \{x : x \in X, x \notin \sigma(x)\}$ is not $\sigma(y)$ for any $y \in X$.

   b) Deduce that Card $X <$ Card $\mathcal{P}(X)$ for any set X.

6. If $(m_i)_{i \in I}$ is any family of cardinal numbers, define $\sum_{i \in I} m_i$ and $\prod_{i \in I} m_i$ to be the cardinality of $\bigcup_{i \in I} X_i$ and $\prod_{i \in I} X_i$ where the X_i are sets such that Card $X_i = m_i$ for each $i \in I$ and where $X_i \cap X_j = \emptyset$ if $i \neq j$. Now let $I = \mathbb{N}$ and suppose $0 < m_0 < m_1 < m_2 < \cdots$. Argue that $\sum_{i=0}^\infty m_i < \prod_{i=0}^\infty m_i$.

Hint: With the notation above, argue that there is no surjection $\bigcup_{i=0}^{\infty} X_i \to \prod_{i=0}^{\infty} X_i$ as follows: Let $f : \bigcup_{i=0}^{\infty} X_i \to \prod_{i=0}^{\infty} X_i$ be a function. Let $x_0 \in X_0$ be any element. Since $m_n < m_{n+1}$, the function $x \mapsto \pi_{n+1}(f(x))$ from X_n to X_{n+1} is not surjective and so let $x_{n+1} \in X_{n+1}$, $x_{n+1} \in \pi_{n+1}(f(X_n))$ for $n \geq 0$. Then argue that $(x_n)_{n \geq 0}$ is not in the image of f.

7. a) Noting that $\sum_{n=0}^{\infty} \aleph_n \leq \aleph_0 \cdot \aleph_\omega$, argue that $\sum_{n=0}^{\infty} \aleph_n = \aleph_\omega$.

   b) Argue that $\prod_{n=0}^{\infty} \aleph_n \leq \aleph_\omega^{\aleph_0}$

   c) Use (a), (b) and problem 6 to deduce that $\aleph_\omega < \aleph_\omega^{\aleph_0}$.
   Note. If m, n are any cardinal numbers and if Card $X = m$, Card $Y = n$, then m^n is defined to be Card X^Y.

8. a) Let α and β be ordinals and let X and Y be well ordered sets such that Ord $X = \alpha$ and Ord $Y = \beta$. Suppose that $f : X \to Y$ is an injective function which preserves order. Argue that $\alpha \leq \beta$ and that $\alpha = \beta$ if and only if f is a bijection.

   b) If α and β are ordinal numbers, show that $\beta \leq \alpha + \beta$ and that $\beta = \alpha + \beta$ if and only if $\alpha = 0$.

9. a) Let X be any well ordered set and $(\alpha_x)_{x \in X}$ be a family of ordinals indexed by X. For each $x \in X$, let Y_x be a well ordered set such that Ord $Y_x = \alpha_x$ and such that $Y_x \cap Y_{x'} = \emptyset$ if $x \neq x'$. Order $Y = \bigcup_{x \in X} Y_x$ so that if $y \in Y_x$ and $y' \in Y_{x'}$ where $x < x'$ then $y < y'$ and so that if $y, y' \in Y_x$ (for some x) then $y \leq y'$ if and only if this holds in the original order on Y_x. Argue that Y is well ordered with this order.

   b) With this Y, Ord Y is denoted $\sum_{x \in X} \alpha_x$. Show that for each $\bar{x} \in X$, $\alpha_{\bar{x}} \leq \sum_{x \in X} \alpha_x$.

1.2   Modules

Throughout this book, R will denote an associative ring with 1.

We will assume that the reader is familiar with modules and their elementary properties. By an R-module M, we shall mean a *unitary left R-module*, that is, an abelian group M with a map $R \times M \to M$, denoted $(r, x) \mapsto rx$, such that for all $x, y \in M, r, s \in R$

$$r(x + y) = rx + ry$$
$$(r + s)x = rx + sx$$
$$(rs)x = r(sx)$$
$$1x = x \text{ where } 1 \in R.$$

If $(rs)x = r(sx)$ is replaced by $(sr)x = r(sx)$, then M is said to be a *right R-module* and we denote the image of (r, x) by xr and so $(sr)x = r(sx)$ becomes $(sr)x = (xs)r$.

Also recall that if R and S are rings, then an abelian group M is said to be an (R, S)-bimodule, denoted $_RM_S$, if M is a left R-module and right S-module and the structures are compatible, that is, $(rx)s = r(xs)$ for all $r \in R$, $s \in S$, $x \in M$. In particular, any ring R is naturally an (R, R)-bimodule.

We will also assume familiarity with R-homomorphisms, R-submodules $S \subset M$, quotient modules M/S, direct products $\prod_{i \in I} M_i$, and direct sums $\bigoplus_{i \in I} M_i$. If $M_i = M$ for each i, then $\prod_{i \in I} M_i$ and $\bigoplus_{i \in I} M_i$ will be denoted by M^I and $M^{(I)}$ respectively.

Definition 1.2.1. An R-module F is said to be *free* if it is a direct sum of copies of R, or equivalently, if it has a basis.

Proposition 1.2.2. *Every R-module is a quotient of a free R-module.*

Proof. Let M be an R-module and $\{x_i : i \in I\}$ be a set of generators of M. Then $R^{(I)}$ is a free R-module. Define a map $\varphi : R^{(I)} \to M$ by $\varphi((r_i)_{i \in I}) = \sum_{i \in I} r_i x_i$. Then φ is onto and so $M \cong R^{(I)}/\operatorname{Ker}\varphi$.   □

Corollary 1.2.3. *An R-module is finitely generated if and only if it is a quotient of R^n for some integer $n > 0$.*

Definition 1.2.4. If M and N are R-modules, then by $\operatorname{Hom}_R(M, N)$ we mean all the R-homomorphisms from M to N. Clearly $\operatorname{Hom}_R(M, N)$ is an abelian group under addition.

Now suppose M is an R-module and N is an (R, S)-bimodule. Let $s \in S$, $f \in \operatorname{Hom}_R(M, N)$ and define $fs : M \to N$ by $(fs)(x) = f(x)s$. Then clearly $f(s_1 + s_2) = fs_1 + fs_2$, $(f + g)s = fs + gs$, $(fs)t = f(st)$, and $f \cdot 1 = f$ for all $f, g \in \operatorname{Hom}_R(M, N)$, $s_1, s_2, s, t, 1 \in S$. That is, $\operatorname{Hom}_R(M, N)$ is a right S-module. Similarly, if M is an (R, S)-bimodule and N is an R-module, then $\operatorname{Hom}_R(M, N)$ is a left S-module. In particular, if $_RM_S$ and $_RN_T$ are modules, then $\operatorname{Hom}_R(M, N)$ is an (S, T)-bimodule. Likewise, given modules $_SM_R$, $_TN_R$, then $\operatorname{Hom}_R(M, N)$ is a (T, S)-bimodule.

Proposition 1.2.5. *If M is an R-module, then the map $\varphi : \operatorname{Hom}_R(R, M) \to M$ defined by $\varphi(f) = f(1)$ is an R-isomorphism.*

Proof. This is left to the reader.   □

Proposition 1.2.6. *Let $(N_i)_{i \in I}$ be a family of R-modules and $\pi_j : \prod_I N_i \to N_j$ for each j be the projection map. Then the map*

$$\varphi : \operatorname{Hom}_R\left(M, \prod_I N_i\right) \to \prod_I \operatorname{Hom}(M, N_i)$$

defined by $\varphi(f) = (\pi_i \circ f)_I$ is an isomorphism.

Proof. φ is clearly an R-homomorphism. Suppose $(f_i)_I \in \prod_I \text{Hom}(M, N_i)$. Then f_i is a map from M to N_i for each i. So we can define a map $f : M \to \prod N_i$ by $f(x) = (f_i(x))_I$. f is clearly an R-homomorphism. Furthermore, $\pi_j \circ f(x) = \pi_j((f_i(x))_I) = f_j(x)$ for all $x \in M$ and so $\pi_j \circ f = f_j$ for each j. Hence $\varphi(f) = (\pi_i \circ f)_I = (f_i)_I$. That is, φ is onto.

Now suppose $\varphi(f) = 0$. Then $(\pi_j \circ f)(x) = \pi_j(f(x)) = 0$ for each j and each $x \in M$. But then $f(x) = 0$ for all $x \in M$. That is, $f = 0$ and so φ is one-to-one. $\square$

A similar proof gives the following.

Proposition 1.2.7. *Let* $(N_i)_{i \in I}$ *be a family of R-modules and* $e_j : N_j \to \bigoplus_I N_i$ *be the jth embedding. Then the map*

$$\varphi : \text{Hom}_R \left(\bigoplus_I N_i, M \right) \to \prod_I \text{Hom}_R(N_i, M)$$

defined by $\varphi(f) = (f \circ e_i)_I$ *is an isomorphism.*

Definition 1.2.8. If M, M', N, N' are R-modules and $f : M' \to M$, $g : N \to N'$ are R-homomorphisms, then define a map $\varphi : \text{Hom}(M, N) \to \text{Hom}(M', N')$ by $\varphi(h) = ghf$. φ is denoted by $\text{Hom}(f, g)$. We have that $\text{Hom}(f, g)(h_1 + h_2) = \text{Hom}(f, g)(h_1) + \text{Hom}(f, g)(h_2)$, that is, $\text{Hom}(f, g)$ is *additive*. Furthermore, in the situation $_R M_S, _R M'_S, _R N, _R N'$, if $f : M' \to M$ is an (R, S)-homomorphism and $g : N \to N'$ is an R-homomorphism, then $\text{Hom}(f, g)$ is an S-homomorphism between the two left S-modules.

If $M'' \xrightarrow{f'} M' \xrightarrow{f} M$ and $N \xrightarrow{g} N' \xrightarrow{g'} N''$ are homomorphisms, then it is easy to see that $\text{Hom}(f', g') \circ \text{Hom}(f, g) = \text{Hom}(f \circ f', g' \circ g)$.

The maps $\text{Hom}(f, \text{id}_N)$, $\text{Hom}(\text{id}_M, g)$ are denoted by $\text{Hom}(f, N)$, $\text{Hom}(M, g)$ respectively. We note that if $f : M' \to M$ is an R-homomorphism, then we have a homomorphism of abelian groups $\text{Hom}(f, N) : \text{Hom}(M, N) \to \text{Hom}(M', N)$. Similarly, for a map $g : N \to N'$, we get a map $\text{Hom}(M, g) : \text{Hom}(M, N) \to \text{Hom}(M, N')$.

Definition 1.2.9. If $f \in \text{Hom}_R(M, N)$ where M and N are R-modules, then the *kernel* of f, denoted $\text{Ker } f$, is defined as usual. The *cokernel* of f, denoted $\text{Coker } f$, is defined to be $N/\text{Im } f$ where $\text{Im } f$ denotes the *image* of f.

Definition 1.2.10. A sequence of R-modules and R-homomorphisms

$$\cdots \to M_2 \to M_1 \xrightarrow{\partial_1} M_0 \xrightarrow{\partial_0} M_{-1} \xrightarrow{\partial_{-1}} M_{-2} \to \cdots$$

is said to be *exact* at M_i if $\text{Im } \partial_{i+1} = \text{Ker } \partial_i$. The sequence is said to be *exact* if it is exact at each M_i. It is easy to see that a sequence $0 \to A \xrightarrow{f} B$ of R-modules is exact if and only if f is one-to-one, and a sequence $B \xrightarrow{g} C \to 0$ is exact if and only if g

is onto. An exact sequence of the form $0 \to M' \xrightarrow{f} M \xrightarrow{g} M'' \to 0$ is said to be a
short exact sequence. In this case, Coker $f = M/\text{Im } f \cong M''$.

Remark 1.2.11. Using Proposition 1.2.2, one can construct an exact sequence $\cdots \to$
$F_1 \to F_0 \to M \to 0$ with each F_i free for any R-module M. This sequence is called
a *free resolution* of M.

Proposition 1.2.12. *The following statements hold.*

(1) *If* $0 \to N' \xrightarrow{f} N \xrightarrow{g} N''$ *is an exact sequence of R-modules, then for each
R-module M the sequence*

$$0 \to \text{Hom}_R(M, N') \xrightarrow{\text{Hom}(M,f)} \text{Hom}_R(M, N) \xrightarrow{\text{Hom}(M,g)} \text{Hom}(M, N'')$$

is also exact.

(2) *If* $M' \xrightarrow{f} M \xrightarrow{g} M'' \to 0$ *is exact, then for each module N the sequence*

$$0 \to \text{Hom}(M'', N) \xrightarrow{\text{Hom}(g,N)} \text{Hom}(M, N) \xrightarrow{\text{Hom}(f,N)} \text{Hom}(M', N)$$

is exact.

Proof. (1) Let $\sigma \in \text{Hom}(M, N')$ be such that $\text{Hom}(M, f)(\sigma) = 0$. Then $f\sigma = 0$
and so $\sigma = 0$ since f is one-to-one. Hence $\text{Hom}(M, f)$ is one-to-one.

We now show exactness at $\text{Hom}(M, N)$. Let $\sigma \in \text{Hom}(M, N')$. Then
$\text{Hom}(M, g) \circ \text{Hom}(M, f)(\sigma) = gf\sigma$. But if $x \in M$, then $f(\sigma(x)) \in \text{Im } f =$
$\text{Ker } g$. So $\text{Hom}(M, g) \circ \text{Hom}(M, f)(\sigma) = gf\sigma = 0$. Thus $\text{Im}(\text{Hom}(M, f)) \subset$
$\text{Ker}(\text{Hom}(M, g))$. Now let $\tau \in \text{Ker}(\text{Hom}(M, g))$. Then $\text{Hom}(M, g)(\tau) = g\tau = 0$.
So $\text{Im } \tau \subset \text{Ker } g = \text{Im } f$. Hence let σ be the map from M to N' defined by
$\sigma = f^{-1}\tau$. Then $\sigma \in \text{Hom}(M, N')$ is such that $\text{Hom}(M, f)(\sigma) = f\sigma = \tau$. That is,
$\tau \in \text{Im}(\text{Hom}(M, f))$. Thus we have exactness at $\text{Hom}(M, N)$.

(2) follows similarly.   □

Proposition 1.2.13 (Snake lemma). *Suppose*

$$
\begin{array}{ccccccc}
M' & \xrightarrow{f} & M & \xrightarrow{g} & M'' & \longrightarrow & 0 \\
\downarrow{\sigma'} & & \downarrow{\sigma} & & \downarrow{\sigma''} & & \\
0 & \longrightarrow & N' & \xrightarrow{f'} & N & \xrightarrow{g'} & N''
\end{array}
$$

*is a commutative diagram (that is, $f'\sigma' = \sigma f$ and $g'\sigma = \sigma''g$) of R-modules with
exact rows. Then there is an exact sequence*

$$\text{Ker } \sigma' \xrightarrow{\bar{f}} \text{Ker } \sigma \to \text{Ker } \sigma'' \xrightarrow{d} \text{Coker } \sigma' \to \text{Coker } \sigma \xrightarrow{\bar{g}'} \text{Coker } \sigma''$$

Furthermore, if f is one-to-one, then $\bar{f}$ is also one-to-one, and if g' is onto, then $\bar{g}'$ is onto.

Proof. The proof for exactness is routine once we define the map d.

Let $x'' \in \operatorname{Ker} \sigma''$. Choose $x \in M$ such that $g(x) = x''$. Then $g' \circ \sigma(x) = \sigma'' \circ g(x) = \sigma''(x'') = 0$. So $\sigma(x) \in \operatorname{Ker} g' = \operatorname{Im} f'$. Thus $\sigma(x) = f'(y')$ for some $y' \in N'$. So define $d : \operatorname{Ker} \sigma'' \to \operatorname{Coker} \sigma'$ by $d(x'') = y' + \operatorname{Im} \sigma'$. Then d is a well-defined homomorphism. □

Definition 1.2.14. An exact sequence $0 \to M' \xrightarrow{f} M \xrightarrow{g} M'' \to 0$ of R-modules is said to be *split exact*, or we say the sequence *splits*, if Im f is a direct summand of M.

Proposition 1.2.15. *Let* $0 \to M' \xrightarrow{f} M \xrightarrow{g} M'' \to 0$ *be an exact sequence of R-modules. Then the following are equivalent.*

(1) *The sequence is split exact.*

(2) *There exists an R-homomorphism $f' : M \to M'$ such that $f' \circ f = \operatorname{id}_{M'}$.*

(3) *There exists an R-homomorphism $g'' : M'' \to M$ such that $g \circ g'' = \operatorname{id}_{M''}$.*

Proof. This is left to the reader. □

Definition 1.2.16. Let M be a right R-module, N a left R-module, and G an abelian group. Then a map $\sigma : M \times N \to G$ is said to be *balanced* (or *bilinear*) if it is additive in both variables (*biadditive*), that is,

$$\sigma(x + x', y) = \sigma(x, y) + \sigma(x', y),$$
$$\sigma(x, y + y') = \sigma(x, y) + \sigma(x, y'),$$
$$\sigma(xr, y) = \sigma(x, ry)$$

for all $x, x' \in M, y, y' \in N, r \in R$.

Definition 1.2.17. $\sigma : M \times N \to G$ is said to be a *universal balanced map* or we say σ *solves* the *"universal mapping problem"* for G if for every abelian group G' and balanced map $\sigma' : M \times N \to G'$, there exists a unique map $h : G \to G'$ such that $\sigma' = h\sigma$.

Definition 1.2.18. A *tensor product* of a right R-module M and left R-module N is an abelian group T together with a universal balanced map $\sigma : M \times N \to T$.

If $\sigma : M \times N \to T$, $\sigma' : M \times N \to T'$ are both universal balanced maps, then we can complete the diagram

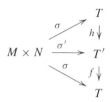

to a commutative diagram. But then $fh = \mathrm{id}_T$. Similarly, $hf = \mathrm{id}_{T'}$ and thus h is an isomorphism. Thus tensor products are unique up to isomorphism. We will thus speak of the tensor product of M_R and $_R N$, and will denote it by $M \otimes_R N$ or simply $M \otimes N$.

Theorem 1.2.19. *The tensor product of M_R and $_R N$ exists.*

Proof. Let F be the free abelian group with base $M \times N$, that is,

$$F = \left\{ \sum_i m_i(x_i, y_i) : m_i \in \mathbb{Z}, (x_i, y_i) \in M \times N \right\} \cong \mathbb{Z}^{(M \times N)}$$

Let S be the subgroup of F generated by elements of F of the form

$$(x + x', y) - (x, y) - (x', y), (x, y + y') - (x, y) - (x, y'), (rx, y) - (x, ry)$$

where $x, x' \in M$, $y, y' \in N$, $r \in R$. Define a map $\sigma : M \times N \to F/S$ by $\sigma(x, y) = (x, y) + S$. Then σ is clearly balanced. Now let $\sigma' : M \times N \to G'$ be a balanced map into an abelian group G'. But F is free on $M \times N$. So there is a unique homomorphism $h' : F \to G'$ that extends σ', that is, $h'(x, y) = \sigma'(x, y)$. But clearly $S \subset \mathrm{Ker}\, h'$ since σ' is balanced. So we get a unique induced map $h : F/S \to G'$ such that $\sigma' = h\sigma$. Thus $F/S = M \otimes_R N$.   □

Remark 1.2.20. We see from the proof above that F/S is generated as an abelian group by cosets $(x, y) + S$. We denote $(x, y) + S$ by $x \otimes y$. So $M \otimes_R N$ is generated as an abelian group by the elements $x \otimes y$. Since $-(x \otimes y) = (-x) \otimes y$, the elements of $M \otimes_R N$ are of the form $\sum x_i \otimes y_i$. Furthermore, if $x, x' \in M$, $y, y' \in N$, and $r \in R$, then

$$(x + x') \otimes y = x \otimes y + x' \otimes y,$$
$$x \otimes (y + y') = x \otimes y + x \otimes y',$$
$$(xr) \otimes y = x \otimes (ry).$$

Proposition 1.2.21. $M_R \otimes R \cong M$ *for every right R-module M, and $R \otimes_R N \cong N$ for every left R-module N.*

Proof. The map $M \times R \to M$ given by $(x, r) \mapsto xr$ is balanced and so there is a unique homomorphism $h : M \otimes_R R \to M$ such that $h(x \otimes r) = xr$. But $h' : M \to M \otimes_R R$ given by $h'(x) = x \otimes 1$ is a group homomorphism and $hh' = \mathrm{id}_M$. Moreover, $M \otimes_R R$ is generated by $x \otimes 1$, $x \in M$ and so easily $h'h = \mathrm{id}_{M \otimes_R R}$. Thus $M \otimes_R R \cong M$. $\square$

Proposition 1.2.22. *Let $(M_i)_I$ be a family of right R-modules and N a left R-module. Then*

$$(\oplus_I M_i) \otimes_R N \cong \oplus_I (M_i \otimes N).$$

Proof. The map $(\oplus M_i) \times N \to \oplus_I (M_i \otimes N)$ given by $((x_i)_I, y) \mapsto (x_i \otimes y)_I$ is balanced and so we have a unique homomorphism $h : (\oplus_I M_i) \otimes N \to \oplus_I (M_i \otimes N)$ such that $h((x_i)_I \otimes y) = (x_i \otimes y)_I$. Similarly one gets a unique homomorphism $h' : \oplus_I (M_i \otimes N) \to (\oplus_I M_i) \otimes N$ given by $h'((x_i \otimes y_i)_I) = \sum_I e_i(x_i) \otimes y_i$. It is easy to see that $h' = h^{-1}$. $\square$

We note that with the appropriate hypothesis there is an isomorphism

$$M \otimes_R (\oplus N_i) \cong \oplus_I (M \otimes N_i).$$

Proposition 1.2.23. *Let $f : M_R \to M'_R$ and $g : {}_R N \to {}_R N'$ be homomorphisms. Then there is a unique homomorphism $h : M \otimes_R N \to M' \otimes_R N'$ such that $h(x \otimes y) = f(x) \otimes g(y)$.*

Proof. We consider the following commutative diagram

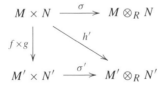

where σ, σ' are the universal balanced maps $(x, y) \mapsto x \otimes y$, $(x', y') \mapsto x' \otimes y'$, respectively, and $h' = \sigma'(f \times g)$. But h' is balanced. So there exists a unique homomorphism $h : M \otimes_R N \to M' \otimes_R N'$ such that $h(x \otimes y) = f(x) \otimes g(y)$. $\square$

Remark 1.2.24. The map $h : M \otimes_R N \to M' \otimes_R N'$ in the proposition above is denoted by $f \otimes g$. Now suppose $f' : M'_R \to M''_R$, $g' : {}_R N' \to {}_R N''$ are R-homomorphisms. Then we get a map $f' \otimes g' : M' \otimes_R N' \to M'' \otimes_R N''$ and it is easy to see that $(f' \otimes g') \circ (f \otimes g) = f'f \otimes g'g$ by evaluating the maps on a generator $x \otimes y \in M \otimes N$. We also note that $\mathrm{id}_M \otimes \mathrm{id}_N : M \otimes N \to M \otimes N$ is clearly the identity on $M \otimes N$, and if $f : M \to M'$, $g : N \to N'$ are isomorphisms, then $f \otimes g$ is an isomorphism and $(f \otimes g)^{-1} = f^{-1} \otimes g^{-1}$.

Proposition 1.2.25. *If* $N' \xrightarrow{f} N \xrightarrow{g} N'' \to 0$ *is an exact sequence of left R-modules, then for each right R-module M, the sequence* $M \otimes_R N' \xrightarrow{\mathrm{id}_M \otimes f}$ $M \otimes_R N \xrightarrow{\mathrm{id}_M \otimes g} M \otimes_R N'' \to 0$ *is also exact.*

Proof. Let $x \otimes y'' \in M \otimes N''$. Then there is a $y \in N$ such that $g(y) = y''$ and so $x \otimes y \in M \otimes N$ is such that $(\mathrm{id}_M \otimes g)(x \otimes y) = x \otimes y''$. That is, $M \otimes_R N \to M \otimes_R N'' \to 0$ is exact.

Clearly, $\mathrm{Im}(\mathrm{id}_M \otimes f) \subset \mathrm{Ker}(\mathrm{id}_M \otimes g)$ since $\mathrm{Im}\, f \subset \mathrm{Ker}\, g$. But $\mathrm{Im}(\mathrm{id}_M \otimes f) \subset \mathrm{Ker}(\mathrm{id}_M \otimes g)$ means that we have an induced commutative diagram

$$
\begin{array}{ccccccc}
M \otimes_R N' & \xrightarrow{\;\mathrm{id}_M \otimes f\;} & M \otimes N & \longrightarrow & (M \otimes N)/\mathrm{Im}(\mathrm{id}_M \otimes f) & \longrightarrow & 0 \\
& & \Big\downarrow{\scriptstyle \mathrm{id}_M \otimes g} & \nearrow {\scriptstyle h} & & & \\
& & M \otimes N''. & & & &
\end{array}
$$

If we can show that h is an isomorphism, then we would be done for then $\mathrm{Ker}(\mathrm{id}_M \otimes g) \subset \mathrm{Im}(\mathrm{id}_M \otimes f)$. So we now define an inverse of h. Define a map $\sigma : M \times N'' \to (M \otimes N)/\mathrm{Im}(\mathrm{id}_M \otimes f)$ by $\sigma(x, y'') = x \otimes y + \mathrm{Im}(\mathrm{id}_M \otimes f)$ where $y'' = g(y)$. σ is a well-defined balanced map. So there is a unique homomorphism $h' : M \otimes N'' \to (M \otimes N)/\mathrm{Im}(\mathrm{id}_M \otimes f)$ given by $h'(x \otimes y'') = x \otimes y + \mathrm{Im}(\mathrm{id}_M \otimes f)$. Then one easily checks that $h' = h^{-1}$. $\square$

Now let I be a right ideal of R and M be an R-module. Then IM, the set of all finite sums $\sum_{i=1}^n r_i x_i$, $r_i \in I$, $x_i \in M$, is a subgroup of M. With this notation, we have the following.

Corollary 1.2.26. *Let I be a right ideal of R and M be a left R-module. Then*

$$(R/I) \otimes_R M \cong M/IM.$$

Proof. We consider the exact sequence $0 \to I \to R \to R/I \to 0$. But $I \otimes M \xrightarrow{f} M \to R/I \otimes M \to 0$ is exact by Propositions 1.2.21 and 1.2.25, and $\mathrm{Im}\, f = \{\sum_i r_i x_i : r_i \in I, x \in M\} = IM$. Hence the result follows. $\square$

Now suppose M is an (S, R)-bimodule and N is a left R-module. Then the tensor product $M \otimes_R N$ is a left S-module with $s(x \otimes y) = (sx) \otimes y$ for all $x \in M, y \in N$, $s \in S$. If M is a right R-module and N is an (R, T)-bimodule, then $M \otimes_R N$ is a right T-module with $(x \otimes y)t = x \otimes (yt)$ for all $x \in M$, $y \in N$, $t \in T$. If M is an (S, R)-bimodule and N is an (R, T)-bimodule, then $M \otimes_R N$ is an (S, T)-bimodule with $s(x \otimes y)t = (sx) \otimes (yt)$.

We finally note that the associativity property holds for tensor products. That is, given M_R, $_R N_S$, $_S P$, we have that $(M \otimes_R N) \otimes_S P \cong M \otimes_R (N \otimes_S P)$. One begins by constructing a balanced map $(M \otimes_R N) \times P \to M \otimes_R (N \otimes_S P)$. To do this let $z \in P$ and note that the function $M \times N \to M \otimes_R (N \otimes_S P)$ which maps (x, y) to $x \otimes (y \otimes z)$ is balanced and gives a map $M \otimes_R N \to M \otimes_R (N \otimes_S P)$ mapping $x \otimes y$ to $x \otimes (y \otimes z)$. Using these maps (for all $z \in P$), we get a map $(M \otimes_R N) \times P \to M \otimes_R (N \otimes_S P)$ which maps $(x \otimes y, z)$ to $x \otimes (y \otimes z)$. This map is balanced and so gives a map $(M \otimes_R N) \otimes_S P \to M \otimes_R (N \otimes_S P)$ which maps $(x \otimes y) \otimes z$ to $x \otimes (y \otimes z)$. Similarly a map $M \otimes_R (N \otimes_S P) \to (M \otimes_R N) \otimes_S P$ can be constructed which maps $x \otimes (y \otimes z)$ to $(x \otimes y) \otimes z$. Clearly the two maps are inverses of one another.

Definition 1.2.27. The intersection of all maximal left ideals of a ring R is called the *Jacobson radical* of R and is denoted rad(R). An R-module M is said to be *simple* if it is isomorphic to $R/\mathfrak{m}$ for some maximal left ideal $\mathfrak{m}$ of R, or equivalently, has no submodules different from 0 and itself. Thus it is easy to see that rad$(R) = \{r \in R : rM = 0$ for every simple left R-module $M\}$. So rad(R) is a two-sided ideal of R. Moreover, rad(R) consists precisely of elements $r \in R$ such that $1 - sr$ is invertible for all $s \in R$. For if $r \in$ rad(R), then $r \in \mathfrak{m}$ for each maximal left ideal $\mathfrak{m}$ of R and so $sr \in \mathfrak{m}$ for all $s \in R$. But then $1 - sr \notin \mathfrak{m}$ for each maximal left ideal $\mathfrak{m}$ and every $s \in R$. Hence $1 - sr$ is invertible for if not then the left-ideal $R(1 - sr)$ would be contained in some maximal left ideal of R. Conversely, if $r \notin$ rad(R), then $r \notin \mathfrak{m}$ for some maximal left ideal $\mathfrak{m}$. But then $Rr + \mathfrak{m} = R$ and so there is an $s \in R$ such that $1 - sr \in \mathfrak{m}$. That is, $1 - sr$ is not invertible.

In particular, if $r \in$ rad(R), then $1 - r$ is invertible.

Proposition 1.2.28 (Nakayama lemma). *Let M be an R-module and I be a subgroup of the additive group of R such that either* (a) *I is nilpotent (that is, $I^n = 0$ for some $n \geq 1$), or* (b) *$I \subset$ rad(R) and M is finitely generated. Then $IM = M$ implies $M = 0$.*

Proof. (a) is trivial for $M = IM = I^2 M = \cdots = 0$. (b) Suppose $M \neq 0$. Then let $\{x_1, \ldots, x_n\}$ be a minimal set of generators of M. So $x_1 = \sum_{i=1}^{n} r_i x_i$ for some $r_i \in I$ since $M = IM$. But $1 - r_1$ is invertible. Thus $x_1 \in Rx_2 + Rx_3 + \cdots + Rx_n$ which contradicts the minimality of $\{x_1, \ldots, x_n\}$. □

Corollary 1.2.29. *Let M be an R-module, N a submodule of M, and I a subgroup of the additive group of R such that either* (a) *I is nilpotent, or* (b) *$I \subset$ rad(R) and M is finitely generated. Then $IM + N = M$ implies $M = N$.*

Proof. We note that $I(M/N) = (IM + N)/N$ and so we apply Nakayama lemma to M/N. □

Proposition 1.2.30. *If M is a nonzero finitely generated R-module and $I \subset$ rad(R) is a right ideal, then $(R/I) \otimes_R M \neq 0$.*

Proof. This easily follows from Nakayama lemma and Corollary 1.2.26. □

Exercises

1. Prove Proposition 1.2.7.

2. Let M be an R-module and $(N_i)_{i \in I}$ be a family of R-modules. Consider the homomorphism of abelian groups

$$\varphi : \bigoplus_{i \in I} \operatorname{Hom}_R(M, N_i) \to \operatorname{Hom}_R \Big(M, \bigoplus_{i \in I} N_i\Big)$$

which maps $(f_i)_{i \in I}$ to f where $f(x) = (f_i(x))_{i \in I}$

   a) Argue that φ is an isomorphism if I is finite or if M is finitely generated.

   b) Find an example where φ is not an isomorphism.

3. If N is an R-module and $(M_i)_{i \in I}$ is a family of R-modules, define

$$\varphi : \bigoplus_{i \in I} \operatorname{Hom}_R(M_i, N) \to \operatorname{Hom}_R \Big(\prod_{i \in I} M_i, N\Big)$$

   by $\varphi(f_i)_{i \in I} = f$ where $f((x_i)_{i \in I}) = \sum_{i \in I} f_i(x_i)$

   a) Argue that φ is an isomorphism if I is finite.

   b) Find an example where φ is not an isomorphism.

4. a) Let $x \in M$ where M is an R-module. For every R-module N consider the homomorphism $\operatorname{Hom}_R(M, N) \to N$ of abelian groups that maps f to $f(x)$. Argue that this is an isomorphism for all N if and only if M is isomorphic to R under an isomorphism that maps x to 1.

   b) Find an example of an M with two different elements $x, y \in M$ such that each of the maps $f \mapsto f(x)$ and $f \mapsto f(y)$ from $\operatorname{Hom}_R(M, N)$ to N are isomorphisms for all N.

5. Prove the second part of Proposition 1.2.12.

6. Complete the proof of Proposition 1.2.13.

7. Prove that if $M \xrightarrow{f} N$ and $N \xrightarrow{g} M$ are such that $g \circ f = \operatorname{id}_M$, then $N \cong \operatorname{Im} f \oplus \operatorname{Ker} g$.

8. Prove Proposition 1.2.15.

9. If $I \subset R$ is a two sided ideal of the ring R and if M is a right R/I-module, then M is a right R-module with $x \cdot r = x \cdot (r + I)$ for $x \in M$, $r \in R$. Similarly any left R/I-module N is a left R-module. In this situation argue that

$$M \otimes_R N \cong M \otimes_{R/I} N.$$

10. Let M be a right R-module and let $x \in M$. For every left R-module N consider the homomorphism $\varphi : N \to M \otimes_R N$ of abelian groups with $\varphi(y) = x \otimes y$. Argue that φ is an isomorphism for all N if and only if M is isomorphic to the right R-module R under an isomorphism that maps x to 1.

11. a) Let $I \subset R$ be a left ideal. Consider the properties: (i) for every module M and submodule $S \subset M$, $IM + S = M$ implies $S = M$ and (ii) for any module M, $IM = 0$ implies $M = 0$. Argue that (i) and (ii) are equivalent.

    b) Let the left ideal $I \subset R$ have properties (i) and (ii) above. Let $F = R \oplus R \oplus \cdots$ and let $S \subset F$ be the submodule generated by $x_1 = (1, -r_1, 0, 0, \dots)$, $x_2 = (0, 1, -r_2, 0, \dots)$, $x_3 = (0, 0, 1, -r_3, 0, \dots) \dots$. Show that $I(F/S) = F/S$ and hence deduce that $F/S = 0$, that is, $S = F$. So $(1, 0, 0, 0, \dots) \in S$ and thus $(1, 0, 0, \dots)$ can be written as a sum $s_1 x_1 + \cdots + s_n x_n$ for some $n \geq 1$ and $s_1, s_2, \dots, s_n \in R$. Now solve for $s_1, s_2, \dots, s_n$ and deduce that $r_1 r_2 \dots r_n = 0$.

12. a) Let $(M_i)_{i \in I}$ be a family of R-modules. Argue that $\bigoplus_{i \in I} M_i$ is finitely generated if and only if each M_i is finitely generated and $M_i = 0$ except for a finite number of $i \in I$.

    b) Find an example of a ring R and a family $(M_i)_{i \in I}$ of R-modules with $M_i \neq 0$ for an infinite number of $i \in I$ such that $\prod_{i \in I} M_i$ is a finitely generated R-module.

1.3   Categories and functors

Definition 1.3.1. A *category* C consists of the following.

1) A class of objects, denoted $Ob(C)$.

2) For any pair $A, B \in Ob(C)$, a set denoted $\mathrm{Hom}_C(A, B)$ with the property that $\mathrm{Hom}_C(A, B) \cap \mathrm{Hom}_C(A', B') = \emptyset$ whenever $(A, B) \neq (A', B')$. $\mathrm{Hom}_C(A, B)$ is called the *set of morphisms* from A to B.
   If $f \in \mathrm{Hom}_C(A, B)$ we write $f : A \to B$ and say f is a *morphism* of C from A to B.

3) A *composition* $\mathrm{Hom}_C(B, C) \times \mathrm{Hom}_C(A, B) \to \mathrm{Hom}_C(A, C)$ for all objects A, B, C, denoted $(g, f) \mapsto gf$ (or $g \circ f$), satisfying the following properties:

   i) for each $A \in Ob(C)$, there is an *identity morphism* $\mathrm{id}_A \in \mathrm{Hom}_C(A, A)$ such that $f \circ \mathrm{id}_A = \mathrm{id}_B \circ f = f$ for all $f \in \mathrm{Hom}(A, B)$,
   ii) $h(gf) = (hg)f$ for all $f \in \mathrm{Hom}_C(A, B), g \in \mathrm{Hom}_C(B, C)$ and $h \in \mathrm{Hom}_C(C, D)$.

Examples include categories **Sets**, **Ab**, **Top**, and $_R$**Mod** whose objects are respectively, sets, abelian groups, topological spaces, and left R-modules, and morphisms are functions, group homomorphisms, continuous maps, and R-homomorphisms, respectively, with the usual compositions.

Now let $Mor(\mathsf{C})$ denote the set of all morphisms of C. Then

$$Mor(\mathsf{C}) = \bigcup_{A,B \in Ob(\mathsf{C})} \mathrm{Hom}(A, B).$$

If $f : A \to B$ is a morphism in C, then f is said to be an *isomorphism* if there is a morphism $g : B \to A$ in C such that $fg = \mathrm{id}_B$ and $gf = \mathrm{id}_A$. Clearly, g is unique if it exists and is denoted by f^{-1}. f is said to be a *monomorphism* (*epimorphism*) if for every morphisms $g, h : C \to A$, $(g, h : B \to C)$ in C, $fg = fh$ $(gf = hf)$ implies $g = h$.

Definition 1.3.2. If C and C' are categories, then C' is said to be a *subcategory* of C if

1) $Ob(\mathsf{C}') \subset Ob(\mathsf{C})$, $Mor(\mathsf{C}') \subset Mor(\mathsf{C})$ and $\mathrm{Hom}_{\mathsf{C}'}(A', B') = \mathrm{Hom}_{\mathsf{C}}(A', B') \cap Mor(\mathsf{C}')$.

2) For any $A' \in Ob(\mathsf{C}')$, the identity morphisms on A' in C and C' are the same, and if $f' \in \mathrm{Hom}_{\mathsf{C}'}(A', B')$, $g' \in \mathrm{Hom}_{\mathsf{C}'}(B', C')$, then the map $g' \circ f'$ is the same in C' as in C.

Definition 1.3.3. A subcategory C' of C is said to be a *full subcategory* if $\mathrm{Hom}_{\mathsf{C}'}(A, B) = \mathrm{Hom}_{\mathsf{C}}(A, B)$ for all $A, B \in Ob(\mathsf{C}')$.

We note that for any category C and any subclass S of $Ob(\mathsf{C})$, there is a unique full subcategory C' of C with $Ob(\mathsf{C}') = S$. **Ab** and the category of compact spaces are full subcategories of the category of groups **Grp** and **Top**, respectively.

Definition 1.3.4. If C and D are categories, then we say that we have a *functor* $F : \mathsf{C} \to \mathsf{D}$ if we have

1) a function $Ob(\mathsf{C}) \to Ob(\mathsf{D})$ (denoted F),

2) functions $\mathrm{Hom}_{\mathsf{C}}(A, B) \to \mathrm{Hom}_{\mathsf{D}}(F(A), F(B))$ (also denoted F) such that

    i) if $f \in \mathrm{Hom}_{\mathsf{C}}(A, B)$, $g \in \mathrm{Hom}_{\mathsf{C}}(B, C)$, then $F(gf) = F(g)F(f)$, and

    ii) $F(\mathrm{id}_A) = \mathrm{id}_{F(A)}$ for each $A \in Ob(\mathsf{C})$.

A functor is sometimes called a *covariant functor*.

A function $Ob(\mathsf{C}) \to Ob(\mathsf{D})$ is said to be *functorial* if it agrees with a functor from C to D (usually in some obvious way).

Example 1.3.5.

1. We have the identify functor $\mathrm{id}_{\mathsf{C}} : \mathsf{C} \to \mathsf{C}$.

2. Define $F : \mathbf{Grp} \to \mathbf{Ab}$ by $F(G) = G/G'$ where G' is the commutator subgroup of G. Then it is easy to see that F is a covariant functor.

3. Let **Top*** denote the category of topological spaces with a base point. Define $F : \textbf{Top}^* \to \textbf{Grp}$ by $F(X) = \pi_1(x)$, the fundamental group. Then continuous maps get mapped to group homomorphisms and F satisfies the conditions of a functor.

4. Let M be a left R-module. Define $F : {}_R\textbf{Mod} \to \textbf{Ab}$ by $F(N) = \text{Hom}_R(M, N)$ and for $f \in \text{Hom}(N', N)$, define $F(f) : \text{Hom}(M, N') \to \text{Hom}(M, N)$ by $F(f)(h) = fh$. Then F is a covariant functor. This functor is denoted by $\text{Hom}_R(M, -)$.

5. Similarly, if M is a right R-module, we can define a function $F : {}_R\textbf{Mod} \to \textbf{Ab}$ by $F(N) = M \otimes_R N$ and for $f \in \text{Hom}(N', N)$, define $F(f) : M \otimes_R N' \to M \otimes_R N$ by $F(f)(x \otimes y') = x \otimes f(y')$. Then F is again a covariant functor. This functor is denoted by $M \otimes_R -$.

Definition 1.3.6. We say that we have a *contravariant functor* $F : \textsf{C} \to \textsf{D}$ if we have

1) a function $Ob(\textsf{C}) \to Ob(\textsf{D})$ (denoted F)

2) functions $\text{Hom}_\textsf{C}(A, B) \to \text{Hom}_\textsf{D}(F(B), F(A))$ (also denoted F) such that

    i) if $f \in \text{Hom}_\textsf{C}(A, B)$, $g \in \text{Hom}_\textsf{C}(B, C)$, then $F(gf) = F(f)F(g)$, and

    ii) $F(\text{id}_A) = \text{id}_{F(A)}$ for each $A \in Ob(\textsf{C})$.

Example 1.3.7.

1. Let M be an R-module. Define $F : {}_R\textbf{Mod} \to \textbf{Ab}$ by $F(N) = \text{Hom}_R(N, M)$ and for $f \in \text{Hom}(N', N)$, define $F(f) : \text{Hom}(N, M) \to \text{Hom}(N', M)$ by $F(f)(h) = hf$. Then F is a contravariant functor and is denoted by $\text{Hom}(-, M)$.

2. Define a function $F : \textbf{Top} \to \textbf{Ab}$ by $F(X) = H^n(X, G)$, the nth cohomology group of the topological space X with coefficients in G. Then F is a contravariant functor.

3. Let $\textsf{C}$ be the category of finite dimensional Galois extensions of k. Then define a function $F : \textsf{C} \to \textbf{Grp}$ by $F(K) = \mathcal{G}(K/k)$, the Galois group of K over k. Then F is a contravariant functor.

Definition 1.3.8. If $\textsf{C}$ and $\textsf{D}$ are categories, the *product* $\textsf{C} \times \textsf{D}$ of $\textsf{C}$ and $\textsf{D}$ is the category whose class of objects is $Ob(\textsf{C}) \times Ob(\textsf{D})$ and where $\text{Hom}_{\textsf{C} \times \textsf{D}}((A, D), (B, E)) = \text{Hom}_\textsf{C}(A, B) \times \text{Hom}_\textsf{D}(D, E)$ with $(g, k) \circ (f, h) = (g \circ f, k \circ h)$ where $f : A \to B$ and $g : B \to C$ are in $\textsf{C}$ and $h : D \to E$ and $k : E \to F$ are in $\textsf{D}$.

Definition 1.3.9. If $\textsf{C}$ is any category, we define $\textsf{C}^0$ (the *category opposite* $\textsf{C}$) to be the category such that $Ob(\textsf{C}^0) = Ob(\textsf{C})$, and $\text{Hom}_{\textsf{C}^0}(B, A) = \text{Hom}_\textsf{C}(A, B)$ where for $g : C \to B$, $f : B \to A$ in $\textsf{C}^0$, $f \circ g$ is defined to be the morphism $g \circ f$ of $\textsf{C}$. So then a contravariant functor $T : \textsf{C} \to \textsf{D}$ is simply a functor $T : \textsf{C}^0 \to \textsf{D}$ or a functor $T : \textsf{C} \to \textsf{D}^0$.

Definition 1.3.10. A functor of the form $F : \mathsf{C} \times \mathsf{D} \to \mathsf{E}$ is called a *functor of two variables*. A functor $F : \mathsf{C}^0 \times \mathsf{D} \to \mathsf{E}$ is said to be a functor $\mathsf{C} \times \mathsf{D} \to \mathsf{E}$ which is contravariant in the first and covariant in the second variable. For example, $\mathrm{Hom}(-, -) :$ $_R\mathbf{Mod} \times {_R}\mathbf{Mod} \to \mathbf{Ab}$ is a functor of two variables which is contravariant in the first and covariant in the second variable. The functor $- \otimes_R - : \mathbf{Mod}_R \times {_R}\mathbf{Mod} \to \mathbf{Ab}$ is covariant in both variables where $\mathbf{Mod}_R$ denotes the category of right R-modules.

Definition 1.3.11. If $F, G : \mathsf{C} \to \mathsf{D}$ are functors, then by a *natural transformation* from F to G we mean a function $\sigma : Ob(\mathsf{C}) \to Mor(\mathsf{D})$ with $\sigma(A) : F(A) \to G(A)$ such that for any $f \in Mor(\mathsf{C})$ there is a commutative diagram

$$
\begin{array}{ccc}
F(A) & \xrightarrow{\ \sigma(A)\ } & G(A) \\
\downarrow{\scriptstyle F(f)} & & \downarrow{\scriptstyle G(f)} \\
F(B) & \xrightarrow[\sigma(B)]{} & G(B)
\end{array}
$$

that is, $\sigma(B)F(f) = G(f)\sigma(A)$.

We denote the natural transformation σ by $\sigma : F \to G$.

Suppose $F, G, H : \mathsf{C} \to \mathsf{D}$ are functors and $\sigma : F \to G, \tau : G \to H$ are natural transformations. Then we can form the composition diagrams to get the following commutative diagram

$$
\begin{array}{ccccc}
F(A) & \xrightarrow{\ \sigma(A)\ } & G(A) & \xrightarrow{\ \tau(A)\ } & H(A) \\
\downarrow{\scriptstyle F(f)} & & \downarrow{\scriptstyle G(f)} & & \downarrow{\scriptstyle H(f)} \\
F(B) & \xrightarrow{\ \sigma(B)\ } & G(B) & \xrightarrow{\ \tau(B)\ } & H(B)
\end{array}
$$

for any morphism $f : A \to B$ in C.

We can also form a category of functors from C to D, denoted D^{C}, where objects are functors and morphisms are natural transformations $\sigma : F \to G$. Two functors F, G in this category are said to be *isomorphic* if there are natural transformations $\sigma : F \to G, \tau : G \to F$ such that $\tau \circ \sigma = \mathrm{id}_F$ (identity transformation) and $\sigma \circ \tau = \mathrm{id}_G$. It is easy to show that F and G are isomorphic if and only if for each $A \in Ob(\mathsf{C})$, $\sigma(A)$ is an isomorphism in D.

Definition 1.3.12. A category C is said to be *additive* if $\mathrm{Hom}_{\mathsf{C}}(A, B)$ is an abelian group such that if $f, f_1, f_2 \in \mathrm{Hom}_{\mathsf{C}}(A, B)$, $g, g_1, g_2 \in \mathrm{Hom}_{\mathsf{C}}(B, C)$ then

$$g(f_1 + f_2) = gf_1 + gf_2 \quad \text{and} \quad (g_1 + g_2)f = g_1 f + g_2 f.$$

We note that if C is an additive category, then $\mathrm{Hom}_{\mathsf{C}}(A, B) \neq \emptyset$ since the zero morphism is always in $\mathrm{Hom}_{\mathsf{C}}(A, B)$. This is denoted 0_{AB} or simply 0. Easily, $_R\mathbf{Mod}$ is an additive category.

Definition 1.3.13. If C, D are additive categories, then a functor $F : C \to D$ is said to be *additive* if for all $f, g \in \mathrm{Hom}_C(A, B)$, $F(f + g) = F(f) + F(g)$.

We note that the composition of additive functors is also additive. Furthermore, if F is additive, then $F(0_{AB}) = 0_{F(A)F(B)}$ and $F(-f) = -F(f)$. For example, let M be an (R, S)-bimodule. Then $\mathrm{Hom}_R(M, -) : {}_R\mathbf{Mod} \to {}_S\mathbf{Mod}$ is an additive covariant functor while $\mathrm{Hom}_R(-, M) : {}_R\mathbf{Mod} \to \mathbf{Mod}_S$ is an additive contravariant functor. Similarly $- \otimes_R M$ and $M \otimes_S -$ are additive covariant functors.

Definition 1.3.14. By a *product* of a family $(A_i)_{i \in I}$ where $A_i \in Ob(C)$ we mean an object A of C together with morphisms $\pi_i : A \to A_i$ such that for each $B \in Ob(C)$ and morphisms $f_i : B \to A_i$ there is a unique morphism $f : B \to A$ such that $\pi_i \circ f = f_i$ for all $i \in I$. A is unique up to isomorphism and is denoted by $\prod_{i \in I} A_i$.

Dually, a *coproduct* of a family $(A_i)_{i \in I}$ of objects in C is an object A in C together with morphisms $e_i : A_i \to A$ such that for each $B \in Ob(C)$ and morphisms $f_i : A_i \to B$, there exists a unique morphism $f : A \to B$ such that $f \circ e_i = f_i$ for all $i \in I$. Again a coproduct is unique up to isomorphism and is denoted by $\coprod A_i$. If the category C is additive, then the coproduct is called a *direct sum* and is denoted $\oplus_I A_i$.

Definition 1.3.15. Let $f : A \to B$ be a morphism in C. Then a *kernel* of f, denoted $ker\, f$, is a morphism $k : K \to A$ such that $fk = 0$ and for each morphism $g : C \to A$ with $fg = 0$, there exists a unique morphism $h : C \to K$ such that $g = kh$. K is denoted by $\mathrm{Ker}\, f$. It is easy to see that $ker\, f$ is a unique monomorphism and that a morphism f is a monomorphism if and only if $\mathrm{Ker}\, f = 0$.

Dually, a *cokernel* of f, denoted $coker\, f$, is a morphism $p : B \to C$ such that $pf = 0$ and for each morphism $g : B \to D$ with $gf = 0$, there exists a unique morphism $h : C \to D$ such that $hp = g$. C is denoted $\mathrm{Coker}\, f$. $coker\, f$ is a unique epimorphism and a morphism f is an epimorphism if and only if $\mathrm{Coker}\, f = 0$.

Suppose that in an additive category C all morphisms have kernels and cokernels. Then a morphism $f : A \to B$ gives rise to $\mathrm{Ker}\, f \xrightarrow{ker\, f} A \xrightarrow{f} B \xrightarrow{coker\, f} \mathrm{Coker}\, f$. Then since $f \circ ker\, f = 0$ we get a decomposition $A \to \mathrm{Coker}\,(ker\, f) \to B$ of f. Since $coker\, f \circ f = 0$ and $A \to \mathrm{Coker}(ker\, f)$ is an epimorphism we get that $\mathrm{Coker}(ker\, f) \to B \to \mathrm{Coker}\, f$ is 0. So then $\mathrm{Coker}(ker\, f) \to B$ factors as $\mathrm{Coker}(ker\, f) \to \mathrm{Ker}(coker\, f) \to B$ giving us a morphism $\mathrm{Coker}(ker\, f) \to \mathrm{Ker}(coker\, f)$.

Definition 1.3.16. An additive category C is said to be an *abelian category* if it satisfies the following conditions

1) C has products (and coproducts),

2) every morphism in C has a kernel and a cokernel, and

3) for every morphism $f : A \to B$, the map $\mathrm{Coker}(ker\, f) \to \mathrm{Ker}(coker\, f)$ as above is an isomorphism.

Examples of abelian categories include $_R\mathbf{Mod}$ and $\mathbf{Mod}_R$.

Definition 1.3.17. If C and D are abelian categories, then a functor $F : \mathsf{C} \to \mathsf{D}$ is said to be *left exact* if for every short exact sequence $0 \to A \to B \to C \to 0$ in C the sequence $0 \to F(A) \to F(B) \to F(C)$ is exact in D. F is said to be *right exact* if $F(A) \to F(B) \to F(C) \to 0$ is exact. If F is contravariant, then it is *left exact* if $0 \to F(C) \to F(B) \to F(A)$ is exact and *right exact* if $F(C) \to F(B) \to F(A) \to 0$ is exact. F is said to be an *exact functor* if it is both left and right exact.

It follows from Proposition 1.2.12 that functors $\mathrm{Hom}(M, -)$ and $\mathrm{Hom}(-, N)$ are left exact and from Proposition 1.2.25 that tensor product functors are right exact. As an application, we have the following result.

Theorem 1.3.18. *Let R be a commutative ring and $T : _R\mathbf{Mod} \to \mathbf{Ab}$ be a contravariant left exact functor which converts sums to products. Then for some R-module D, $T(M) \cong \mathrm{Hom}_R(M, D)$, that is, T is isomorphic to $\mathrm{Hom}_R(-, D)$.*

Proof. Let $D = T(R)$. Then D can be viewed as an R-module. For if M is an R-module, then for each $x \in M$, the map $f_x : R \to M$ defined by $f_x(r) = rx$ is an R-homomorphism. But T is contravariant. So we have an R-homomorphism $T(f_x) \in \mathrm{Hom}(T(M), T(R))$. But then we have a well-defined map $\sigma : R \times T(R) \to T(R)$ given by $\sigma(r, x) = T(f_r)(x)$. σ gives $T(R)$ an R-module structure where we denote $\sigma(r, x)$ by rx noting that $T(f_r)T(f_s) = T(f_{rs})$ for all $r, s \in R$ since $f_{rs} = f_s f_r$ and T is contravariant and so $(rs)x = \sigma(rs, x) = T(f_{rs})(x) = T(f_r)T(f_s)(x) = r(sx)$.

We now define a map $g_M : T(M) \to \mathrm{Hom}(M, T(R))$ by $g_M(x)(a) = T(f_a)(x)$. Then g_M is a natural transformation and g_R is an isomorphism on $T(R)$. But T and $\mathrm{Hom}(-, T(R))$ convert sums to products. So g_M is an isomorphism for a free R-module M. Now let M be an R-module and consider the presentation $F_1 \to F_0 \to M \to 0$ of M where F_0, F_1 are free. Since T and $\mathrm{Hom}(-, D)$ are both left exact we have the following commutative diagram

$$
\begin{array}{ccccccc}
0 & \longrightarrow & T(M) & \longrightarrow & T(F_0) & \longrightarrow & T(F_1) \\
 & & \downarrow{\scriptstyle g_M} & & \downarrow{\scriptstyle g_{F_0}} & & \downarrow{\scriptstyle g_{F_1}} \\
0 & \longrightarrow & \mathrm{Hom}(M, D) & \longrightarrow & \mathrm{Hom}(F_0, D) & \longrightarrow & \mathrm{Hom}(F_1, D)
\end{array}
$$

with exact rows. But the last two vertical maps are isomorphisms by the above. So g_M is also an isomorphism and we are done.   □

Definition 1.3.19. Let C be an abelian category. Then a *pushout* of the diagram

$$A \xrightarrow{\ f\ } B$$
$$g \downarrow$$
$$C$$

in C is an object D together with morphisms $h : B \to D$ and $k : C \to D$ such that $kg = hf$ and if

$$A \xrightarrow{\ f\ } B$$
$$g \downarrow \qquad \qquad g' \downarrow$$
$$C \xrightarrow{\ f'\ } D'$$

is any commutative diagram in C, then there is a unique morphism $D \to D'$ such that the diagram

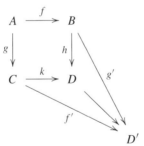

is commutative.

The diagram

$$A \xrightarrow{\ f\ } B$$
$$g \downarrow \qquad \qquad h \downarrow$$
$$C \xrightarrow{\ k\ } D$$

in the above is called a *pushout diagram*.

Dually, a *pullback diagram* is a commutative diagram

$$P \xrightarrow{\ h\ } A$$
$$k \downarrow \qquad \qquad f \downarrow$$
$$B \xrightarrow{\ g\ } C$$

such that if

is any commutative diagram in C, then there is a unique morphism $\sigma : P' \to P$ such that $h\sigma = h'$ and $k\sigma = k'$.

In this case P with morphisms h, k is called a *pullback* of morphisms $f : A \to C$ and $g : B \to C$.

It is now easy to see that pullbacks and pushouts are unique up to isomorphism.

Example 1.3.20.

1. In the above, if we set $D = (C \oplus B)/K$ where $K = \{(g(a), -f(a)) : a \in A\}$, and let $h(b) = (0, b) + K$, $k(c) = (c, 0) + K$, then D with morphisms h, k is a pushout.

2. If we set $P = \{(b, a) \in B \oplus A : g(b) = f(a)\}$ and let h and k be projection morphisms, then P with morphisms h, k is a pullback.

Exercises

1. a) Let C and D be categories. Argue that there is a category whose objects are the functors $F : C \to D$ and whose morphisms are the natural transformations $\sigma : F \to G$ (where $F, G : C \to D$ are functors).

   b) If the category D has products, argue that the category described in a) does too.

2. a) Let C be the category of commutative rings. Argue that C admits finite sums (for commutative rings R and S argue that $R \otimes_{\mathbb{Z}} S$ can be made into a commutative ring. Then consider the functions $r \mapsto r \otimes 1$ and $s \mapsto 1 \otimes s$ from R and S to $R \otimes_{\mathbb{Z}} S$).

   b) If R and S are subrings of $\mathbb{Q}$, argue that $R \otimes_{\mathbb{Z}} S$ is isomorphic to a subring of $\mathbb{Q}$.

   c) Find a sequence of commutative rings $R_0, R_1, R_2, \ldots$ no two of which are isomorphic such that $\mathbb{Q}$ is a sum of $R_0, R_1, R_2, \ldots$ in C.

3. Let C be a category and consider the category of functors $C \to$ **Sets**. Argue that any natural transformation $\mathrm{Hom}_C(X, -) \to \mathrm{Hom}_C(Y, -)$ is given by a morphism $Y \to X$ of C. Deduce that $\mathrm{Hom}_C(X, -)$ and $\mathrm{Hom}_C(Y, -)$ are isomorphic functors if and only if X and Y are isomorphic objects of C.

4. a) For a ring R and any R-modules M and N, define a relation on $\operatorname{Hom}_R(M, N)$ so that f and g are related if and only if $f - g : M \to N$ can be factored through a free R-module F, that is, $f - g$ can be written as a composition $M \to F \to N$. Check that this is an equivalence relation on $\operatorname{Hom}_R(M, N)$. For $f \in \operatorname{Hom}_R(M, N)$, let $[f]$ be the equivalence class of f. For $f : M \to N$ and $g : N \to P$, check that the definition of $[g] \circ [f]$ as $[g \circ f]$ is well-defined.

b) Argue that there is an additive category C whose objects are the R-modules M and whose morphisms are the $[f] : M \to N$ where $f \in \operatorname{Hom}_R(M, N)$.

c) Argue that in C of (b), M and $M \oplus F$ are isomorphic objects when F is a free R-module.

5. Verify Example 1.3.20.

6. Prove that in the pullback and pushout diagrams in Example 1.3.20, if f is one-to-one (onto), then so is k.

7. Let C be an abelian category. Argue that

$$
\begin{array}{ccc}
P & \xrightarrow{\ 0\ } & 0 \\
{\scriptstyle k}\downarrow & & \downarrow{\scriptstyle 0} \\
B & \xrightarrow{\ g\ } & C
\end{array}
$$

is a pullback diagram if and only if $P = \operatorname{Ker} g$. Now state and prove a dual result and conclude that cokernels are pushouts.

1.4 Complexes of modules and homology

Definition 1.4.1. By a *(chain) complex* C of R-modules we mean a sequence

$$
\mathsf{C} : \cdots \to C_2 \xrightarrow{\partial_2} C_1 \xrightarrow{\partial_1} C_0 \xrightarrow{\partial_0} C_{-1} \xrightarrow{\partial_{-1}} C_{-2} \to \cdots
$$

of R-modules and R-homomorphisms such that $\partial_{n-1} \circ \partial_n = 0$ for all $n \in \mathbb{Z}$. C is denoted by $((C_n), (\partial_n))$. If F is a covariant additive functor into some category of modules, then $F(\mathsf{C}) : \cdots \to F(C_2) \xrightarrow{F(\partial_2)} F(C_1) \xrightarrow{F(\partial_1)} F(C_0) \xrightarrow{F(\partial_0)} F(C_{-1}) \to \cdots$ is also a complex. Similarly if F is a contravariant additive functor, then the sequence $F(\mathsf{C}) : \cdots \to F(C_{-1}) \xrightarrow{F(\partial_0)} F(C_0) \xrightarrow{F(\partial_1)} F(C_1) \to \cdots$ is also a chain complex.

Let $\mathsf{C}' = ((C_n'), (\partial_n'))$ be another complex of R-modules. Then by a *map (or chain map)* $\mathbf{f} : \mathsf{C} \to \mathsf{C}'$ we mean a sequence of maps $f_n : C_n \to C_n'$ such that the diagram

$$
\begin{array}{ccc}
C_n & \xrightarrow{\partial_n} & C_{n-1} \\
{\scriptstyle f_n}\downarrow & & \downarrow{\scriptstyle f_{n-1}} \\
C_n' & \xrightarrow{\partial_n'} & C_{n-1}'
\end{array}
$$

is commutative for each $n \in \mathbb{Z}$. $\mathbf{f}$ is denoted by (f_n).

We note that if $\mathbf{g} = (g_n) : \mathbf{C} \to \mathbf{C}'$ is another map, then $(f_n + g_n) : \mathbf{C} \to \mathbf{C}'$ is a map. If $\mathbf{h} : \mathbf{C}' \to \mathbf{C}''$ is another map of complexes, then we can define a map $\mathbf{hf} : \mathbf{C} \to \mathbf{C}''$ by $\mathbf{hf} = (h_n f_n)$. So we obviously get a category, the category of complexes of R-modules which is denoted by **Comp**. It is not hard to see that **Comp** is an abelian category.

Definition 1.4.2. If $\mathbf{C} = ((C_n), (\partial_n))$ is a complex, then Im $\partial_{n+1} \subset$ Ker ∂_n. The nth *homology module* of $\mathbf{C}$ is defined to be Ker ∂_n/Im ∂_{n+1} and is denoted by $H_n(\mathbf{C})$. So $H_n(\mathbf{C}) = 0$ if and only if $\mathbf{C}$ is exact at C_n. Ker ∂_n, Im ∂_{n+1} are usually denoted by $Z_n(\mathbf{C})$, $B_n(\mathbf{C})$ and their elements are called *n-cycles, n-boundaries* respectively.

Definition 1.4.3. A chain complex of the form $\mathbf{C} : \cdots \to C^{-2} \to C^{-1} \xrightarrow{\partial^{-1}} C^0 \xrightarrow{\partial^0} C^1 \xrightarrow{\partial^1} C^2 \to \cdots$ is called a *cochain complex*. In this case, $\partial^n \circ \partial^{n-1} = 0$ for all $n \in \mathbb{Z}$. Ker ∂^n, Im ∂^{n-1} are denoted by $Z^n(\mathbf{C})$ and $B^n(\mathbf{C})$ respectively and their elements are called *n-cocycles, n-coboundaries,* respectively. Ker ∂^n/Im $\partial^{(n-1)}$ is called the nth *cohomology module* and is denoted by $H^n(\mathbf{C})$. We note that a cochain complex is simply a chain complex with C^i replaced by C_{-i} and ∂^i by ∂_{-i}. Consequently, we will only consider chain complexes in this section.

Now suppose $\mathbf{f} : \mathbf{C} \to \mathbf{C}'$ is a chain map. Then we have a commutative diagram

$$
\begin{array}{ccccc}
C_{n+1} & \xrightarrow{\partial_{n+1}} & C_n & \xrightarrow{\partial_n} & C_{n-1} \\
\downarrow{f_{n+1}} & & \downarrow{f_n} & & \downarrow{f_{n-1}} \\
C'_{n+1} & \xrightarrow{\partial'_{n+1}} & C'_n & \xrightarrow{\partial'_n} & C'_{n-1} .
\end{array}
$$

If $x \in$ Ker ∂_n, then $\partial'_n(f_n(x)) = f_{n-1}(\partial_n(x)) = 0$ and so $f_n(x) \in$ Ker ∂'_n. Hence we get an induced map Ker $\partial_n \to$ Ker ∂'_n. Furthermore, suppose $x \in$ Im ∂_{n+1}. Then $x = \partial_{n+1}(y)$, $y \in C_{n+1}$. So $\partial'_{n+1}(f_{n+1}(y)) = f_n(\partial_{n+1}(y)) = f_n(x)$. That is, $f_n(x) \in$ Im ∂'_{n+1}. So we consider the composition Ker $\partial_n \to$ Ker $\partial'_n \to$ Ker ∂'_n/Im $\partial'_{n+1} = H_n(\mathbf{C}')$. This composition maps Im ∂_{n+1} onto zero by the above. So we get an induced map

$$
H_n(\mathbf{C}) = \text{Ker } \partial_n/\text{Im } \partial_{n+1} \to \text{Ker } \partial'_n/\text{Im } \partial'_{n+1} = H_n(\mathbf{C}')
$$

given by $x + $ Im $\partial_{n+1} \mapsto f_n(x) + $ Im ∂'_{n+1}. This map is denoted by $H_n(\mathbf{f})$.

We note that if $\mathbf{g} : \mathbf{C}' \to \mathbf{C}''$ is another chain map, then $H_n(\mathbf{g}) : H_n(\mathbf{C}') \to H_n(\mathbf{C}'')$ maps $x' + $ Im ∂'_{n+1} onto $g_n(x') + $ Im ∂''_{n+1}. Hence $H_n(\mathbf{g})H_n(\mathbf{f}) = H_n(\mathbf{gf})$. Also easily $H_n(\text{id}_{\mathbf{C}}) = \text{id}_{H_n(\mathbf{C})}$ and if $\mathbf{f}_1, \mathbf{f}_2 : \mathbf{C} \to \mathbf{C}'$ are chain maps, then $H_n(\mathbf{f}_1 + \mathbf{f}_2) = H_n(\mathbf{f}_1) + H_n(\mathbf{f}_2)$. Thus we have the following result.

Theorem 1.4.4. $H_n :$ **Comp** $\to {}_R$**Mod** *defined by* $H_n(\mathbf{C}) = $ Ker ∂_n/Im ∂_{n+1} *is an additive covariant functor for each* $n \in \mathbb{Z}$.

Definition 1.4.5. A complex $\mathbf{C}' = ((C'_n), (\partial'_n))$ is said to be a *subcomplex* of a complex $\mathbf{C} = ((C_n), (\partial_n))$ if $C'_n \subset C_n$ and ∂_n agrees with ∂'_n on C'_n (so necessarily $\partial_n(C'_n) \subset C'_{n-1}$). In this case, we can form a complex $((C_n/C'_n), (\bar{\partial}_n))$ where $\bar{\partial}_n : C_n/C'_n \to C_{n-1}/C'_{n-1}$ is the induced map given by $\bar{\partial}_n(x + C'_n) = \partial_n(x) + C'_{n-1}$. This complex is called the *quotient complex* and is denoted by $\mathbf{C}/\mathbf{C}'$.

Definition 1.4.6. If $\mathbf{f} : \mathbf{C}' \to \mathbf{C}$, $\mathbf{g} : \mathbf{C} \to \mathbf{C}''$ are chain maps, then we say that $\mathbf{C}' \xrightarrow{\mathbf{f}} \mathbf{C} \xrightarrow{\mathbf{g}} \mathbf{C}''$ is an *exact sequence* if $C'_n \xrightarrow{f_n} C_n \xrightarrow{g_n} C''_n$ is exact for each $n \in \mathbb{Z}$.

Now let $\mathbf{C} = \cdots \to C_{n+1} \xrightarrow{\partial_{n+1}} C_n \xrightarrow{\partial_n} C_{n-1} \to \cdots$ be a chain complex. Then $Z(\mathbf{C}) = \cdots \to \mathrm{Ker}\,\partial_{n+1} \xrightarrow{0} \mathrm{Ker}\,\partial_n \xrightarrow{0} \mathrm{Ker}\,\partial_{n-1} \to \cdots$ is a subcomplex of $\mathbf{C}$. Similarly, we have a complex $B(\mathbf{C}) = \cdots \to \mathrm{Im}\,\partial_{n+1} \xrightarrow{0} \mathrm{Im}\,\partial_n \xrightarrow{0} \mathrm{Im}\,\partial_{n-1} \to \cdots$.

So if $0 \to \mathbf{C}' \to \mathbf{C} \to \mathbf{C}'' \to 0$ is an exact sequence of complexes, then we have a commutative diagram

$$
\begin{array}{ccccccccc}
0 & \longrightarrow & C'_{n+1} & \longrightarrow & C_{n+1} & \longrightarrow & C''_{n+1} & \longrightarrow & 0 \\
 & & \downarrow{\scriptstyle \partial_{n+1}'} & & \downarrow{\scriptstyle \partial_{n+1}} & & \downarrow{\scriptstyle \partial_{n+1}''} & & \\
0 & \longrightarrow & C'_n & \longrightarrow & C_n & \longrightarrow & C''_n & \longrightarrow & 0
\end{array}
$$

with exact rows which by the Snake lemma (Proposition 1.2.13) gives an exact sequence $0 \to Z_{n+1}(\mathbf{C}') \to Z_{n+1}(\mathbf{C}) \to Z_{n+1}(\mathbf{C}'') \to C'_n/B'_n(\mathbf{C}) \to C_n/B_n(\mathbf{C}) \to C''_n/B_n(\mathbf{C}'') \to 0$ for each $n \in \mathbb{Z}$. So we have a commutative diagram

$$
\begin{array}{ccccccccc}
0 & \longrightarrow & C'_{n+1} & \longrightarrow & C_{n+1} & \longrightarrow & C''_{n+1} & \longrightarrow & 0 \\
 & & \downarrow & & \downarrow & & \downarrow & & \\
0 & \longrightarrow & Z_n(\mathbf{C}') & \longrightarrow & Z_n(\mathbf{C}) & \longrightarrow & Z_n(\mathbf{C}'') & &
\end{array}
$$

with exact rows and an induced commutative diagram

$$
\begin{array}{ccccccc}
C'_{n+1}/B_{n+1}(\mathbf{C}') & \longrightarrow & C_{n+1}/B_{n+1}(\mathbf{C}) & \longrightarrow & C''_{n+1}/B_{n+1}(\mathbf{C}'') & \longrightarrow & 0 \\
\downarrow & & \downarrow & & \downarrow & & \\
\end{array}
$$

$$
\begin{array}{ccccccc}
0 & \longrightarrow & Z_n(\mathbf{C}') & \longrightarrow & Z_n(\mathbf{C}) & \longrightarrow & Z_n(\mathbf{C}'') \\
\end{array}
$$

with exact rows where the vertical maps are given by $\partial'_{n+1}, \partial_{n+1}, \partial''_{n+1}$ respectively. We now apply the Snake lemma again to get an exact sequence

$$
\begin{aligned}
\cdots &\to Z_{n+1}(\mathbf{C}')/B_{n+1}(\mathbf{C}') \to Z_{n+1}(\mathbf{C})/B_{n+1}(\mathbf{C}) \to Z_{n+1}(\mathbf{C}'')/B_{n+1}(\mathbf{C}'') \\
&\to Z_n(\mathbf{C}')/B_n(\mathbf{C}') \to Z_n(\mathbf{C})/B_n(\mathbf{C}) \to Z_n(\mathbf{C}'')/B_n(\mathbf{C}'') \to \cdots
\end{aligned}
$$

So we have proved the following result.

Theorem 1.4.7. *If* $0 \to \mathbf{C}' \to \mathbf{C} \to \mathbf{C}'' \to 0$ *is an exact sequence of complexes, then there is an exact sequence*

$$\cdots \to H_{n+1}(\mathbf{C}') \to H_{n+1}(\mathbf{C}) \to H_{n+1}(\mathbf{C}'') \to H_n(\mathbf{C}') \to H_n(\mathbf{C}) \to \cdots$$

for each $n \in \mathbb{Z}$.

Definition 1.4.8. The homomorphism $H_{n+1}(\mathbf{C}'') \to H_n(\mathbf{C}')$ is called the *connecting homomorphism* associated with the exact sequence $0 \to \mathbf{C}' \to \mathbf{C} \to \mathbf{C}'' \to 0$ and the sequence $\cdots \to H_{n+1}(\mathbf{C}') \to H_{n+1}(\mathbf{C}) \to H_{n+1}(\mathbf{C}'') \to H_n(\mathbf{C}') \to H_n(\mathbf{C}) \to H_n(\mathbf{C}'') \to \cdots$ is called the *long exact sequence*. Clearly, a map of the exact sequence $0 \to \mathbf{C}' \to \mathbf{C} \to \mathbf{C}'' \to 0$ into an exact sequence $0 \to \bar{\mathbf{C}}' \to \bar{\mathbf{C}} \to \bar{\mathbf{C}}'' \to 0$ gives rise to a map of the long exact sequence associated with the first into that associated with the second.

Definition 1.4.9. Let $\mathbf{C} = ((C_n), (\partial_n))$ and $\mathbf{C}' = ((C_n), (\partial_n'))$ be chain complexes of R-modules and $\mathbf{f} = (f_n)$, $\mathbf{g} = (g_n)$ be maps from $\mathbf{C}$ to $\mathbf{C}'$. Then $\mathbf{f}$ is said to be *homotopic* to $\mathbf{g}$, denoted $\mathbf{f} \sim \mathbf{g}$, if there are maps $s_n : C_n \to C_{n+1}'$ such that for every $n \in \mathbb{Z}$,

$$f_n - g_n = \partial_{n+1}' s_n + s_{n-1} \partial_n.$$

$\mathbf{s} = (s_n)_{n \in \mathbb{Z}}$ is called a *chain homotopy* between $\mathbf{f}$ and $\mathbf{g}$.

We note that $\mathbf{f} \sim \mathbf{f}$ (let $s_n = 0$) and if $\mathbf{f} \sim \mathbf{g}$ then $\mathbf{g} \sim \mathbf{f}$ (use $-s_n$'s). Now suppose $\mathbf{f} \sim \mathbf{g}$ and $\mathbf{g} \sim \mathbf{h}$ with homotopies $\mathbf{s} = (s_n)$ and $\mathbf{t} = (t_n)$ respectively. Then $\mathbf{f} \sim \mathbf{h}$ by adding $f_n - g_n = \partial_{n+1}' s_n + s_{n-1} \partial_n$ to $g_n - h_n = \partial_{n+1}' t_n + t_{n-1} \partial_n$ and using $s_n + t_n$. Thus $\sim$ is an equivalence relation and we let $[\mathbf{f}]$ denote the equivalence class of $\mathbf{f}$. $[\mathbf{f}]$ is called a *homotopy class* of $\mathbf{f}$.

Proposition 1.4.10. *Let* $\mathbf{C}, \mathbf{C}', \mathbf{C}''$ *be complexes and* $\mathbf{f}, \mathbf{g} : \mathbf{C} \to \mathbf{C}'$ *and* $\mathbf{h} : \mathbf{C}' \to \mathbf{C}''$ *be chain maps. If* $\mathbf{f} \sim \mathbf{g}$, *then* $\mathbf{h}\mathbf{f} \sim \mathbf{h}\mathbf{g}$.

Proof. Let $\mathbf{s}$ be a chain homotopy between $\mathbf{f}$ and $\mathbf{g}$. Then with obvious notation, $f_n - g_n = \partial_{n+1}' s_n + s_{n-1} \partial_n$ and so

$$
\begin{aligned}
h_n f_n - h_n g_n &= h_n (f_n - g_n) \\
&= h_n \partial_{n+1}' s_n + h_n s_{n-1} \partial_n \\
&= \partial_{n+1}'' h_{n+1} s_n + h_n s_{n-1} \partial_n.
\end{aligned}
$$

Thus $t_n : C_n \to C_{n+1}''$ given by $t_n = h_{n+1} s_n$ gives a chain homotopy between $\mathbf{h}\mathbf{f}$ and $\mathbf{h}\mathbf{g}$. $\square$

Similarly, we have the following result.

Proposition 1.4.11. *Let* $\mathbf{C}, \mathbf{C}', \mathbf{C}''$ *be complexes and* $\mathbf{f} : \mathbf{C} \to \mathbf{C}'$ *and* $\mathbf{h}, \mathbf{k} : \mathbf{C}' \to \mathbf{C}''$ *be maps. If* $\mathbf{h} \sim \mathbf{k}$, *then* $\mathbf{h}\mathbf{f} \sim \mathbf{k}\mathbf{f}$.

Corollary 1.4.12. *If* $\mathbf{f} \sim \mathbf{g}$ *and* $\mathbf{h} \sim \mathbf{k}$, *then* $\mathbf{fh} \sim \mathbf{gk}$.

Hence if $\mathbf{f} : \mathbf{C} \to \mathbf{C}'$ and $\mathbf{g} : \mathbf{C}' \to \mathbf{C}''$ are maps of complexes, then we can define $[\mathbf{g}][\mathbf{f}] = [\mathbf{gf}]$. By the above, this is well-defined. So we get the category of chain complexes of R-modules with objects the chain complexes as usual but such that morphisms are homotopy classes $[\mathbf{f}]$ where $\mathbf{f}$ is a map of complexes.

Proposition 1.4.13. *If* $\mathbf{f}, \mathbf{g} : \mathbf{C} \to \mathbf{C}'$ *are homotopic, then* $H_n(\mathbf{f}) = H_n(\mathbf{g})$ *for each* n.

Proof. Let $\mathbf{s} = (s_n)$ be the homotopy connecting $\mathbf{f}$ and $\mathbf{g}$ so that $f_n - g_n = \partial'_{n+1} s_n + s_{n-1} \partial_n$ and let $x \in \operatorname{Ker} \partial_n$. Then $f_n(x) - g_n(x) = \partial'_{n+1}(s_n(x)) + s_{n-1}(\partial_n(x)) = \partial'_{n+1}(s_n(x)) \in \operatorname{Im} \partial'_{n+1}$. That is, $f_n(x) + \operatorname{Im} \partial'_{n+1} = g_n(x) + \operatorname{Im} \partial'_{n+1}$. Hence $H_n(\mathbf{f}) = H_n(\mathbf{g})$. $\square$

A basic tool will be the following elementary result about complexes (essentially involving the mapping cone of a morphism of complexes, that is, the complex obtained in the following proposition).

Proposition 1.4.14. *Let*

$$\cdots \longrightarrow D_2 \xrightarrow{d_2} D_1 \xrightarrow{d_1} D_0 \xrightarrow{d_0} D_{-1} \longrightarrow D_{-2} \longrightarrow \cdots$$
$$\downarrow{\rho_2} \quad\quad \downarrow{\rho_1} \quad\quad \downarrow{\rho_0} \quad\quad \downarrow{\rho_{-1}} \quad\quad \downarrow{\rho_{-2}}$$
$$\cdots \longrightarrow C_2 \xrightarrow{d_2} C_1 \xrightarrow{d_1} C_0 \xrightarrow{d_0} C_{-1} \longrightarrow C_{-2} \longrightarrow \cdots$$

be a commutative diagram where the rows are complexes. Form the complex $\cdots \to C_2 \oplus D_1 \to C_1 \oplus D_0 \to C_0 \oplus D_{-1} \to \cdots$ *where the map* $C_{n+1} \oplus D_n \to C_n \oplus D_{n-1}$ *is the map* $(x, y) \mapsto (dx + (-1)^n \rho_n(y), dy)$ *(it is immediate that this is a complex). Then this complex is exact at* $C_n \oplus D_{n-1}$ *if the complex* $\cdots \to C_1 \to C_0 \to C_{-1} \to \cdots$ *is exact at* C_n *and* $\cdots \to D_1 \to D_0 \to D_{-1} \to \cdots$ *is exact at* D_{n-1}.

Proof. By diagram chasing. $\square$

Remark 1.4.15. We see that the construction of the single complex from the diagram above is compatible with the application of any covariant additive functor. We will also apply this result to diagrams involving complexes of finite length where we substitute 0 for all missing terms. This result implies that if both rows of the diagram are exact then so is the associated single complex.

We will also use the fact that if $\mathbf{C}$ is an exact complex and $\mathbf{S} \subset \mathbf{C}$ is a subcomplex, then the quotient complex $\mathbf{C}/\mathbf{S}$ is exact if and only if $\mathbf{S}$ is exact. This follows from applying Theorem 1.4.7 to the exact sequence $0 \to \mathbf{S} \to \mathbf{C} \to \mathbf{C}/\mathbf{S} \to 0$ of complexes.

Proposition 1.4.16. *Given a commutative diagram*

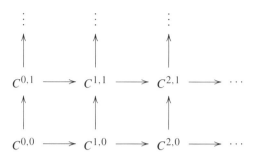

of R-modules, suppose that each row and each column is exact. For each $n \geq 0$, let $D^n = \text{Ker}(C^{0,n} \to C^{1,n})$ and let $E^n = \text{Ker}(C^{n,0} \to C^{n,1})$. Then noting that for each n we get induced maps $D^n \to D^{n+1}$ and $E^n \to E^{n+1}$ such that

$$\mathbf{D} = \cdots \to 0 \to 0 \to D^0 \to D^1 \to D^2 \to \cdots$$

and

$$\mathbf{E} = \cdots \to 0 \to 0 \to E^0 \to E^1 \to E^2 \to \cdots$$

are complexes, we have that $H^n(\mathbf{D}) \cong H^n(\mathbf{E})$ for all n.

Proof. This is an exercise in diagram chasing. One starts by using the diagram to define a (well-defined) homomorphism $H^n(\mathbf{D}) \to H^n(\mathbf{E})$. Then it is simple to check that the corresponding homomorphism $H^n(\mathbf{E}) \to H^n(\mathbf{D})$ is its inverse.   □

We note that in this statement we used upper indices in order to avoid using negative indices.

Exercises

1. a) Let M be an R-module and $\bar{\mathbf{M}}$ denote the complex

$$\cdots \to 0 \to 0 \to M \xrightarrow{\text{id}} M \to 0 \to 0 \to \cdots$$

with M in the 0-th place. For any complex $\mathbf{C} = ((C_n), (\partial_n))$ of R-modules argue that the set of chain maps $\bar{\mathbf{M}} \to \mathbf{C}$ is in bijective correspondence with the set $\text{Hom}_R(M, C_0)$.

b) Argue that every chain map $\bar{\mathbf{M}} \to \mathbf{C}$ is homotopic to the zero chain map.

2. Let $\mathbf{C}$ be a complex of R-modules and F be an exact additive functor of R-modules. Prove that $H_n(F(\mathbf{C})) \cong F(H_n(\mathbf{C}))$ for each $n \in \mathbb{Z}$.

3. Argue that **Comp** admits sums and products.

4. a) Let $\mathbf{C}$ be a complex of R-modules and let $x \in Z_0(\mathbf{C})$. Show that we can define a complex

$$\mathbf{C}' = \cdots \to C_2 \to C_1 \oplus R \to C_0 \to C_{-1} \to \cdots$$

in such a way that $\mathbf{C}$ is a subcomplex of $\mathbf{C}'$ and so that $x \in B_0(\mathbf{C}')$.

b) Generalize (a) and argue that for any $\mathbf{C}$, $\mathbf{C}$ is a subcomplex of a complex $\mathbf{C}'$ such that $Z_n(\mathbf{C}) \subset B_n(\mathbf{C}')$ for all n.

c) Use (b) to argue that every complex $\mathbf{C}$ is a subcomplex of an exact complex.

5. Prove Proposition 1.4.14.

6. Complete the proof of Proposition 1.4.16.

1.5 Direct and inverse limits

Definition 1.5.1. Let I be a *directed set*, that is, I is a partially ordered set such that for any $i, j \in I$ there is a $k \in I$ with $i, j \leq k$. Let $\{M_i\}_{i \in I}$ be a family of R-modules and suppose for each pairs $i, j \in I$ with $i \leq j$ there is an R-homomorphism $f_{ji} : M_i \to M_j$ such that

1) $f_{ii} = \mathrm{id}_{M_i}$ for each $i \in I$

2) if $i \leq j \leq k$, then $f_{kj} \circ f_{ji} = f_{ki}$.

Then we say that the R-modules M_i together with the homomorphisms f_{ji} form a *direct* (or *inductive*) *system* which is denoted $((M_i), (f_{ji}))$.

Definition 1.5.2. The *direct (inductive) limit* of a direct system $((M_i), (f_{ji}))$ of R-modules is an R-module M with R-homomorphisms $g_i : M_i \to M$ for $i \in I$ with $g_i = g_j \circ f_{ji}$ whenever $i \leq j$ and such that if $(N, \{h_i\})$ is another such family, then there is a unique R-homomorphism $f : M \to N$ such that $f \circ g_i = h_i$ for all $i \in I$. It is easy to check that the direct limit defined above is unique up to isomorphism. The direct limit $(M, \{g_i\})$ is denoted by $\varinjlim M_i$.

Theorem 1.5.3. *The inductive limit of an inductive system of R-modules always exists.*

Proof. Let $((M_i), (f_{ji}))$ be an inductive system of R-modules and U be the disjoint union of the M_i. Define a relation on U by $x_i \sim x_j$ if there exists $k \geq i, j$ such that $f_{ki}(x_i) = f_{kj}(x_j)$ where $x_i \in M_i$, $x_j \in M_j$. Then $\sim$ is an equivalence relation. Now let M be the set of equivalence classes under this relation and let $[x]$ denote the equivalence class of x. Define operations on M by $r[x_i] = [rx_i]$ if $r \in R$ and $[x_i] + [x_j] = [y_k + y_k']$ where $k \geq i, j$ and $y_k = f_{ki}(x_i)$, $y_k' = f_{kj}(x_j)$. Then M is an R-module. Now define maps $g_i : M_i \to M$ by $g_i(x_i) = [x_i]$. Then it is easy to see that $(M, \{g_i\})$ is the direct limit. $\qquad \square$

Remark 1.5.4.

1. Let S be the submodule of $\oplus_I M_i$ generated by $e_j \circ f_{ji}(x_i) - e_i(x_i)$ where $x_i \in M_i$ and $e_i : M_i \to \oplus_I M_i$ is the ith embedding. Then the map $\tau :$ $\lim_{\to} M_i \to \oplus_I M_i / S$ defined by $\tau([x_i]) = e_i(x_i) + S$ is an isomorphism.

2. Let $\mathcal{F} = ((M_i), (f_{ji}))$ and $\mathcal{F}' = ((M_i'), (f_{ji}'))$ be inductive systems over the same set I. Then a map $T : \mathcal{F} \to \mathcal{F}'$ is a family of R-homomorphisms $\tau_i : M_i \to M_i'$ such that $f_{ji}' \circ \tau_i = \tau_j \circ f_{ji}$ whenever $i < j$. So T induces an R-homomorphism $\tau : \lim_{\to} M_i \to \lim_{\to} M_i'$ defined by $\tau([x_i]) = [\tau_i(x_i)]$. τ is denoted $\lim_{\to} \tau_i$.

3. If $J \subset I$ is a subset such that for every $i \in I$ there is a $j \in J$ such that $i \leq j$, then we say J is a *cofinal subset* of I. If this is the case then we have an induced inductive system over J. From the proof of Theorem 1.5.3 we easily see that the inductive limit of the induced system over J is isomorphic to the limit of the original system over I.

Example 1.5.5.

1. Let M be an R-module. For any directed set I, set $M_i = M$ for each $i \in I$ and $f_{ji} = \mathrm{id}_M$ for all $i \leq j$. Then $((M), (\mathrm{id}_M))$ is an inductive system called the *constant inductive system*. In this case, $\lim_{\to} M_i = M$.

2. Let I be a directed set and M be an R-module and $\{M_i\}_{i \in I}$ be a family of submodules of M such that for each pair $i, j \in I$, there is a $k \in I$ with $M_i + M_j \subseteq M_k$. Set $i \leq j$ if $M_i \subseteq M_j$ and let $f_{ji} : M_i \to M_j$ be the inclusion. Then $((M_i), (f_{ji}))$ is an inductive system and $\lim_{\to} M_i = \cup_I M_i$. If the M_i are all the finitely generated submodules of M, the condition is satisfied and $\cup M_i = M$. Hence every R-module is the direct limit of its finitely generated submodules.

Theorem 1.5.6. *Let* $\mathcal{F}' = ((M_i'), (f_{ji}'))$, $\mathcal{F} = ((M_i), (f_{ji}))$, $\mathcal{F}'' = ((M_i''), (f_{ji}''))$ *be inductive systems over* I *and suppose there are maps* $\mathcal{F}' \overset{\{\sigma_i\}}{\longrightarrow} \mathcal{F} \overset{\{\tau_i\}}{\longrightarrow} \mathcal{F}''$ *such that* $M_i' \overset{\sigma_i}{\to} M_i \overset{\tau_i}{\to} M_i''$ *is exact for each* $i \in I$, *that is,* $\mathcal{F}' \overset{\{\sigma_i\}}{\longrightarrow} \mathcal{F} \overset{\{\tau_i\}}{\longrightarrow} \mathcal{F}''$ *is exact, then*
$$\lim_{\to} M_i' \overset{\lim_{\to} \sigma_i}{\longrightarrow} \lim_{\to} M_i \overset{\lim_{\to} \tau_i}{\longrightarrow} \lim_{\to} M_i'' \text{ is exact.}$$

Proof. Let $[x] \in \lim_{\to} M_i$ where $x \in M_i$ is such that $(\lim_{\to} \tau_i)([x]) = 0$. So there is a $j \geq i$ such that $f_{ji}'' \circ \tau_i(x) = 0$. But $f_{ji}'' \circ \tau_i = \tau_j \circ f_{ji}$. Therefore there is an $x' \in M_j'$ such that $f_{ji}(x) = \sigma_j(x')$. But $x \sim f_{ji}(x)$. So $[x] = [\sigma_j(x')] = (\lim_{\to} \sigma_i)([x'])$ and thus $[x] \in \mathrm{Im}(\lim_{\to} \sigma_i)$. It is now easy to see that $\lim_{\to} \tau_i \circ \lim_{\to} \sigma_i = 0$ and hence we are done. $\qquad\qquad \square$

Theorem 1.5.7. *Let N be a left R-module and $\mathcal{F} = ((M_i), (f_{ji}))$ be an inductive system of right R-modules. Then*

$$\lim_{\rightarrow}(M_i \otimes_R N) \cong (\lim_{\rightarrow} M_i) \otimes_R N$$

Proof. We note that $((M_i \otimes N), (f_{ji} \otimes \mathrm{id}_N))$ is an inductive system. So we have for each i a homomorphism $g_i \otimes \mathrm{id}_N : M_i \otimes N \to \lim_{\rightarrow} M_i \otimes N$ where $(\lim_{\rightarrow} M_i, \{g_i\})$ is the inductive limit. This induces a homomorphism $g : \lim_{\rightarrow}(M_i \otimes N) \to (\lim_{\rightarrow} M_i) \otimes N$ given by $g([x_i \otimes y]) = g_i(x_i) \otimes y = [x_i] \otimes y$. We claim that g is an isomorphism. For given $y \in N$, we have a map $h_{i,y} : M_i \to \lim_{\rightarrow}(M_i \otimes N)$ defined by $h_{i,y}(x_i) = [x_i \otimes y]$. This induces a map $h_y : \lim_{\rightarrow} M_i \to \lim_{\rightarrow}(M_i \otimes N)$ given by $h_y([x_i]) = [x_i \otimes y]$. But then h_y is balanced on $(\lim_{\rightarrow} M_i) \times N$. So we have a map $h : (\lim_{\rightarrow} M_i) \otimes N \to \lim_{\rightarrow}(M_i \otimes N)$ given by $h([x_i] \otimes y) = h_y([x_i]) = [x_i \otimes y]$. One then easily checks that h is the inverse of g. $\square$

Definition 1.5.8. Let I be a directed set and $\{M_i\}_{i \in I}$ a family of R-modules. Suppose for each $i, j \in I$ with $i \leq j$, there is an R-homomorphism $f_{ij} : M_j \to M_i$ such that

1) $f_{ii} = \mathrm{id}_{M_i}$ for each $i \in I$

2) if $i \leq j \leq k$, then $f_{ij} \circ f_{jk} = f_{ik}$,

then $((M_i), (f_{ij}))$ is called an *inverse system* of the R-modules M_i indexed by I.

Definition 1.5.9. The *inverse* (or *projective*) *limit* of an inverse system $((M_i), (f_{ij}))$ is an R-module M with R-homomorphisms $g_i : M \to M_i$ for $i \in I$ with $g_i = f_{ij} \circ g_j$ whenever $i \leq j$ and such that if $(N, \{h_i\})$ is also such a family, then there is a unique R-homomorphism $f : N \to M$ such that $h_i = g_i \circ f$ for all $i \in I$. The inverse limit $(M, \{g_i\})$ is denoted $\lim_{\leftarrow} M_i$ and is unique up to isomorphism.

Theorem 1.5.10. *The projective limit of an inverse system always exists.*

Proof. Let $((M_i), (f_{ij}))$ be an inverse system. Then for each $i \in I$, let $\pi_i : \prod M_i \to M_i$ be the ith projection map. We set $M = \{(x_i)_I \in \prod M_i : x_i = f_{ij}(x_j) \text{ whenever } i \leq j\}$ and define $g_i : M \to M_i$ by $g_i = \pi_{i|M}$. Then $(M, \{g_i\})$ is an inverse limit. $\square$

Remark 1.5.11. Let $\mathcal{F} = ((M_i), (f_{ij}))$ and $\mathcal{F}'' = ((M_i''), (f_{ij}''))$ be inverse systems over I. Then a map $T : \mathcal{F} \to \mathcal{F}''$ is a family of R-homomorphisms $\tau_i : M_i \to M_i''$ such that $\tau_i \circ f_{ij} = f_{ij}'' \circ \tau_j$ whenever $i < j$. Thus T induces an R-homomorphism $\tau : \lim_{\leftarrow} M_i \to \lim_{\leftarrow} M_i''$ defined by $\tau((x_i)_I) = (\tau_i(x_i))_I$ where $x_i = f_{ij}(x_j)$ and $\tau_i(x_i) = \tau_i \circ f_{ij}(x_j) = f_{ij}'' \circ \tau_j(x_j) = f_{ij}''(\tau_j(x_j))$ whenever $i \leq j$. τ is denoted by $\lim_{\leftarrow} \tau_i$.

Example 1.5.12.

1. If $((M), (\mathrm{id}_M))$ is the constant inverse system, then $\varprojlim M_i = M$.

2. Let $\{M_i\}_{i \in I}$ be a family of submodules of an R-module M that is ordered by reverse inclusion, that is, $i \le j$ implies $M_i \supseteq M_j$. Then $\{M_i\}_{i \in I}$ together with the reverse inclusions form an inverse system and $\varprojlim M_i \cong \cap_I M_i$.

Theorem 1.5.13. *Let* $\mathcal{F}' = ((M_i'), (f_{ij}'))$, $\mathcal{F} = ((M_i), (f_{ij}))$, $\mathcal{F}'' = ((M_i''), (f_{ij}''))$, *be inverse systems over the same directed set and suppose there are maps* $\mathcal{F}' \xrightarrow{\{\sigma_i\}} \mathcal{F} \xrightarrow{\{\tau_i\}} \mathcal{F}''$ *such that* $0 \to M_i' \xrightarrow{\sigma_i} M_i \xrightarrow{\tau_i} M_i''$ *is exact for each* i, *then the induced sequence*

$$0 \to \varprojlim M_i' \xrightarrow{\varprojlim \sigma_i} \varprojlim M_i \xrightarrow{\varprojlim \tau_i} \varprojlim M_i''$$

is exact. If furthermore the set of indices is $\mathbb{N}$ *and if the maps* f_{ij}' *are surjective, then when* $0 \to M_i' \to M_i \to M_i'' \to 0$ *are exact for each* i, *the induced sequence*

$$0 \to \varprojlim M_i' \to \varprojlim M_i \to \varprojlim M_i'' \to 0$$

is exact.

Proof. We use the description of the inverse limit given in Theorem 1.5.10. Then since $\tau_i \circ \sigma_i = 0$ for each i we see that $\varprojlim \tau_i \circ \varprojlim \sigma_i = 0$. If $(x_i) \in \mathrm{Ker}(\varprojlim \tau_i)$, then $\tau_i(x_i) = 0$ for each i. By the exactness of $0 \to M_i' \xrightarrow{\sigma_i} M_i \xrightarrow{\tau_i} M_i''$ there is a $x_i' \in M_i'$ such that $\sigma_i(x_i') = x_i$. If $i \le j$ then $f_{ij}(xj) = x_i$ and so $f_{ij}(\sigma_j(x_j')) = \sigma_i(x_i')$. But $f_{ij}(\sigma_j(x_j')) = \sigma_i(f_{ij}'(x_j'))$. Since σ_i is an injection, $f_{ij}'(x_j') = x_i'$. Thus $(x_i') \in \varprojlim M_i'$ and then clearly $\varprojlim \sigma_i(x_i') = (x_i)$. So $(x_i) \in \mathrm{Im}(\varprojlim \sigma_i)$. Hence the sequence is exact at $\varprojlim M_i$. It is easy to see that $\varprojlim \sigma_i$ is one-to-one.

We now let the index set be $\mathbb{N}$ and consider the module $M = \prod M_i$ and the map $\psi_{\mathcal{F}} : M \to M$ defined by $\psi_{\mathcal{F}}((x_i)) = (x_i - f_{i,i+1}(x_{i+1}))$. Then $\mathrm{Ker}\, \psi_{\mathcal{F}} = \varprojlim M_i$. We can similarly consider modules $M' = \prod M_i'$ and $M'' = \prod M_i''$ and define maps $\psi_{\mathcal{F}'}$ and $\psi_{\mathcal{F}''}$ respectively. So we have the following commutative diagram

$$
\begin{array}{ccccccccc}
0 & \longrightarrow & M' & \longrightarrow & M & \longrightarrow & M'' & \longrightarrow & 0 \\
& & \downarrow{\scriptstyle \psi_{\mathcal{F}'}} & & \downarrow{\scriptstyle \psi_{\mathcal{F}}} & & \downarrow{\scriptstyle \psi_{\mathcal{F}''}} & & \\
0 & \longrightarrow & M' & \longrightarrow & M & \longrightarrow & M'' & \longrightarrow & 0
\end{array}
$$

with exact rows. Then we have an exact sequence $0 \to \mathrm{Ker}\, \psi_{\mathcal{F}'} \to \mathrm{Ker}\, \psi_{\mathcal{F}} \to \mathrm{Ker}\, \psi_{\mathcal{F}''} \to \mathrm{Coker}\, \psi_{\mathcal{F}'} \to \mathrm{Coker}\, \psi_{\mathcal{F}} \to \mathrm{Coker}\, \psi_{\mathcal{F}''} \to 0$ by Proposition 1.2.13. But if f_{ij}' are surjective, then $\psi_{\mathcal{F}'}$ is surjective and so $\mathrm{Coker}\, \psi_{\mathcal{F}'} = 0$. Thus we are done. $\qquad\square$

We now state the following result.

Theorem 1.5.14. *If N is an R-module, then*

(1) $\operatorname{Hom}(N, \varprojlim M_i) \cong \varprojlim \operatorname{Hom}(N, M_i)$,

(2) $\operatorname{Hom}(\varinjlim M_i, N) \cong \varprojlim \operatorname{Hom}(M_i, N)$.

Proof. This is left to the reader.   □

Exercises

1. If A is an abelian group, A is said to be *torsion* if every element of A has finite order and *torsion free* if every element but 0 has infinite order. Given an inductive system $((A_i), (f_{ji}))$ of abelian groups, argue that $\varinjlim A_i$ is torsion (torsion free) if every A_i is torsion (torsion free).

2. Find an inductive system $((A_i), (f_{ji}))$ of abelian groups with $A_i \cong \mathbb{Z}$ for all i and such that $\varinjlim A_i \cong \mathbb{Q}$.

3. a) Let $((M_i), (f_{ji}))$ be an inductive system of R-modules and let X be a set. Define an inductive system $((M_i^{(X)}), (f_{ji}^{(X)}))$ and argue that

$$\varinjlim_i M_i^{(X)} \cong (\varinjlim M_i)^{(X)}.$$

b) If we consider the inductive system $((M_i^X), (f_{ij}^X))$, show that

$$\varinjlim_i M_i^X \ncong (\varinjlim M_i)^X$$

can occur.

Hint: Let p be a prime and let $\mathbb{Z}(p^\infty)$ be the subgroup of $\mathbb{Q}/\mathbb{Z}$ consisting of all $\frac{m}{p^n} + \mathbb{Z}$, $m \in \mathbb{Z}, n \geq 0$. Consider the inductive system described by the diagram

$$\mathbb{Z}(p^\infty) \xrightarrow{p} \mathbb{Z}(p^\infty) \xrightarrow{p} \mathbb{Z}(p^\infty) \to \cdots .$$

Argue that the inductive limit of this system is 0 but that the limit of the system $\mathbb{Z}(p^\infty)^{\mathbb{N}} \to \mathbb{Z}(p^\infty)^{\mathbb{N}} \to \mathbb{Z}(p^\infty)^{\mathbb{N}} \to \cdots$ is not 0.

c) Argue that the limit of the latter system in (b) is torsion free.

4. Prove part (1) of Remark 1.5.4.

5. Let $\{\mathbf{C}^i : i \in I\}$ be a family of complexes. Prove that if I is a directed set, then $H_n(\varinjlim \mathbf{C}^i) \cong \varinjlim H_n(\mathbf{C}^i)$ for all $n \in \mathbb{Z}$.

6. a) Let p be a prime. Consider the inverse system

$$\cdots \to \mathbb{Z}/(p^3) \to \mathbb{Z}/(p^2) \to \mathbb{Z}/(p^1) \to \mathbb{Z}/(p^0)$$

with all maps canonical surjections. The inverse limit is denoted $\hat{\mathbb{Z}}_p$. Argue that $\hat{\mathbb{Z}}_p$ is uncountable.

b) Consider the inverse system $\cdots \to \mathbb{Z} \overset{id}{\to} \mathbb{Z} \overset{id}{\to} \mathbb{Z}$. Use this system and the system of (a) to argue that the conclusions of Theorem 1.5.13 can fail to be true if we drop the hypothesis that the maps f_{ij} are surjective.

7. Prove Theorem 1.5.14.

1.6 I-adic topology and completions

Throughout this section, R will be a commutative ring.

Definition 1.6.1. Let I be an ideal of R and M be an R-module. Then $M \supset IM \supset I^2 M \supset I^3 M \supset \cdots$ and so we have R-homomorphisms $f_{ij} : M/I^j M \to M/I^i M$ defined by $f_{ij}(x + I^j M) = x + I^i M$ whenever $i \leq j$. Thus $((M/I^i M), (f_{ij}))$ is an inverse system over $\mathbb{Z}_+$ and so has the projective limit $\varprojlim M/I^i M$. We note that

$$\varprojlim M/I^i M = \{(x_1 + IM, \ldots) : x_i + I^i M = f_{i,i+1}(x_{i+1} + I^{i+1} M)\}$$

$$= \{(x_1 + IM, x_2 + I^2 M, \ldots) : x_i \equiv x_{i+1} \bmod I^i M\}.$$

It is easy to see that $\{x + I^i M\}_{i \geq 0}$ is a basis for a topology on M. For clearly it covers M and if $z \in (x + I^i M) \cap (y + I^j M)$, then $z + I^k M = (x + I^i M) \cap (y + I^j M)$ where $k = \max(i, j)$. The topology generated by $\{x + I^i M\}$ is called the I-adic topology of M. It is easy to see that in this topology, addition and scalar multiplication are continuous, and if $M = R$, then multiplication is also continuous so that R is a topological ring.

We note that if N is a submodule of M, then the decreasing sequences $(I^i M \cap N)$ and $((I^i M + N)/N)$ determine the *subspace topology* of N and the *quotient topology* of M/N induced by the I-adic topology of M, respectively.

Proposition 1.6.2. *M is Hausdorff if and only if $\cap I^i M = 0$*

Proof. Suppose M is Hausdorff. Let $x \in \cap I^i M$, $x \neq 0$. Then there exist i, j such that $(x + I^i M) \cap I^j M = \emptyset$. But then $x \notin I^j M$, a contradiction. Conversely, suppose $\cap I^i M = 0$ and $x \neq y$. Then there is a k such that $x + I^k M \neq y + I^k M$ for otherwise $x - y \in I^i M$ for each i and so $x = y$. So $(x + I^k M) \cap (y + I^k M) = \emptyset$. $\square$

Remark 1.6.3. If N is a submodule of M, then the *closure* of N in M with respect to the I-adic topology of M is defined by $\bar{N} = \cap(I^i M + N)$. So M/N is Hausdorff with respect to the quotient topology if and only if $\cap(I^i M + N) = N$ and thus if and only if N is closed in M with respect to the I-adic topology.

Definition 1.6.4. A sequence $\{x_n\}$ of elements of an R-module M is said to be a *cauchy sequence* in the I-adic topology if given any nonnegative integer k there exists a nonnegative integer n_0 such that $x_{i+1} - x_i \in I^k M$ whenever $i \geq n_0$. $\{x_n\}$ is said to be *convergent* if there is an $x \in M$ such that given any k there is an n_0 such that $x_n - x \in I^k M$ whenever $n \geq n_0$. x is called a *limit* of the sequence $\{x_n\}$. We note that the limit is unique if M is Hausdorff (that is, if $\cap I^i M = 0$ by the proposition above) and that every convergent sequence is a cauchy sequence.

An R-module M is said to be *complete* in its I-adic topology if every cauchy sequence in M converges. Now let C be the set of all cauchy sequences in M in the I-adic topology. Define addition and scalar multiplication on C by $\{x_n\} + \{y_n\} = \{x_n + y_n\}$ and $r\{x_n\} = \{rx_n\}$ where $r \in R$. Then C is an R-module. Now let C_0 be the subset of C consisting of those cauchy sequences that converge to zero. Then C_0 is a submodule of C. The quotient R-module C/C_0 is called the *I-adic completion* of M and is denoted by $\hat{M}$. So we have the following result.

Proposition 1.6.5. *Let $\varphi : M \to \hat{M}$ be defined by $\varphi(x) = \{x\} + C_0$. Then*

(1) *φ is an R-homomorphism and $\text{Ker } \varphi = \cap I^i M$.*

(2) *$\varphi(x) = \{x_n\} + C_0$ if and only if x is a limit of $\{x_n\}$.*

(3) *φ is a monomorphism if and only if M is Hausdorff.*

(4) *φ is an epimorphism if and only if M is complete.*

Proof. (1) φ is clearly a homomorphism. Now let $\varphi(x) = 0$. Then $\{x\}$ is in C_0 and so $x \in \cap I^i M$. On the other hand, if $x \in \cap I^i M$, then $x \in I^i M$ for each i and so $\{x\}$ converges to zero. That is, $\{x\}$ is in C_0 and so $\varphi(x) = C_0$. Thus we are done.

(2) We simply note that $\varphi(x) = \{x_n\} + C_0$ if and only if $\{x\} - \{x_n\}$ is in C_0 if and only if $\{x - x_n\}$ converges to zero and if and only if $\{x_n\}$ converges to x.

(3) follows from Proposition 1.6.2 and part (1) above.

(4) φ is an epimorphism if and only if for each $\{x_n\}$ in C, there is an $x \in M$ such that $\varphi(x) = \{x_n\} + C_0$ if and only if each $\{x_n\}$ converges by part (2) above. $\square$

Remark 1.6.6. We see that if $\varphi : M \to \hat{M}$ is an epimorphism, then M is complete and $M/\cap I^i M \cong \hat{M}$. Furthermore, φ is an isomorphism if and only if M is Hausdorff and complete. In this case $M \cong \hat{M}$.

Theorem 1.6.7. $\hat{M} \cong \varprojlim M/I^i M$.

Proof. Let $(x_1 + IM, x_2 + I^2 M, \dots) \in \varprojlim M/I^i M$. Then $x_{i+1} - x_i \in I^i M$ for each $i \geq 1$ and so $\{x_n\}$ is a cauchy sequence. Now define a map $\sigma : \varprojlim M/I^i M \to \hat{M}$ by $\sigma((x_1 + IM, x_2 + I^2 M, \dots)) = \{x_n\} + C_0$. Then σ is well-defined for if $x_n + I^n M = y_n + I^n M$, then $x_n - y_n \in I^n M$ for each n and thus $\{x_n - y_n\}$ converges to zero. So $\{x_n - y_n\} \in C_0$ and hence $\{x_n\} + C_0 = \{y_n\} + C_0$. Clearly, σ is a homomorphism.

Now let $\{x_n\} \in C$. Then for each n there is an integer $s(n)$ such that $x_{i+1} - x_i \in I^n M$ whenever $i \geq s(n)$. So $x_i + I^n M = x_j + I^n M$ for $i, j \geq s(n)$. Thus we have a well-defined map $\tau' : C \to \varprojlim M/I^i M$ given by $\tau'(\{x_n\}) = (x_i + I^n M)$ where $i \geq s(n)$. τ' is an R-homomorphism. But if $\{x_n\}$ is in C_0, then $x_i \in I^i M$ for all sufficiently large i. So $\tau'(\{x_i\}) = (x_i + I^n M)$ where $x_i \in I^i M \subset I^n M$ for sufficiently large i. Thus $\tau'(\{x_n\}) = (0)$ and so τ' induces a homomorphism $\tau : \hat{M} \to \varprojlim M/I^i M$ defined by $\tau(\{x_n\} + C_0) = (x_i + I^n M)$ where $i \geq s(n)$. It is now standard to argue that $\tau = \sigma^{-1}$. $\square$

In view of Theorem 1.6.7, we will refer to $\varprojlim M/I^i M$ as the *I-adic completion* of M and simply write $\hat{M} = \varprojlim M/I^i M$.

Example 1.6.8.

1. Let $M = R = \mathbb{Z}$ and $I = (p)$, p a prime. In this case $\mathbb{Z}/(p^{i+1}) \to \mathbb{Z}/(p^i)$ is the natural map $r + (p^{i+1}) \mapsto r + (p^i)$ and $\hat{\mathbb{Z}}_p = \varprojlim \mathbb{Z}/(p^i) = \{(r_1 + (p), r_2 + (p^2), \dots) : r_i \cong r_{i+1} \bmod p^i\}$. We note that $(r_1 + (p), r_2 + (p^2), \dots)$ has a unique representation $(a_0 + (p), a_0 + a_1 p + (p^2), a_0 + a_1 p + a_2 p^2 + (p^3), \dots)$ with $0 \leq a_i < p$. Hence to each element in $\hat{\mathbb{Z}}_p$, we can associate a unique *p-adic number* $\sum_{i=0}^{\infty} a_i p^i$ where $0 \leq a_i < p$. $\hat{\mathbb{Z}}_p$ is called the *ring of p-adic integers*.

2. Let $R = S[x_1, \dots, x_n]$ with S any ring and $I = \langle x_1, \dots, x_n \rangle$. Then $\hat{R} = S[[x_1, \dots, x_n]]$.

Exercises

1. Prove Remark 1.6.3.

2. If $I \subset R$ is a finitely generated ideal of the commutative ring R, prove that

   a) I^n is finitely generated for each $n \geq 1$.

   b) If $(M_s)_{s \in S}$ is a family of R-modules, then $I \prod_{s \in S} M_s = \prod_{s \in S}(I M_s)$.

   c) Let $M = \prod_{s \in S} M_s$ and $\hat{M}$ be the I-adic completion of M and $\hat{M}_s$ that of M_s for each s. Show that $\hat{M} \cong \prod_{s \in S} \hat{M}_s$.

3. Using the hypotheses and notation of (2), let $N = \bigoplus_{s \in S} M_s$ and argue that $\hat{N} \ncong \bigoplus_{s \in S} \hat{M}_s$ can happen.

   Hint: Let $R = \mathbb{Z}$, $S = \mathbb{N}$, $M_n = \mathbb{Z}/(p^n)$, and $I = (p)$. Argue that $\hat{N}$ is not countable.

4. Let $n = p_1 p_2 \ldots p_s$ where $p_1, p_2, \ldots, p_s$ are distinct primes of $\mathbb{Z}$. Use the fact that $\mathbb{Z}/(n^k) \cong \mathbb{Z}/(p_1^k) \oplus \cdots \oplus \mathbb{Z}/(p_s^k)$ for $k \geq 1$ to argue that if $\hat{\mathbb{Z}}$ is the completion of $\mathbb{Z}$ with the (n)-adic topology, then $\hat{\mathbb{Z}} = \hat{\mathbb{Z}}_{p_1} \times \hat{\mathbb{Z}}_{p_2} \times \cdots \times \hat{\mathbb{Z}}_{p_s}$.

5. Let $I \subset R$ be an ideal of the ring R. Suppose that R is complete with respect to the I-adic topology. Prove that for every $r \in I$, $1 + r$ is a unit of R.

6. Let $J \subset I \subset R$ be ideals of R such that R is complete with respect to the I-adic topology.

   a) Argue that R is complete with respect to the J-adic topology.

   b) If the R-module R/J is Hausdorff with respect to the I-adic topology, argue that R/J is complete with respect to the I/J-topology.

7. Prove Example 1.6.8(2).

Chapter 2

Flat Modules, Chain Conditions and Prime Ideals

2.1 Flat modules

We start with the following

Definition 2.1.1. An R-module P is said to be *projective* if given an exact sequence $A \xrightarrow{\psi} B \to 0$ of R-modules and an R-homomorphism $f : P \to B$, there exists an R-homomorphism $g : P \to A$ such that $f = \psi \circ g$. Thus every free module is projective. So every R-module M has a *projective resolution*, that is, an exact sequence $\cdots \to P_1 \to P_0 \to M \to 0$ with each P_i projective by Remark 1.2.11.

Theorem 2.1.2. *The following are equivalent for an R-module P.*

(1) *P is projective.*

(2) $\mathrm{Hom}(P, -)$ *is right exact.*

(3) *Every exact sequence $0 \to A \to B \to P \to 0$ is split exact.*

(4) *P is a direct summand of a free R-module.*

Proof. (1) $\Rightarrow$ (2) is clear.

(2) $\Rightarrow$ (3) follows from Proposition 1.2.15 since if $B \xrightarrow{\sigma} P \to 0$ is exact then $\mathrm{Hom}(P, B) \to \mathrm{Hom}(P, P) \to 0$ is exact and so $B \to P \to 0$ splits.

(3) $\Rightarrow$ (4) follows from Proposition 1.2.2 easily.

(4) $\Rightarrow$ (1). Let P be a direct summand of a free R-module F. Then there is a map $s : F \to P$ such that $s \circ i = \mathrm{id}_P$ where $i : P \to F$ is the inclusion. Now let $A \xrightarrow{\psi} B \to 0$ be exact and $f : P \to B$ be an R-homomorphism. Then there is a map $g : F \to A$ such that $\psi \circ g = f \circ s$. But then $\psi \circ g \circ i = f \circ s \circ i = f$. Thus P is projective. $\qquad\square$

Definition 2.1.3. An R-module F is said to be *flat* if given any exact sequence $0 \to A \to B$ of right R-modules, the tensored sequence $0 \to A \otimes_R F \to B \otimes_R F$ is exact.

Proposition 2.1.4. *The direct sum $\bigoplus_{i \in I} F_i$ is flat if and only if each F_i is flat.*

Proof. This easily follows by Proposition 1.2.22.   □

Corollary 2.1.5. *Every projective module is flat.*

Proof. Let P be projective. Then P is a summand of a free by Theorem 2.1.2. But R is a flat R-module and so every free R-module is flat by Proposition 2.1.4 above. Thus P is a direct summand of a flat module and hence is flat again by Proposition 2.1.4.□

Definition 2.1.6. It follows from the above that every R-module has a *flat resolution*, that is, an exact sequence $\cdots \to F_1 \to F_0 \to M \to 0$ with each F_i flat.

Proposition 2.1.7. *If F is a flat R-module and I is a right ideal of R, then $I \otimes F \cong IF$.*

Proof. We consider the exact sequence $0 \to I \to R$. Then $0 \to I \otimes F \to F$ is exact. But the image of $I \otimes F$ in F under this embedding is IF. So we are done. □

Let $\cdots \to P_1 \to P_0 \to M \to 0$ be a projective resolution of a right R-module M and N be a left R-module. Then the ith homology module of the complex $\cdots \to P_1 \otimes N \to P_0 \otimes N \to 0$ is denoted $\mathrm{Tor}_i^R(M, N)$. Note that $\mathrm{Tor}_0^R(M, N) = M \otimes N$ since $P_1 \otimes N \to P_0 \otimes N \to M \otimes N \to 0$ is exact, and $\mathrm{Tor}_i(M, F) = 0$ for all flat left R-modules F. $\mathrm{Tor}_i^R(M, N)$ can also be computed using a projective resolution of N and is independent of the projective resolutions used, and moreover given an exact sequence $0 \to M' \to M \to M'' \to 0$ of right R-modules there exists a long exact sequence $\cdots \to \mathrm{Tor}_1(M'', N) \to M' \otimes N \to M \otimes N \to M'' \otimes N \to 0$ (see Chapter 8 for details). Using such a long exact sequence and Theorem 1.5.7, one can easily show by induction that Tor commutes with direct limits (see Exercise 4).

Theorem 2.1.8. *The following are equivalent for an R-module F.*

(1) F *is flat.*

(2) $- \otimes_R F$ *is left exact.*

(3) $\mathrm{Tor}_i^R(M, F) = 0$ *for all right R-modules M and for all $i \geq 1$.*

(4) $\mathrm{Tor}_1^R(M, F) = 0$ *for all right R-modules M.*

(5) $\mathrm{Tor}_1^R(M, F) = 0$ *for all finitely generated right R-modules M.*

Proof. $(1) \Rightarrow (2) \Rightarrow (3)$ is clear from the above while $(3) \Rightarrow (4) \Rightarrow (5)$ is trivial.
    $(5) \Rightarrow (4)$ since Tor commutes with direct limits.
    $(4) \Rightarrow (1)$. We consider the exact sequence $0 \to A \to B \to C \to 0$ of right R-modules. Then we have an exact sequence $0 = \mathrm{Tor}_1^R(C, F) \to A \otimes_R F \to B \otimes_R F \to C \otimes_R F \to 0$. Thus F is flat.   □

Remark 2.1.9. Since Tor commutes with direct limits, we also see that any direct limit of flat R-modules is also flat.

We now recall the following natural identity and leave its proof as an exercise.

Theorem 2.1.10. *Let R and S be rings, A a left S-module, B an (R, S)-bimodule, and C a left R-module. Then the map*

$$\varphi : \mathrm{Hom}_S(A, \mathrm{Hom}_R(B, C)) \to \mathrm{Hom}_R(B \otimes_S A, C)$$

defined by $\varphi(f)(b \otimes a) = (f(a))(b)$ where $f \in \mathrm{Hom}_S(A, \mathrm{Hom}_R(B, C))$, $a \in A$, $b \in B$, is an isomorphism.

Theorem 2.1.11. *Let R and S be commutative rings and let $R \to S$ be a ring homomorphism that makes S into a flat left R-module. If M and N are R-modules, then*

$$\mathrm{Tor}_i^R(M, N) \otimes_R S \cong \mathrm{Tor}_i^S(M \otimes_R S, N \otimes_R S)$$

for all $i \geq 0$.

Proof. Let $\cdots \to P_1 \to P_0 \to M \to 0$ be a projective resolution of M. Then $\cdots \to P_1 \otimes_R S \to P_0 \otimes_R S \to M \otimes_R S \to 0$ is exact. But $P_i \otimes_R S$ is a projective S-module for each i. For if $A \to B \to 0$ is an exact sequence of S-modules, then $\mathrm{Hom}_S(S, A) \to \mathrm{Hom}_S(S, B) \to 0$ is an exact sequence of R-modules. But then $\mathrm{Hom}_R(P_i, \mathrm{Hom}_S(S, A)) \to \mathrm{Hom}_R(P_i, \mathrm{Hom}_S(S, B)) \to 0$ is exact since P_i is projective. Hence $\mathrm{Hom}_S(P_i \otimes_R S, A) \to \mathrm{Hom}_S(P_i \otimes_R S, B) \to 0$ is exact by the theorem above. So $\cdots \to P_1 \otimes_R S \to P_0 \otimes_R S \to M \otimes_R S \to 0$ is a projective resolution of the S-module $M \otimes_R S$. Thus if $\mathbf{P_\bullet}$ denotes the *deleted complex* $\cdots \to P_1 \to P_0 \to 0$ of M, then since $- \otimes_R S$ commutes with homology (see Exercise 2 of Section 1.4) we have that

$$\begin{aligned}
\mathrm{Tor}_i^R(M, N) \otimes_R S &\cong H_i(\mathbf{P_\bullet} \otimes_R N) \otimes_R S \\
&\cong H_i((\mathbf{P_\bullet} \otimes_R N) \otimes_R S) \\
&\cong H_i((\mathbf{P_\bullet} \otimes_R S) \otimes_S (N \otimes_R S)) \\
&\cong \mathrm{Tor}_i^S(M \otimes_R S, N \otimes_R S).
\end{aligned}$$

$\square$

Definition 2.1.12. An R-module F is said to be *faithfully flat* if $0 \to A_R \to B_R$ is an exact sequence of R-modules if and only if $0 \to A \otimes_R F \to B \otimes_R F$ is exact. It is easy to see that every free R-module is faithfully flat.

Lemma 2.1.13. *The following are equivalent for a left R-module F.*

(1) *F is faithfully flat.*

(2) *F is flat and for any right R-module N, $N \otimes F = 0$ implies $N = 0$*

(3) *F is flat and $\mathfrak{m}F \neq F$ for every maximal right ideal $\mathfrak{m}$ of R.*

Proof. (1) $\Rightarrow$ (2). We consider the sequence $0 \to N \to 0$. $0 \to N \otimes F \to 0$ is exact by assumption and so $0 \to N \to 0$ is exact by (1).

(2) $\Rightarrow$ (3). This follows from the fact that $F/\mathfrak{m}F \cong (R/\mathfrak{m}) \otimes_R F$.

(3) $\Rightarrow$ (2). Suppose $N \neq 0$ and let $x \in N, x \neq 0$. Then $xR = R/I$ for some right ideal I of R. Let $\mathfrak{m}$ be a maximal right ideal containing I. Then $F \neq \mathfrak{m}F \supset IF$ by assumption. So $xR \otimes F = F/IF \neq 0$. But F is flat and so $0 \to xR \otimes F \to N \otimes F$ is exact. Hence $N \otimes F \neq 0$.

(2) $\Rightarrow$ (1). Suppose $0 \to A \xrightarrow{f} B$ is a sequence of R-modules. If $0 \to A \otimes F \to B \otimes F$ is exact, then $(\mathrm{Ker}\, f) \otimes F = 0$ since F is flat. So $\mathrm{Ker}\, f = 0$ by assumption and hence $0 \to A \to B$ is exact. $\square$

Exercises

1. Give an example of a projective module that is not free.
   Hint: Consider $R = \mathbb{Z}/6\mathbb{Z} = \mathbb{Z}/2\mathbb{Z} \oplus \mathbb{Z}/3\mathbb{Z}$.

2. Prove that if $f : M \to P$ is an epimorphism with P projective, then $M \cong \mathrm{Ker}\, f \oplus P$.

3. Prove Proposition 2.1.4.

4. Let $((M_i), (f_{ji}))$ be inductive system of right R-modules and N be a left R-module. Prove that $\varinjlim \mathrm{Tor}_n^R(M_i, N) \cong \mathrm{Tor}_n^R(\varinjlim M_i, N)$. Hint: Use a *dimension shifting* argument, that is, an induction argument on n involving the long exact sequence corresponding to a short exact sequence $0 \to K \to P \to N \to 0$ with P projective.

5. Prove that if every finitely generated submodule of M is flat, then M is flat.

6. Prove that if R is an integral domain, then every flat R-module is *torsion free* (that is, $rx = 0$ for $r \in R, x \in M$ implies $r = 0$ or $x = 0$). Conclude that $\mathbb{Z}/n\mathbb{Z}$ is not a flat $\mathbb{Z}$-module for $n \geq 2$.

7. Prove that the quotient field of an integral domain R is a flat R-module.

8. Consider the exact sequence $0 \to M' \to M \to M'' \to 0$ of R-modules. Prove that if M'' is flat, then M' is flat if and only if M is flat. Give an example to show that M' and M can be flat without M'' being flat.

9. Let M be an $(R\text{-}S)$-bimodule and N be a left S-module. Prove that if M is a flat R-module and N is a flat S-module, then $M \otimes_S N$ is a flat R-module.

10. (Schanuel's lemma) If $0 \to K_1 \to P_1 \to M \to 0$ and $0 \to K_2 \to P_2 \to M \to 0$ are exact with P_1, P_2 projective, then $K_1 \oplus P_2 \cong K_2 \oplus P_1$.

11. Prove Theorem 2.1.10.

12. Show that the $\mathbb{Z}$-module $\mathbb{Q}$ is flat but not faithfully flat.

13. Prove that if R' is a subring of R, then R/R' is a flat R'-module if and only if R is a faithfully flat R'-module.

2.2 Localization

In this section, R will denote a commutative ring.

Definition 2.2.1. Let S be a *multiplicative subset* of R, that is, $1 \in S$ and S is closed under multiplication. Then the *localization* of R with respect to S, denoted by $S^{-1}R$, is the set of all equivalence classes (a, s) with $a \in R$, $s \in S$ under the equivalence relation $(a, s) \sim (b, t)$ if there is an $s' \in S$ such that $(at - bs)s' = 0$. It is easy to check that this relation is indeed an equivalence relation. The equivalence class (a, s) is denoted by a/s.

We now define addition and multiplication on $S^{-1}R$ by

$$a/s + b/t = (at + bs)/st$$
$$(a/s)(b/t) = ab/st.$$

These operations are well-defined and $S^{-1}R$ is then a commutative ring with identity.

Remark 2.2.2. We note that $S^{-1}R = 0$ if and only if $0 \in S$. The map $\varphi : R \to S^{-1}R$ defined by $\varphi(a) = a/1$ is a homomorphism with $\mathrm{Ker}\,\varphi = \{a \in R : as' = 0 \text{ for some } s' \in S\}$. As a consequence, if S has no zero divisors, then φ is monic. Moreover, if R is a domain, then $S^{-1}R$ is the quotient field of R when S is the set of all nonzero elements of R.

Definition 2.2.3. Let $S \subset R$ be a multiplicative set and M be an R-module. Then the *localization* of M with respect to S, denoted $S^{-1}M$ is defined as for $S^{-1}R$. $S^{-1}M$ is an abelian group under addition and is an $S^{-1}R$-module via $(a/s) \cdot (x/t) = ax/st$.

Proposition 2.2.4. *Let $S \subset R$ be a multiplicative set. Then:*

(1) *If $f : M \to N$ is an R-module homomorphism, then $S^{-1}f : S^{-1}M \to S^{-1}N$ defined by $(S^{-1}f)(x/s) = f(x)/s$ is an $S^{-1}R$-module homomorphism.*

(2) *If $M' \to M \to M''$ is exact at M, then $S^{-1}M' \to S^{-1}M \to S^{-1}M''$ is exact at $S^{-1}M$.*

(3) *If $N \subset M$ are R-modules, then $S^{-1}(M/N) \cong S^{-1}(M)/S^{-1}(N)$.*

(4) *If M is an R-module, then $S^{-1}R \otimes_R M \cong S^{-1}M$.*

(5) *$S^{-1}R$ is a flat R-module.*

Proof. The proof of (1) and (2) are left to the reader.

(3) follows from (2) by considering the exact sequence $0 \to N \to M \to M/N \to 0$.

(4) Define a map $\varphi : S^{-1}R \otimes_R M \to S^{-1}M$ by $\varphi((a/s) \otimes x) = (ax)/s$. Then φ is a well-defined $S^{-1}R$-homomorphism. φ is clearly onto. Now suppose $(ax)/s = 0$.

Then there is an $s' \in S$ such that $as'x = 0$. So $(a/s) \otimes x = (as'/ss') \otimes x = (1/ss') \otimes as'x = 0$. Thus φ is one-to-one.

(5) follows from parts (3) and (4). $\qquad \square$

Remark 2.2.5. It is now easy to see that if M is a free (projective) R-module, then $S^{-1}M$ is a free (projective) $S^{-1}R$-module, and that if M is a finitely generated R-module, then $S^{-1}M$ is also such as an $S^{-1}R$-module. Moreover since $S^{-1}M \cong S^{-1}R \otimes M$, if M is a flat R-module, then it is easy to check that $S^{-1}M$ is a flat $S^{-1}R$-module.

Lemma 2.2.6. *Let $S \subset R$ be a multiplicative set. If J is an ideal of $S^{-1}R$, then $J = IS^{-1}R = S^{-1}I$ for some ideal I of R.*

Proof. Let $I = J \cap R$ (or, more precisely, I is the inverse image of J under $R \to S^{-1}R$). Then I is an ideal of R. Clearly $IS^{-1}R \subset J$. Now let $a = r/s \in J$. Then $a = (r/1)(1/s)$. So it suffices to show that $r \in I$. For then $a \in IS^{-1}R$. But $r/1 = (a/1)(s/1) \in J$ and so $r \in J \cap R = I$. Thus $J = IS^{-1}R$. But $IS^{-1}R \cong S^{-1}R \otimes I$ by Proposition 2.1.7 since $S^{-1}R$ is a flat R-module by Proposition 2.2.4. So $IS^{-1}R \cong S^{-1}I$ again by Proposition 2.2.4. $\qquad \square$

Proposition 2.2.7. *Let $S \subset R$ be a multiplicative set. Then there is a one-to-one order preserving correspondence between the prime ideals of $S^{-1}R$ and the prime ideals of R disjoint from S given by $S^{-1}\mathfrak{p} \leftrightarrow \mathfrak{p}$.*

Proof. Let J be a prime ideal of $S^{-1}R$, and let $\mathfrak{p} = J \cap R$. Then $\mathfrak{p}$ is a prime ideal of R. But then $J = \mathfrak{p}S^{-1}R = S^{-1}\mathfrak{p}$ by Lemma 2.2.6. If $\mathfrak{p} \cap S \neq \emptyset$, then $1 \in S^{-1}\mathfrak{p} = J$, a contradiction. Hence $\mathfrak{p} \cap S = \emptyset$.

Now suppose $\mathfrak{p}$ is a prime ideal of R disjoint from S. We claim that $S^{-1}\mathfrak{p}$ is a prime ideal. But $1 \notin S^{-1}\mathfrak{p}$ since $\mathfrak{p} \cap S = \emptyset$. Moreover, if $(a/s) \cdot (b/t) \in S^{-1}\mathfrak{p}$ with $s, t \in S$, then $(a/s) \cdot (b/t) = c/r$ for some $c \in \mathfrak{p}$, $r \in S$. So there is an $s' \in S$ such that $(abr - stc)s' = 0$. But $stcs' \in \mathfrak{p}$. So $abrs' \in \mathfrak{p}$ where $rs' \in S$. But then $ab \in \mathfrak{p}$ and so $a \in \mathfrak{p}$ or $b \in \mathfrak{p}$. That is, $a/s \in S^{-1}\mathfrak{p}$ or $b/s \in S^{-1}\mathfrak{p}$. Hence $S^{-1}\mathfrak{p}$ is a prime ideal of $S^{-1}R$. $\qquad \square$

Definition 2.2.8. Let $\mathfrak{p}$ be a prime ideal of R. Then $S = R - \mathfrak{p}$ is a multiplicative subset of R. In this case $S^{-1}R$, $S^{-1}M$, and $S^{-1}f$ are denoted by $R_{\mathfrak{p}}$, $M_{\mathfrak{p}}$, and $f_{\mathfrak{p}}$ respectively where M is an R-module. We say that $M_{\mathfrak{p}}$ is the *localization* of M at $\mathfrak{p}$.

As a consequence of Proposition 2.2.7, we have the following result.

Theorem 2.2.9. *Let $\mathfrak{p}$ be a prime ideal of R. Then there is a one-one order preserving correspondence between the prime ideals of $R_{\mathfrak{p}}$ and the prime ideals of R contained in $\mathfrak{p}$.*

Remark 2.2.10. Let $\mathfrak{p}$ be a prime ideal of R. Then $\mathfrak{p}R_\mathfrak{p}$ is a prime ideal of $R_\mathfrak{p}$ from the above. But if J is an ideal of $R_\mathfrak{p}$, then $J = IR_\mathfrak{p}$ where I is an ideal of R such that $I \cap (R - \mathfrak{p}) = \emptyset$. So $I \subset \mathfrak{p}$ and hence $J = IR_\mathfrak{p} \subset \mathfrak{p}R_\mathfrak{p}$. Thus $\mathfrak{p}R_\mathfrak{p}$ is the maximal ideal of $R_\mathfrak{p}$.

We recall that a ring R is said to be *local* if it has only one maximal ideal. So the localization of R at a prime ideal $\mathfrak{p}$ is a local ring with maximal ideal $\mathfrak{p}R_\mathfrak{p}$. The field $R_\mathfrak{p}/\mathfrak{p}R_\mathfrak{p}$ is called the *residue field* of $R_\mathfrak{p}$ and is denoted by $k(\mathfrak{p})$.

Exercises

1. Prove that the relation in Definition 2.2.1 is indeed an equivalence relation.

2. Prove that the operations in Definition 2.2.1 are well-defined.

3. Prove parts (1), (2), (5) of Proposition 2.2.4.

4. Prove Remark 2.2.5.

5. Let M_1, M_2 be submodules of M. Prove that

   a) $S^{-1}(M_1 + M_2) = S^{-1}M_1 + S^{-1}M_2$
   b) $S^{-1}(M_1 \cap M_2) = S^{-1}M_1 \cap S^{-1}M_2$

6. Let $((M_i), (f_{ji}))$ be a direct system of R-modules. Prove that

$$\lim_{\rightarrow} S^{-1}M_i \cong S^{-1} \lim_{\rightarrow} M_i.$$

7. Let M and N be R-modules. Prove that

   a) $M = 0$ if and only if $M_\mathfrak{m} = 0$ for all maximal ideals $\mathfrak{m}$ of R.
   b) An R-homomorphism $f : M \rightarrow N$ is monic(epic) if and only if $f_\mathfrak{m} : M_\mathfrak{m} \rightarrow N_\mathfrak{m}$ is monic(epic) for all maximal ideals $\mathfrak{m}$ of R.
   c) M is a flat R-module if and only if $M_\mathfrak{m}$ is flat for all maximal ideals $\mathfrak{m}$ of R.

2.3 Chain conditions

In this section, we introduce some of the basic concepts concerning artinian and noetherian rings and modules.

Definition 2.3.1. An R-module M is said to be *noetherian* (*artinian*) if every ascending (descending) chain of submodules of M terminates.

Remark 2.3.2. It is easy to see that an R-module M is noetherian (artinian) if and only if every nonempty set of submodules of M has a maximal (minimal) element. For suppose there is a nonempty set of submodules of M that has no maximal element. Let M_1 be an element of this set. Then M_1 is not maximal. So there is an element M_2 in the set such that $M_1 \subsetneq M_2$. Repeat the argument to get a chain of submodules $M_1 \subsetneq M_2 \subsetneq M_3 \cdots$ of M that never terminates. The converse is easy to see.

The proof for artinian modules is similar.

Proposition 2.3.3. *An R-module M is noetherian if and only if every submodule of M is finitely generated.*

Proof. Suppose there is a submodule S of M that is not finitely generated. Let $x_1 \in S$. Then $S \neq \langle x_1 \rangle$. So let $x_2 \in S - \langle x_1 \rangle$. Then $\langle x_1 \rangle \subsetneq \langle x_1, x_2 \rangle \subsetneq S$. Repeat the process to get a strictly increasing chain of R-modules that never terminates.

Now suppose $M_1 \subset M_2 \subset \cdots$ is an ascending chain of submodules of M. Then $S = \bigcup_{i=1}^{\infty} M_i$ is a submodule of M and so is finitely generated by assumption. Let $S = \langle x_1, x_2, \ldots, x_n \rangle$ for some $x_1, \ldots, x_n \in M$. Then each $x_i \in M_{m_i}$ for some m_i. So let $m = \max(m_1, m_2, \ldots, m_n)$. Then $x_1, \ldots, x_n \in M_m$ and so $S = \langle x_1, \ldots, x_n \rangle \subset M_m \subset S$. Thus $M_m = S$ and hence the chain terminates. □

Definition 2.3.4. A ring R is said to be *left (right) noetherian (artinian)* if it is noetherian (artinian) as a left (right) module over itself. Noetherian (artinian) will always mean left noetherian (left artinian).

We now have the following

Corollary 2.3.5. *A ring R is noetherian if and only if every left ideal of R is finitely generated.*

Lemma 2.3.6. *Let $0 \to M' \to M \to M'' \to 0$ be exact with $M' \subset M$, $M'' = M/M'$ with obvious maps. Suppose S_1, S_2 are submodules of M such that $S_1 \subset S_2$ and $S_1 \cap M' = S_2 \cap M'$. If $(M' + S_1)/M' = (M' + S_2)/M'$, then $S_1 = S_2$.*

Proof. This is left to the reader. □

Proposition 2.3.7. *Let $0 \to M' \to M \to M'' \to 0$ be an exact sequence of R-modules. Then M is noetherian (artinian) if and only if M' and M'' are noetherian (artinian).*

Proof. We can suppose the sequence is as above. If M is noetherian (artinian), then clearly so are M' and M''. Now suppose $\{M_i\}$ form an ascending (descending) chain of submodules of M. Then $\{M_i \cap M'\}$ and $\{(M' + M_i)/M'\}$ form an ascending (descending) chain of submodules of M', M'' respectively. If these chains terminate, then so does the chain for $\{M_i\}$ by Lemma 2.3.6 above. Thus the result follows. □

Remark 2.3.8. It now easily follows that if R is noetherian (artinian), then every quotient R/I is noetherian (artinian) and if R is furthermore commutative then every localization $S^{-1}R$ is also noetherian (artinian).

Corollary 2.3.9. *A finite direct sum of noetherian (artinian) R-modules is also noetherian (artinian).*

Proof. Let $\{M_i\}_{i=1}^n$ be a family of noetherian R-modules. Then we consider the exact sequence $0 \to M_n \to \bigoplus_{i=1}^n M_i \to \bigoplus_{i=1}^{n-1} M_i \to 0$. Then $\bigoplus_{i=1}^n M_i$ is noetherian by induction on n. Similarly for artinian.   □

Corollary 2.3.10. *A finitely generated module over a noetherian (artinian) ring is noetherian (artinian). In particular, if R is noetherian, then an R-module M is noetherian if and only if M is finitely generated.*

Proof. A finitely generated R-module M is a quotient of R^n for some n. But R^n is noetherian (artinian) by the corollary above. So M is noetherian (artinian) by Proposition 2.3.7. The second part now follows by Proposition 2.3.3.   □

Corollary 2.3.11. *A ring R is noetherian if and only if every submodule of a finitely generated R-module is finitely generated.*

Proof. Let S be a submodule of a finitely generated R-module M. Then S is finitely generated by Proposition 2.3.3 since M is noetherian by the corollary above. The converse is clear from Corollary 2.3.5.   □

Remark 2.3.12. It is now easy to see that if R is noetherian, then every finitely generated R-module M has a free resolution $\cdots \to F_1 \to F_0 \to M \to 0$ where each F_i is finitely generated.

Theorem 2.3.13 (Hilbert Basis Theorem). *If R is noetherian, then so is $R[x]$.*

Proof. Suppose $R[x]$ is not noetherian. Then let I be a left ideal of $R[x]$ that is not finitely generated. Let f_1 be a nonzero polynomial in I of minimal degree. Then $\langle f_1 \rangle \subsetneq I$ and so let f_2 be a nonzero polynomial in $I - \langle f_1 \rangle$ with minimal degree. Then $\langle f_1 \rangle \subsetneq \langle f_1, f_2 \rangle$ and $\deg f_1 \leq \deg f_2$. Repeat this procedure to get ideals $\langle f_1 \rangle \subsetneq \langle f_1, f_2 \rangle \subsetneq \langle f_1, f_2, f_3 \rangle \subsetneq \cdots$ in I such that $\deg f_i \leq \deg f_{i+1}$.

Now let a_i be the leading coefficient of f_i. Then $\langle a \rangle \subset \langle a_1, a_2 \rangle \subset \cdots$ is an ascending chain of ideals of R. It now suffices to show that $\langle a_1, \ldots, a_k \rangle \subsetneq \langle a_1, a_2, \ldots, a_{k+1} \rangle$. Suppose not. Then $a_{k+1} = \sum_{i=1}^k r_i a_i$ for some $r_i \in R$. One then considers the polynomial $h = f_{k+1} - \sum_{i=1}^k r_i x^{n_{k+1}-n_i} f_i$ where $n_i = \deg f_i$. Note that $h \notin \langle f_1, \ldots, f_k \rangle$ for otherwise $f_{k+1} \in \langle f_1, \ldots, f_k \rangle$, a contradiction. But $r_i x^{n_{k+1}-n_i} f_i = r_i x^{n_{k+1}-n_i}(a_i x^{n_i} + \cdots) = r_i a_i x^{n_{k+1}} +$ lower terms. So $h = f_{k+1} - \sum_i r_i a_i x^{n_{k+1}} -$ lower terms $= f_{k+1} - a_{k+1} x^{n_{k+1}} -$ lower terms. So $h \in I - \langle f_1, f_2, \ldots, f_k \rangle$ and

$\deg h < n_{k+1}$, a contradiction. Hence $\langle a_1, \ldots, a_k \rangle \subsetneq \langle a_1, \ldots, a_{k+1} \rangle$ and so R is not noetherian.   □

Corollary 2.3.14. *If R is noetherian, then so is $R[x_1, \ldots, x_n]$.*

Proof. By induction on n.   □

Remark 2.3.15. Likewise, $R[[x_1, \ldots, x_n]]$ is noetherian whenever R is.

Definition 2.3.16. A chain of R-submodules $M = M_0 \supset M_1 \supset \cdots \supset M_n = 0$ is said to be a *composition series* of M if M_i / M_{i+1} is a simple R-module for each i, that is, $M_i / M_{i+1} \cong R / \mathfrak{m}_i$ for some maximal left ideal $\mathfrak{m}_i$ of R.

The length of a composition series of M does not depend on the choice of the series (in the sense of Jordan–Hölder theorem), and every chain of submodules of M can be refined to a composition series. The common *length* of the composition series of M is denoted $\text{length}_R M$ or simply $\text{length } M$.

Theorem 2.3.17. *An R-module M has finite length if and only if M is artinian and noetherian.*

Proof. If M has finite length, then any composition series is finite and so all the chains are stationary. Conversely suppose M is artinian and noetherian. Since M is noetherian, it has a maximal proper submodule M_1. We note that M/M_1 is simple. But M_1 is noetherian since M is. So let M_2 be a maximal proper submodule of M_1. Repeat this procedure to get a strictly descending chain $M = M_0 \supset M_1 \supset M_2 \supset \cdots$ of submodules of M such that M_i / M_{i+1} is simple. But M is artinian and so the chain stops. Hence $\text{length } M < \infty$.   □

Definition 2.3.18. An R-module M is said to be *semisimple* if it is a direct sum of simple modules. A module M is semisimple if and only if every submodule of M is a direct summand(see Exercise 6). Thus every submodule and homomorphic image of a semisimple is also semisimple. It easily follows from the definition that the direct sum of semisimple modules is also semisimple. A ring R is said to be *semisimple* if it semisimple as an R-module.

Lemma 2.3.19. *The following are equivalent for a semisimple R-module M.*

(1) *M is artinian.*

(2) *M is noetherian.*

(3) *M is a direct sum of finitely many simple modules.*

(4) *M is a finitely generated R-module.*

In particular, a semisimple ring R is a direct sum of finitely many simple modules.

Proof. (1), (2) $\Rightarrow$ (3). If M is a direct sum of infinitely many simple modules, then M has ascending and descending chains of submodules of M that are not stationary.

(3) $\Rightarrow$ (1), (2). We simply note that a simple module is of finite length.

(3) $\Rightarrow$ (4) is trivial since simple modules are cyclic.

(4) $\Rightarrow$ (3). Let $x_1, x_2, \ldots, x_n$ be generators of M and $M = \bigoplus_I S_i$ where S_i are simple submodules of M. But then there are finitely many simple submodules, say $S_{i_1}, \ldots, S_{i_m}$ such that each $x_k \in S_{i_1} + \cdots + S_{i_m}$. So $M \subseteq \bigoplus_{j=1}^m S_{i_j}$ and we are done. $\qquad\square$

Proposition 2.3.20. *Suppose R is a ring such that $R/\operatorname{rad}(R)$ is semisimple and $\operatorname{rad}(R)$ is nilpotent. Then an R-module M is noetherian if and only if M is artinian.*

Proof. Let $J = \operatorname{rad}(R)$. Then $J^n = 0$ for some n. We then consider the descending chain $M \supset JM \supset J^2M \supset \cdots \supset J^{n-1}M \supset J^nM = 0$. The quotient modules $J^iM/J^{i+1}M$, $0 \le i \le n - 1$, can be viewed as R/J-modules. But R/J is a semisimple ring by assumption. So each $J^iM/J^{i+1}M$ is a semisimple R/J-module since it is a homomorphic image of a free R/J-module. Thus if M is artinian or noetherian, then each $J^iM/J^{i+1}M$ is of finite length as an R/J-module by Lemma 2.3.19, and thus as an R-module. But then M is of finite length by Proposition 2.3.7.$\square$

Lemma 2.3.21. *A ring R is semisimple if and only if R is artinian and $\operatorname{rad}(R) = 0$. In particular, if R is artinian, then $R/\operatorname{rad}(R)$ is semisimple.*

Proof. If R is semisimple, then R is a direct sum of finitely many simple modules and so R is artinian and $\operatorname{rad}(R) = 0$. Conversely suppose R is artinian and $\operatorname{rad}(R) = 0$. Consider the set S of all finite intersections of maximal ideals of R. Then S has a minimal element, say I, by Remark 2.3.2. So if $\mathfrak{m}$ is a maximal ideal of R, then $\mathfrak{m} \cap I = I$ by the minimality of I and so $I \subset \mathfrak{m}$. Thus $I \subset \operatorname{rad}(R)$. But then $I = 0$ since $\operatorname{rad}(R) = 0$. Hence there are finitely many maximal ideals, say $\mathfrak{m}_1, \ldots, \mathfrak{m}_n$ such that $\bigcap_{i=1}^n \mathfrak{m}_i = I = 0$. But the map $\varphi : R \to \prod_{i=1}^n R/\mathfrak{m}_i$ defined by $\varphi(r) = (r + \mathfrak{m}_i)$ has $\operatorname{Ker} \varphi = \bigcap_{i=1}^n \mathfrak{m}_i = 0$. So φ embedds R into a semisimple module. Thus R is semisimple.

Since $\operatorname{rad}(R/\operatorname{rad}(R)) = 0$, we see that the second part follows by applying the lemma to $R/\operatorname{rad}(R)$. $\qquad\square$

Proposition 2.3.22. *If R is artinian, then $R/\operatorname{rad}(R)$ is semisimple and $\operatorname{rad}(R)$ is nilpotent.*

Proof. The first part follows from the lemma above.

Now let $J = \operatorname{rad}(R)$ and consider the descending chain $J \supset J^2 \supset J^3 \supset \cdots$. Then $J^n = J^{n+1}$ for some n since R is artinian. Suppose $J^n \ne 0$. Then let I be the minimal left ideal such that $J^nI \ne 0$. Then $J^n(JI) = J^{n+1}I = J^nI \ne 0$. But $JI \subset J$ and I is a minimal left ideal such that $J^nI \ne 0$. So $JI = I$. But I is a principal ideal. So $I = 0$ by Nakayama lemma, a contradiction. Hence $J^n = 0$. $\qquad\square$

Corollary 2.3.23. *An artinian local ring is complete.*

Proof. Let $\mathfrak{m}$ be the maximal ideal of R. Then $\mathfrak{m}$ is nilpotent by the proposition above. So $\hat{R} = \varprojlim R/\mathfrak{m}^i = R$. $\square$

Corollary 2.3.24. *If R is artinian, then an R-module M is noetherian if and only if M is artinian.*

Proof. This follows from Propositions 2.3.20 and 2.3.22. $\square$

Corollary 2.3.25. *A ring R is artinian if and only if* length $_R R < \infty$.

As another consequence of Propositions 2.3.20 and 2.3.22, we get the following characterization of artinian rings.

Theorem 2.3.26. *A ring R is artinian if and only if $R/\operatorname{rad}(R)$ is semisimple, $\operatorname{rad}(R)$ is nilpotent, and R is noetherian.*

Exercises

1. Prove Remark 2.3.2 for artinian modules.

2. Prove Lemma 2.3.6.

3. Prove that if R is a commutative noetherian ring, then so is its localization $S^{-1}R$.

4. Prove Remark 2.3.15.

5. Let M be a noetherian R-module and $\varphi : M \to M$ be a homomorphism. Prove that if φ is surjective, then φ is an automorphism.

6. Prove that a module M is semisimple if and only if every submodule of M is a direct summand (Rotman [160, page 15]).

7. Suppose an R-module M has length n. Prove that every composition series of M has length n and every chain of R-submodules of M can be refined to a composition series.

8. Prove that if $0 \to M' \to M \to M'' \to 0$ is an exact sequence of R-modules, then length $M = $ length $M' + $ length M''.

2.4 Prime ideals and primary decomposition

Throughout this section, R will denote a commutative ring.

Definition 2.4.1. If M is an R-module, then the *annihilator* of M, denoted $\operatorname{Ann}(M)$, is defined by $\operatorname{Ann}(M) = \{r \in R : rx = 0 \text{ for all } x \in M\}$. The annihilator of an element $x \in M$ is defined by $\operatorname{Ann}(x) = \{r \in R : rx = 0\}$. $\operatorname{Ann}(M)$ is an ideal of R. Moreover, if I is an ideal of R such that $I \subset \operatorname{Ann}(M)$, then M is an R/I-module via scalar multiplication $(r + I)x = rx$. This is well-defined for if $r + I = s + I$, then $r - s \in I \subset \mathrm{Ann}(M)$ and so $(r - s)x = 0$. In particular, we have that M is always an $R/\mathrm{Ann}(M)$-module.

Definition 2.4.2. Let M be an R-module. A prime ideal $\mathfrak{p}$ is said to be an *associated prime ideal* of M if $\mathfrak{p} = \mathrm{Ann}(x)$ for some $x \in M$. It is easy to see that this is equivalent to M containing a cyclic submodule isomorphic to $R/\mathfrak{p}$. The set of associated prime ideals of M is denoted by $\mathrm{Ass}(M)$.

Proposition 2.4.3. *If R is noetherian and M is an R-module, then $M = 0$ if and only if $\mathrm{Ass}(M) = \emptyset$.*

Proof. If $M = 0$ then $\mathrm{Ass}(M) = \emptyset$. Let $M \neq 0$ and $x \in M$, $x \neq 0$. If $\mathrm{Ann}(x)$ is a prime ideal we are through. If not let $rs \in \mathrm{Ann}(x)$ with $r, s \notin \mathrm{Ann}(x)$. Then $rx \neq 0$ and $s \in \mathrm{Ann}(rx)$. So $\mathrm{Ann}(x) \subsetneq \mathrm{Ann}(rx)$. If $\mathrm{Ann}(rx)$ is not a prime ideal then we can repeat the procedure. If the procedure did not stop we would contradict the fact that R is noetherian. Hence the procedure stops and we see that $\mathrm{Ass}(M) \neq \emptyset$. □

Remark 2.4.4. From the proof we see that for $x \in M$, $x \neq 0$, $\mathrm{Ann}(x) \subset \mathfrak{p}$ for some $\mathfrak{p} \in \mathrm{Ass}(M)$. Hence $\bigcup_{\mathfrak{p}\in\mathrm{Ass}(M)} \mathfrak{p}$ is the set of all *zero divisors* on M, that is, all $r \in R$ with $rx = 0$ for an $x \in M$, $x \neq 0$.

Proposition 2.4.5. *Let R be noetherian, M an R-module, and $\mathfrak{p}$ a prime ideal of R. Then $\mathfrak{p} \in \mathrm{Ass}(M)$ if and only if $\mathfrak{p}R_\mathfrak{p} \in \mathrm{Ass}_{R_\mathfrak{p}}(M_\mathfrak{p})$.*

Proof. If $\mathfrak{p} \in \mathrm{Ass}(M)$, then $R/\mathfrak{p} \cong Rx$ for some $x \in M$, $x \neq 0$. So $R/\mathfrak{p}$ is isomorphic to a submodule of M. Thus $R_\mathfrak{p}/\mathfrak{p}R_\mathfrak{p}$ is isomorphic to a submodule of $M_\mathfrak{p}$. Hence $\mathfrak{p}R_\mathfrak{p} \in \mathrm{Ass}_{R_\mathfrak{p}}(M_\mathfrak{p})$. Conversely, if $\mathfrak{p}R_\mathfrak{p} \in \mathrm{Ass}_{R_\mathfrak{p}}(M_\mathfrak{p})$, then $\mathfrak{p}R_\mathfrak{p} = \mathrm{Ann}_{R_\mathfrak{p}}(\frac{x}{t})$ where $\frac{x}{t} \in M_\mathfrak{p}$ for some $x \in M$ and $t \in R - \mathfrak{p}$. Since $\mathfrak{p}$ is finitely generated, let $\mathfrak{p} = \langle a_1, a_2, \ldots, a_n \rangle$. Then $\frac{a_i}{1} \cdot \frac{x}{t} = 0$ for each i. So there is an $r_i \in R - \mathfrak{p}$ such that $r_i a_i x = 0$ for each i. Now set $r = r_1 r_2 \ldots r_n$. Then $rax = 0$ for all $a \in \mathfrak{p}$. Thus $\mathfrak{p} \subset \mathrm{Ann}_R(rx)$. If $a \in \mathrm{Ann}_R(rx)$, then $arx = 0$ and so $\frac{a}{1} \cdot \frac{x}{t} = 0$. But then $\frac{a}{1} \in \mathfrak{p}R_\mathfrak{p}$. Consequently $a \in \mathfrak{p}$. Thus $\mathrm{Ann}_R(rx) \subset \mathfrak{p}$. Hence $\mathfrak{p} = \mathrm{Ann}_R(rx)$ and so $\mathfrak{p} \in \mathrm{Ass}_R(M)$. □

Definition 2.4.6. The *spectrum* of R, denoted $\mathrm{Spec}\, R$, is the set of all prime ideals of R. The set of maximal ideals is called a *maximal spectrum* of R and is denoted by $\mathrm{mSpec}\, R$.

Lemma 2.4.7. *Let R be a noetherian ring. If $M \neq 0$ is a finitely generated R-module, then there exists a chain $0 = M_0 \subset M_1 \subset \cdots \subset M_{n-1} \subset M_n = M$ of submodules of M such that for each $1 \leq i \leq n$, $M_i/M_{i-1} \cong R/\mathfrak{p}_i$ for some $\mathfrak{p}_i \in \mathrm{Spec}\, R$.*

Proof. Let $\mathfrak{p}_1 \in \mathrm{Ass}(M)$. Then $R/\mathfrak{p}_1$ is isomorphic to a submodule of M. That is, there is a submodule M_1 of M such that $M_1 \cong R/\mathfrak{p}_1$. If $M_1 = M$, then we are done. Otherwise let $\mathfrak{p}_2 \in \mathrm{Ass}(M/M_1)$. Then there is a submodule M_2 of M containing M_1 such that $M_2/M_1 \cong R/\mathfrak{p}_2$. One then repeats this procedure to get the required submodules noting that the process stops since M is noetherian. □

Lemma 2.4.8. *Let* $0 \to M' \to M \to M'' \to 0$ *be an exact sequence of R-modules, then*
$$\mathrm{Ass}(M) \subset \mathrm{Ass}(M') \cup \mathrm{Ass}(M'').$$

Proof. Let $\mathfrak{p} \in \mathrm{Ass}(M)$. Then $R/\mathfrak{p}$ is isomorphic to a submodule Rx of M for some $x \in M$, $x \neq 0$. If $Rx \cap M' \neq 0$, let $y \in Rx \cap M'$, $y \neq 0$. Then $\mathfrak{p} = \mathrm{Ann}(x)$ is equal to $\mathrm{Ann}(y)$ since $\mathfrak{p}$ is a prime ideal. Thus $\mathfrak{p} \in \mathrm{Ass}(M')$. If $Rx \cap M' = 0$, then the image of Rx in M'' is $(Rx + M')/M' \cong Rx \cong R/\mathfrak{p}$. Thus $\mathfrak{p} \in \mathrm{Ass}(M'')$. □

Theorem 2.4.9. *If R is noetherian and M is a finitely generated R-module, then* $\mathrm{Ass}(M)$ *is finite.*

Proof. We consider the chain $0 = M_0 \subset M_1 \subset \cdots \subset M_{n-1} \subset M_n = M$ of Lemma 2.4.7. Then we have short exact sequences $0 \to M_{i-1} \to M_i \to M_i/M_{i-1} \to 0$ for $i = 1, 2, \ldots, n$. So $\mathrm{Ass}(M_i) \subset \mathrm{Ass}(M_{i-1}) \cup \mathrm{Ass}(M_i/M_{i-1})$ by Lemma 2.4.8 above. Thus inductively,

$$\mathrm{Ass}(M) \subset \mathrm{Ass}(M_n/M_{n-1}) \cup \mathrm{Ass}(M_{n-1}/M_{n-2}) \cup \cdots \cup \mathrm{Ass}(M_2/M_1) \cup \mathrm{Ass}(M_1).$$

But each $M_i/M_{i-1} \cong R/\mathfrak{p}_i$ for some $\mathfrak{p}_i \in \mathrm{Spec}\, R$ by Lemma 2.4.7. So $\mathrm{Ass}(M) \subset \{\mathfrak{p}_1, \mathfrak{p}_2, \ldots, \mathfrak{p}_n\}$ since $\mathrm{Ass}(R/\mathfrak{p}) = \{\mathfrak{p}\}$ for each $\mathfrak{p} \in \mathrm{Spec}\, R$. Hence we are done. □

Definition 2.4.10. The *support* of an R-module M, denoted $\mathrm{Supp}(M)$, is the set of all prime ideals $\mathfrak{p}$ of R such that $M_\mathfrak{p} \neq 0$. If $M \neq 0$ we see that $\mathrm{Supp}(M) \neq \emptyset$. For let $x \in M$, $x \neq 0$, then $\mathrm{Ann}(x) \subset \mathfrak{p}$ for $\mathfrak{p}$ a maximal ideal of R. Then $\mathfrak{p}$ is a prime ideal of R. But $\frac{x}{1} \neq 0$ in $M_\mathfrak{p}$ and so $\mathfrak{p} \in \mathrm{Supp}(M)$. Furthermore, if $0 \to M' \to M \to M'' \to 0$ is an exact sequence R-modules, then $\mathrm{Supp}(M) = \mathrm{Supp}(M') \cup \mathrm{Supp}(M'')$.

Remark 2.4.11. If R is noetherian and $\mathfrak{p} \in \mathrm{Ass}(M)$, then $\mathfrak{p}R_\mathfrak{p} \in \mathrm{Ass}(M_\mathfrak{p})$ by Proposition 2.4.5. So $R_\mathfrak{p}/\mathfrak{p}R_\mathfrak{p}$ is isomorphic to a submodule of $M_\mathfrak{p}$. Hence $M_\mathfrak{p} \neq 0$ and so $\mathfrak{p} \in \mathrm{Supp}(M)$. Thus $\mathrm{Ass}(M) \subset \mathrm{Supp}(M)$.

Theorem 2.4.12. *Let R be noetherian and M be an R-module. If $\mathfrak{p}$ is a minimal element in* $\mathrm{Supp}(M)$, *then* $\mathfrak{p} \in \mathrm{Ass}(M)$.

Proof. Let $\mathfrak{p}$ be a minimal element in $\mathrm{Supp}(M)$. By Proposition 2.4.5, it suffices to prove the result for a local ring R with maximal ideal $\mathfrak{p}$ and a nonzero R-module M. Since $\mathfrak{p}$ is minimal, we further assume that $M_\mathfrak{q} = 0$ for all prime ideals $\mathfrak{q}$ contained in $\mathfrak{p}$. So $\mathrm{Supp}(M) = \{\mathfrak{p}\}$. But $\mathrm{Ass}(M) \subset \mathrm{Supp}(M)$ by the remark above. So $\mathfrak{p} \in \mathrm{Ass}(M)$ since $\mathrm{Ass}(M) \neq \emptyset$. □

Definition 2.4.13. The *height* (ht) of a prime ideal $\mathfrak{p}$ is the supremum of the lengths s of strictly decreasing chains $\mathfrak{p} = \mathfrak{p}_0 \supset \mathfrak{p}_1 \supset \cdots \supset \mathfrak{p}_{s-1} \supset \mathfrak{p}_s$ of prime ideals of R.

The *Krull dimension* of R, denoted dim R, is defined by

$$\dim R = \sup\{\text{ht } \mathfrak{p} : \mathfrak{p} \in \text{Spec } R\}.$$

It follows from the definitions above that ht $\mathfrak{p} + \dim R/\mathfrak{p} \leq \dim R$ and ht $\mathfrak{p} = \dim R_\mathfrak{p}$.

If dim $R = 0$, then every prime ideal of R is minimal, and if R is a principal ideal domain which not a field, then dim $R = 1$.

Definition 2.4.14. Now let V be a subset of Spec R. Then the *Krull dimension* of V, denoted dim V, is defined to be the supremum of the lengths of strictly decreasing chains $\mathfrak{p}_0 \supset \mathfrak{p}_1 \supset \cdots \supset \mathfrak{p}_{s-1} \supset \mathfrak{p}_s$ of prime ideals of V. In particular, the Krull dimension of R is dim Spec R. The *dimension* of an R-module M, denoted dim M, is defined by dim $M = \dim \text{Supp}(M)$. So dim $M \leq \dim R$.

Remark 2.4.15. If M is finitely generated, then $\text{Supp}(M) = \{\mathfrak{p} \in \text{Spec } R : \text{Ann}(M) \subset \mathfrak{p}\}$. For if $M = m_1 R + m_2 R + \cdots + m_n R$ for some $m_1, m_2, \ldots, m_n \in M$, then $\mathfrak{p} \in \text{Supp}(M)$ if and only if there is an i such that $\frac{m_i}{1} \neq 0$ in $M_\mathfrak{p}$. But this means that there is an i such that $\text{Ann}(m_i) \subset \mathfrak{p}$. But this holds if and only if $\text{Ann}(M) = \bigcap_{i=1}^n \text{Ann}(m_i) \subset \mathfrak{p}$. Hence if M is finitely generated, then dim $M = \dim R/\text{Ann}(M)$.

We now recall the following.

Definition 2.4.16. The *radical* of an ideal I of R, denoted $\sqrt{I}$, is defined by $\sqrt{I} = \{r \in R : r^n \in I \text{ for some } n > 0\}$. We note that $I \subset \sqrt{I}$. If $I = 0$, then $\sqrt{I}$ is called the *nilradical*. It is easy to see that the nilradical is the set of all nilpotent elements of R.

Proposition 2.4.17. $\sqrt{I}$ *is the intersection of all prime ideals containing* I.

Proof. Let $\mathfrak{p}$ be a prime ideal containing I. If $r \in \sqrt{I}$, then $r^n \in I \subset \mathfrak{p}$ and so $r \in \mathfrak{p}$. Hence $\sqrt{I} \subset \bigcap_{\mathfrak{p} \supset I} \mathfrak{p}$. Now let $x \notin \sqrt{I}$. Then $x^n \notin I$ for each $n \geq 0$. So $S = \{1, x, x^2, \ldots\}$ is a multiplicative set disjoint from I. Then the set of ideals J such that $J \supset I$ and $J \cap S = \emptyset$ has a maximal element $\mathfrak{q}$ by Zorn's lemma. We claim that $\mathfrak{q}$ is a prime ideal. We first note that if $x \notin \mathfrak{q}$, then $(\mathfrak{q} + Rx) \cap S \neq \emptyset$ for otherwise $\mathfrak{q} + Rx$ would contradict the maximality of $\mathfrak{q}$. So $x \in \mathfrak{q}$ if and only if $(\mathfrak{q} + Rx) \cap S = \emptyset$. Thus $x_1 \notin \mathfrak{q}$, $x_2 \notin \mathfrak{q}$ implies that $(\mathfrak{q} + Rx_i) \cap S \neq \emptyset$. So $((\mathfrak{q} + Rx_1)(\mathfrak{q} + Rx_2)) \cap S \neq \emptyset$. But $(\mathfrak{q} + Rx_1)(\mathfrak{q} + Rx_2) \subset (\mathfrak{q} + Rx_1x_2)$. So $(\mathfrak{q} + Rx_1x_2) \cap S \neq \emptyset$ and thus $x_1x_2 \notin \mathfrak{q}$. So $\mathfrak{q}$ is a prime ideal. Hence $x \notin \bigcap_{\mathfrak{p} \supset I} \mathfrak{p}$. Thus $\sqrt{I} = \bigcap_{\mathfrak{p} \supset I} \mathfrak{p}$. $\square$

Corollary 2.4.18. *The nilradical of* R *is the intersection of all prime ideals of* R.

Definition 2.4.19. An ideal I of R is said to be *primary* if $ab \in I$ and $a \notin I$ implies that $b^n \in I$ for some integer $n \geq 1$. It is easy to see that I is primary if and only if every zero divisor of R/I is nilpotent.

Remark 2.4.20. If I is a primary ideal, then $\sqrt{I}$ is a prime ideal. For if $ab \in \sqrt{I}$, then $a^n b^n \in I$ for some $n > 0$. If $a \notin \sqrt{I}$, then $a^n \notin I$. But I is primary. So $(b^n)^m \in I$ for some $m > 0$. Hence $b \in \sqrt{I}$ and we are done. It follows from Proposition 2.4.17 that if I is primary, then $\sqrt{I}$ is the smallest prime ideal containing I.

Definition 2.4.21. If I is a primary ideal and $\mathfrak{p} = \sqrt{I}$, then I is said to be $\mathfrak{p}$-*primary*.

Lemma 2.4.22. *If $\sqrt{I}$ is a maximal ideal, then I is primary. In particular, if $\mathfrak{m}$ is a maximal ideal, then $\mathfrak{m}^n$ is $\mathfrak{m}$-primary for each $n > 0$.*

Proof. Let $\mathfrak{m} = \sqrt{I}$. Then since $\sqrt{I}$ is the intersection of prime ideals $\mathfrak{p}$ of R containing I, we have that $I \subset \mathfrak{m} \subset \mathfrak{p}$. But $\mathfrak{m}$ is maximal. So $\mathrm{Spec}(R/I) = \{\mathfrak{m}/I\}$. But then $\bar{x} \in \mathfrak{m}/I$ implies $\bar{x}$ is nilpotent and $\bar{x} \notin \mathfrak{m}/I$ implies that it is a unit. So if $x + I$ is a zero divisor of R/I, then $x \in \mathfrak{m}$ and so $x + I$ is nilpotent. Hence zero divisors of R/I are nilpotent. That is, I is primary. The second part is now clear since $\sqrt{\mathfrak{m}^n} = \mathfrak{m}$. □

Lemma 2.4.23. *If R is noetherian and I is an ideal of R, then $(\sqrt{I})^n \subset I$ for some $n > 0$.*

Proof. Since R is noetherian, let $\sqrt{I} = \langle r_1, \ldots, r_s \rangle$. Then $r_i^{n_i} \in I$ for some $n_i > 0$. Let $n = (n_1 - 1) + (n_2 - 1) + \cdots + (n_s - 1) + 1$. Then $(\sqrt{I})^n$ is generated by monomials $r_1^{m_1} r_2^{m_2} \ldots r_s^{m_s}$ where $n = \sum_{i=1}^{s} m_i$ and $m_i \geq n_i$ for some i. Thus $r_1^{m_1} r_2^{m_2} \ldots r_s^{m_s} \in I$ and so $(\sqrt{I})^n \subset I$. □

Proposition 2.4.24. *If R is noetherian, then the nilradical is nilpotent.*

Proof. We simply let $I = 0$ in Lemma 2.4.23 above. □

Proposition 2.4.25. *Let R be noetherian, $\mathfrak{m}$ a maximal ideal of R, and I an ideal of R. Then I is $\mathfrak{m}$-primary if and only if $\mathfrak{m}^n \subset I \subset \mathfrak{m}$ for some $n > 0$.*

Proof. If I is $\mathfrak{m}$-primary, then $\sqrt{I} = \mathfrak{m}$ and so $I \subset \mathfrak{m}$ since $I \subset \sqrt{I}$. Thus the conclusion follows from Lemma 2.4.23. Conversely, $\sqrt{\mathfrak{m}^n} = \mathfrak{m}$ and so $\mathfrak{m}^n \subset I \subset \mathfrak{m}$ implies that $\mathfrak{m} = \sqrt{\mathfrak{m}^n} \subset \sqrt{I} \subset \sqrt{\mathfrak{m}} = \mathfrak{m}$. □

Remark 2.4.26. Let R be noetherian and M be a finitely generated R-module. The number of minimal elements of $\mathrm{Supp}(M)$ is finite since the sets of minimal elements of $\mathrm{Supp}(M)$ and $\mathrm{Ass}(M)$ are the same by Theorem 2.4.12 and $\mathrm{Ass}(M)$ is finite by Theorem 2.4.9. Such elements are called *isolated associated primes* of M while the remaining primes in $\text{Ass}(M)$ are said to be *embedded*. So by Remark 2.4.15, the isolated associated primes of M are precisely the minimal prime ideals that contain $\text{Ann}(M)$. So $\sqrt{\text{Ann}(M)} = \bigcap_{\mathfrak{p} \supset \text{Ann}(M)} \mathfrak{p} = \bigcap_{i=1}^{s} \mathfrak{p}'_i$ where $\mathfrak{p}'_1, \dots, \mathfrak{p}'_s$ are isolated associated primes of M. Elements of $\text{Ass}(R/I)$ are sometimes called *prime divisors* of I and so isolated associated primes of R/I are referred to as *minimal prime divisors* of I. Hence minimal prime divisors of I are precisely the minimal prime ideals that contain I.

Theorem 2.4.27. *A ring R is artinian if and only if R is noetherian and* $\dim R = 0$.

Proof. If R is artinian, then R is noetherian by Theorem 2.3.26. Now let $\mathfrak{p}$ be a prime ideal of R and $\bar{r} \in R/\mathfrak{p}, \bar{r} \neq 0$. Then $\langle \bar{r} \rangle^n = \langle \bar{r} \rangle^{n+1}$ for some n since $R/\mathfrak{p}$ is artinian. So $\bar{r}^n = \bar{r}^{n+1} \cdot \bar{s}$ for some $\bar{s} \in R/\mathfrak{p}$. But then $1 = \bar{r} \cdot \bar{s}$ since $R/\mathfrak{p}$ is an integral domain. That is, $R/\mathfrak{p}$ is a field and so $\mathfrak{p}$ is maximal. Hence every prime ideal is maximal and so $\dim R = 0$.

Conversely, if $\dim R = 0$, then each $\mathfrak{p} \in \text{Spec } R$ is both minimal and maximal. But R is noetherian. So by Remark 2.4.26, there are only finitely many minimal divisors of the zero ideal, say $\mathfrak{p}_1, \mathfrak{p}_2, \dots, \mathfrak{p}_r$. Hence $\mathfrak{p}_1, \mathfrak{p}_2, \dots, \mathfrak{p}_r$ are the maximal ideals of R. Thus $\text{rad}(R) = \bigcap_{i=1}^{r} \mathfrak{p}_i = \sqrt{0}$. So $\text{rad}(R)$ is nilpotent by Proposition 2.4.24. But $R/\text{rad}(R) = R/\bigcap_{i=1}^{r} \mathfrak{p}_i$ is isomorphic to the semisimple R-module $\prod_{i=1}^{r} R/\mathfrak{p}_i$. Hence R is artinian by Theorem 2.3.26.   □

As an application, we have the following result.

Theorem 2.4.28 (Principal Ideal Theorem). *Let R be noetherian and $\mathfrak{p}$ be a minimal prime ideal containing a principal ideal $I \neq R$. Then* $\text{ht } \mathfrak{p} \leq 1$.

Proof. We first note that $\text{ht } \mathfrak{p} = \dim R_{\mathfrak{p}}$ by Definition 2.4.13 and $\mathfrak{p} R_{\mathfrak{p}}$ is a minimal prime ideal of the principal ideal $I R_{\mathfrak{p}}$. Thus we may assume that R is a local ring with maximal ideal $\mathfrak{m}$ such that $\mathfrak{m}$ is minimal over a principal ideal I of $R_{\mathfrak{p}}$.

Now let $\mathfrak{q}$ be a prime ideal such that $\mathfrak{q} \subsetneq \mathfrak{m}$. We then consider the ideals $\mathfrak{q}^i R_{\mathfrak{q}}$ of $R_{\mathfrak{q}}$ and set $\mathfrak{q}^{(i)}$ to be the preimage of $\mathfrak{q}^i R_{\mathfrak{q}}$ under the natural map $R \to R_{\mathfrak{q}}$. Then $I + \mathfrak{q}^{(i+1)} \subset I + \mathfrak{q}^{(i)}$ for each $i > 0$ and so we get a descending chain of ideals of R. But $\mathfrak{m}/I$ is the only prime ideal of R/I since $\mathfrak{m}$ is minimal over I. Hence $\dim R/I = 0$ and so R/I is artinian by the theorem above. Therefore there is an $n > 0$ such that $I + \mathfrak{q}^{(n+1)} = I + \mathfrak{q}^{(n)}$.

We now claim that if $I = \langle a \rangle$, then $\mathfrak{q}^{(n)} = a\mathfrak{q}^{(n)} + \mathfrak{q}^{(n+1)}$. Clearly $a\mathfrak{q}^{(n)} + \mathfrak{q}^{(n+1)} \subseteq \mathfrak{q}^{(n)}$. Now let $x \in \mathfrak{q}^{(n)}$. Then since $I + \mathfrak{q}^{(n+1)} = I + \mathfrak{q}^{(n)}$ we have that $x = ra + x'$ with $x' \in \mathfrak{q}^{(n+1)}, r \in R$, and $ra \in \mathfrak{q}^{(n)}$. But $\mathfrak{m}$ is minimal over I and $\mathfrak{q} \subsetneq \mathfrak{m}$. So $a \notin \mathfrak{q}$. But $\mathfrak{q}^n R_{\mathfrak{q}}$ is $\mathfrak{q}R_{\mathfrak{q}}$-primary by Lemma 2.4.22 and hence easily its preimage $\mathfrak{q}^{(n)}$ is $\mathfrak{q}$-primary. So $a^i \notin \mathfrak{q}^{(n)}$ for any $i > 0$ and hence $r \in \mathfrak{q}^{(n)}$. Thus $ra \in a\mathfrak{q}^{(n)}$ and hence $\mathfrak{q}^{(n)} = a\mathfrak{q}^{(n)} + \mathfrak{q}^{(n+1)}$.

But $a \in \mathfrak{m}$ and so $\mathfrak{q}^{(n)} = \mathfrak{q}^{(n+1)}$ by Corollary 1.2.29 and so $\mathfrak{q}^n R_{\mathfrak{q}} = \mathfrak{q}^{n+1} R_{\mathfrak{q}}$ over $R_{\mathfrak{q}}$. Hence $\mathfrak{q}^n R_{\mathfrak{q}} = 0$ by Nakayama lemma (Proposition 1.2.28). Thus $\mathfrak{q}R_{\mathfrak{q}}$

is nilpotent and so $R_{\mathfrak{q}}$ is artinian by Proposition 2.3.20. But then $\dim R_{\mathfrak{q}} = 0$ by Theorem 2.4.27 above. Hence ht $\mathfrak{q} = 0$ for all primes $\mathfrak{q} \subsetneq \mathfrak{m}$. That is, ht $\mathfrak{m} \le 1$. □

Remark 2.4.29. Since minimal prime ideals of R consist of only zero divisors of R by Remarks 2.4.4 and 2.4.26, we see that if a is not a zero divisor of R, then any minimal prime ideal $\mathfrak{p}$ containing $I = \langle a \rangle$ is not a minimal prime ideal of R. Hence ht $\mathfrak{p} \ge 1$. But then ht $\mathfrak{p} = 1$ by the theorem above.

Theorem 2.4.30 (Generalized Krull Principal Ideal Theorem). *Let R be noetherian and $\mathfrak{p}$ be a minimal prime ideal containing an ideal I generated by n elements. Then* ht $\mathfrak{p} \le n$.

Proof. By induction on n. The case $n = 1$ is Theorem 2.4.28 above. We may again assume R is local with maximal ideal $\mathfrak{m}$ which is minimal over an ideal I generated by n elements, say $a_1, a_2, \ldots, a_n$. Suppose ht $\mathfrak{m} > n$. Then there is a descending chain of prime ideals $\mathfrak{m} = \mathfrak{p}_0 \supset \mathfrak{p}_1 \supset \mathfrak{p}_2 \supset \cdots \supset \mathfrak{p}_n$. We may assume that there is no prime ideal $\mathfrak{p}'$ such that $\mathfrak{p}_1 \subsetneq \mathfrak{p}' \subsetneq \mathfrak{m}$. So I is not contained in $\mathfrak{p}_1$ because of the minimality of $\mathfrak{m}$. Thus some a_i, say a_1, is not an element of $\mathfrak{p}_1$. We note that $\mathfrak{m}$ is minimal over $\mathfrak{p}_1 + \langle a_1 \rangle$ and so $\sqrt{\mathfrak{p}_1 + \langle a_1 \rangle} = \mathfrak{m}$ by Remark 2.4.20. Hence there is an $t > 0$ such that $\mathfrak{m}^t \subset \mathfrak{p}_1 + \langle a_1 \rangle$. So for each $i = 2, 3, \ldots, n$, $a_i^t = b_i + r_i a_1$ where $b_i \in \mathfrak{p}_1, r_i \in R$. Now set $J = \langle b_2, \ldots, b_n \rangle$. Then $J \subset \mathfrak{p}_1$. But ht $\mathfrak{p}_1 \ge n$. So by the induction hypothesis, $\mathfrak{p}_1$ is not minimal over J since J is generated by $n - 1$ elements. Hence there is a prime ideal $\mathfrak{q}$ such that $J \subset \mathfrak{q} \subsetneq \mathfrak{p}_1$. It is clear from $a_i^t = b_i + r_i a_1$ above that $\mathfrak{q} + \langle a_1 \rangle$ contains a power of I. But then $\mathfrak{m}$ is a minimal prime ideal of $\mathfrak{q} + \langle a_1 \rangle$ by minimality of $\mathfrak{m}$. So the ideal $\mathfrak{m}/\mathfrak{q}$ of $R/\mathfrak{q}$ is minimal over the principal ideal $(\mathfrak{q} + \langle a_1 \rangle)/\mathfrak{q}$. Hence $\mathrm{ht}(\mathfrak{m}/\mathfrak{q}) \le 1$ by Theorem 2.4.28 above. But $R/\mathfrak{q}$ has a chain of prime ideals $\mathfrak{m}/\mathfrak{q} \supset \mathfrak{p}_1/\mathfrak{q} \supset 0$ of length 2, a contradiction. □

Corollary 2.4.31. *If R is noetherian, then every prime ideal of R has finite height. In particular, the Krull dimension of a semilocal ring is finite.*

Proof. The first part easily follows from the theorem above. For the second part, we simply recall that a ring is *semilocal* if it has finitely many maximal ideals, and so the Krull dimension of a semilocal ring is the maximum of the heights of finitely many maximal ideals and hence is finite by the above. □

Corollary 2.4.32. *A noetherian ring satisfies the descending chain condition on its prime ideals.*

We now prove a converse of the Generalized Krull Principal Ideal Theorem.

Theorem 2.4.33. *Let R be noetherian and $\mathfrak{p}$ be a prime ideal of R of height n. Then there exist elements $a_1, a_2, \ldots, a_n$ in $\mathfrak{p}$ such that $\mathfrak{p}$ is minimal over $I = \langle a_1, \ldots, a_n \rangle$.*

Proof. If $n = 0$, there is nothing to prove. So we assume $n \geq 1$. By Theorem 2.4.12, R has a finite number of minimal prime ideals, say $\mathfrak{p}_1, \mathfrak{p}_2, \ldots, \mathfrak{p}_r$. But ht $\mathfrak{p} \geq 1$. So $\mathfrak{p}$ is not contained in any $\mathfrak{p}_i$ and thus $\mathfrak{p} \not\subset \bigcup_{i=1}^{r} \mathfrak{p}_i$. So let $a_1 \in \mathfrak{p} - \bigcup_{i=1}^{r} \mathfrak{p}_i$ and set $\bar{R} = R/\langle a_1 \rangle$, $\bar{\mathfrak{p}} = \mathfrak{p}/\langle a_1 \rangle$. Then dim $\bar{R} \leq n - 1$ and so by the induction hypothesis there exist a sequence $\bar{a}_2, \ldots, \bar{a}_n$ in $\bar{\mathfrak{p}}$ such that $\bar{\mathfrak{p}}$ is minimal over the ideal $\langle \bar{a}_2, \ldots, \bar{a}_n \rangle$ in $\bar{R}$. But $\bar{a}_i = a_i + \langle a_1 \rangle$ for some $a_i \in \mathfrak{p}$, $i = 2, \ldots, n$. So $\mathfrak{p}$ is minimal over I. □

We now generalize the notion of primary ideals to modules.

Definition 2.4.34. A submodule N of an R-module M is said to be a *primary submodule* if $N \neq M$ and $xy \in N$ and $x \notin N$ implies $y^n M \subset N$ for some $n > 0$. It is easy to see that N is a primary submodule of M if and only if every zero divisor r of M/N is *nilpotent* for M/N, that is, $r^n(M/N) = 0$ for some $n > 0$, or equivalently $r \in \sqrt{\operatorname{Ann}(M/N)}$.

Remark 2.4.35. We note that if M is a finitely generated R-module, then $\sqrt{\operatorname{Ann}(M)} = \bigcap \mathfrak{p}$ over primes $\mathfrak{p}$ containing $\operatorname{Ann}(M)$ by Proposition 2.4.17. But then $\sqrt{\operatorname{Ann}(M)} = \bigcap \mathfrak{p}$ over $\mathfrak{p} \in \operatorname{Supp} M$ by Remark 2.4.15. So if M is finitely generated, then a submodule N of M is a primary submodule if and only if each zero divisor of M/N is an element of $\bigcap_{\mathfrak{p} \in \operatorname{Supp}(M/N)} \mathfrak{p}$.

Proposition 2.4.36. *Let R be noetherian and M be a finitely generated R-module. Then a submodule N of M is primary if and only if $\operatorname{Ass}(M/N) = \{\mathfrak{p}\}$ for some $\mathfrak{p} \in \operatorname{Spec} R$. In this case, $\operatorname{Ann}(M/N)$ is a primary ideal of R and $\sqrt{\operatorname{Ann}(M/N)} = \mathfrak{p}$.*

Proof. If $\operatorname{Ass}(M/N) = \{\mathfrak{p}\}$, then $\mathfrak{p}$ is the only minimal element of $\operatorname{Supp}(M/N)$ by Theorem 2.4.12. Hence $\sqrt{\operatorname{Ann}(M/N)} = \mathfrak{p}$. If r is a nonzero divisor of M/N, then $r \in \mathfrak{p}$ by Remark 2.4.4 and so $r \in \sqrt{\operatorname{Ann}(M/N)}$. So the conclusion follows from Definition 2.4.34.

Conversely, suppose N is a primary submodule of M. Then $\bigcup_{\mathfrak{p} \in \operatorname{Ass}(M/N)} \mathfrak{p} = \sqrt{\operatorname{Ann}(M/N)}$. But minimal elements of $\operatorname{Ass}(M/N)$ and $\operatorname{Supp}(M/N)$ coincide. So

$$\bigcap_{\mathfrak{p} \in \operatorname{Ass}(M/N)} \mathfrak{p} = \bigcap_{\mathfrak{p} \in \operatorname{Supp}(M/N)} \mathfrak{p} = \sqrt{\operatorname{Ann}(M/N)} = \bigcup_{\mathfrak{p} \in \operatorname{Ass}(M/N)} \mathfrak{p}.$$

But then $\operatorname{Ass}(M/N) = \{\mathfrak{p}\}$.

We now show that $\operatorname{Ann}(M/N)$ is primary. Let $ab \in \operatorname{Ann}(M/N)$ and $a \notin \operatorname{Ann}(M/N)$. Then $ab(M/N) = 0$ and $a(M/N) \neq 0$. Thus b is a zero-divisor for M/N and so $b \in \sqrt{\operatorname{Ann}(M/N)}$. That is, $b^n \in \operatorname{Ann}(M/N)$ for some $n > 0$. So $\operatorname{Ann}(M/N)$ is a primary ideal and moreover $\sqrt{\operatorname{Ann}(M/N)} = \mathfrak{p}$ from the above. □

Definition 2.4.37. If N is a primary submodule of M and $\operatorname{Ass}(M/N) = \{\mathfrak{p}\}$, then we say that N is a $\mathfrak{p}$-*primary* submodule of M.

Lemma 2.4.38. *If R is noetherian, then the intersection of a finite number of $\mathfrak{p}$-primary submodules of an R-module is also $\mathfrak{p}$-primary.*

Proof. It suffices to prove the result for two $\mathfrak{p}$-primary submodules N_1, N_2 of an R-module M. We consider the obvious exact sequence $0 \to M/(N_1 \cap N_2) \to M/N_1 \oplus M/N_2$. Then $\mathrm{Ass}(M/(N_1 \cap N_2)) \subset \mathrm{Ass}(M/N_1 \oplus M/N_2) \subset \mathrm{Ass}(M/N_1) \cup \mathrm{Ass}(M/N_2) = \{\mathfrak{p}\}$ by Lemma 2.4.8 and so we are done. $\square$

Definition 2.4.39. A submodule N of M is said to be *irreducible* if $N = N_1 \cap N_2$ where N_1, N_2 are submodules of M implies $N = N_1$ or $N = N_2$. It is easy to see that N is an irreducible submodule of M if and only if 0 is irreducible in M/N.

Proposition 2.4.40. *Let R be noetherian. Then every irreducible proper submodule of a finitely generated R-module is primary.*

Proof. Let N be an irreducible submodule of a finitely generated R-module M with $N \neq M$. By Proposition 2.4.36, it suffices to show that $\mathrm{Ass}(M/N)$ consists of a single prime ideal. Suppose to the contrary $\mathrm{Ass}(M/N)$ has two distinct prime ideals $\mathfrak{p}_1$ and $\mathfrak{p}_2$. Then M/N has distinct submodules A and B such that $A \cong R/\mathfrak{p}_1$, $B \cong R/\mathfrak{p}_2$. But then $A \cap B \cong R/\mathfrak{p}_1 \cap R/\mathfrak{p}_2 = 0$. So it follows from the definition above that $A = 0$ or $B = 0$, a contradiction. Thus the result follows. $\square$

Proposition 2.4.41. *Let M be a noetherian R-module. Then every proper submodule N of M is an intersection of finitely many irreducible submodules of M.*

Proof. Let C be the set of all proper submodules A of M that are not a finite intersection of irreducible submodules of M. We claim that $C = \emptyset$. For if not, then C has a maximal element A_0. But A_0 is not irreducible and so $A_0 = A \cap B$ for some submodules A, B of M with $A_0 \neq A$, $A_0 \neq B$. So A_0 is strictly contained in A and B. Thus A, $B \notin C$. Hence A, B are finite intersections of irreducible submodules and so is A_0, a contradiction. $\square$

Definition 2.4.42. A *primary decomposition* of a submodule N of M is the finite intersection $N = N_1 \cap N_2 \cap \cdots \cap N_r$ where each N_i is a primary submodule of M. A primary decomposition $N = \bigcap_{i=1}^{r} N_i$ is said to be *reduced* if

1) N_i is $\mathfrak{p}_i$-primary for $i = 1, 2, \ldots, r$ implies $\mathfrak{p}_i \neq \mathfrak{p}_j$ for $i, j = 1, \ldots, r$.

2) $N_1 \cap \cdots \cap N_{i-1} \cap N_{i+1} \cap \cdots \cap N_r \not\subseteq N_i$ for $i = 1, \ldots, r$.

We note that given any primary decomposition, we can get a reduced one by combining the N_i's with the same prime ideal $\mathfrak{p}_i$ using Lemma 2.4.38 and by dropping redundant $\mathfrak{p}_i$'s one by one. So the two propositions above give the following important result.

Theorem 2.4.43. *Let R be noetherian and M be a finitely generated R-module. Then every proper submodule N of M has a reduced primary decomposition. Furthermore, if $N = N_1 \cap N_2 \cap \cdots \cap N_r$ is a reduced primary decomposition of N with $\mathrm{Ass}(M/N_i) = \{\mathfrak{p}_i\}$, then $\mathrm{Ass}(M/N) = \{\mathfrak{p}_1, \ldots, \mathfrak{p}_r\}$ and $\sqrt{\mathrm{Ann}(M/N)} = \bigcap_{i=1}^{s} \mathfrak{p}_i'$ where $\mathfrak{p}_1', \mathfrak{p}_2', \ldots, \mathfrak{p}_s'$ are the minimal elements in $\{\mathfrak{p}_1, \ldots, \mathfrak{p}_r\}$. The decomposition of N therefore depends only on N and M.*

Proof. The first part follows from Proposition 2.4.40 and 2.4.41 and the remarks above. We now embed M/N into $\bigoplus_{i=1}^{r} M/N_i$. Then $\mathrm{Ass}(M/N) \subset \bigcup_{i=1}^{r} \mathrm{Ass}(M/N_i) = \{\mathfrak{p}_1, \ldots, \mathfrak{p}_r\}$. Conversely, let $N' = N_1 \cap \cdots \cap N_{i-1} \cap N_{i+1} \cap \cdots \cap N_r$. Then $N'/N \cong N'/(N' \cap N_i) \cong (N' + N_i)/N_i \subset M/N_i$. So $\mathrm{Ass}(N'/N) \subset \mathrm{Ass}(M/N_i) = \{\mathfrak{p}_i\}$. That is, $\mathrm{Ass}(N'/N) = \{\mathfrak{p}_i\}$. But $N'/N \subset M/N$. So $\mathfrak{p}_i \in \mathrm{Ass}(M/N)$. Thus $\{\mathfrak{p}_1, \ldots, \mathfrak{p}_r\} \subset \mathrm{Ass}(M/N)$. Hence $\mathrm{Ass}(M/N) = \{\mathfrak{p}_1, \ldots, \mathfrak{p}_r\}$. The last part follows from Remark 2.4.26. $\qquad\square$

Corollary 2.4.44. *Let R be noetherian. Then every proper ideal I of R has a reduced primary decomposition $I = I_1 \cap I_2 \cap \cdots \cap I_r$ where each I_i is $\mathfrak{p}_i$-primary. Furthermore, $\mathrm{Ass}(R/I) = \{\mathfrak{p}_1, \ldots, \mathfrak{p}_r\}$ and $\sqrt{I} = \bigcap_{i=1}^{s} \mathfrak{p}_i'$ where $\mathfrak{p}_1', \ldots, \mathfrak{p}_s'$ are the minimal elements in $\mathrm{Ass}(R/I)$.*

Exercises

1. Let $\mathfrak{p}$ be a prime ideal of R. Prove that $\mathfrak{p} \in \mathrm{Ass}(M)$ if and only if M contains a submodule isomorphic to $R/\mathfrak{p}$.

2. Let $\mathfrak{p}$ be a prime ideal of a noetherian ring R. Prove that $\mathrm{Ass}_R(R/\mathfrak{p}) = \{\mathfrak{p}\}$.

3. Prove that if $0 \to M' \to M \to M'' \to 0$ is an exact sequence R-modules, then $\mathrm{Supp}(M) = \mathrm{Supp}(M') \cup \mathrm{Supp}(M'')$.

4. Let R be noetherian and M be a finitely generated R-module. Prove that the set of minimal elements of $\mathrm{Ass}_R(M)$ and $\mathrm{Supp}(M)$ coincide.

5. Let R be noetherian and M be a finitely generated R-module. Prove that the following are equivalent.

   a) $\dim M = 0$.

   b) $R/\mathrm{Ann}(M)$ is an artinian ring.

   c) M is of finite length.

   d) Every $\mathfrak{p} \in \mathrm{Ass}(M)$ is a maximal ideal of R.

   e) Every $\mathfrak{p} \in \mathrm{Supp}(M)$ is a maximal ideal of R.

6. Let R be noetherian and M be an R-module of finite length. Prove that $\mathrm{Ass}(M) = \mathrm{Supp}(M)$.

7. Prove that an ideal I of R is primary if and only if every zero divisor of R/I is nilpotent.

8. Prove that if $\mathfrak{p} \in \operatorname{Spec} R$, then $\sqrt{\mathfrak{p}^n} = \mathfrak{p}$.

9. Show that if $\mathfrak{p} \in \operatorname{Spec} R$, then a power $\mathfrak{p}^n$ is not necessarily a primary ideal even though $\sqrt{\mathfrak{p}^n}$ is a prime ideal.
   Hint: Consider $R = k[x, y, z]/\langle z^2 - xy \rangle$ where k is a field and let $\bar{x}, \bar{y}, \bar{z}$ be images of x, y, z in R. Then show that $\mathfrak{p} = \langle \bar{x}, \bar{z} \rangle \in \operatorname{Spec} R$ and $\mathfrak{p}^2$ is not a primary ideal.

2.5 Artin–Rees lemma and Zariski rings

In this section, all rings are commutative.

A ring S is said to be an *R-algebra* if there is a ring homomorphism $\varphi : R \to S$. It is easy to see that S is an R-module via $rs = \varphi(r)s$. For example, every ring is a $\mathbb{Z}$-algebra.

Definition 2.5.1. A *graded ring* is a ring R together with subgroups R_n of the additive group of R, $n \geq 0$, such that $R = \bigoplus_{n \geq 0} R_n$ and $R_m R_n \subset R_{m+n}$ for all $m, n \geq 0$. So in this case $R_0 R_0 \subset R_0$ and thus a graded ring R is an R_0-algebra. It is easy to see that if R is a graded ring, then $R_+ = \bigoplus_{n > 0} R_n$ is an ideal of R and $R/R_+ \cong R_0$.

Now let R be a graded ring. Then a *graded R-module* is an R-module M together with subgroups M_n of M, $n \geq 0$, such that $M = \bigoplus_{n \geq 0} M_n$ and $R_m M_n \subset M_{m+n}$ for all $m, n \geq 0$. Each element $x \in M_n$ is said to be *homogeneous* of degree n.

We state and prove the next result for completeness.

Proposition 2.5.2. *Let R be a graded ring. Then R is noetherian if and only if R_0 is noetherian and $R = R_0[x_1, \ldots, x_r]$ for some $x_1, \ldots, x_r \in R$.*

Proof. If R is noetherian, then $R_0 \cong R/R_+$ is also noetherian. Now since R_+ is an ideal of R, $R_+ = \langle x_1, \ldots, x_r \rangle$ for some $x_1, \ldots, x_r \in R$. Clearly, $R_0[x_1, \ldots, x_r] \subseteq R$. To show $R \subseteq R_0[x_1, \ldots x_r]$, we argue by induction that for each $n \geq 0$, $R_n \subseteq R_0[x_1, \ldots, x_r]$. The case $n = 0$ is trivial. Now suppose $n > 0$ and $R_k \subseteq R_0[x_1, \ldots, x_r]$ for all $k \leq n - 1$. Assume each x_i is homogeneous of degree α_i. If $y \in R_n$, then $y \in R_+$ and so $y = \sum_{i=1}^{r} a_i x_i$ where $a_i \in R_{n-\alpha_i}$ taking $R_{n-\alpha_i} = 0$ if $\alpha_i > n$. But $\alpha_i > 0$. So $n - \alpha_i \leq n - 1$ and thus each $a_i \in R_0[x_1, \ldots, x_r]$ by the induction hypothesis. Thus $y \in R_0[x_1, \ldots, x_r]$ and hence $R = R_0[x_1, \ldots, x_r]$.

The converse follows from the Hilbert basis theorem. $\square$

Definition 2.5.3. Let M be an R-module. Then a decreasing sequence (M_n) of sub-modules of M is called a *filtration* of M. If I is an ideal of R, then the filtration (M_n) of M is said to be an *I-filtration* if $I M_n \subset M_{n+1}$. An I-filtration of M is said to be *stable*, or according to Bourbaki, *I-good*, if there is an integer n_0 such that $I M_n = M_{n+1}$ for all $n > n_0$. It is clear that the filtration $M = I^0 M \supset I M \supset I^2 M \supset \cdots$ is a stable I-filtration. We recall that this filtration determines the I-adic topology of M generated by $\{x + I^n M\}$.

Now let x be an indeterminate, then $R' = R + Ix + I^2x^2 + \cdots$ is a graded subring of the polynomial ring $R[x]$. Furthermore,

$$M' = M + (IM)x + (I^2M)x^2 + \cdots$$

is a subgroup of $M \otimes_R R[x]$ noting that $M' = \sum_{n \geq 0} M_n \otimes Rx^n$ where $M_n = I^n M$. But

$$(I^m x^m)(M_n \otimes_R Rx^n) \subset I^m M_n \otimes_R Rx^{m+n} \subset M_{m+n} \otimes Rx^{m+n}.$$

So M' is a graded R'-module. With this notation, we have the following result.

Lemma 2.5.4. *Let I be an ideal of R and (M_n) be an I-filtration of an R-module M such that each M_n is a finitely generated submodule of M. Then the filtration is stable if and only if M' is a finitely generated R'-module.*

Proof. From the above, M' is a graded R'-module. Suppose $M' = \langle y_1, y_2, \dots, y_r \rangle$ where $y_i \in M'_{n_i} = M_{n_i} \otimes Rx^{n_i}$. We note that each $y_i = m_i \otimes x^{n_i}$ for some $m_i \in M_{n_i}$. Now let $n_0 = \max\{n_i\}$, $i = 1, \dots, r$. If $n \geq n_0$ and $m \in M_n$, then $m \otimes x^n = \sum_i a_i (m_i \otimes x^{n_i})$ where $a_i \in R'$. But then we may assume $a_i = b_i x^{n-n_i}$ where $b_i \in I^{n-n_i}$. So $m \otimes x^n = (\sum_i b_i m_i) \otimes x^n$ and therefore $m = \sum_i b_i m_i \in I^{n-n_0} M_{n_0}$. Hence if $n \geq n_0$, $M_n \subset I^{n-n_0} M_{n_0}$. But clearly $I^{n-n_0} M_{n_0} \subset M_n$. Hence $M_n = I^{n-n_0} M_{n_0}$. But then $M_n = I M_{n-1}$ whenever $n > n_0$. That is, (M_n) is stable.

Now suppose $M_n = I M_{n-1}$ for $n > n_0$. If $n \leq n_0$, let $M_n = \langle y_{n_1}, \dots, y_{n_{r_n}} \rangle$. Then $M_n \otimes_R Rx^n = \langle y_{n_1} \otimes x^n, \dots, y_{n_{r_n}} \otimes x^n \rangle$ as an R-module for each $n \leq n_0$. If $n > n_0$, then $M_n \otimes_R Rx^n = I M_{n-1} \otimes Rx^n = \cdots = I^{n-n_0} M_{n_0} \otimes Rx^n$. Thus M' is generated by $\{y_{n_j} \otimes x^n\}$ for $0 \leq n \leq n_0$ and $1 \leq j \leq r_n$ as an R'-module. $\square$

Lemma 2.5.5. *If R is noetherian, then so is R'.*

Proof. Since $R' = R + Ix + I^2x^2 + \cdots$, we see that $R'/R_0 \cong R_+ = Ix + I^2x^2 + \cdots$ where $R_0 = R$. But I is finitely generated. So $I = \langle a_1, \dots, a_r \rangle$. But then $R_+ = \langle a_1 x, a_2 x, \dots a_r x \rangle$. Thus $R' = R[a_1 x, \dots, a_r x]$ as in the proof of Proposition 2.5.2 above. Hence R' is noetherian. $\square$

Theorem 2.5.6 (Artin–Rees lemma). *Let R be a noetherian ring, I an ideal of R, M a finitely generated R-module and N a submodule of M. If (M_n) is a stable I-filtration of M, then $(M_n \cap N)$ is also a stable I-filtration. In particular, there exists an integer r such that*

$$(I^n M) \cap N = I^{n-r}((I^r M) \cap N)$$

for all $n \geq r$.

Proof. We have $I(M_n \cap N) \subset I M_n \cap I N \subset M_{n+1} \cap N$. So $(M_n \cap N)$ is an I-filtration which defines a graded R'-module $N' = \sum_{n \geq 0} (M_n \cap N) \otimes Rx^n$ which is an R'-submodule of M'. But (M_n) is stable. So M' is a finitely generated R'-module by Lemma 2.5.4. Hence N' is a finitely generated R'-module since R' is noetherian by Lemma 2.5.5 above. But then $(M_n \cap N)$ is stable again by Lemma 2.5.4.

In particular, if we set $M_n = I^n M$, then $((I^n M) \cap N)$ is a stable I-filtration since $(I^n M)$ is. So there is an integer r such that $I((I^r M) \cap N) = (I^{r+1} M) \cap N$. Thus if $n \geq r$, then

$$I^{n-r}((I^r M) \cap N) = I^{n-r-1}((I^{r+1} M) \cap N) = \cdots = (I^n M) \cap N. \qquad \square$$

Theorem 2.5.7 (Krull Intersection Theorem). *Let R be noetherian, I an ideal of R, M a finitely generated R-module and $N = \bigcap_{n \geq 0} I^n M$. Then $N = IN$.*

Proof. There exists an integer r such that $N = (I^n M) \cap N = I^{n-r}((I^r M) \cap N) \subset IN \subset N$ by the Artin–Rees lemma. Hence $IN = N$. $\qquad \square$

Corollary 2.5.8. *Let R be noetherian, I an ideal of R, and M a finitely generated R-module. If $I \subset \mathrm{rad}(R)$, then M is Hausdorff and every submodule of M is closed with respect to the I-adic topology on M.*

Proof. By Proposition 1.6.2, to show M is Hausdorff it suffices to show that $\bigcap_{n \geq 0} I^n M = 0$. So let $N = \bigcap_{n \geq 0} I^n M$. Then $N = IN$ by the theorem above. But $I \subset \mathrm{rad}(R)$. So $N = 0$ by Nakayama lemma. Now if N is a submodule of M, then M/N is Hausdorff with respect to the quotient topology. Hence N is closed in M by Remark 1.6.3. $\qquad \square$

Lemma 2.5.9. *Stable I-filtrations of an R-module M determine the same topology on M, namely the I-adic topology on M.*

Proof. Let $M_n = I^n M$. Then (M_n) is a stable I-filtration of M that determines the I-adic topology on M. Now let (M_n') be a stable I-filtration. Then $I M_n' \subseteq M_{n+1}'$ and so $M_n = I^n M \subseteq M_n'$ since $M_0' = M$. Thus $M_{n+r} = I^r M_n \subset M_{n+r}'$ for all $n \geq 0$. But there is an integer r such that $I M_n' = M_{n+1}'$ for all $n \geq r$. So $M_{n+r}' = I^n M_r' \subseteq I^n M = M_n$ for all $n \geq 0$. Hence (M_n') and $(I^n M)$ induce the same topology on M. $\qquad \square$

Theorem 2.5.10. *Let R be a noetherian ring, I an ideal of R, M a finitely generated R-module, and N a submodule of M. Then the I-adic topology of N coincides with the subspace topology induced by the I-adic topology of M.*

Proof. We simply note that $(I^n N)$ is a stable I-filtration of N. But $((I^n M) \cap N)$ is also a stable I-filtration of N by the Artin–Rees lemma (Theorem 2.5.6). So the result follows from Lemma 2.5.9 above. $\qquad \square$

Theorem 2.5.11. *Let R be noetherian, I an ideal of R, and $0 \to M' \xrightarrow{\varphi} M \xrightarrow{\psi} M'' \to 0$ be an exact sequence of finitely generated R-modules. Then the sequence of I-adic completions*

$$0 \to \hat{M}' \to \hat{M} \to \hat{M}'' \to 0$$

is also exact.

Proof. The filtration $(I^n M)$ determines the I-adic topology on M. So the filtration $(\varphi^{-1}(I^n M)) = ((I^n M) \cap M')$ determines the I-adic topology on M' by the theorem above and $(\psi(I^n M)) = ((I^n M + M')/M')$ determines the I-adic topology on M''. Thus we consider the exact sequence

$$0 \to M'/((I^n M) \cap M') \to M/I^n M \to M''/\psi(I^n M) \to 0.$$

But the natural maps $M'/(I^{n+1} M) \cap M' \to M'/(I^n M) \cap M'$ are clearly surjective. So taking inverse limits gives the exact sequence

$$0 \to \varprojlim M'/(I^n M) \cap M' \to \varprojlim M/I^n M \to \varprojlim M''/\psi(I^n M) \to 0$$

by Theorem 1.5.13. But then the result follows by Theorem 1.6.7.   □

Corollary 2.5.12. *If $0 \to M' \to M \to M'' \to 0$ is an exact sequence of finitely generated R-modules, then $(M/M')^\wedge \cong \hat{M}/\hat{M}'$.*

If we set $M' = I^n M$, then $M'' = M/I^n M$ has the discrete topology and so $\hat{M}'' = M''$. Hence we have the following.

Corollary 2.5.13. $\widehat{I^n M}$ *is an $\hat{R}$-submodule of $\hat{M}$ and $\hat{M}/\widehat{I^n M} \cong M/I^n M$.*

Theorem 2.5.14. *Let R be a noetherian ring, I an ideal of R and M a finitely generated R-module. If $\hat{M}$, $\hat{R}$ denote the I-adic completions of M and R respectively, then*

$$\hat{R} \otimes_R M \cong \hat{M}.$$

In particular, if R is complete, then so is M.

Proof. By Remark 2.3.12, M has an exact sequence $F_1 \to F_0 \to M \to 0$ with F_1, F_0 finitely generated and free. So we have the following commutative diagram

$$
\begin{array}{ccccccc}
\hat{R} \otimes_R F_1 & \longrightarrow & \hat{R} \otimes_R F_0 & \longrightarrow & \hat{R} \otimes_R M & \longrightarrow & 0 \\
\downarrow & & \downarrow & & \downarrow & & \\
\hat{F}_1 & \longrightarrow & \hat{F}_0 & \longrightarrow & \hat{M} & \longrightarrow & 0
\end{array}
$$

with exact rows. But the first two vertical maps are isomorphisms. So $\hat{R} \otimes_R M \cong \hat{M}$.□

Corollary 2.5.15. *If R is noetherian and $\hat{R}$ is the I-adic completion of R, then*

(1) *$\hat{R}$ is a flat R-algebra.*

(2) *$I\hat{R} \cong I \otimes_R \hat{R} \cong \hat{I}$.*

(3) *The topology of $\hat{R}$ is the $\hat{I}$-adic topology.*

(4) *$\hat{I} \subset \text{rad}(\hat{R})$.*

Proof. (1) Let M be a finitely generated R-module. Then there is an exact sequence $0 \to K \to P \to M \to 0$ with P projective and K, P finitely generated. So there is an exact sequence $0 \to \text{Tor}_1^R(\hat{R}, M) \to \hat{R} \otimes_R K \to \hat{R} \otimes_R P \to \hat{R} \otimes_R M \to 0$. But $0 \to \hat{K} \to \hat{P} \to \hat{M} \to 0$ is exact by Theorem 2.5.11. So $\text{Tor}_1^R(\hat{R}, M) = 0$ by Theorem 2.5.14. Hence $\hat{R}$ is a flat R-algebra by Theorem 2.1.8.

(2) Since $\hat{R}$ is flat, $I\hat{R} \cong I \otimes_R \hat{R}$ by Proposition 2.1.7. So the result follows from the theorem above.

(3) Since $I\hat{R} \cong \hat{I}$, the topology of $\hat{R}$ is determined by $(I^n \hat{R}) = (\hat{I}^n)$.

(4) We note that $\hat{R}$ is complete in its $\hat{I}$-adic topology. So if $x \in \hat{I}$, then $(1-x)^{-1} = 1 + x + x^2 + \cdots$ converges in $\hat{R}$. Thus $1 - xy$ is a unit in $\hat{R}$ for all $y \in \hat{R}$. So $x \in \text{rad}(\hat{R})$ by Definition 1.2.27. That is, $\hat{I} \subset \text{rad}(\hat{R})$. $\square$

Corollary 2.5.16. *If R is noetherian, then the I-adic completion $\hat{R}$ is also noetherian.*

Proof. Let $I = \langle a_1, a_2, \ldots, a_r \rangle$. Suppose $S = R[x_1, \ldots, x_r]$ and $J = \sum_{i=1}^r (x_i - a_i)S$. Then $S/J \cong R$ and so S/J is an S-algebra. Furthermore, if $I_1 = \sum_{i=1}^r x_i S$, then the I_1-adic topology on the S-algebra S/J coincides with the I-adic topology on R and so with respect to these topologies, we have $(S/J)^\wedge \cong \hat{R}$. But $(S/J)^\wedge \cong \hat{S}/\hat{J} \cong \hat{S}/J\hat{S} \cong R[[x_1, \ldots, x_r]]/(x_1 - a_1, \ldots, x_r - a_r)$. So the result follows since $R[[x_1, \ldots, x_r]]$ is noetherian. $\square$

Lemma 2.5.17. *Let $\varphi : R \to S$ be a ring homomorphism and S be a faithfully flat R-module, that is, S is a faithfully flat R-algebra. Then*

(1) *If M is an R-module, then the map $\bar{\varphi} : M \to M \otimes_R S$ defined by $\bar{\varphi}(x) = x \otimes 1$ is a monomorphism. In particular φ is a monomorphism.*

(2) *If I is an ideal of R, then $IS \cap R = I$.*

(3) *The map $\psi : \text{Spec } S \to \text{Spec } R$ defined by $\psi(\mathfrak{p}) = \varphi^{-1}(\mathfrak{p}) = \mathfrak{p} \cap R$ is surjective.*

(4) *If $\mathfrak{m}$ is a maximal ideal of R, then there exists a maximal ideal $\mathfrak{m}'$ of S such that $\mathfrak{m}' \cap R = \mathfrak{m}$, that is, $\mathfrak{m}'$ lies over $\mathfrak{m}$.*

Proof. (1) Suppose $x \in M$, $x \neq 0$. Then $0 \neq Rx \otimes_R S \subset M \otimes_R S$ since S is faithfully flat. So $Rx \otimes_R S = (x \otimes 1)S \neq 0$ and thus $x \otimes 1 \neq 0$.

(2) We simply note that $R/I \to R/I \otimes_R S = S/IS$ is an embedding by part (1) above since $R/I \otimes_R S$ is a faithfully flat R/I-module. So $I = IS \cap R$ by Lemma 2.1.13.

(3) Let $\mathfrak{q} \in \operatorname{Spec} R$. Then $S \otimes_R R_{\mathfrak{q}} = S_{\mathfrak{q}}$ is a faithfully flat $R_{\mathfrak{q}}$-module. So $S_{\mathfrak{q}} \neq \mathfrak{q} S_{\mathfrak{q}}$ by again the lemma above. Therefore there exists a maximal ideal $\mathfrak{m}$ of $S_{\mathfrak{q}}$ that contains $\mathfrak{q} S_{\mathfrak{q}}$. So $\mathfrak{m} \cap R_{\mathfrak{q}} \supset \mathfrak{q} R_{\mathfrak{q}}$. But $\mathfrak{q} R_{\mathfrak{q}}$ is maximal. So $\mathfrak{m} \cap R_{\mathfrak{q}} = \mathfrak{q} R_{\mathfrak{q}}$. We now let $\mathfrak{p} = \mathfrak{m} \cap S$. Then $\mathfrak{p} \in \operatorname{Spec} S$ and $\psi(\mathfrak{p}) = \mathfrak{p} \cap R = (\mathfrak{m} \cap S) \cap R = \mathfrak{m} \cap R = (\mathfrak{m} \cap R_{\mathfrak{q}}) \cap R = \mathfrak{q} R_{\mathfrak{q}} \cap R = \mathfrak{q}$.

(4) Since $\mathfrak{m} \in \operatorname{Spec} R$, we have that there is a $\mathfrak{p} \in \operatorname{Spec} S$ such that $\mathfrak{p} \cap R = \mathfrak{m}$ by part (3) above. Now let $\mathfrak{m}'$ be a maximal ideal of S containing $\mathfrak{p}$. Then $\mathfrak{m}' \cap R \supset \mathfrak{p} \cap R = \mathfrak{m}$. But $\mathfrak{m}$ is maximal. So $\mathfrak{m}' \cap R = \mathfrak{m}$.   □

Theorem 2.5.18. *Let R be noetherian and I be an ideal of R. Then the following are equivalent.*

(1) $I \subset \operatorname{rad}(R)$.

(2) *Every finitely generated R-module is Hausdorff with respect to the I-adic topology.*

(3) *If M is a finitely generated R-module, then every submodule of M is closed with respect to the I-adic topology on M.*

(4) *Every ideal of R is closed with respect to the I-adic topology.*

(5) *Every maximal ideal of R is closed with respect to the I-adic topology.*

(6) *The I-adic completion $\hat{R}$ is a faithfully flat R-module.*

Proof. $(1) \Rightarrow (2) \Rightarrow (3)$ by the proof of Corollary 2.5.8.

$(3) \Rightarrow (4) \Rightarrow (5)$ is trivial.

$(5) \Rightarrow (6)$. $\hat{R}$ is a flat R-module by Corollary 2.5.15. Now let $\mathfrak{m}$ be a maximal ideal of R. Then $\mathfrak{m}\hat{R} = \hat{\mathfrak{m}}$ is the closure of $\mathfrak{m}$ in $\hat{R}$. But $\mathfrak{m}$ is closed. So $\mathfrak{m}\hat{R} \cap R = \mathfrak{m}$ and thus $\mathfrak{m}\hat{R} \neq \hat{R}$. That is, $\hat{R}$ is faithfully flat by Lemma 2.1.13.

$(6) \Rightarrow (1)$. Let $\mathfrak{m}$ be a maximal ideal of R. Then by Lemma 2.5.17, there exists a maximal ideal $\mathfrak{m}'$ of $\hat{R}$ such that $\mathfrak{m}' \cap R = \mathfrak{m}$. But $\hat{I} \subset \operatorname{rad}(\hat{R})$ by Corollary 2.5.15. So $\hat{I} \subset \mathfrak{m}'$. Hence $I \subset \hat{I} \cap R \subset \mathfrak{m}' \cap R = \mathfrak{m}$. Thus $I \subset \operatorname{rad}(R)$.   □

Definition 2.5.19. A *Zariski ring* is a noetherian ring R with an I-adic topology that satisfies the equivalent conditions of Theorem 2.5.18 above. In this book, we will from time to time be concerned with an important class of Zariski rings, namely, noetherian local rings $(R, \mathfrak{m}, k)$ with the $\mathfrak{m}$-adic topology. In this case, $\hat{R}$ is a local ring with maximal ideal $\mathfrak{m}\hat{R}$ and residue field $\hat{R}/\mathfrak{m}\hat{R} \cong \widehat{(R/\mathfrak{m})} = k$.

Theorem 2.5.20. *Let R be a semilocal ring and $\mathfrak{m}_1, \mathfrak{m}_2, \ldots, \mathfrak{m}_r$ be its maximal ideals. If $I = \operatorname{rad}(R)$, then the I-adic completion $\hat{R}$ is a direct product of local rings $\hat{R}_{\mathfrak{m}_i}$. That is,*

$$\hat{R} \cong \hat{R}_{\mathfrak{m}_1} \times \hat{R}_{\mathfrak{m}_2} \times \cdots \times \hat{R}_{\mathfrak{m}_r}.$$

Proof. $I = \mathfrak{m}_1\mathfrak{m}_2\ldots\mathfrak{m}_r = \bigcap_{i=1}^r \mathfrak{m}_i$ since $I = \mathrm{rad}(R)$. So for each $n \geq 0$, $I^n = \mathfrak{m}_1^n\mathfrak{m}_2^n\ldots\mathfrak{m}_r^n$ where $\mathfrak{m}_i^n$'s are pairwise coprime. Hence $R/I^n \cong R/\mathfrak{m}_1^n \times R/\mathfrak{m}_2^n \times \cdots \times R/\mathfrak{m}_r^n$. But $R/\mathfrak{m}_i^n$ is local and so $R/\mathfrak{m}_i^n = (R/\mathfrak{m}_i^n)_{\mathfrak{m}_i} = R\mathfrak{m}_i/(\mathfrak{m}_i R_{\mathfrak{m}_i})^n$. Thus $\lim_{\leftarrow} R/\mathfrak{m}_i^n = \hat{R}_{\mathfrak{m}_i}$. But $\hat{R} = \lim_{\leftarrow} R/I^n$. So the result follows. $\qquad\square$

Exercises

1. Prove that if R is a graded ring, then $R_+ = \bigoplus_{n>0} R_n$ is an ideal of R.

2. Let R be a noetherian ring and $I = \mathrm{rad}(R)$. Prove that $\bigcap_{n>0} I^n = 0$.

3. If R is noetherian and $\hat{R}$ is its I-adic completion, prove that $\widehat{(I^n)} \cong (\hat{I})^n$.

4. Let R be a Zariski ring and $\hat{R}$ be its completion. Prove that

   a) $R \subset \hat{R}$ and $I\hat{R} \cap R = I$ for any ideal I of R.

   b) There is a bijective map $\psi : \mathrm{mSpec}\,R \to \mathrm{mSpec}\,\hat{R}$ given by $\psi(\mathfrak{m}) = \mathfrak{m}\hat{R}$ where $\mathfrak{m}\hat{R} \cap R = \mathfrak{m}$.

   c) If R is a local ring, then $\hat{R}$ is also a local ring.

5. Prove that if R is a noetherian ring, then $R[[x_1,\ldots,x_n]]$ is a faithfully flat R-module.

6. Let R be noetherian, I be an ideal of R such that $I \subset \mathrm{rad}(R)$, M and N be finitely generated R-modules, and $\hat{R}$, $\hat{M}$, $\hat{N}$ denote I-adic completions. Argue that an R-homomorphism $f : M \to N$ is an isomorphism if and only if $\hat{f} : \hat{M} \to \hat{N}$ is an isomorphism.

7. Let R be noetherian, I be an ideal of R and M, N be R-modules. Prove that $\mathrm{Tor}_i^R(M, N)^\wedge \cong \mathrm{Tor}_i^{\hat{R}}(\hat{M}, \hat{N})$ for all $i \geq 0$ where $^\wedge$ denotes the I-adic completion.

Chapter 3

Injective and Flat Modules

3.1 Injective modules

We recall the following

Definition 3.1.1. An R-module E is said to be *injective* if given R-modules $A \subset B$ and a homomorphism $f : A \to E$, there exists a homomorphism $g : B \to E$ such that $g_{|A} = f$.

Theorem 3.1.2. *The following are equivalent for an R-module E.*

(1) *E is injective.*

(2) *$\mathrm{Hom}(-, E)$ is right exact.*

(3) *E is a direct summand of every R-module containing E.*

Proof. (1) $\Rightarrow$ (2) is clear.

(2) $\Rightarrow$ (3). We consider the exact sequence $0 \to E \to B \to C \to 0$ of R-modules. Then $\mathrm{Hom}(B, E) \to \mathrm{Hom}(E, E) \to 0$ is exact and so E is a direct summand of B.

(3) $\Rightarrow$ (1). Let $A \subset B$ be R-modules. Then we consider the pushout diagram

$$
\begin{array}{ccc}
0 \longrightarrow A & \overset{i}{\longrightarrow} & B \\
\downarrow{\scriptstyle f} & & \downarrow{\scriptstyle f'} \\
E & \overset{j}{\longrightarrow} & C
\end{array}
$$

of Example 1.3.20. But then j is one-to-one and thus $0 \to E \overset{j}{\to} C$ is split exact. So there is a map $s : C \to E$ such that $s \circ j = \mathrm{id}_E$. Then $g = s \circ f'$ is an extension of f since $g \circ i = s \circ f' \circ i = s \circ j \circ f = f$. Hence E is injective. $\square$

Theorem 3.1.3 (Baer's Criterion). *An R-module E is injective if and only if for all ideals I of R, every homomorphism $f : I \to E$ can be extended to R.*

Proof. Let $A \subset B$ be R-modules and $f : A \to E$ be a homomorphism. Now let $\mathcal{C}$ be the collection of all pairs (C, g) such that $A \subset C \subset B$ and $g_{|A} = f$. Then $\mathcal{C} \neq \emptyset$ since $(A, f) \in \mathcal{C}$. Now partially order $\mathcal{C}$ by $(C, g) \leq (C', g')$ if $C \subset C'$ and $g'_{|C} = g$. Then $\mathcal{C}$ is an inductive system and hence has a maximal element (C_0, g_0) by Zorn's lemma.

Suppose $C_0 \neq B$. Then let $x \in B - C_0$ and set $I = \{r \in R : rx \in C_0\}$. Then I is a left ideal of R. Define a map $h : I \to E$ by $h(r) = g_0(rx)$. Then h is a homomorphism and thus can be extended to $h' : R \to E$ by assumption. We now define a map $\bar{g} : C_0 + Rx \to E$ by $\bar{g}(c_0 + rx) = g_0(c_0) + h'(r)$. If $c_0 + rx = c'_0 + r'x$, then $c_0 - c'_0 = (r' - r)x$ and so $r' - r \in I$. Thus $g_0(c_0 - c'_0) = g_0((r' - r)x) = h(r' - r) = h'(r' - r)$ and so $g_0(c_0) + h'(r) = g_0(c'_0) + h'(r')$. Hence $\bar{g}$ is a well-defined homomorphism. Furthermore $\bar{g}(a) = g_0(a) = f(a)$ for all $a \in A$ and so $(C_0 + Rx, \bar{g}) \in \mathcal{C}$. This contradicts the maximality of (C_0, g_0) since $C_0 \subsetneq C_0 + Rx$. Hence $C_0 = B$ and we are done. $\square$

Theorem 3.1.4. *Let R be a principal ideal domain. Then an R-module M is injective if and only if it is divisible.*

Proof. Let $x \in M$ and $r \in R$ be a nonzero divisor. Then we define a map $f : \langle r \rangle \to M$ by $f(sr) = sx$. f is a well-defined homomorphism since r is a nonzero divisor. If M is injective, then we can extend the map f to a map $g : R \to M$ such that $x = f(r) = g(r) = rg(1)$. Thus M is divisible. Conversely, let I be an ideal of R and $f : I \to M$ be an R-homomorphism. By Baer's Criterion, it suffices to extend f to R for $I \neq 0$. But R is a principal ideal domain and so $I = \langle s \rangle$ for some $s \in R, s \neq 0$. If M is divisible, then there is $x \in M$ such that $f(s) = sx$. Now define an R-homomorphism $g : R \to M$ by $g(r) = rx$. Then $g_{|I} = f$ for if $r' \in R$, then $g(r's) = r'sx = r'f(s) = f(r's)$. $\square$

Corollary 3.1.5. *Every abelian group can be embedded in an injective abelian group.*

Proof. Let G be an abelian group. Then $G = (\oplus \mathbb{Z})/S \subset (\oplus \mathbb{Q})/S$. But $(\oplus \mathbb{Q})/S$ is divisible since $\mathbb{Q}$ is and so we are done by the theorem above. $\square$

Proposition 3.1.6. *If $R \to S$ is a ring homomorphism and if E is an injective left R module, then $\mathrm{Hom}_R(S, E)$ is an injective left S module.*

Proof. Note that S is an (R, S)-bimodule. Let $A \subset B$ be a submodule of the left S-module B. Then by Theorem 2.1.10, $\mathrm{Hom}_S(A, \mathrm{Hom}_R(S, E)) \cong \mathrm{Hom}_R(S \otimes_S A, E) \cong \mathrm{Hom}_R(A, E)$ and likewise for $\mathrm{Hom}_S(B, \mathrm{Hom}_R(S, E))$. So we have that $\mathrm{Hom}_S(B, \mathrm{Hom}_R(S, E)) \to \mathrm{Hom}_S(A, \mathrm{Hom}_R(S, E)) \to 0$ is exact since $\mathrm{Hom}_R(B, E) \to \mathrm{Hom}_R(A, E) \to 0$ is exact. Hence $\mathrm{Hom}_R(S, E)$ is injective. $\square$

We note that it follows from the above that $\mathrm{Hom}_{\mathbb{Z}}(R, G)$ is an injective left R-module for any ring R when G is a divisible abelian group.

Theorem 3.1.7. *Every R-module can be embedded in an injective R-module.*

Proof. Let M be an R-module. Then M can be embedded into an injective abelian group G by Corollary 3.1.5. But M can be embedded in $\operatorname{Hom}_{\mathbb{Z}}(R, G)$ by the map $\varphi : M \to \operatorname{Hom}_{\mathbb{Z}}(R, G)$ defined by $\varphi(x)(r) = rx$ since $\varphi(x) = 0$ implies $x = \varphi(x)(1) = 0$. Hence we are done by Proposition 3.1.6 above. □

Remark 3.1.8. It follows from the theorem above that every R-module N has an exact sequence $0 \to N \to E^0 \to E^1 \to \cdots$ with each E^i injective. This sequence is called an *injective resolution* of N.

Let $\cdots \to P_1 \to P_0 \to M \to 0$ be a projective resolution of a left R-module M and consider the deleted projective resolution $\cdots \to P_1 \to P_0 \to 0$. Then the ith cohomology module of the complex $0 \to \operatorname{Hom}(P_0, N) \to \operatorname{Hom}(P_1, N) \to \cdots$ is denoted $\operatorname{Ext}^i_R(M, N)$. Note that $\operatorname{Ext}^0_R(M, N) = \operatorname{Hom}(M, N)$ since $0 \to \operatorname{Hom}(M, N) \to \operatorname{Hom}(P_0, N) \to \operatorname{Hom}(P_1, N)$ is exact. $\operatorname{Ext}^i_R(M, N)$ can also be computed using a deleted injective resolution of N and is independent of the projective and injective resolutions used, and moreover given an exact sequence $0 \to M' \to M \to M'' \to 0$ there exists a long exact sequence $0 \to \operatorname{Hom}(M'', N) \to \operatorname{Hom}(M, N) \to \operatorname{Hom}(M', N) \to \operatorname{Ext}^1(M'', N) \to \cdots$ (see Chapter 8 for details).

We can now characterize injective modules as follows.

Theorem 3.1.9. *The following are equivalent for an R-module E.*

(1) *E is injective.*

(2) $\operatorname{Ext}^i(M, E) = 0$ *for all R-modules M and for all $i \geq 1$.*

(3) $\operatorname{Ext}^1(M, E) = 0$ *for all R-modules M.*

(4) $\operatorname{Ext}^i(R/I, E) = 0$ *for all ideals I of R and for all $i \geq 1$.*

(5) $\operatorname{Ext}^1(R/I, E) = 0$ *for all ideals I of R.*

Proof. (1) $\Rightarrow$ (2). Let $\cdots \to P_1 \to P_0 \to M \to 0$ be a projective resolution of M. Then $0 \to \operatorname{Hom}(M, E) \to \operatorname{Hom}(P_0, E) \to \operatorname{Hom}(P_1, E) \to \cdots$ is exact since E is injective and so (2) follows.

(2) $\Rightarrow$ (3) $\Rightarrow$ (5) and (2) $\Rightarrow$ (4) $\Rightarrow$ (5) are trivial.

(5) $\Rightarrow$ (1) follows from Baer's Criterion (Theorem 3.1.3). □

Corollary 3.1.10. *A product of R-modules $\prod_{i \in I} E_i$ is injective if and only if each E_i is injective.*

Definition 3.1.11. If M is a submodule of an injective R-module E, then $M \subset E$ is called an *injective extension* of M. It therefore follows from Theorem 3.1.7 that every R-module has an injective extension.

Definition 3.1.12. Let $A \subset B$ be R-modules. Then B is said to be an *essential extension* of A if for each submodule N of B, $N \cap A = 0$ implies $N = 0$. In this case, A is said to be an *essential submodule* of B.

We note that if $A \subset B \subset C$ are modules and C is an essential extension of A, then B is an essential extension of A and C is an essential extension of B. If $A \subset B$ is a direct summand, then B is an essential extension of A if and only if $A = B$.

Definition 3.1.13. An injective module E which is an essential extension of an R-module M is said to be an *injective envelope* of M.

Theorem 3.1.14. *Every R-module has an injective envelope which is unique up to isomorphism.*

Proof. Embed an R-module M into an injective R-module E by Theorem 3.1.7, and let $\mathcal{C}$ be the collection of all essential extensions of M in E. $\mathcal{C} \neq \emptyset$ since $M \in \mathcal{C}$. Partially order $\mathcal{C}$ by inclusion. Then $\mathcal{C}$ is an inductive system and so has a maximal element E' by Zorn's lemma. We claim that E' is a maximal essential extension of M. For let E'' be an essential extension of M that contains E'. Then we have a commutative diagram

since E is injective. But $\operatorname{Ker} \varphi \cap E' = 0$ and $E' \subset E''$ is an essential extension. So $\operatorname{Ker} \varphi = 0$ and thus φ is an embedding. Therefore, $\varphi(E'')$ is an essential extension of M contained in E, that is, $\varphi(E'') \in \mathcal{C}$. Thus $\varphi(E'') = E'$ and so $E'' = E'$. We now want to argue E' is injective. We do this by arguing E' is a direct summand of E. Consider the submodules $S \subset E$ with $E' \cap S = 0$. Using Zorn's lemma we see that there is a maximal such S. We claim then that $(E' + S)/S \subset E/S$ is an essential extension. For if T/S is a nonzero submodule of E/S then $S \subsetneq T$. But by the maximality of S, $E' \cap T \neq 0$. So $((E'+S)/S) \cap (T/S) = ((E'+S) \cap T)/S = ((E' \cap T)+S)/S \neq 0$. The canonical isomorphism $(E' + S)/S \to E'$ can be extended to a necessarily injective $f : E/S \to E$. Since $(E' + S)/S$ is essential in E/S we get $f((E' + S)/S) = E'$ is essential in $f(E/S)$. But then by the maximality of E', $f(E/S) = E'$, that is, $f((E'+S)/S) = f(E/S)$. Since f is injective this implies $(E'+S)/S = E/S$. This means that $E' + S = E$. Since $E' \cap S = 0$ we get that E' is a direct summand of E and so is injective.

Now suppose E', E'' are injective envelopes of M. Then since E'' is injective, the inclusion map $M \to E''$ can be extended to a map $\varphi : E' \to E''$. But $M \subset E'$ is an essential extension. So φ is an embedding as in the above. Thus $\varphi(E')$ is an injective extension of M and $\varphi(E')$ is a direct summand of E''. But then $\varphi(E') = E''$ since M is essential in E'' and so φ is an isomorphism.   □

Remark 3.1.15. We can construct an exact sequence $0 \to M \to E^0 \to E^1 \to \cdots$ with each E^i injective using injective envelopes by the theorem above. This sequence is called the *minimal injective resolution* of M.

Notation. The injective envelope of an R-module M is denoted by $E(M)$. We easily see that if $M \subset E$ with E injective then E contains an injective envelope of M (just extend the identity $M \to E$ to $E(M) \to E$).

Lemma 3.1.16. *Let R be left noetherian, M be a finitely generated R-module, and $\varinjlim N_j$ be a direct limit of R-modules. Then*

$$\operatorname{Ext}^i_R(M, \varinjlim N_j) \cong \varinjlim \operatorname{Ext}^i_R(M, N_j)$$

for all $i \geq 0$.

Proof. By Remark 2.3.12, M has an exact sequence $\cdots \to F_1 \to F_0 \to M \to 0$ with each F_i finitely generated and free. We consider the complex

$$0 \to \operatorname{Hom}(M, \varinjlim N_j) \to \operatorname{Hom}(F_0, \varinjlim N_j) \to \operatorname{Hom}(F_1, \varinjlim N_j) \to \cdots$$

and note that $\operatorname{Hom}(F_i, \varinjlim N_j) \cong \oplus N_j \cong \varinjlim \operatorname{Hom}(F_i, N_j)$. Thus the result follows since $\varinjlim$ commutes with homology. □

Theorem 3.1.17. *The following are equivalent for a ring R.*

(1) *R is left noetherian.*

(2) *Every direct limit of injective R-modules is injective.*

(3) *Every direct sum of injective R-modules is injective.*

Proof. (1) $\Rightarrow$ (2). Let $E = \varinjlim E_j$ where each E_j is an injective R-module, and I be a left ideal of R. Then $\operatorname{Ext}^1(R/I, E) = \varinjlim \operatorname{Ext}^1(R/I, E_j)$ by the lemma above. Hence $\operatorname{Ext}^1(R/I, E) = 0$. Thus E is injective by Theorem 3.1.9.

(2) $\Rightarrow$ (3) is trivial since a direct sum is a direct limit of the finite sums which are injective.

(3) $\Rightarrow$ (1). Suppose R is not noetherian. Then there exists a strictly ascending chain $I_1 \subset I_2 \subset \cdots$ of left ideals of R that never stops. Let $I = \bigcup_{i=1}^\infty I_i$. Then I is an ideal of R and $I/I_i \neq 0$ for each i. Now let f_i be the composition of the natural map $\tau_i : I \to I/I_i$ and the inclusion $I/I_i \subset E(I/I_i)$. Then define a map $f : I \to \bigoplus_{i=1}^\infty E(I/I_i)$ by $f(a) = (f_i(a))$. We note that for each $a \in I$, $\tau_i(a) = 0$ for sufficiently large i. So $f(I)$ is indeed contained in the direct sum.

If $\bigoplus_{i=1}^\infty E(I/I_i)$ is injective, then f extends to a map $g : R \to \bigoplus_{i=1}^\infty E(I/I_i)$. Now let $\pi_j : \bigoplus_{i=1}^\infty E(I/I_i) \to E(I/I_j)$ be the projection map. Then for sufficiently large i, $\pi_i \circ g(1) = 0$. So for each $a \in I$, $f_i(a) = \pi_i \circ f(a) = \pi_i \circ g(a) =$

$a(\pi_i \circ g)(1) = 0$ for sufficiently large i. So for such i, the map $f_i : I \rightarrow E(I/I_i)$ is a zero homomorphism, a contradiction. Hence $\bigoplus_{i=1}^{\infty} E(I/I_i)$ is not injective and the result follows. □

Theorem 3.1.18 (Eakin–Nagata Theorem). *If $R \subset S$ is a subring of the commutative ring S and if S is a finitely generated R-module, then R is noetherian if and only if S is noetherian.*

Proof. If R is noetherian then S is a noetherian R-module by Corollary 2.3.10 and so clearly S is a noetherian ring.

Now suppose S is noetherian. Let $(E_i)_{i \in I}$ be an arbitrary family of injective S-modules. By Theorem 3.1.17 above it suffices to prove that $\oplus E_i$ is injective. Let $\oplus E_i \subset E$ be an injective envelope. Since $\oplus E_i$ is essential in E (as R-modules), we claim $\text{Hom}_R(S, \oplus E_i)$ is essential in $\text{Hom}_R(S, E)$ as S-modules. For if $f \in \text{Hom}_R(S, E)$, $f \neq 0$, then since S is a finitely generated R-module and $\oplus E_i$ is essential in E, there is an $r \in R$ such that $rf(S) \subset \oplus E_i$ with $rf(S) \neq 0$. Thus $rf \neq 0$ and $rf \in \text{Hom}_R(S, \oplus E_i)$. Hence $\text{Hom}_R(S, \oplus E_i)$ is essential in $\text{Hom}_R(S, E)$ as R-modules and so also as S-modules.

Now since S is a finitely generated R-module, $\text{Hom}_R(S, \oplus E_i) \cong \oplus \text{Hom}_R(S, E_i)$ (naturally). By Proposition 3.1.6, $\text{Hom}_R(S, E_i)$ is an injective S-module and since S is noetherian, $\oplus \text{Hom}_R(S, E_i)$ and so also $\text{Hom}_R(S, \oplus E_i)$ are injective S-modules by the preceding theorem.

But by the above $\text{Hom}_R(S, \oplus E_i)$ is essential in $\text{Hom}_R(S, E)$. By Theorem 3.1.2 $\text{Hom}(S, \oplus E_i)$ is a direct summand of $\text{Hom}_R(S, E)$ and so in fact $\text{Hom}_R(S, \oplus E_i) = \text{Hom}_R(S, E)$.

It only remains to show that this equality implies $\oplus E_i = E$, that is, that $\oplus E_i$ is injective. But if $x \in E$, the function $R \rightarrow E$ that maps r to rx has an R-linear extension $f : S \rightarrow E$. So $f \in \text{Hom}_R(S, E) = \text{Hom}_R(S, \oplus E_i)$. But then $f(1) = x \in \oplus E_i$. Hence $\oplus E_i = E$. □

Definition 3.1.19. The sum of all simple submodules of M is called the *socle* of M and is denoted by $\text{Soc}(M)$. If M has no simple submodules, we set $\text{Soc}(M) = 0$. Clearly $\text{Soc}(M)$ is the largest semisimple submodule of M and $\text{Soc}(M) = M$ if and only if M is semisimple. It is also easy to see that $\text{Soc}(M) = \{x \in M : \text{Ann}(x)$ is the intersection of finitely many maximal left ideals of $R\}$. In particular, if R is local with maximal ideal $\mathfrak{m}$, then $\text{Soc}(M) = \{x \in M : \text{Ann}(x) = \mathfrak{m}\}$ is a vector space over the residue field $k = R/\mathfrak{m}$.

Proposition 3.1.20. *Let $A \subset B$ be R-modules. Then $\text{Soc}(A) \subset \text{Soc}(B)$ and equality holds if B is an essential extension of A.*

Proof. $\text{Soc}(A) \subset \text{Soc}(B)$ is trivial. Now let S be a simple submodule of B. When B is an essential extension of A, then $S \cap A \neq 0$ since $A \subset B$ is essential. So $S \cap A = S$ and thus $S \subset A$. □

Corollary 3.1.21. *If M is an R-module, then* $\mathrm{Soc}(M) = \mathrm{Soc}(E(M))$.

Proposition 3.1.22. *If M is an artinian R-module, then M is an essential extension of* $\mathrm{Soc}(M)$.

Proof. Let N be a nonzero submodule of M. Then N is artinian and so the collection of all nonzero submodules of N has a minimal element, say S. S is a simple submodule and so $N \cap \mathrm{Soc}(M) \neq 0$. Hence $\mathrm{Soc}(M)$ is essential in M.   □

Exercises

1. Let R be an integral domain. Prove that a torsion free R-module is injective if and only if it is divisible.

2. Let R be an integral domain and K be its field of fractions. Prove that a torsion free R-module is divisible if and only if it is a vector space over K. Conclude that every torsion free R-module can be embedded in a vector space over K, and hence in particular every finitely generated torsion free R-module can be embedded in a finitely generated free R-module.

3. Prove that if R is a commutative noetherian ring, then an R-module E is injective if and only if $\mathrm{Ext}^1(R/\mathfrak{p}, E) = 0$ for all $\mathfrak{p} \in \mathrm{Spec}\, R$.
   Hint: Use Lemma 2.4.7 and Baer's Criterion.

4. Let $(M_i)_{i \in I}$ be a family of R-modules and N be an R-module. Prove

   a) $\mathrm{Ext}^n(N, \prod_I M_i) \cong \prod_I \mathrm{Ext}^n(N, M_i)$ for all $n \geq 0$.
   b) $\mathrm{Ext}^n(\bigoplus_I M_i, N) \cong \prod_I \mathrm{Ext}^n(M_i, N)$ for all $n \geq 0$.

   Hint: By induction on n using the long exact sequence corresponding to a short exact sequence $0 \to N \to E \to C \to 0$ with E injective and Propositions 1.2.6 and 1.2.7.

5. Prove Corollary 3.1.10.

6. Let $A \subset B \subset C$ be modules. Prove

   a) C is an essential extension of A if and only if C is an essential extension of B and B is an essential extension of A.

   b) If B and C are both essential extensions of A, then C is an essential extension of B.

7. Prove that an R-module M is injective if and only if it has no proper essential extension.
   Hint: If M has no essential extension, let $M \subset E$ be an injective extension and $S \subset E$ be maximal with respect to $M \cap S = 0$ (by Zorn's lemma). Argue that $M \cong (M + S)/S \subset E/S$ is essential.

8. Let M be a submodule of E. Prove that the following are equivalent.

    a) E is an injective envelope of M.

    b) E is a maximal essential extension of M.

    c) E is a minimal injective extension of M.

9. Let R be an integral domain. Show that $E(R)$ is its quotient field.

10. Prove that the following are equivalent.

    a) R is semisimple.

    b) Every R-module is semisimple.

    c) Every R-module is injective.

    d) Every exact sequence $0 \to A \to B \to C \to 0$ of R-modules is split exact.

    e) Every R-module is projective.

11. Let $(M_i)_{i \in I}$ be a family of R-modules. Prove

    a) If N_i is an essential extension of M_i for each i, then $\bigoplus_{i \in I} N_i$ is an essential extension of $\bigoplus_{i \in I} M_i$.

    b) $\mathrm{Soc}(\bigoplus_{i \in I} M_i) = \bigoplus_{i \in I} \mathrm{Soc}(M_i)$.

    c) If I is finite or R is noetherian, then $E(\bigoplus_{i \in I} M_i) \cong \bigoplus_{i \in I} E(M_i)$.

12. Prove that $\mathrm{Soc}(M) = \{x \in M : \mathrm{Ann}(x) \text{ is the intersection of finitely many maximal left ideals of } R\}$.

13. Prove that if R is a local ring with maximal ideal $\mathfrak{m}$ and residue field k, then $\mathrm{Soc}(M)$ is a vector space over k.

14. State and prove Schanuel's lemma for injective modules.

3.2 Natural identities, flat modules, and injective modules

We start with the following

Theorem 3.2.1. *Let R and S be rings, A a left R-module, and B an (S, R)-bimodule. If C is an injective left S-module, then*

$$\mathrm{Ext}^i_R(A, \mathrm{Hom}_S(B, C)) \cong \mathrm{Hom}_S(\mathrm{Tor}^R_i(B, A), C)$$

for all $i \geq 0$.

Proof. The case $i = 0$ is the natural identity in Theorem 2.1.10. Now let $\mathbf{P}_\bullet$ denote a deleted projective resolution of A. Then

$$H^i(\mathrm{Hom}_R(\mathbf{P}_\bullet, \mathrm{Hom}_S(B, C))) \cong H^i(\mathrm{Hom}_S(B \otimes_R \mathbf{P}_\bullet, C))$$
$$\cong \mathrm{Hom}_S(H_i(B \otimes_R \mathbf{P}_\bullet, C))$$

since H^i commutes with $\mathrm{Hom}_S(-, C)$ if C is injective (see Exercise 2 of Section 1.4). So we are done. We also note that this theorem easily follows from Theorem 8.2.11 of Chapter 8.   □

Definition 3.2.2. An R-module M is said to be *finitely presented* if there is an exact sequence $F_1 \to F_0 \to M \to 0$ where F_0 and F_1 are finitely generated free R-modules.

Remark 3.2.3. It is easy to see that an R-module M is finitely presented if and only if there is an exact sequence $0 \to K \to F \to M \to 0$ where F and K are finitely generated R-modules and F is free. In particular, every finitely presented R-module is finitely generated and the converse holds if R is left noetherian by Corollary 2.3.11.

Lemma 3.2.4. *Let R and S be commutative rings, S be a flat R-algebra and M, N be R-modules. If M is finitely presented, then*

$$\mathrm{Hom}_R(M, N) \otimes_R S \cong \mathrm{Hom}_S(M \otimes_R S, N \otimes_R S).$$

Proof. Since M is finitely presented, we have the following commutative diagram with exact rows

$$
\begin{array}{ccccccc}
0 & \longrightarrow & \mathrm{Hom}(M, N) \otimes S & \longrightarrow & \mathrm{Hom}(F_0, N) \otimes S & \longrightarrow & \mathrm{Hom}(F_1, N) \otimes S \\
 & & \varphi \downarrow & & \varphi \downarrow & & \varphi \downarrow \\
0 & \to & \mathrm{Hom}(M \otimes S, N \otimes S) & \to & \mathrm{Hom}(F_0 \otimes S, N \otimes S) & \to & \mathrm{Hom}(F_1 \otimes S, N \otimes S)
\end{array}
$$

where the maps φ are given by $\varphi(f \otimes s)(x \otimes t) = s(f(x) \otimes t)$. But the last two vertical maps are isomorphisms since F_0 and F_1 are free and finitely generated. Hence the first φ is also an isomorphism.   □

Theorem 3.2.5. *Let R and S be commutative rings, S be a flat R-algebra, and M, N be R-modules. If R is noetherian and M is finitely generated, then*

$$\mathrm{Ext}^i_R(M, N) \otimes_R S \cong \mathrm{Ext}^i_S(M \otimes_R S, N \otimes_R S)$$

for all $i \geq 0$.

Proof. By Remark 2.3.12, M has a projective resolution $\cdots \to P_1 \to P_0 \to M \to 0$ with each P_i finitely generated (and hence finitely presented). But then $\cdots \to P_1 \otimes_R S \to P_0 \otimes_R S \to M \otimes_R S \to 0$ is a projective resolution of the S-module $M \otimes_R S$. So as in Theorem 2.1.11,

$$
\begin{aligned}
\mathrm{Ext}^i_R(M, N) \otimes_R S &\cong H^i(\mathrm{Hom}_R(\mathbf{P}_\bullet, N)) \otimes_R S \\
&\cong H^i(\mathrm{Hom}_R(\mathbf{P}_\bullet, N) \otimes_R S) \\
&\cong H^i(\mathrm{Hom}_S(\mathbf{P}_\bullet \otimes_R S, N \otimes_R S)) \\
&\cong \mathrm{Ext}^i_S(M \otimes_R S, N \otimes_R S).
\end{aligned}
$$

□

Corollary 3.2.6. *Let R, M, and N be as in the theorem above. Then*

$$\operatorname{Ext}^i_R(M, N)_{\mathfrak{p}} \cong \operatorname{Ext}^i_{R_{\mathfrak{p}}}(M_{\mathfrak{p}}, N_{\mathfrak{p}})$$

for all $\mathfrak{p} \in \operatorname{Spec} R$ and all $i \geq 0$.

Definition 3.2.7. An injective R-module E is said to be an *injective cogenerator* for R-modules if for each R-module M and nonzero element $x \in M$, there is $f \in \operatorname{Hom}_R(M, E)$ such that $f(x) \neq 0$. This is equivalent to the condition that $\operatorname{Hom}_R(M, E) \neq 0$ for any module $M \neq 0$. For if $x \in M$, $x \neq 0$, any $g \in \operatorname{Hom}_R(Rx, E)$ with $g \neq 0$ has $g(x) \neq 0$. And such a g has an extension $f \in \operatorname{Hom}_R(M, E)$. The group $\mathbb{Q}/\mathbb{Z}$ is an injective cogenerator for abelian groups. Hence if M is a nonzero right R-module, then the *character module M^+* of M defined by $M^+ = \operatorname{Hom}_{\mathbb{Z}}(M, \mathbb{Q}/\mathbb{Z})$ is a nonzero left R-module. Moreover, if M is any left R-module, then $\operatorname{Hom}_R(M, R^+) \cong M^+$ by Theorem 2.1.10. Hence $R^+ = \operatorname{Hom}_{\mathbb{Z}}(R, \mathbb{Q}/\mathbb{Z})$ is an injective cogenerator for left R-modules since R^+ is injective by Proposition 3.1.6. Thus there exists an injective cogenerator for R-modules for any ring R.

Lemma 3.2.8. *Let R and S be rings and E be an injective cogenerator for R-modules. Then a sequence $0 \to A \xrightarrow{\varphi} B \xrightarrow{\psi} C \to 0$ of (R, S)-bimodules is exact if and only if the sequence $0 \to \operatorname{Hom}_R(C, E) \xrightarrow{\psi^*} \operatorname{Hom}_R(B, E) \xrightarrow{\varphi^*} \operatorname{Hom}_R(A, E) \to 0$ of left S-modules is exact.*

Proof. The 'only if' part is clear since E is an injective R-module. For the 'if' part, we show that $\operatorname{Im} \varphi = \operatorname{Ker} \psi$. Suppose $\operatorname{Im} \varphi \not\subset \operatorname{Ker} \psi$. Then choose $b \in \operatorname{Im} \varphi - \operatorname{Ker} \psi$. So $\psi(b) \neq 0$. But $\psi(b) \in C$. So there is an $f \in \operatorname{Hom}_R(C, E)$ such that $f(\psi(b)) \neq 0$ since E is an injective cogenerator. But $b = \varphi(a)$ for some $a \in A$. Thus $f \circ \psi \circ \varphi \neq 0$. But then $(\varphi^* \circ \psi^*)(f) \neq 0$, a contradiction. So $\operatorname{Im} \varphi \subset \operatorname{Ker} \psi$. Now suppose $\operatorname{Im} \varphi \subsetneqq \operatorname{Ker} \psi$. Then let $b \in \operatorname{Ker} \psi - \operatorname{Im} \varphi$. So $b + \operatorname{Im} \varphi$ is nonzero in $B/\operatorname{Im} \varphi$. Thus there is an $f \in \operatorname{Hom}_R(B/\operatorname{Im} \varphi, E)$ such that $f(b + \operatorname{Im} \varphi) \neq 0$. Hence the composite map $g : B \xrightarrow{\tau} B/\operatorname{Im} \varphi \xrightarrow{f} E$ where τ is the natural homomorphism is such that $g(b) \neq 0$. But $\varphi^*(g) = g \circ \varphi = 0$ since $f(\operatorname{Im} \varphi) = 0$. So $g \in \operatorname{Ker} \varphi^* = \operatorname{Im} \psi^*$. That is, $g = \psi^*(h) = h \circ \psi$ for some $h \in \operatorname{Hom}_R(C, E)$. But $b \in \operatorname{Ker} \psi$. So $g(b) = h(\psi(b)) = 0$, a contradiction since $g(b) \neq 0$. $\square$

Theorem 3.2.9. *The following are equivalent for an (R, S)-bimodule F.*

(1) *F is a flat left R-module.*

(2) *$\operatorname{Hom}_S(F, E)$ is an injective right R-module for all injective right S-modules E.*

(3) *$\operatorname{Hom}_S(F, E)$ is an injective right R-module for any injective cogenerator E for right S-modules.*

Proof. $(1) \Rightarrow (2)$. Let I be a right ideal of R. If F is flat, then $0 \to I \otimes F \to R \otimes F$ is an exact sequence of right S-modules. But then $\operatorname{Hom}_S(R \otimes_R F, E_S) \to \operatorname{Hom}_S(I \otimes_R F, E_S) \to 0$ is exact for any injective right S-module E. Hence by Theorem 2.1.10, $\operatorname{Hom}_R(R, \operatorname{Hom}_S(F, E)) \to \operatorname{Hom}_R(I, \operatorname{Hom}_S(F, E)) \to 0$ is exact. Thus $\operatorname{Hom}_S(F, E)$ is injective by Baer's Criterion.

$(2) \Rightarrow (3)$ is trivial.

$(3) \Rightarrow (1)$. Let $A \subset B$ be right R-modules. By (3), $\operatorname{Hom}_R(B, \operatorname{Hom}_S(F, E)) \to \operatorname{Hom}_R(A, \operatorname{Hom}_S(F, E)) \to 0$ is exact. So $\operatorname{Hom}_S(B \otimes_R F, E) \to \operatorname{Hom}_S(A \otimes_R F, E) \to 0$ is exact. But then $0 \to A \otimes_R F \to B \otimes_R F$ is exact by Lemma 3.2.8 above. $\square$

Theorem 3.2.10. *The following are equivalent for an R-module F.*

(1) F *is flat.*

(2) *The character module F^+ is an injective right R-module.*

(3) $\operatorname{Tor}_1^R(R/I, F) = 0$ *for all finitely generated right ideals I of R.*

(4) $0 \to I \otimes_R F \to F$ *is exact for all finitely generated right ideals of R.*

Proof. $(1) \Leftrightarrow (2)$ follows from Theorem 3.2.9 above.

$(1) \Rightarrow (3) \Rightarrow (4)$ is trivial.

$(4) \Rightarrow (2)$. Every ideal is a direct limit of finitely generated ideals and direct limits preserve exact sequences. Hence (4) means that $0 \to I \otimes_R F \to F$ is exact for all right ideals I of R. So $F^+ \to (I \otimes_R F)^+ \to 0$ is exact. But then $\operatorname{Hom}_R(R, F^+) \to \operatorname{Hom}_R(I, F^+) \to 0$ is exact for all ideals I of R. So (2) follows by Baer's Criterion. $\square$

We now consider yet another natural identity.

Theorem 3.2.11. *Let R and S be rings, A be a finitely presented right S-module, B an (R, S)-bimodule, and C an injective left R-module. Then the natural homomorphism*

$$\tau : A \otimes_S \operatorname{Hom}_R(B, C) \to \operatorname{Hom}_R(\operatorname{Hom}_S(A, B), C)$$

defined by $\tau(a \otimes f)(g) = f(g(a))$ is an isomorphism where $a \in A$, $f \in \operatorname{Hom}(B, C)$, and $g \in \operatorname{Hom}(A, B)$. If A is a finitely presented left R-module, B an (R, S)-bimodule, and C and injective right S-module, then

$$\operatorname{Hom}_S(B, C) \otimes_R A \cong \operatorname{Hom}_S(\operatorname{Hom}_R(A, B), C)$$

where the isomorphism is given by $\tau(f \otimes a)(g) = f(g(a))$.

Proof. We consider the exact sequence $F_1 \to F_0 \to A \to 0$ with F_0, F_1 finitely generated and free. Then we have the following commutative diagram

$$
\begin{array}{ccccccc}
F_1 \otimes \operatorname{Hom}(B, C) & \longrightarrow & F_0 \otimes \operatorname{Hom}(B, C) & \longrightarrow & A \otimes \operatorname{Hom}(B, C) & \longrightarrow & 0 \\
\downarrow & & \downarrow & & \downarrow{\scriptstyle \tau} & & \\
\operatorname{Hom}(\operatorname{Hom}(F_1, B), C) & \longrightarrow & \operatorname{Hom}(\operatorname{Hom}(F_0, B), C) & \longrightarrow & \operatorname{Hom}(\operatorname{Hom}(A, B), C) & \longrightarrow & 0
\end{array}
$$

with exact rows. But the first two vertical maps are isomorphisms. So τ is an isomorphism. The second isomorphism follows similarly.   □

As an application, we have the following result.

Proposition 3.2.12. *A finitely presented flat R-module is projective.*

Proof. Let F be a finitely presented flat right R-module and $B \to C \to 0$ be an exact sequence of right R-modules. We want to show that $\text{Hom}_R(F, B) \to \text{Hom}_R(F, C) \to 0$ is exact, or equivalently $0 \to \text{Hom}_R(F, C)^+ \to \text{Hom}_R(F, B)^+$ is exact by Lemma 3.2.8. But then we have the following commutative diagram

$$
\begin{array}{ccccc}
0 & \longrightarrow & F \otimes_R C^+ & \longrightarrow & F \otimes_R B^+ \\
& & \downarrow & & \downarrow \\
0 & \longrightarrow & \text{Hom}_R(F, C)^+ & \longrightarrow & \text{Hom}_R(F, B)^+
\end{array}
$$

where the first row is exact since F is flat. But the vertical maps are isomorphisms by Theorem 3.2.11 above since F is finitely presented. Hence the second row is also exact and thus we are done.   □

Theorem 3.2.13. *Let R and S be rings. If R is left noetherian, A a finitely presented left R-module, B an (R, S)-bimodule, and C an injective right S-module, then*

$$\text{Tor}_i^R(\text{Hom}_S(B, C), A) \cong \text{Hom}_S(\text{Ext}_R^i(A, B), C)$$

for all $i \geq 0$.

Proof. This follows from Theorem 3.2.11 as in the proof of Theorem 3.2.1 since A has a deleted projective resolution with each term finitely presented.   □

Theorem 3.2.14. *Let A be a finitely presented left R-module, B and (R, S)-bimodule and C a left flat S-module. Then the natural map $\tau : \text{Hom}_R(A, B) \otimes_S C \to \text{Hom}_R(A, B \otimes_S C)$ defined by $\tau(f \otimes c)(a) = f(a) \otimes c$ is an isomorphism.*

Proof. The proof is similar to that of Theorem 3.2.11.   □

Theorem 3.2.15. *Let R, S, A, B and C be as in Theorem 3.2.14 above. If R is left noetherian, then*
$$\text{Ext}_R(A, B) \otimes_S C \cong \text{Ext}_R^i(A, B \otimes_S C)$$

for all $i \geq 0$.

Proof. This is also similar to the proofs of Theorems 3.2.1 and 3.2.13.   □

Theorem 3.2.16. *Let R be left noetherian. Then the following are equivalent for an (R, S)-bimodule E.*

(1) E is an injective left R-module.

(2) $\mathrm{Hom}_S(E, E')$ is a flat right R-module for all injective right S-modules E'.

(3) $\mathrm{Hom}_S(E, E')$ is a flat right R-module for any injective cogenerator E' for right S-modules.

(4) $E \otimes_S F$ is an injective left R-module for all flat left S-modules F.

(5) $E \otimes_S F$ is an injective left R-module for any faithfully flat left S-module F.

Proof. (1) $\Rightarrow$ (2). Let I be a left ideal of R. Then I is finitely presented since R is noetherian. But E is injective. So

$$0 \to \mathrm{Hom}_S(\mathrm{Hom}_R(I, E), E') \to \mathrm{Hom}_S(\mathrm{Hom}_R(R, E), E')$$

being exact means $0 \to \mathrm{Hom}_S(E, E') \otimes_R I \to \mathrm{Hom}_S(E, E')$ is exact by Theorem 3.2.11. Hence (2) follows by Proposition 3.2.10.

(2) $\Rightarrow$ (3) and (4) $\Rightarrow$ (5) are trivial.

(3) $\Rightarrow$ (1) follows by reversing the proof of (1) $\Rightarrow$ (2) and using the fact that E' is an injective cogenerator.

(1) $\Rightarrow$ (4). $\mathrm{Hom}_R(R, E) \otimes_S F \to \mathrm{Hom}_R(I, E) \otimes_S F \to 0$ is exact for an ideal I of R since $_R E$ is injective. But then $\mathrm{Hom}_R(R, E \otimes_S F) \to \mathrm{Hom}_R(I, E \otimes_S F) \to 0$ is exact by Theorem 3.2.14 since $_S F$ is flat. Hence (4) follows.

(5) $\Rightarrow$ (1) follows by reversing the proof of (1) $\Rightarrow$ (4) above and using the fact that $_S F$ is faithfully flat. $\square$

Corollary 3.2.17. *Let R be left noetherian. Then a left R-module E is injective if and only if the character module E^+ is a flat right R-module.*

Definition 3.2.18. An R-module M is said to have *injective dimension* at most n, denoted $\mathrm{inj\,dim} \le n$, if there is an injective resolution $0 \to M \to E^0 \to E^1 \to \cdots \to E^n \to 0$. If n is the least, then we set $\mathrm{inj\,dim}\, M = n$. The *flat dimension* and *projective dimension* of an R-module are defined similarly using flat and projective resolutions, respectively. These are denoted $\mathrm{flat\,dim}\, M$ and $\mathrm{proj\,dim}\, M$, respectively. We note that $\mathrm{flat\,dim}\, M \le \mathrm{proj\,dim}\, M$ and equality holds if R is left noetherian and M is finitely generated.

Now using Theorems 3.2.1, 3.2.13, and 3.2.15, we get the following results of Ishikawa.

Theorem 3.2.19. *Let M be an (R, S)-bimodule and E an injective cogenerator for right S-modules. Then $\mathrm{flat\,dim}\, {}_R M = \mathrm{inj\,dim}\, {}_R \mathrm{Hom}_S(M, E)$. If furthermore R is left noetherian, then $\mathrm{inj\,dim}\, {}_R M = \mathrm{flat\,dim}\, {}_R \mathrm{Hom}_S(M, E)$.*

Theorem 3.2.20. *Let M be an (R, S)-bimodule and F be a faithfully flat left S-module. If R is left noetherian, then $\mathrm{inj\,dim}\, {}_R M = \mathrm{inj\,dim}\, {}_R M \otimes_S F$.*

Now let M be a right R-module and (A_λ) be a family of left R-modules over some index set Λ. Then we define a map

$$\tau : M \otimes_R \prod_\lambda A_\lambda \to \prod_\lambda M \otimes_R A_\lambda$$

by $\tau(x \otimes (a_\lambda)) = (x \otimes a_\lambda)$. If $A_\lambda = R$ for each λ, then we have a map $\tau : M \otimes R^\Lambda \to M^\Lambda$ given by $\tau(x \otimes (r_\lambda)) = (xr_\lambda)$. It is easy to see that τ is an isomorphism when M is finitely generated and free.

Lemma 3.2.21. *The following are equivalent for a right R-module M.*

(1) *M is finitely generated.*

(2) *$\tau : M \otimes \prod_\Lambda A_\lambda \to \prod_\Lambda M \otimes A_\lambda$ is an epimorphism for every family $(A_\lambda)_\Lambda$ of left R-modules.*

(3) *$\tau : M \otimes R^\Lambda \to M^\Lambda$ is an epimorphism for any set Λ,*

(4) *$\tau : M \otimes R^M \to M^M$ is an epimorphism.*

Proof. (1) $\Rightarrow$ (2). Let $0 \to K \to F \to M \to 0$ be exact with F finitely generated and free. Then we have the following commutative diagram

$$
\begin{array}{ccccccc}
K \otimes \prod_\Lambda A_\lambda & \longrightarrow & F \otimes \prod_\Lambda A_\lambda & \longrightarrow & M \otimes \prod_\Lambda A_\lambda & \longrightarrow & 0 \\
\downarrow{\scriptstyle \tau_K} & & \downarrow{\scriptstyle \tau_F} & & \downarrow{\scriptstyle \tau_M} & & \\
\prod_\Lambda K \otimes A_\lambda & \longrightarrow & \prod_\Lambda F \otimes A_\lambda & \longrightarrow & \prod_\Lambda M \otimes A_\lambda & \longrightarrow & 0
\end{array}
$$

with exact rows. But τ_F is an isomorphism. So τ_M is onto.

(2) $\Rightarrow$ (3) $\Rightarrow$ (4) is trivial.

(4) $\Rightarrow$ (1). Let $(x_x)_M \in M^M$. Then since τ is onto, $(x_x)_M = \tau(\sum_{i=1}^n x_i \otimes (r_{ix}))$ where $x_i \in M$, $r_{ix} \in R$. So $(x_x)_M = (\sum_{i=1}^n x_i r_{ix})$. Hence $x = \sum_{i=1}^n x_i r_{ix}$ and thus $x_1, x_2, \ldots, x_n$ are generators of M. $\qquad\square$

Theorem 3.2.22. *The following are equivalent for a right R-module M.*

(1) *M is finitely presented.*

(2) *$\tau : M \otimes \prod_\Lambda A_\lambda \to \prod_\Lambda M \otimes A_\lambda$ is an isomorphism for every family $(A_\lambda)_\Lambda$ of left R-modules.*

(3) *$\tau : M \otimes R^\Lambda \to M^\Lambda$ is an isomorphism for any set Λ.*

Proof. (1) $\Rightarrow$ (2). Let $F_1 \to F_0 \to M \to 0$ be exact with F_0, F_1 finitely generated and free. Then we get a commutative diagram similar to the one in the proof of the lemma above where τ_{F_1} and τ_{F_0} are isomorphisms. Thus τ_M is an isomorphism.

(2) $\Rightarrow$ (3) is trivial.

(3) $\Rightarrow$ (1). M is finitely generated by the lemma above. So let $0 \to K \to F \to M \to 0$ be exact with F finitely generated and free. It now suffices to show that K is finitely generated. But for any Λ, we have a commutative diagram

$$
\begin{array}{ccccccccc}
K \otimes R^\Lambda & \longrightarrow & F \otimes R^\Lambda & \longrightarrow & M \otimes R^\Lambda & \longrightarrow & 0 \\
\downarrow{\scriptstyle \tau_K} & & \downarrow{\scriptstyle \tau_F} & & \downarrow{\scriptstyle \tau_M} & & \\
0 \longrightarrow & K^\Lambda & \longrightarrow & F^\Lambda & \longrightarrow & M^\Lambda & \longrightarrow & 0
\end{array}
$$

with exact rows where τ_F and τ_M are isomorphisms. So τ_K is onto and hence K is finitely generated again by Lemma 3.2.21. $\qquad\square$

Definition 3.2.23. A ring R is said to be *right coherent* if every finitely generated right ideal of R is finitely presented. It follows from Remark 3.2.3 that every right noetherian is right coherent.

We are now in position to prove the following characterization of coherent rings.

Theorem 3.2.24. *The following are equivalent for a ring R.*

(1) *R is right coherent.*

(2) *Every product of flat left R-modules is flat.*

(3) *R^Λ is a flat left R-module for any set Λ.*

(4) *Every finitely generated submodule of a finitely presented right R-module is finitely presented.*

Proof. (1) $\Rightarrow$ (2). Let $(F_\lambda)_\Lambda$ be a family of flat left R-modules. If I is a finitely generated right ideal of R, then $I \otimes \prod_\lambda F_\lambda \cong \prod_\lambda I \otimes F_\lambda$ by Theorem 3.2.22. But $I \otimes F_\lambda \subset F_\lambda$ since F_λ is flat. So we have an embedding $I \otimes \prod_\lambda F_\lambda \hookrightarrow \prod_\lambda F_\lambda$ for each finitely generated right ideal I of R. Hence $\prod_\lambda F_\lambda$ is flat by Theorem 3.2.10.

(2) $\Rightarrow$ (3) and (4) $\Rightarrow$ (1) are trivial.

(3) $\Rightarrow$ (4). Let M be a finitely presented right R-module and N be a finitely generated submodule of M. We now consider the following commutative diagram

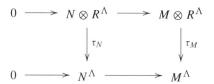

with exact rows. But τ_M is an isomorphism by Theorem 3.2.22. So τ_N is one-to-one. But N is finitely generated. So τ_N is surjective by Lemma 3.2.21. Hence τ_N is an isomorphism and thus N is finitely presented. $\qquad\square$

Remark 3.2.25. Suppose M is a finitely presented right R-module. Then there is an exact sequence $0 \to K \to F_0 \to M \to 0$ with F_0 and K finitely generated and F_0 free. If R is right coherent, then K is finitely presented by the theorem above. Thus continuing in this manner, we see that if R is right coherent then every finitely presented right R-module M has a free resolution $\cdots \to F_1 \to F_0 \to M \to 0$ with each F_i finitely generated and free.

Theorem 3.2.26. *Let R be right coherent, M be a finitely presented right R-module, and $(A_\lambda)_\Lambda$ be a family of left R- modules. Then*

$$\mathrm{Tor}_n^R\left(M, \prod_\lambda A_\lambda\right) \cong \prod_\lambda \mathrm{Tor}_n^R(M, A_\lambda)$$

for all $n \geq 0$.

Proof. The case $n = 0$ is Theorem 3.2.22 and the rest is left as an exercise.   □

Remark 3.2.27. Lemma 3.1.16 and Theorems 3.2.5, 3.2.13, and 3.2.15, hold without the noetherian hypothesis if we assume that the left R-module M (or A) has a projective resolution $\cdots \to P_1 \to P_0 \to M \to 0$ with each P_i finitely presented. Hence by Remark 3.2.25 above, these results together with Theorems 3.2.19 and 3.2.20 hold if we assume R is right coherent.

Exercises

1. Prove Remark 3.2.3.

2. Prove that the group $\mathbb{Q}/\mathbb{Z}$ is an injective cogenerator for abelian groups.

3. A ring is said to be *left semihereditary* if its finitely generated left ideals are projective. Prove that if R is a left semihereditary ring, then every finitely generated submodule of a free R-module is a direct sum of finitely many finitely generated left ideals.

4. Prove that the following are equivalent for a ring R.

   a) R is left semihereditary.
   b) Every finitely generated submodule of a projective R-module is projective.
      Moreover if R is a domain, then the above statements are equivalent to
   c) Every finitely generated torsion free R-module is projective.

5. A semihereditary integral domain is called a *Prüfer domain*. Prove that if R is a Prüfer domain, then an R-module is flat if and only if it is torsion free.
   Hint: Use Exercises 5 and 6 of Section 2.1.

6. Prove Theorem 3.2.13.

7. Prove Theorem 3.2.14.

8. Prove Theorem 3.2.15.

9. Prove Theorems 3.2.19 and 3.2.20.

10. Prove Theorem 3.2.26.

11. Prove Remark 3.2.27.

12. Let R be a commutative noetherian ring, I be an ideal of R, and M, N be R-modules. Prove that if M is finitely generated, then

$$\mathrm{Ext}^i_R(M, N)^\wedge \cong \mathrm{Ext}^i_{\hat{R}}(\hat{M}, \hat{N})$$

for all $i \geq 0$ where $^\wedge$ denotes the I-adic completion.

3.3 Injective modules over commutative noetherian rings

In this section, R will denote a commutative noetherian ring. We start with the following result.

Proposition 3.3.1. *Let $S \subset R$ be a multiplicative set. If $A \subset B$ is an essential extension of R-modules, then so is $S^{-1}A \subset S^{-1}B$ as R-modules (and so also as $S^{-1}R$-modules).*

Proof. Let N be a nonzero finitely generated submodule of $S^{-1}B$. Then $N = S^{-1}B'$ for some finitely generated submodule B' of B. Suppose $N \cap S^{-1}A = 0$. Then $S^{-1}(B' \cap A) = N \cap S^{-1}A = 0$. Thus since $B' \cap A$ is finitely generated, $t(B' \cap A) = 0$ for some $t \in S$. Now let $I = Rt$. Then by Artin–Rees lemma, there is an r such that

$$(I^n B') \cap (B' \cap A) = I^{n-r}((I^r B') \cap (B' \cap A)) = 0$$

for all $n > r$ since $I(B' \cap A) = 0$. But $(I^n B') \cap A = (I^n B') \cap (B' \cap A)$. So $(I^n B') \cap A = 0$, a contradiction since $I^n B' \neq 0$ and $A \subset B$ is an essential extension. Hence $S^{-1}A$ is an essential submodule of $S^{-1}B$.   □

Proposition 3.3.2. *Let $S \subset R$ be a multiplicative set. If E is an injective R-module, then $S^{-1}E$ is an injective $S^{-1}R$-module.*

Proof. Let J be an ideal of $S^{-1}R$. Then $J = S^{-1}I$ for some ideal I of R by Lemma 2.2.6. So

$$\mathrm{Ext}^1_{S^{-1}R}((S^{-1}R)/J, S^{-1}E) \cong \mathrm{Ext}^1_{S^{-1}R}(S^{-1}(R/I), S^{-1}E)$$
$$\cong S^{-1}\mathrm{Ext}^1_R(R/I, E)$$

by Theorem 3.2.5. So the result follows from Baer's Criterion (Theorem 3.1.3).   □

Theorem 3.3.3. $S^{-1}E_R(M) \cong E_{S^{-1}R}(S^{-1}M)$ *for any R-module M.*

Proof. $S^{-1}E_R(M)$ is an injective $S^{-1}R$-module by the proposition above. But then $S^{-1}E_R(M)$ is an essential injective extension of $S^{-1}M$ by Proposition 3.3.1. Thus $S^{-1}E_R(M)$ is an injective envelope of $S^{-1}M$. □

Remark 3.3.4. It follows from the above that if $0 \to M \to E^0 \to E^1 \to \cdots$ is an injective resolution of an R-module M, then $0 \to S^{-1}M \to S^{-1}E^0 \to S^{-1}E^1 \to \cdots$ is an injective resolution of the $S^{-1}R$-module $S^{-1}M$. Similarly for the minimal injective resolution of M.

Definition 3.3.5. An R-module M is said to be *indecomposable* if there are no nonzero submodules M_1 and M_2 of M such that $M = M_1 \oplus M_2$.

Lemma 3.3.6. *An injective R-module M is indecomposable if and only if it is the injective envelope of each of its nonzero submodules.*

Proof. Let N be a nonzero submodule of M. Then $M \cong E(N) \oplus N'$ for some R-module N'. Thus $N' = 0$ since M is indecomposable. Conversely, suppose $M = M_1 \oplus M_2$. If $M_1 \neq 0$, then $M_1 \subseteq M$ is an essential extension by assumption. But $M_1 \cap M_2 = 0$. So $M_2 = 0$ and we are done. □

Theorem 3.3.7. *The following properties hold.*

(1) $E(R/\mathfrak{p})$ *is indecomposable for all* $\mathfrak{p} \in \operatorname{Spec} R$.

(2) *If E is an indecomposable injective R-module, then* $E \cong E(R/\mathfrak{p})$ *for some* $\mathfrak{p} \in \operatorname{Spec} R$.

Proof. (1) Suppose there are nonzero submodules E_1 and E_2 of $E(R/\mathfrak{p})$ such that $E(R/\mathfrak{p}) = E_1 \oplus E_2$. Then $E_i \cap R/\mathfrak{p} \neq 0$ for $i = 1, 2$ since $R/\mathfrak{p} \subset E(R/\mathfrak{p})$ is an essential extension. So let $x_i \in E_i \cap R/\mathfrak{p}$ be a nonzero element. Then $x_1 x_2 \in (E_1 \cap R/\mathfrak{p}) \cap (E_2 \cap R/\mathfrak{p})$. But $(E_1 \cap R/\mathfrak{p}) \cap (E_2 \cap R/\mathfrak{p}) = 0$. So x_1, x_2 are nonzero elements in $R/\mathfrak{p}$ such that $x_1 x_2 = 0$. This contradicts the fact that $R/\mathfrak{p}$ is a domain. Hence $E(R/\mathfrak{p})$ is indecomposable.

(2) Let $\mathfrak{p} \in \operatorname{Ass}(E)$. Then $R/\mathfrak{p}$ is isomorphic to a submodule of E. Thus $E \cong E(R/\mathfrak{p})$ by Lemma 3.3.6 above. □

Theorem 3.3.8. *Let $\mathfrak{p}, \mathfrak{q} \in \operatorname{Spec} R$. Then*

(1) *If $r \in R - \mathfrak{p}$, then r is an automorphism on* $E(R/\mathfrak{p})$.

(2) $E(R/\mathfrak{p}) \cong E(R/\mathfrak{q})$ *if and only if* $\mathfrak{p} = \mathfrak{q}$.

(3) $\operatorname{Ass} E(R/\mathfrak{p}) = \{\mathfrak{p}\}$.

(4) *If $x \in E(R/\mathfrak{p})$, then there exists a positive integer t such that* $\mathfrak{p}^t x = 0$.

(5) $\operatorname{Hom}(E(R/\mathfrak{p}), E(R/\mathfrak{q})) \neq 0$ *if and only if* $\mathfrak{p} \subset \mathfrak{q}$.

(6) *If $S \subset R$ is a multiplicative set, then*

$$S^{-1} E(R/\mathfrak{p}) = \begin{cases} E(R/\mathfrak{p}) & \text{if } S \cap \mathfrak{p} = \emptyset \\ 0 & \text{if } S \cap \mathfrak{p} \neq \emptyset. \end{cases}$$

(7) $\operatorname{Hom}_{R_\mathfrak{p}}(k(\mathfrak{p}), E(R/\mathfrak{p})) \cong k(\mathfrak{p})$.

Proof. (1) Let $\varphi : E(R/\mathfrak{p}) \to E(R/\mathfrak{p})$ be the map multiplication by r. Then φ is one-to-one on $R/\mathfrak{p}$. So $\operatorname{Ker} \varphi \cap R/\mathfrak{p} = 0$. But $R/\mathfrak{p} \subset E(R/\mathfrak{p})$ is essential. So $\operatorname{Ker} \varphi = 0$ and thus φ is one-to-one. But then $\varphi(E(R/\mathfrak{p}))$ is an injective summand of $E(R/\mathfrak{p})$. So φ is an isomorphism since $E(R/\mathfrak{p})$ is indecomposable by Theorem 3.3.7.

(2) Suppose $\mathfrak{p} \neq \mathfrak{q}$. Let $\mathfrak{p} \not\subset \mathfrak{q}$. Then $r \in \mathfrak{p} - \mathfrak{q}$ is an automorphism on $E(R/\mathfrak{q})$ but not on $E(R/\mathfrak{p})$. So $E(R/\mathfrak{q}) \not\cong E(R/\mathfrak{p})$.

(3) Let $\mathfrak{q} \in \operatorname{Ass}(E(R/\mathfrak{p}))$, then $R/\mathfrak{q}$ is isomorphic to a submodule of $E(R/\mathfrak{p})$ and so $E(R/\mathfrak{p}) \cong E(R/\mathfrak{q})$. Hence $\mathfrak{p} = \mathfrak{q}$ by (2).

(4) Let $x \in E(R/\mathfrak{p})$, $x \neq 0$. Then $Rx \cong R/\operatorname{Ann}(x)$. But $\operatorname{Ass} E(R/\mathfrak{p}) = \{\mathfrak{p}\}$ by (3). So $\operatorname{Ass}(Rx) = \{\mathfrak{p}\}$. But then $\mathfrak{p}$ is the unique minimal element in $\operatorname{Supp}(Rx)$. But $\operatorname{Supp}(Rx) = \{\mathfrak{q} \in \operatorname{Spec} R : \operatorname{Ann}(x) \subset \mathfrak{q}\}$ by Remark 2.4.15. Hence $\mathfrak{p}$ is the radical of $\operatorname{Ann}(x)$. That is, $\operatorname{Ann}(x)$ is $\mathfrak{p}$-primary and we are done.

(5) If $\mathfrak{p} \subset \mathfrak{q}$, then we have a map $R/\mathfrak{p} \xrightarrow{\varphi} R/\mathfrak{q}$ induced by the inclusion $\mathfrak{p} \subset \mathfrak{q}$. Now embed $R/\mathfrak{q}$ into $E(R/\mathfrak{q})$. Then the composition of φ and the inclusion $R/\mathfrak{q} \subset E(R/\mathfrak{q})$ can be extended to a nonzero map in $\operatorname{Hom}(E(R/\mathfrak{p}), E(R/\mathfrak{q}))$.

Now let $\varphi \in \operatorname{Hom}(E(R/\mathfrak{p}), E(R/\mathfrak{q}))$ be nonzero. Then let $x \in E(R/\mathfrak{p})$ be such that $\varphi(x) \neq 0$. If $r \in \mathfrak{p}$, then $r^t x = 0$ for some $t > 0$ by (4) above. So $r^t \in \operatorname{Ann}(x)$. But $\operatorname{Ann}(\varphi(x)) \subset \mathfrak{q}$ by (3). Therefore $\operatorname{Ann}(x) \subset \operatorname{Ann}(\varphi(x)) \subset \mathfrak{q}$. So $r^t \in \mathfrak{q}$ and thus $r \in \mathfrak{q}$. Hence $\mathfrak{p} \subset \mathfrak{q}$.

(6) This follows from parts (1) and (4).

(7) $E(R/\mathfrak{p}) = E(R/\mathfrak{p})_\mathfrak{p}$ by (6) above. So $E(R/\mathfrak{p}) \cong E_{R_\mathfrak{p}}(k(\mathfrak{p}))$ by Theorem 3.3.3. So $\operatorname{Hom}_{R_\mathfrak{p}}(k(\mathfrak{p}), E(R/\mathfrak{p})) \cong \operatorname{Hom}_{R_\mathfrak{p}}(k(\mathfrak{p}), E_{R_\mathfrak{p}}(k(\mathfrak{p}))) \cong k(\mathfrak{p})$. □

Remark 3.3.9. We see from Theorems 3.3.7 and 3.3.8 above that there is a bijective correspondence between the prime ideals $\mathfrak{p}$ of R and the indecomposable injective modules given by $\mathfrak{p} \leftrightarrow E(R/\mathfrak{p})$.

Theorem 3.3.10. *Every injective R-module E is a direct sum of indecomposable R-modules. This decomposition is unique in the sense that for each $\mathfrak{p} \in \operatorname{Spec} R$, the number $\mu_\mathfrak{p}$ of summands isomorphic to $E(R/\mathfrak{p})$ depends only on $\mathfrak{p}$ and E. In fact, $\mu_\mathfrak{p} = \dim_{k(\mathfrak{p})} \operatorname{Hom}(k(\mathfrak{p}), E_\mathfrak{p})$ and $E = \bigoplus_{\mathfrak{p} \in \operatorname{Spec} R} E(R/\mathfrak{p})^{(X_\mathfrak{p})}$ where $\operatorname{Card} X_\mathfrak{p} = \mu_\mathfrak{p}$.*

Proof. We assume $E \neq 0$. Let $\mathcal{C}$ be the class of submodules of E that are direct sums of indecomposable injective modules. $\mathcal{C} \neq \emptyset$ since E has an indecomposable injective summand. For if $\mathfrak{p} \in \operatorname{Ass}(E)$, then $E(R/\mathfrak{p})$ is an indecomposable summand of E. Now partially order $\mathcal{C}$ by inclusion. Then $\mathcal{C}$ is an inductive system and so it has a maximal element E_0. But R is noetherian. So E_0 is an injective R-module by Theorem 3.1.17. Thus $E \cong E_0 \oplus E'$ for some injective E'. If $E' = 0$, then we are done. If not, let $\mathfrak{p} \in \text{Ass}(E')$. So $E' \cong E(R/\mathfrak{p}) \oplus E''$. But then $E_0 \oplus E(R/\mathfrak{p})$ contradicts the maximality of E_0. Hence $E' = 0$.

Now let $E = \bigoplus_{i \in I} E_i$ where the E_i are indecomposable. Then $E_i \cong E(R/\mathfrak{p})$ for some $\mathfrak{p} \in \text{Spec } R$ by Theorem 3.3.7. But if $\mathfrak{p}, \mathfrak{q} \in \text{Spec } R$, then $E(R/\mathfrak{q})_{\mathfrak{p}} = 0$ if $\mathfrak{q} \not\subset \mathfrak{p}$ by Theorem 3.3.8. So for each $\mathfrak{p} \in \text{Spec } R$, $E_{\mathfrak{p}} = (\bigoplus_i E_i)_{\mathfrak{p}} = \bigoplus_{\mathfrak{q} \subset \mathfrak{p}} E(R/\mathfrak{q})_{\mathfrak{p}}$. But if $\mathfrak{q} \neq \mathfrak{p}$, then $r \in \mathfrak{p} - \mathfrak{q}$ is an automorphism on $E(R/\mathfrak{q})$ and is zero on $k(\mathfrak{p})$ and so $\text{Hom}_{R_{\mathfrak{p}}}(k(\mathfrak{p}), E(R/\mathfrak{q})_{\mathfrak{p}}) = 0$. Thus using Lemma 3.1.16 and Theorem 3.3.8, we have

$$\text{Hom}_{R_{\mathfrak{p}}}(k(\mathfrak{p}), E_{\mathfrak{p}}) \cong \text{Hom}_{R_{\mathfrak{p}}}\left(k(\mathfrak{p}), \bigoplus_{\mathfrak{q} \subset \mathfrak{p}} E(R/\mathfrak{q})_{\mathfrak{p}}\right)$$

$$\cong \bigoplus_{\mathfrak{q} \subset \mathfrak{p}} \text{Hom}_{R_{\mathfrak{p}}}(k(\mathfrak{p}), E(R/\mathfrak{q})_{\mathfrak{p}})$$

$$\cong \bigoplus_{\mathfrak{p}} \text{Hom}_{R_{\mathfrak{p}}}(k(\mathfrak{p}), E(R/\mathfrak{p}))$$

$$\cong k(\mathfrak{p})^{(X_{\mathfrak{p}})}$$

where Card $X_{\mathfrak{p}}$ is the number of copies of $E(R/\mathfrak{p})$ in E. □

Corollary 3.3.11. *If M is an R-module, then $E(M) \cong \oplus E(R/\mathfrak{p})^{(X_{\mathfrak{p}})}$ over $\mathfrak{p} \in \text{Spec } R$ where* Card $X_{\mathfrak{p}} = \dim_{k(\mathfrak{p})} \text{Hom}(k(\mathfrak{p}), M_{\mathfrak{p}}) = \dim_{k(\mathfrak{p})} \text{Hom}(R/\mathfrak{p}, M)_{\mathfrak{p}}$. *In particular, if M is finitely generated, then* Card $X_{\mathfrak{p}} < \infty$ *for each $\mathfrak{p} \in \text{Spec } R$.*

Proof. It suffices to show that $\text{Hom}(k(\mathfrak{p}), M_{\mathfrak{p}}) = \text{Hom}(k(\mathfrak{p}), E_{\mathfrak{p}})$ where $E = E(M)$. But clearly $\text{Hom}(k(\mathfrak{p}), M_{\mathfrak{p}}) \subset \text{Hom}(k(\mathfrak{p}), E_{\mathfrak{p}})$. Now let $0 \neq f \in \text{Hom}(k(\mathfrak{p}), E_{\mathfrak{p}})$. Then f is one-to-one since $k(\mathfrak{p})$ is simple and so Ker $f = 0$. Therefore $f(k(\mathfrak{p}))$ is a simple submodule of $E_{\mathfrak{p}}$. But $M_{\mathfrak{p}} \subset E_{\mathfrak{p}}$ is essential and so $f(k(\mathfrak{p})) \cap M_{\mathfrak{p}} = f(k(\mathfrak{p}))$. Hence $f(k(\mathfrak{p})) \subset M_{\mathfrak{p}}$ and thus $\text{Hom}(k(\mathfrak{p}), E_{\mathfrak{p}}) \subset \text{Hom}(k(\mathfrak{p}), M_{\mathfrak{p}})$. Now the result follows from the theorem above. □

Theorem 3.3.12. *Let F be a flat R-module and $\mathfrak{p} \in \text{Spec } R$. Then $F \otimes_R E(k(\mathfrak{p})) \cong E(k(\mathfrak{p}))^{(X)}$ for some set X.*

Proof. $F \otimes_R E(k(\mathfrak{p}))$ is an injective $R_{\mathfrak{p}}$-module by Theorem 3.2.16 and so $F \otimes_R E(k(\mathfrak{p})) \cong \oplus E(k(\mathfrak{q}))^{(X_{\mathfrak{q}})}$ over $\mathfrak{q} \subset \mathfrak{p}$. If $\mathfrak{q} \subsetneq \mathfrak{p}$, let $r \in \mathfrak{p}$, $r \notin \mathfrak{q}$. For each $w \in F \otimes E(k(\mathfrak{p}))$, $w \neq 0$, there exists an $n > 0$ such that $r^n w = 0$ by Theorem 3.3.8. If $w \in \oplus E(k(\mathfrak{q}))^{(X_{\mathfrak{q}})}$, then multiplication by r^n is an automorphism of $\oplus E(k(\mathfrak{q}))^{(X_{\mathfrak{q}})}$ and thus $r^n w \neq 0$. Hence $\mathfrak{p} = \mathfrak{q}$. □

Theorem 3.3.13. *If E and E' are injective R-modules, then*

$$\text{Hom}_R(E, E') \cong \prod \text{Hom}(E(k(\mathfrak{p})), E(k(\mathfrak{p}))^{(X_{\mathfrak{p}})})$$

for some sets $X_{\mathfrak{p}}$, $\mathfrak{p} \in \text{Spec } R$.

Proof. E can be written as $\oplus E(k(\mathfrak{p}))^{(X_\mathfrak{p})}$ over $\mathfrak{p} \in \operatorname{Spec} R$ by Theorem 3.3.10 above. So

$$\operatorname{Hom}(E, E') \cong \prod \operatorname{Hom}(E(k(\mathfrak{p})), E')^{X_\mathfrak{p}} \cong \prod \operatorname{Hom}(E(k(\mathfrak{p})), E'^{X_\mathfrak{p}})$$

by Propositions 1.2.6 and 1.2.7. Hence we consider the modules $\operatorname{Hom}(E(k(\mathfrak{p})), G)$ for G injective. We have

$$\operatorname{Hom}(E(k(\mathfrak{p})), G) \cong \operatorname{Hom}(E(k(\mathfrak{p})) \otimes R_\mathfrak{p}, G) \cong \operatorname{Hom}(E(k(\mathfrak{p})), \operatorname{Hom}(R_\mathfrak{p}, G)).$$

But $\operatorname{Hom}(R_\mathfrak{p}, G)$ is injective over $R_\mathfrak{p}$. So $\operatorname{Hom}(R_\mathfrak{p}, G) \cong \oplus E(k(\mathfrak{q}))^{(X_\mathfrak{q})}$ over $\mathfrak{q} \subset \mathfrak{p}$. But if $\mathfrak{q} \subsetneq \mathfrak{p}$, then $\operatorname{Hom}(E(k(\mathfrak{p})), E(k(\mathfrak{q}))) = 0$. Hence $\operatorname{Hom}(E(k(\mathfrak{p})), G) \cong \operatorname{Hom}(E(k(\mathfrak{p})), E(k(\mathfrak{p}))^{(X_\mathfrak{p})})$. □

Remark 3.3.14. The $R_\mathfrak{p}$-module $\operatorname{Hom}(E(k(\mathfrak{p})), E(k(\mathfrak{p}))^{(X_\mathfrak{p})})$ is denoted by $T_\mathfrak{p}$ and so $\operatorname{Hom}_R(E, E') \cong \prod_{\mathfrak{p} \in \operatorname{Spec} R} T_\mathfrak{p}$. In the next section, we will use Matlis duality to show that $T_\mathfrak{p}$ is the $\mathfrak{p}R_\mathfrak{p}$-adic completion of a free $R_\mathfrak{p}$-module (Theorem 3.4.1 (7)).

Exercises

1. Let E be an injective R-module. Prove that the following are equivalent.

   a) E is indecomposable.
   b) 0 is not the intersection of two nonzero submodules of E.
   c) $\operatorname{Hom}_R(E, E)$ is a local ring.

2. Prove part 6 of Theorem 3.3.8

3. Prove Remark 3.3.9.

4. Prove that $\bigoplus_{\mathfrak{m} \in \operatorname{mSpec}} E(R/\mathfrak{m})$ is an injective cogenerator of R.

3.4   Matlis duality

Throughout this section, R will denote a commutative local noetherian ring with maximal ideal $\mathfrak{m}$ and residue field k. M^v will denote the *Matlis dual* $\operatorname{Hom}_R(M, E(k))$ of the R-module M. There is a natural homomorphism $\varphi : M \to M^{vv}$ defined by $\varphi(x)(f) = f(x)$ for $x \in M$, $f \in M^v$. We call this map the *canonical homomorphism*. We will say that an R-module M is *Matlis reflexive* (or simply *reflexive*) if $M \cong M^{vv} = (M^v)^v$ under the canonical homomorphism $M \to M^{vv}$.

We start with the following result.

Theorem 3.4.1. *Let $\hat{R}$ be the $\mathfrak{m}$-adic completion of R. Then*

(1) *$E(k)$ is an injective cogenerator for R.*

(2) *The canonical map $\varphi : M \to M^{vv}$ is an embedding.*

(3) *If M is an R-module of finite length, then* length M = length M^v *and M is reflexive.*

(4) $\hat{R} \otimes_R E(k) \cong E(k)$.

(5) $E(k) \cong E_{\hat{R}}(\hat{R}/\hat{\mathfrak{m}})$ *as an $\hat{R}$-module.*

(6) $\hat{R} \cong \operatorname{Hom}_R(E(k), E(k))$.

(7) $\operatorname{Hom}_R(E(k), E(k)^{(X)}) \cong \widehat{R^{(X)}}$, *the $\mathfrak{m}$-adic completion of a free R-module.*

(8) *If M is a finitely generated R-module, then* $\hat{M} \cong M^{vv}$.

(9) $E(k)$ *is artinian as an R and $\hat{R}$-module.*

Proof. (1) Let M be an R-module and $x \in M$, $x \neq 0$. Then $Rx \cong R/\operatorname{Ann}(x)$ and so we have a nonzero map $g : Rx \to E(k)$ which is obtained from the canonical map $R/\operatorname{Ann}(x) \to k = R/\mathfrak{m}$ and the inclusion $k \subset E(k)$. But g can be extended to a map $f \in M^v$. So $f(x) \neq 0$. Thus $E(k)$ is an injective cogenerator.

(2) Let $x \in M$, $x \neq 0$. Then there is an $f \in M^v$ such that $f(x) \neq 0$ by part (1). So $\varphi(x)(f) = f(x) \neq 0$. Thus φ is one-to-one.

(3) By induction on length M. If length $M = n$, then M has a submodule N of length $n - 1$ where M/N is a simple R-module. So we have an exact sequence $0 \to k \to M^v \to N^v \to 0$ where $(M/N)^v = k^v = k$. But length $N^v = n - 1$ by the induction hypothesis. Hence length $M^v = n$. Thus length M = length M^{vv}. So the embedding $\varphi : M \to M^{vv}$ of part (2) above is an isomorphism.

(4) Let $S_n = \{x \in E(k) : \mathfrak{m}^n x = 0\}$. Then $E(k) = \bigcup S_n$ by Theorem 3.3.8. But $S_1 \subset S_2 \subset \cdots$ and so $E(k) = \varinjlim S_n$. Now we consider the exact sequence

$$0 \to \mathfrak{m}^n \hat{R} \to \hat{R} \to \hat{R}/\mathfrak{m}^n \hat{R} \to 0. \text{ Then } \mathfrak{m}^n \hat{R} \otimes S_n \to \hat{R} \otimes S_n \to (\hat{R}/\mathfrak{m}^n \hat{R}) \otimes S_n \to 0$$

is exact. But $\mathfrak{m}^n \hat{R} \otimes S_n \to \hat{R} \otimes S_n$ is the zero map and $\hat{R}/\mathfrak{m}^n \hat{R} \cong R/\mathfrak{m}^n$ by Corollary 2.5.13. So $\hat{R} \otimes S_n \cong (\hat{R}/\mathfrak{m}^n \hat{R}) \otimes S_n \cong R/\mathfrak{m}^n \otimes S_n \cong S_n$. Hence taking limits gives the result.

(5) By part (4), $E(k)$ is an $\hat{R}$-module and so $E(k)$ is an injective $\hat{R}$-module that contains $k \cong \hat{R}/\hat{\mathfrak{m}}$ by Theorem 3.2.16. Now it is easy to show that $E_{\hat{R}}(k)$ is an indecomposable $\hat{R}$-module. Hence $E_{\hat{R}}(k)$ is an injective envelope of $\hat{R}/\hat{\mathfrak{m}}$ as an $\hat{R}$-module.

(6) Let S_n be as in the above. Then $\operatorname{Hom}(R/\mathfrak{m}^n, E(k)) \cong S_n$ and so

$$\operatorname{Hom}_R(E(k), E(k)) = \operatorname{Hom}_R(\varinjlim S_n, E(k)) = \varprojlim \operatorname{Hom}(S_n, E(k))$$

$$= \varprojlim (R/\mathfrak{m}^n)^{vv} = \varprojlim R/\mathfrak{m}^n = \hat{R}.$$

(7) We again let S_n be as in the above and note that $S_n \cong \operatorname{Hom}(R/\mathfrak{m}^n, E(k))$ is of

finite length. So

$$\operatorname{Hom}_R(E(k), E(k)^{(X)}) \cong \varprojlim \operatorname{Hom}(S_n, E(k)^{(X)})$$
$$\cong \varprojlim \operatorname{Hom}(S_n, E(k))^{(X)}$$
$$\cong \varprojlim (R/\mathfrak{m}^n)^{(X)}$$
$$\cong \widehat{R^{(X)}}.$$

(8) If M is finitely generated, then

$$M^{vv} \cong \operatorname{Hom}(M^v, E(k))$$
$$\cong M \otimes_R \operatorname{Hom}(E(k), E(k)) \text{ by Theorem 3.2.11}$$
$$\cong M \otimes_R \hat{R} \text{ by part (6) above}$$
$$\cong \hat{M} \text{ by Theorem 2.5.14.}$$

(9) By part (4), $E(k)$ is an $\hat{R}$-module and the $\hat{R}$-submodules of $E(k)$ coincide with the R-submodules. Hence we may assume R is complete.

Let $E(k) = N_0 \supset N_1 \supset \cdots$ be a descending chain of submodules of $E(k)$. Then $\operatorname{Hom}(N_i, E(k)) = R/I_i$ for some ideal I_i of R and so we have an ascending chain $I_0 \subset I_1 \subset I_2 \subset \cdots$ of ideals of R. But R is noetherian, so there is an integer n_0 such that $I_i = I_{i+1}$ for all $i \geq n_0$. So $R/I_i = R/I_{i+1}$ for all $i \geq n_0$. We claim that this implies that $N_i = N_{i+1}$ for all $i \geq n_0$. For if $N_i \neq N_{i+1}$ and $R/I_i = R/I_{i+1}$, then $N_i/N_{i+1} \neq 0$ with $\operatorname{Hom}(N_i/N_{i+1}, E(k)) = 0$, a contradiction since $E(k)$ is an injective cogenerator by part (1) above.   □

Proposition 3.4.2. *An artinian local ring R with residue field k is self injective if and only if $\dim_k \operatorname{Soc}(R) = 1$.*

Proof. If $\dim_k \operatorname{Soc}(R) = 1$, then k is an essential submodule of R by Proposition 3.1.22 and so an extension $R \to E(k)$ of the embedding $k \to E(k)$ is an injection. But since the lengths of R and $E(k)$ are the same by Corollary 2.3.25 and Theorem 3.4.1 above, $R \to E(k)$ is an isomorphism and so R is injective.

Conversely, if R is injective then $\dim_k \operatorname{Soc}(R) = 1$. For if $k \oplus k \subset R$, then $E(k \oplus k) = E(k) \oplus E(k) \subset R$ and R would not be indecomposable.   □

Theorem 3.4.3. *An R-module M is artinian if and only if it is finitely embedded, that is, $M \subset E(k)^n$ for some $n \geq 1$.*

Proof. $E(k)$ is artinian by Theorem 3.4.1 above and so the "if" part is clear.

Now suppose M is artinian and $M \neq 0$. Then since $E(k)$ is an injective cogenerator, the set of nonzero R-homomorphisms from M to $E(k)$ is nonempty. We now consider the set of all possible maps $f : M \to E(k)^i$ where $i \geq 1$. Since M is artinian, we can find an $f : M \to E(k)^n$ with minimal kernel. We claim that

f is one-to-one. For if not, let $x \in \operatorname{Ker} f$, $x \neq 0$. Then there is a nonzero map $g : M \to E(k)$ such that $g(x) \neq 0$ as in proof of part (1) of the theorem above. Thus $h = (f, g) : M \to E(k)^n \oplus E(k)$ is an R-homomorphism such that $\operatorname{Ker} h \subsetneq \operatorname{Ker} f$. But this contradicts the minimality of $\operatorname{Ker} f$. So f is one-to-one. $\square$

Corollary 3.4.4. *An R-module M is noetherian if and only if M^v is artinian.*

Proof. If M is noetherian, then there is an exact sequence $R^n \to M \to 0$. But then $M^v \subset E(k)^n$ and so M^v is artinian by the theorem above. Conversely, if M^v is artinian, then $M^v \subset E(k)^n$ for some $n \geq 1$. But then we have an exact sequence $(R^n)^{vv} \to M^{vv} \to 0$. Thus the map $R^n \to M \to 0$ is exact since $E(k)$ is an injective cogenerator. That is, M is noetherian. $\square$

Corollary 3.4.5. *If M^v is noetherian, then M is artinian.*

Proof. If M^v is noetherian, then M^{vv} is artinian by the above. But the canonical homomorphism $M \to M^{vv}$ is an embedding. So M is artinian. $\square$

Lemma 3.4.6. *Let R be complete. If an R-module M is noetherian or artinian, then M is reflexive.*

Proof. If M is noetherian, then we have an exact sequence $R^m \to R^n \to M \to 0$. Thus we have the following commutative diagram

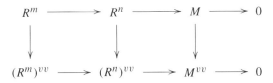

with exact rows. But the first two vertical maps are isomorphisms since $E(k)^v = \hat{R} = R$. So $M \cong M^{vv}$.

Similarly for artinian modules. $\square$

Theorem 3.4.7. *If R is complete, then an R-module M is artinian if and only if M^v is noetherian.*

Proof. Suppose M is artinian, then $M \subset E(k)^n$ for some $n \geq 1$. But then we have an exact sequence $R^n \to M^v \to 0$ since R is complete and so $E(k)^v = R$. Thus M^v is noetherian. The converse follows from Corollary 3.4.5. $\square$

Exercises

1. Let $S_n = \{x \in E(k) : \mathfrak{m}^n x = 0\}$. Prove that $\operatorname{Hom}(R/\mathfrak{m}^n, E(k)) \cong S_n$.

2. Prove that if $0 \to M' \to M \to M'' \to 0$ is an exact sequence of R-modules, then M is reflexive if and only if M', M'' are reflexive.

3. Let N be a submodule of M and set $N^{\perp} = \{\varphi \in M^{\upsilon} : \varphi(N) = 0\}$. Prove that if M is reflexive, then the map $N \mapsto N^{\perp}$ gives a one-to-one correspondence between all submodules of M and all submodules of M^{υ} which reverses inclusions.

   Hint: Note that $N^{\perp} = \mathrm{Ker}(M^{\upsilon} \to N^{\upsilon})$.

4. Prove that if R is complete, then there is a one-to-one correspondence between all submodules of $E(k)$ and all ideals of R given by $N \mapsto \mathrm{Ann}\, N$.

5. Prove Lemma 3.4.6 for artinian modules.

6. Enochs [53, Proposition 1.3]. Let R be complete. Prove that an R-module M is reflexive if and only if it contains a finitely generated R-submodule N such that M/N is artinian.

Chapter 4

Torsion Free Covering Modules

In this chapter, R will denote an integral domain, K will denote its field of fractions, and module will mean an R-module. We recall that an R-module M is said to be *torsion free* if $rx = 0$ for $r \in R$ and $x \in M$ implies $r = 0$ or $x = 0$. M is said to be a *torsion* module if for each $x \in M$, there is an $r \in R$, $r \neq 0$ with $rx = 0$. Each module M has a largest torsion submodule, denoted $t(M)$, and $M/t(M)$ is torsion free. Furthermore, the canonical map $M \to M/t(M)$ is *universal* in the sense that any linear map $M \to G$ where G is a torsion free module, the diagram

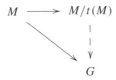

can be completed uniquely to a commutative diagram. Clearly the module $M/t(M)$ and the map $M \to M/t(M)$ are characterized up to isomorphism by these properties.

In general, given an R-module M, there does not exist a torsion free module T and a map $T \to M$ satisfying the dual condition, that is, given any linear map $G \to M$ where G is torsion free, the diagram

can be completed uniquely to a commutative diagram.

However, if the requirements on T and $T \to M$ are weakened, their existence and uniqueness can be proved.

4.1   Existence of torsion free precovers

Definition 4.1.1. A linear map $\varphi : T \to M$ where T is a torsion free module is called a *torsion free cover* of M if

1) for every torsion free module G and linear map $f : G \to M$ there is a linear map $g : G \to T$ such that $\varphi \circ g = f$.

2) Ker φ contains no nonzero pure submodule of T (here S is a *pure submodule* of T means that $aS = aT \cap S$ for all $a \in R$).

Definition 4.1.2. If φ satisfies 1) and maybe not 2) above, it is called a *torsion free precover* of M.

The existence of a torsion free precover of M is a consequence of the following lemmas:

Lemma 4.1.3. *If $\varphi : T \to M$ is a torsion free precover of M and $S \subset M$ is a submodule, then $\varphi^{-1}(S) \to S$ is a torsion free precover.*

Proof. Immediate. □

Lemma 4.1.4. *If M is divisible, then $\varphi : T \to M$ is a torsion free precover (where T is torsion free) if and only if for every linear map $f : K \to M$ (recall that K is the field of fractions of R) there is a linear map $g : K \to T$ with $\varphi \circ g = f$.*

Proof. The condition is clearly necessary. Let $G \to M$ be linear with G torsion free. Since G is a submodule of a torsion free and divisible module G' and $G \to M$ can be extended to $G' \to M$, we can assume G is itself divisible. But then G is the direct sum of modules isomorphic to K and hence the condition is sufficient. □

Lemma 4.1.5. *If M is a divisible R-module, then there is a torsion free precover $\varphi : T \to M$.*

Proof. By the preceding lemma, it suffices to let $T = K^{(X)}$ where $X = \operatorname{Hom}_R(K, M)$ and $\varphi : T \to M$ be the evaluation map defined by $\varphi((x_\sigma)_\sigma) = \sum \sigma(x_\sigma)$, $\sigma \in \operatorname{Hom}(K, M)$. □

Corollary 4.1.6. *Every module M has a torsion free precover.*

Proof. We only need note that M is a submodule of a divisible module and then appeal to Lemmas 4.1.5 and 4.1.3. □

Exercises

1. Prove that if T is torsion free, then S is a pure submodule of T if and only if T/S is torsion free.

2. Prove Lemma 4.1.3.

3. Prove that every torsion free R-module is a submodule of a torsion free divisible R-module.

4. If each $\varphi_\lambda : T_\lambda \to M_\lambda$ for $\lambda \in \Lambda$ is a torsion free precover, argue that $\prod_{\lambda \in \Lambda} \varphi_\lambda :$ $\prod T_\lambda \to \prod M_\lambda$ is also a torsion free precover.

5. Let $n > 1$ be a natural number. Argue that the canonical surjection $\sigma : \mathbb{Z} \to$ $\mathbb{Z}/(n)$ is not a torsion free precover (here $R = \mathbb{Z}$).
   Hint: Let $T \subset \mathbb{Q}$ consist of all rational numbers that can be written a/b with $gcd(b, n) = 1$. Argue that $\mathbb{Z} \to \mathbb{Z}/(n)$ can be extended to a map $\varphi : T \to$ $\mathbb{Z}/(n)$. Then argue that there is no map $f : T \to \mathbb{Z}$ except 0.

6. Argue that a torsion free precover is necessarily surjective.

7. Let $a \in R$ where R is an integral domain and a is not a unit of R. Let $b \in R$ be such that $b + (a)$ is a unit of $R/(a)$. Show that if the canonical surjection $R \to R/(a)$ is a torsion free precover then $(b) + (a) = R$.

8. If $\varphi : T \to M$ is a torsion free precover over some integral domain R and M is torsion free, show that $T \cong M \oplus N$ for some R-module N.

9. Let $\varphi : T \to M$ be a torsion free precover. Show by an example that in general $\varphi[x] : T[x] \to M[x]$ is not a torsion free precover over $R[x]$.

10. If $M = M_1 \oplus M_2$ and M admits a torsion free precover $\varphi : T \to M$, argue that both M_1 and M_2 admit torsion free precovers.

11. If $\varphi : T \to M$ is a torsion free precover and $\varphi = \psi \circ \rho$ with $\rho : T \to U$, $\psi : U \to M$ where U is torsion free, argue that $\psi : U \to M$ is also a torsion free precover.

12. If $\varphi : T \to M$ is a torsion free precover, argue that $\varphi[[x]] : T[[x]] \to M[[x]]$ is a torsion free precover over $R[[x]]$.
    Hint: If G is torsion free over $R[[x]]$ and $n \geq 1$, then $G/x^n G$ is torsion free over R. Use the previous problem to get that

$$T[[x]]/x^n T[[x]] \to M[[x]]/x^n M[[x]]$$

is a torsion free precover over R.

4.2   Existence of torsion free covers

Theorem 4.2.1. *For each module M there exists a torsion free cover $\varphi : T \to M$. Furthermore if $\varphi' : T' \to M$ is also a torsion free cover, then any map $g : T \to T'$ such that $\varphi' \circ g = \varphi$ is an isomorphism.*

Proof. By Corollary 4.1.6, there is a torsion free precover $\varphi : T \to M$. Now let the submodule $S \subset T$ be maximal such that

a) $S \subset \mathrm{Ker}\,\varphi$,

b) S is a pure submodule of T.

Then a quick check shows that the induced map $T/S \to M$ is a torsion free cover of M.

Now let $\varphi : T \to M$ and $\varphi' : T' \to M$ be torsion free covers. If $g : T \to T'$ is such that $\varphi' \circ g = \varphi$, then $\mathrm{Ker}\, g \subset \mathrm{Ker}\, \varphi$ and since T' is torsion free, $\mathrm{Ker}\, g$ is pure in T. Hence by hypothesis, $\mathrm{Ker}\, g = 0$ and so g is an injection. This means Card $T \leq$ Card T' and so reversing the argument we see that Card $T =$ Card T'. We must show g is a surjection.

Now let X be a set such that $X \supset T$, $X \supset T'$ and such that Card $X >$ Card T. Let $\mathcal{F}$ be the set of all pairs (T_0, φ_0) where T_0 is a torsion free module whose elements are elements of X and where $\varphi_0 : T_0 \to M$ is a torsion free cover. Partially order $\mathcal{F}$ by setting $(T_0, \varphi_0) \leq (T_1, \varphi_1)$ if T_0 is a submodule of T_1 and $\varphi_1|T_0 = \varphi_0$. Then $\mathcal{F}$ has maximal elements, for if $\mathcal{C}$ is a chain of $\mathcal{F}$, let T^* be the union of the first coordinates of the pairs in $\mathcal{C}$ with the unique structure of a module such that for each $(T_0, \varphi_0) \in \mathcal{C}$, T_0 is a submodule of T^*. Let $\varphi^* : T^* \to M$ be such that $\varphi^*|T_0 = \varphi_0$ for each such (T_0, φ_0). Then clearly $\varphi^* : T^* \to M$ is a torsion free precover. If $N \subset T^*$ is a pure submodule with $N \subset \mathrm{Ker}\, \varphi^*$, then for each $(T_0, \varphi_0) \in \mathcal{C}$, $N \cap T_0$ is a pure submodule of E_0 contained in the kernel of φ_0, so $N \cap T_0 = 0$. Hence $N = 0$ and so (T^*, φ^*) belongs to $\mathcal{F}$ and is an upper bound of $\mathcal{C}$.

So assume (T^*, φ^*) is a maximal element of $\mathcal{F}$. Let $f : T^* \to T$ be such that $\varphi \circ f = \varphi^*$. As above, f is an injection. To show that it is a surjection, let $Y \subset X$ be such that Card $Y =$ Card$(T - f(T^*))$ and such that $T^* \cap Y = \emptyset$. Such a Y is available since Card $X >$ Card $T =$ Card T^*. Let $\bar{T} = T^* \cup Y$ and let $h : \bar{T} \to T$ be a bijection with $h|T^* = f$. Then $\bar{T}$ can be made into a module such that g is an isomorphism. But then $(\bar{T}, \varphi \circ h)$ is an element of $\mathcal{F}$ and $(T^*, \varphi^*) \leq (\bar{T}, \varphi \circ h)$. Since (T^*, φ^*) is maximal, $Y = \emptyset$. So f is a surjection and thus an isomorphism. But then the same argument gives that $g \circ f : T^* \to T \to T'$ is an isomorphism and so that g is an isomorphism. This completes the proof. $\square$

Using the fact that the *Pontryagin dual* of a compact abelian group is torsion free if and only if the group is connected (see [156]), we get the following result.

Corollary 4.2.2. *Every compact abelian group G can be embedded uniquely up to isomorphism in a connected compact abelian group G' in such a way that every continuous homomorphism of G into a connected compact abelian group can be extended into G' and such that G' has no closed connected proper subgroups containing G. G' is uniquely determined up to isomorphism by these properties.*

Exercises

1. Let $n > 1$ and $\sigma : \mathbb{Z} \to \mathbb{Z}/(n)$ be the canonical surjection. Prove that there are an infinite number of morphisms $f : \mathbb{Z} \to \mathbb{Z}$ with $\sigma \circ f = \sigma$ which are not automorphisms of $\mathbb{Z}$. Conclude that σ is not a torsion free cover.

2. Let $\varphi : T \to M$ be a torsion free cover. Show that if U is a torsion free module then $\varphi \oplus \mathrm{id}_U : T \oplus U \to M \oplus U$ is a torsion free cover.

3. Let $a \in R$ where R is an integral domain. Suppose that $R \to R/(a)$ is a torsion free cover. Show that if $b \in R$ and $(b) + (a) = R$, then b is invertible in R.

4. If $\varphi : T \to M$ is a torsion free cover and $\varphi' : T' \to M$ a torsion free precover, argue that T is isomorphic to a direct summand of T'.

5. Let $\varphi_i : T_i \to M_i$ for $i = 1, 2$ be torsion free covers. Show that for every morphism $f : M_1 \to M_2$ there is a morphism $g : T_1 \to T_2$ such that $\varphi_2 \circ g = f \circ \varphi_1$. Then argue that if f is such that g is an isomorphism for some g with $\varphi_2 \circ g = f \circ \varphi_1$, then any morphism $\bar{g} : T_1 \to T_2$ with $\varphi_2 \circ \bar{g} = f \circ \varphi_1$ is an isomorphism.

6. Let $\varphi : T \to M$ be a torsion free precover and let $U = \varphi^{-1}(t(M))$. Note that $U \to t(M)$ is a torsion free precover. Show that if $U \to t(M)$ is a cover, then so is $\varphi : T \to M$.

7. Let $\varphi : T \to M$ be a torsion free cover. Argue that if $L \subset \text{Ker}(\varphi)$ is a direct summand of T, then $L = 0$.

4.3 Examples

The proof of the existence of torsion free covers in the previous section does not give much information about a torsion free cover $\varphi : T \to M$ of a given module M. The next proposition is useful in this respect.

Proposition 4.3.1. *Let M be a module and let $E(M)$ be its injective envelope. Let $T \subset \text{Hom}(K, E(M))$ consist of all σ such that $\sigma(1) \in M$ and let $\varphi : T \to M$ be the evaluation map $\sigma \to \sigma(1)$. Then $\varphi : T \to M$ is a torsion free cover of M.*

Proof. $\text{Hom}(K, E(M))$ is a vector space over K and so T is torsion free. Let G be torsion free and let $f : G \to M$ be linear. Since G is torsion free, $G \cong G \otimes R \to G \otimes K$ is an injection. Hence $G \to M$ can be extended to $G \otimes K \to E(M)$. By the natural isomorphism $\text{Hom}(G \otimes K, E(M)) \cong \text{Hom}(G, \text{Hom}(K, E(M)))$, we get a map $G \to \text{Hom}(K, E(M))$. But then a quick check gives that the image of G is in T and the map $g : G \to T$ is such that $\varphi \circ g = f$.

To argue that no pure submodule of T is in the kernel of φ, let $\sigma \in \text{Ker}\, \varphi, \sigma \neq 0$. Then $\sigma(K) \neq 0$. If $\sigma(\frac{r}{s}) \neq 0$, with $r, s \in R, s \neq 0$, then $\frac{r}{s}\sigma \notin \text{Ker}\, \varphi$. Hence the pure submodule of T generated by σ is not contained in $\text{Ker}\, \varphi$. $\square$

Remark 4.3.2. The first part of the proof of the above proposition gives a quick proof of Corollary 4.1.6. The two proofs are included since the two different approaches may be of interest.

Example 4.3.3. If $R = \mathbb{Z}$, $p \in \mathbb{Z}$ is a prime, and $T \to \mathbb{Z}/(p)$ is a torsion free cover, then $T \cong \hat{\mathbb{Z}}_p$ since by the above $T \subset \text{Hom}(\mathbb{Q}, \mathbb{Z}(p^\infty))$ consists of those σ with $\sigma(1) \in \mathbb{Z}(p)(\cong \mathbb{Z}/(p))$. But then $\sigma(p) = 0$. Hence $T \cong \text{Hom}(\mathbb{Q}/p\mathbb{Z}, \mathbb{Z}(p^\infty))$.

But $\mathbb{Q}/p\mathbb{Z} \cong \oplus\mathbb{Z}(q^\infty)$ (over all positive primes q) and $\operatorname{Hom}(\mathbb{Z}(q^\infty), \mathbb{Z}(p^\infty)) = 0$ if $q \neq p$ and $\operatorname{Hom}(\mathbb{Z}(p^\infty), \mathbb{Z}(p^\infty)) \cong \hat{\mathbb{Z}}_p$. We note that this construction carries through if R is any principal ideal domain and $p \in R$ is a prime. So the natural map $\hat{R}_p \to R/(p)$ is a torsion free cover. For example, if k is a field and we let $R = k[x]$ with $p = x$, then $k[[x]] \to k$ in a torsion free covering where $xk = 0$ and the map is $f \mapsto f(0)$ for $f \in k[[x]]$. It is natural to ask whether the similar map $k[[x, y]] \to k$ over $R = k[x, y]$ is a torsion free cover. But as will be shown below in Corollary 4.3.11, this is not the case.

Now given maps $f : S \to P$ and $g : S \to M$ of modules, we can form the pushout diagram

as in Example 1.3.20. We will be interested in the case when g is the canonical injection. G will then be denoted $M \oplus_f P$. We note that the map $P \to M \oplus_f P$ which maps y onto the coset of $(0, y)$ is an injection in this case. So identifying P with its image gives $M \oplus_f P/P \cong M/S$. Any submodule of $M \oplus_f P$ containing P is of the form $M' \oplus_f P$ where $S \subset M' \subset M$. Also it is easy to see that $f : S \to P$ can be extended to M if and only if P is a direct summand of $M \oplus_f P$.

Proposition 4.3.4. *Let $\varphi : T \to M$ be a torsion free cover and let $S \subset G$ be a pure submodule of a torsion free module G. If a linear map $f : S \to T$ is such that $\varphi \circ f : S \to M$ can be extended to G, then $f : S \to T$ can be extended to G so that the extension $G \to T$ lifts the extension $G \to M$.*

Proof. The hypothesis guarantees a map $G \oplus_f T \to M$ such that $G \to G \oplus_f T \to M$ is the given extension and $T \to G \oplus_f T \to M$ is φ. Since S is pure in G, G/S is torsion free. But $G \oplus_f T/T \cong G/S$ by the above. So $G \oplus_f T$ is torsion free. Therefore, the map $G \oplus_f T \to M$ can be lifted to a map $g : G \oplus_f T \to T$ since $\varphi : T \to M$ is a torsion free cover. Then $\varphi \circ (g|T) = \varphi$ and so by Theorem 4.2.1, $g|T$ is an automorphism of T. Replacing g with $(g|T)^{-1} \circ g$ we see that we can assume $g|T = \operatorname{id}_T$. Then the composition $G \to G \oplus T \to T$ gives the required extension. $\square$

Corollary 4.3.5. *Any linear map $S \to \operatorname{Ker} \varphi$ can be extended to G.*

Proof. The map $S \to \operatorname{Ker} \varphi \subset T$ gives the zero map $S \to M$ and so can be trivially extended to G. If $G \to M$ extends $S \to M$ and lifts the zero map, we get the required extension $G \to \operatorname{Ker} \varphi$. $\square$

Corollary 4.3.6. $\operatorname{Ext}^n(G, \operatorname{Ker} \varphi) = 0$ *and* $\operatorname{Ext}^n(G, T) \to \operatorname{Ext}^n(G, M)$ *is an isomorphism for all torsion free groups G and $n \geq 1$.*

Proof. Let $0 \to S \to F \to G \to 0$ be exact with F a free module. Then it follows from Corollary 4.3.5 that $\text{Hom}(F, \text{Ker}\,\varphi) \to \text{Hom}(S, \text{Ker}\,\varphi) \to 0$ is exact and so $0 \to \text{Ext}^1(G, \text{Ker}\,\varphi) \to \text{Ext}^1(F, \text{Ker}\,\varphi) = 0$ is exact since F is free. Thus $\text{Ext}^1(G, \text{Ker}\,\varphi) = 0$. Since F is free, $\text{Ext}^n(F, \text{Ker}\,\varphi) = 0$ for all $n \geq 1$ and so $\text{Ext}^n(S, \text{Ker}\,\varphi) \to \text{Ext}^{n+1}(G, \text{Ker}\,\varphi)$ is an isomorphism for $n \geq 1$. But S is torsion free. So by the above, $\text{Ext}^1(S, \text{Ker}\,\varphi) = 0$ and hence $\text{Ext}^2(G, \text{Ker}\,\varphi) = 0$. Proceeding in this manner we get $\text{Ext}^n(G, \text{Ker}\,\varphi) = 0$ for all $n \geq 1$. The fact that $\text{Ext}^n(G, T) \to \text{Ext}^n(G, M)$ is an isomorphism for all $n \geq 1$ then follows from the fact that $\text{Ext}^n(G, \text{Ker}\,\varphi) = 0$ for $n \geq 1$ and that $\text{Hom}(G, T) \to \text{Hom}(G, M) \to 0$ is exact. $\square$

Definition 4.3.7. A map of modules $f : M \to M'$ is said to be *neat* if given any submodule S of a module G and any map $\sigma : S \to M$, σ has a proper extension in G whenever $f \circ \sigma$ does. Diagrammatically $f : M \to M'$ being neat means that a commutative diagram

$$
\begin{array}{ccc}
S \hookrightarrow & & H \subset G \\
\sigma \downarrow & & \downarrow \\
M \xrightarrow{\ f\ } & & M'
\end{array}
$$

with $S \neq H$ always guarantees the existence of a commutative diagram

$$
\begin{array}{ccc}
S \hookrightarrow & & H \subset G \\
\sigma \downarrow & \swarrow & \\
M \xrightarrow{\ f\ } & & M'
\end{array}
$$

(with, perhaps a different H with $S \neq H$). It is not hard to see that in order to check whether $M \to M'$ is neat, it suffices to restrict ourselves to the case $G = R$ by using the type of argument one uses to prove Baer's Criterion for the injectivity of a module (see Theorem 3.1.3).

Example 4.3.8. If $R = \mathbb{Z}$ and M' is a subgroup of the group M, then $M' \to M$ is neat if and only if $pM' = (pM) \cap M'$ for all primes p. To show that the condition is necessary, note that any $x \in M' \cap pM$, say $x = py$, gives a commutative diagram

$$
\begin{array}{ccc}
(p) \hookrightarrow & & \mathbb{Z} \\
\downarrow & & \downarrow \\
M' \xrightarrow{\quad} & & M\,.
\end{array}
$$

Then by hypothesis, there is a commutative diagram

showing that $x \in pM'$. In a similar manner we see that the condition is sufficient.

Proposition 4.3.9. *If $\varphi : T \to M$ is a torsion free cover, then φ is neat.*

Proof. Suppose $\sigma : I \to T$ is linear where $I \subset R$ is an ideal and suppose $\varphi \circ \sigma$ has an extension to an ideal J with $I \not\subseteq J \subset R$. Then we have a map $J \oplus_\sigma T \to M$. If $J \oplus_\sigma T$ is torsion free, then $J \oplus_\sigma T \to M$ is a torsion free precover and so T is a direct summand of $J \oplus_\sigma T$ by Problem 4 of Section 4.2 and thus $I \to T$ can be extended to J. Hence suppose $t(J \oplus_\sigma T) \neq 0$. Then $t(J \oplus_\sigma T) + T = I' \oplus_\sigma T$ for $I \not\subseteq I' \subset J$. Since T is torsion free, $T \cap t(J \oplus_\sigma T) = 0$ and so T is a direct summand of $I' \oplus_\sigma T$. This means $I \to T$ can be extended to I'. □

Corollary 4.3.10. *T is injective if and only if M is injective.*

Proof. If T is injective and $I \to M$ is linear for an ideal $I \subset R$, lift to $I \to T$. This can be extended to $R \to T$. Then the composition $R \to T \to M$ gives the required extension of $I \to M$. Conversely, if M is injective, let $\sigma : I \to T$ be linear for $I \subset R$. We can suppose σ cannot be extended in R. But then if $I \neq R$, $\varphi \circ \sigma$ can be extended to R. So there is a proper extension of σ in R since φ is neat. □

Corollary 4.3.11. *If k is a field and $R = k[x, y]$, the map $S \to S(0, 0)$ from $k[[x, y]]$ to k is not a torsion free cover.*

Proof. If this were the case, then consider the map $\sigma : (x) \to k[[x, y]]$ with $\sigma(x) = y$. Since $(x) \to k[[x, y]] \to k$ is the 0 map, it has an extension to $k[x, y]$. If $\sigma : (x) \to k[[x, y]]$ has a proper extension, say τ, then for some nonzero $g(y) \in k[y]$, $g(y)$ is in the domain of τ. But $\sigma(g(y)x) = g(y)\sigma(x) = g(y)y$. Also $\sigma(g(y)x) = \sigma(xg(y)) = x\sigma(g(y))$. But clearly $g(y)y = x\sigma(g(y))$ is impossible. □

Other interesting examples can be found in Cheatham [36], Jenda [123] and Matlis [145].

Exercises

1. Prove that the canonical mapping $\hat{\mathbb{Z}}_p \to \mathbb{Z}/(p^n)$ with $n \geq 1$ is a torsion free cover.

2. Show that any surjective morphism $\hat{\mathbb{Z}}_p \to \mathbb{Z}/(p^n)$ with $n \geq 1$ is a torsion free cover.

3. Show that if $\varphi_i : T_i \to M_i$ for $i = 1, 2$ are torsion free covers, then $\varphi_1 \oplus \varphi_2 : T_1 \oplus T_2 \to M_1 \oplus M_2$ is also a torsion free cover.

4. Give an example of torsion free covers $\varphi_i : T_i \to M_i, i = 1, 2$, and morphisms $g : T_1 \to T_2, f : M_1 \to M_2$ with $\varphi_2 \circ g = f \circ \varphi_1$ where g is an isomorphism but f is not an isomorphism.

5. The canonical homomorphism $\hat{\mathbb{Z}}_p \to \mathbb{Z}/(p)$ is a torsion free cover by Problem 1. So Problem 3 gives a torsion free cover $\varphi : \hat{\mathbb{Z}}_p \oplus \hat{\mathbb{Z}}_p \to \mathbb{Z}/(p) \oplus \mathbb{Z}/(p)$. Prove that the group of linear maps $f : \hat{\mathbb{Z}}_p \oplus \hat{\mathbb{Z}}_p \to \hat{\mathbb{Z}}_p \oplus \hat{\mathbb{Z}}_p$ such that $\varphi \circ f = \varphi$ is isomorphic to the group of 2×2 matrices (a_{ij}) over $\hat{\mathbb{Z}}_p$ such that $a_{ij} \equiv \delta_{ij}$ (mod p) for all i, j.

6. Let R be a local domain and let $a \in R$ where $a \neq 0$ and a is not a unit of R. Show that the canonical surjection $R \to R/(a)$ is a torsion free cover if and only if $\operatorname{Ext}^1(G, R) = 0$ for all torsion free modules G.

7. Use Corollary 4.3.10 to argue that the torsion free cover of every divisible module is divisible if and only if every divisible module is injective.

8. If k is a field and $R = k[x, y, z]$, argue that the map $S \to S(0, 0, 0)$ from $k[[x, y, z]]$ to k is not a torsion free cover.

4.4 Direct sums and products

It is rarely true that given a family $\varphi_i : T_i \to M_i$ of torsion free covers, $\oplus T_i \to \oplus M_i$ is also a torsion free cover. In Section 5 of Chapter 5, we will show that this property is preserved by countable sums under suitable conditions. The next result considers the corresponding question for products.

Theorem 4.4.1. *The following are equivalent for R.*

(1) *Every torsion module $G \neq 0$ has a simple submodule.*

(2) *If $\varphi_i : T_i \to M_i$ is any family of torsion free covers, then $\prod \varphi_i : \prod T_i \to \prod M_i$ is a torsion free cover.*

(3) *A module E is injective if and only if $\operatorname{Ext}^1(S, E) = 0$ for all simple modules S.*

Proof. Recall that K is the field of fractions of R. We first show that (1) is equivalent to each of the following:

(a) R/I has a simple submodule for each ideal $I \neq 0, R$.

(b) K/T has a simple submodule for each submodule T of $K, T \neq 0, K$.

Since every torsion module $G \neq 0$ has a submodule isomorphic to R/I for an ideal $I \neq 0, R$, clearly (1) and (a) are equivalent. (1) $\Rightarrow$ (b) is trivial.

Now suppose (b) holds and let $I \neq 0, R$ be an ideal. By transfinite induction we can define a submodule I_σ of K for each ordinal σ such that

i) $I_0 = I$,

ii) $I_{\sigma+1}/I_\sigma$ is the socle of K/I_σ for each σ,

iii) $I_\sigma = \bigcup I_\tau$ (for $\tau < \sigma$) when σ is a limit ordinal.

Then by (b), $I_\sigma = K$ for some σ. Hence since $I \neq R$, there is a least σ with $I \neq R \cap I_\sigma$. Clearly σ is not a limit ordinal and $\sigma > 0$. Then $I = R \cap I_{\sigma-1}$ and so

$$R \cap I_\sigma / I \rightarrow I_\sigma / I_{\sigma-1}$$

is a nonzero injection. Thus R/I has a simple submodule.

Now we prove (1) $\Rightarrow$ (3). Suppose E is such that $\mathrm{Ext}^1(S, E) = 0$ for all simple modules S. Let $\sigma : I \rightarrow E$ be linear where $I \subset R$ is an ideal. We want to show σ can be extended to R.

Clearly we can assume $I \neq 0, R$. But then R/I has a simple submodule J/I and so $\mathrm{Ext}^1(J/I, E) = 0$. Then the exactness of

$$\mathrm{Hom}(J, E) \rightarrow \mathrm{Hom}(I, E) \rightarrow \mathrm{Ext}^1(J/I, E) = 0$$

implies σ can be extended to J. But $J \supsetneq I$. So by using Zorn's lemma, σ can be extended to R.

(3) $\Rightarrow$ (1) By (b) above it suffices to prove K/T has a simple submodule for $T \neq 0, K$. Suppose not. Then we claim $\mathrm{Ext}^1(S, T) = 0$ for any simple module S. For let $S = R/\mathrm{m}$, m a maximal ideal. Then if $\sigma : \mathrm{m} \rightarrow T$ is linear, there is a linear $\tau : R \rightarrow K$ agreeing with σ on m. If $\tau(R) \not\subset T$, then K/T would have a simple submodule. So $\tau(R) \subset T$. This means

$$\mathrm{Hom}(R, T) \rightarrow \mathrm{Hom}(\mathrm{m}, T) \rightarrow 0$$

is exact. But

$$\mathrm{Hom}(R, T) \rightarrow \mathrm{Hom}(\mathrm{m}, T) \rightarrow \mathrm{Ext}^1(S, T) \rightarrow \mathrm{Ext}^1(R, T)$$

is exact and $\mathrm{Ext}^1(R, T) = 0$. So $\mathrm{Ext}^1(S, T) = 0$. Since S was arbitrary, by (3), T is injective. Since $T \neq 0$, this is possible only if $T = K$. Hence (3) $\Rightarrow$ (b) and so (3) $\Rightarrow$ (1).

(1) $\Rightarrow$ (2) Clearly $\prod \varphi_i : \prod T_i \rightarrow \prod M_i$ is a precover, so we only need to show that $\mathrm{Ker}(\prod \varphi_i)$ contains no nonzero pure submodule of $\prod T_i$. Suppose $(x_i) \in \mathrm{Ker}(\prod \varphi_i)$ generates a pure submodule contained in $\mathrm{Ker}(\prod \varphi_i)$. We want to show $(x_i) = 0$. Let $R \rightarrow \prod T_i$ map 1 to (x_i). We order the extensions of $R \rightarrow \prod T_i$ to submodules T of K in the obvious fashion. $\prod T_i$ torsion free implies there is a unique maximal extension $\sigma : T \rightarrow \prod T_i$. Now $\sigma(T) \subset \mathrm{Ker}(\prod \varphi_i)$ by our hypothesis on (x_i). If $T = K$, then either $\sigma(K) = 0$ in which case $\sigma(1) = (x_i) = 0$ and we are through, or $\sigma(K) \cong K$ in which case $\sigma(K)$ is a divisible submodule of $\prod T_i$ in $\mathrm{Ker}(\prod \varphi_i)$, and in fact for each i, the pure submodule of T_i generated by x_i would be a divisible submodule of T_i contained in $\mathrm{Ker}(\varphi_i)$ which is possible only if $x_i = 0$. But this

contradicts $\sigma(K) \cong K$ since $\sigma(1) = (x_i)$. Hence we suppose $T \neq K$. We have $T \neq 0$ and $R \subset T$. Let U/T be a simple submodule of K/T. Consider the diagram

$$\begin{array}{ccc} T & \longrightarrow & U \\ \downarrow & & \\ T_i & \longrightarrow & M_i \end{array}$$

for each i and note that $T \to M_i$ is the 0 map, which can be extended to the 0 map $U \to M_i$. But φ_i is neat, so $T \to T_i$ can be extended to U, that is, $T \to \prod T_i$ can be extended to U. This contradicts the maximality of T. Hence $(x_i) = 0$.

(2) $\Rightarrow$ (1) Let $T \subset K$, $T \neq 0, K$. We prove (2) $\Rightarrow$ (b), that is, that K/T has a simple submodule. Let (M_i) be a family of modules such that each $M_i = U/T$ for some $U \subset K$ with $T \subsetneq U$ and such that given U, $M_i = U/T$ for some i. Let $\varphi_i : T_i \to M_i$ be a torsion free cover for each i. Given i, if $M_i = U_i/T$, let $\sigma_i : U_i \to T_i$ lift $U_i \to M_i$ (the canonical surjection). Then $(\sigma_i|T)$ gives rise to a map $\sigma : T \to \prod T_i$ with $\sigma(T) \subset \mathrm{Ker}(\prod \varphi_i)$. Hence the composition

$$T \to \prod T_i \to \prod M_i$$

is 0 and can be extended. By hypothesis $\prod T_i \to \prod M_i$ is a torsion free cover and so is neat. Thus $T \to \prod T_i$ has an extension to some $U \subset K$, $T \subsetneq U$. Let $\tau : U \to \prod T_i$ be this extension. Suppose now for some j, $E_j = U_j/T \subset U/T$, that is, $U_j \subset U$. Let $\pi_j : \prod T_i \to T_j$ be the canonical projection. By hypothesis, $\tau|T = \sigma$. But $\pi_j \circ \sigma = \sigma_j|T$. So $\pi_j \circ \tau|T = \sigma_j|T$. Since $\prod T_i$ is torsion free, we get that $\pi_j \circ \tau|U_j = \sigma_j$. But since $\varphi_j \circ \sigma_j : U_j \to U_j/T$ is the canonical surjection we have that

$$\varphi_j \circ \pi_j \circ \tau : U \to U_j/T$$

induces a map $U/T \to U_j/T$ whose restriction to U_j/T is $\mathrm{id}_{U_j/T}$. Hence U_j/T is a direct summand of U/T. Since j was arbitrary with $U_j \subset U$, we see that U/T has every submodule a direct summand and so by a standard argument (since $U/T \neq 0$), U/T has a simple submodule. Thus K/T has a simple submodule. $\square$

Remark 4.4.2. If R is a *Dedekind* domain, that is, every ideal is projective, then R is noetherian and every nonzero prime ideal is maximal (see Jacobson [118] for example). So R satisfies (1)–(3) of the theorem by Exercise 3 of Section 3.1.

If R is not a field and satisfies (1)–(3), then it has Krull dimension 1 for if $\mathfrak{p}$ is a nonzero prime ideal, then $R/\mathfrak{p}$ has a simple submodule only if $\mathfrak{p}$ is maximal. In fact, if R is noetherian (and not a field), R satisfies (1)–(3) if and only if its Krull dimension is 1.

If R is a local domain with maximal ideal $\mathfrak{m}$, then R satisfies (1)–(3) if and only if for every sequence (a_i) of elements of $\mathfrak{m}$ and every ideal $I \subset R$, $a_1 a_2 \ldots a_n \in I$ for some $n \geq 1$.

Exercises

1. If k is a field, argue that $R = k[x, y]$ does not have the property of Theorem 4.4.1.

2. Give an argument for the last comment in Remark 4.4.2.

3. Let k be a field and let $R = k[[x^2, x^3]]$ (so R consists of all formal power series $a_0 + a_1 x + a_2 x^2 + \cdots$ with $a_1 = 0$). Argue that R satisfies the conditions of Theorem 4.4.1.

   Hint: Let $I \subset R$ be a nonzero ideal. Argue that for some $n_0 \geq 2$, I contains all elements of R of the form $a_{n_0} x^{n_0} + a_{n_0+1} x^{n_0+1} + \cdots$. Deduce that if $\mathfrak{m} \subset R$ is generated by x^2 and x^3, then $\mathfrak{m}^{n_0} \subset I$.

4. If I is an infinite set, argue that the canonical morphism $\hat{\mathbb{Z}}^{(I)} \to \mathbb{Z}/(p)^{(I)}$ is not a torsion free precover.

   Hint: Show there is a natural extension of $\hat{\mathbb{Z}}_p^{(I)} \to \mathbb{Z}/(p)^{(I)}$ to a morphism $\hat{\mathbb{Z}}_p^{(I)} + p\hat{\mathbb{Z}}_p^I \to \mathbb{Z}/(p)^{(I)}$. Then use Proposition 4.3.4.

5. If a domain R has the property of Theorem 4.4.1 and if $\varphi : T \to M$ is any torsion free cover, argue that $\mathrm{Hom}(P, T) \to \mathrm{Hom}(P, M)$ is a torsion free cover for any projective R-module P.

Chapter 5

Covers

We note that there is no loss of generality in what follows in assuming that if $\mathcal{F}$ is a class of R-modules and C, D are R-modules such that $C \cong D$ and $C \in \mathcal{F}$, then $D \in \mathcal{F}$. Hence we will always assume that the classes of modules $\mathcal{F}$ are closed under isomorphisms. In the previous chapter, we considered torsion free coverings. In this chapter, we will give a general definition of coverings whose examples will include torsion free covers and the familiar projective covers.

5.1 $\mathcal{F}$-precovers and covers

Definition 5.1.1. Let R be a ring and let $\mathcal{F}$ be a class of R-modules. Then for an R-module M, a morphism $\varphi : C \to M$ where $C \in \mathcal{F}$ is called an $\mathcal{F}$-cover of M if

1) any diagram with $C' \in \mathcal{F}$

can be completed to a commutative diagram and

2) the diagram

can be completed only by automorphisms of C.

So if an $\mathcal{F}$-cover exists, then it is unique up to isomorphism.

If $\varphi : C \to M$ satisfies (1) but may be not (2), then it is called an $\mathcal{F}$-precover of M.

For example, if $\mathcal{F}$ is the class of projective modules, an $\mathcal{F}$-cover (*precover*) is called a *projective cover* (*precover*). Note that this is not the usual definition but can be seen to agree with it. Similar terminology will be used in other situations, for example, if $\mathcal{F}$ is the class of torsion free modules over an integral domain, we get the torsion free covers of Chapter 4 (see Theorem 4.2.1). Note also that if the class $\mathcal{F}$ contains the ring R, then $\mathcal{F}$-precovers are surjective.

We say that a class $\mathcal{F}$ is (*pre*)*covering* if every R-module has an $\mathcal{F}$-(pre)cover. For example, we saw in Chapter 4 that the class of torsion free modules over a domain is covering.

The following proposition illustrates a close relationship between covers and pre-covers.

Proposition 5.1.2. *Let M be an R-module. Then the $\mathcal{F}$-cover of M, if it exists, is a direct summand of any $\mathcal{F}$-precover of M.*

Proof. Let $C \to M$ be the $\mathcal{F}$-cover and $C' \to M$ be an $\mathcal{F}$-precover. Then we have the following commutative diagram

But then $C \to C' \to C$ is an automorphism. So C is a direct summand of C'.   □

Exercises

1. If $\varphi : C \to M$ is an $\mathcal{F}$-cover, argue that there is a bijective correspondence between the set of linear maps $f : C \to C$ such that $\varphi \circ f = \varphi$ and the set $\mathrm{Hom}_R(C, \mathrm{Ker}\,\varphi)$.

2. Let $\mathcal{F}$ be a class of R-modules closed under summands and such that every R-module M has an $\mathcal{F}$-cover. Argue that every such cover is surjective if and only if $\mathcal{F}$ contains all the projective modules.

3. If $M = M_1 \oplus M_2$ for R-modules M_1 and M_2 and if M has an $\mathcal{F}$-precover, argue that both M_1 and M_2 have $\mathcal{F}$-precovers.

4. Let R be an integral domain and $\mathcal{F}$ be the class of torsion R-modules. Show that every R-module has an $\mathcal{F}$-cover which is an injection.

5. An R-submodule A of B is said to be *superfluous* if for each submodule N of B, $N + A = B$ implies $N = 0$. Prove that a map $\varphi : P \to M$ is a projective cover if and only if P is projective and $\mathrm{Ker}\,\varphi$ is superfluous.

6. Let $J = \mathrm{rad}(R)$ and M be a finitely generated R-module. Prove that JM is a superfluous submodule of M. Conclude that the natural map $\varphi : R \to R/J$ is a projective cover.

5.2 Existence of precovers and covers

We start by giving easy necessary and sufficient conditions for the existence of pre-covers.

Lemma 5.2.1. *Let $\mathcal{F}$ be a class of R-modules that is closed under direct sums. Then an R-module M has an $\mathcal{F}$-precover if and only if there is a set I and a family $(C_i)_{i \in I}$ of elements of $\mathcal{F}$ and morphisms $\varphi_i : C_i \to M$ for each $i \in I$ such that any morphism $D \to M$ with $D \in \mathcal{F}$ has a factorization $D \to C_j \overset{\varphi_j}{\to} M$ for some $j \in I$.*

Proof. The only if part is trivial. For the if part, we simply note that $\oplus_{i \in I} C_i \overset{\oplus \varphi_j}{\longrightarrow} M$ is an $\mathcal{F}$-precover. □

Proposition 5.2.2. *If $\mathcal{F}$ is a class of R-modules closed under direct sums, then an R-module M has an $\mathcal{F}$-precover if and only if there is a cardinal number $\aleph_\alpha$ such that any morphism $D \to M$ with $D \in \mathcal{F}$ has a factorization $D \to C \to M$ with $C \in \mathcal{F}$ and $\mathrm{Card}\, C \leq \aleph_\alpha$.*

Proof. If M has an $\mathcal{F}$-precover $C \to M$, then let $\aleph_\alpha = \mathrm{Card}\, C$. Conversely, there is a set X with cardinality $\aleph_\alpha$ such that any morphism $D \to M$ with $D \in \mathcal{F}$ has a factorization $D \to C \to M$ for some $C \in \mathcal{F}$ with $C \subset X$ (as sets). Let $(\varphi_i)_{i \in I}$ give all such morphisms $\varphi_i : C_i \to M$ with $C_i \subset X$ (as sets). Then any morphism $D \to M$ has a factorization $D \to C_j \to M$ for some j. So M has a $\mathcal{F}$-precover by Lemma 5.2.1. □

We now show that if we assume that the class $\mathcal{F}$ is closed under inductive limits, then a module M has a $\mathcal{F}$-cover whenever it has an $\mathcal{F}$-precover. We prove the following

Theorem 5.2.3. *Let R be a ring, $\mathcal{F}$ be a class of R-modules closed under summands, and M be an R-module. Suppose for any well ordered inductive system $((C_\alpha), (\varphi_{\beta\alpha}))$ of modules in $\mathcal{F}$, every map $\lim_{\to} C_\alpha \to M$ can be factored through some $C \to M$ with $C \in \mathcal{F}$. If M has an $\mathcal{F}$-precover, then it has an $\mathcal{F}$-cover.*

Proof. We break the proof into the following three lemmas. □

Lemma 5.2.4. *For each $C \in \mathcal{F}$ and morphism $C \to M$, there exists an $\mathcal{F}$-precover $D \to M$ and a morphism f in the factorization $C \overset{f}{\to} D \to M$ such that for any morphism g in the factorization $D \overset{g}{\to} E \to M$ where $E \to M$ is an $\mathcal{F}$-precover, $\mathrm{Ker}(g \circ f) = \mathrm{Ker}\, f$.*

Proof. Suppose the conclusion is not true. Then any such f does not have the desired property. Hence we can construct a diagram

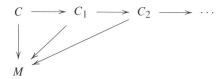

with $\mathrm{Ker}(C \to C_n) \subsetneq \mathrm{Ker}(C \to C_{n+1})$ for each $n \geq 1$ and with each $C_n \to M$ an $\mathcal{F}$-precover. But by assumption, the map $\varinjlim C_n \to M$ can be factored through a map $D \to M$ with $D \in \mathcal{F}$. We can assume $D \to M$ is an $\mathcal{F}$-precover. Let $C_\omega = D$. Then $\mathrm{Ker}(C \to C_n) \subsetneq \mathrm{Ker}(C \to C_\omega)$ for all n, and $C \to C_\omega$ does not have the desired property. So there is an $\mathcal{F}$-precover $C_{\omega+1} \to M$ and a map $C_\omega \to C_{\omega+1}$ in the factorization $C_\omega \to C_{\omega+1} \to M$ such that $\mathrm{Ker}(C \to C_\omega) \subsetneq \mathrm{Ker}(C \to C_{\omega+1})$. Continuing in this fashion, we see that for any ordinal α, we can construct precovers $C_\beta \to M$ for all $\beta < \alpha$ with maps $C \to C_\beta$ so that for $\beta < \nu < \alpha$, $\mathrm{Ker}(C \to C_\beta) \subsetneq \mathrm{Ker}(C \to C_\nu)$. If for each β with $\beta+1 < \alpha$, we choose x_β with $x_\beta \notin \mathrm{Ker}(C \to C_\beta)$, $x_\beta \in \mathrm{Ker}(C \to C_{\beta+1})$, then $x_\beta \neq x_{\beta'}$ for $\beta \neq \beta'$. This implies $\mathrm{Card}\, C \geq \mathrm{Card}\,(\alpha)$ whenever α is infinite. This is impossible since α is arbitrary. $\qquad\square$

Lemma 5.2.5. *There exists an $\mathcal{F}$-precover $D \to M$ such that in any factorization $D \to E \to M$ with $E \to M$ an $\mathcal{F}$-precover, $D \to E$ is an injection.*

Proof. Using the lemma above, for any ordinal γ we construct the diagram in the lemma where for each $\alpha < \gamma$, $C_\alpha \to M$ is an $\mathcal{F}$-precover and such that when $\alpha + 1 < \gamma$,

has the property guaranteed by the lemma. For $\alpha + 1 < \gamma$, we let $D_\alpha \subset C_{\alpha+1}$ be the image of $C_\alpha \to C_{\alpha+1}$. Then by our requirement on $C_\alpha \to C_{\alpha+1}$, we see that if $C \to M$ is any $\mathcal{F}$-precover and if

is commutative then $C_{\alpha+1} \to C$ restricted to D_α is an injection. Hence if we let $\aleph_\lambda = \mathrm{Card}\, C$ for some fixed $\mathcal{F}$-precover $C \to M$, we see that $\mathrm{Card}\, D_\alpha \leq \aleph_\lambda$ for all such α. But when $\alpha < \beta < \beta + 1 < \gamma$ we see that $C_{\alpha+1} \to C_{\beta+1}$ induces an injection $D_\alpha \to D_\beta$. So because of the restriction on the size of the D_α's, we see that if the original γ is large enough, we must be able to find an α and β with $\alpha < \beta < \beta + 1 < \gamma$ such that $D_\alpha \to D_\beta$ is a bijection. But then D_α is a direct summand of $C_{\alpha+1}$ and so is in $\mathcal{F}$. Then noting that any factorization $D_\alpha \to E \to M$ with $E \to M$ an $\mathcal{F}$-precover can be extended to a factorization $C_{\alpha+1} \to E \to M$, we see that $D_\alpha \to E$ is an injection. So we let $D = D_\alpha$ and let $D \to M$ be the restriction of $C_{\alpha+1} \to M$ to D. □

Lemma 5.2.6. *If $\psi : D \to M$ is an $\mathcal{F}$-precover having the property of Lemma 5.2.5, then $\psi : D \to M$ is an $\mathcal{F}$-cover.*

Proof. If every map $D \to D$ completing

is an isomorphism, then we are through. So suppose $C \to C$ is such a map, but not an automorphism. Then for any ordinal α, we can construct a commutative diagram

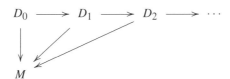

with $D_\beta = D$ for each $\beta < \alpha$ and $D_\beta \to D_{\beta+1}$ not an isomorphism (but an injection) for each $\beta < \alpha$. Now complete

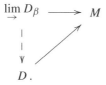

By construction, if $\beta < \upsilon < \alpha$, then $D_\beta \to D_\upsilon$, $D_\upsilon \to D$ are injections but not surjections. This implies $\operatorname{Card} D \geq \operatorname{Card}(\alpha)$ for all infinite α, and so gives a contradiction. □

Corollary 5.2.7. *Let $\mathcal{F}$ be a class of R-modules that is closed under well ordered inductive limits and M be an R-module. If M has an $\mathcal{F}$-precover, then it has an $\mathcal{F}$-cover.*

Proof. In Theorem 5.2.3 we can let $C = \varinjlim C_\alpha$. □

Exercises

1. Let G be any set of modules and $\mathcal{F}$ be the class of modules which are isomorphic to a direct summand of a direct sum of modules in G. Argue that $\mathcal{F}$ is precovering.

2. Let $\mathcal{F}$ be a class such that M has an $\mathcal{F}$-cover and let $\varphi : C \to M$ be an $\mathcal{F}$-precover. Then show that the following are equivalent:

   a) $\varphi : C \to M$ is a cover
   b) any $f : C \to C$ with $\varphi \circ f = \varphi$ is an injection
   c) any $f : C \to C$ with $\varphi \circ f = \varphi$ is a surjection

3. Let $\mathcal{F}$ be a class of R-modules and M_1 and M_2 be two R-modules satisfying the initial hypothesis of Theorem 5.2.3. If $\mathcal{F}$ is closed under finite sums, argue that $M_1 \oplus M_2$ also satisfies this hypothesis.

4. Let $R = \mathbb{Z}$ and $\mathcal{F}$ consist of all direct sums of cyclic groups. Show that every module has an $\mathcal{F}$-precover. Find a module that does not have an $\mathcal{F}$-cover.

5.3   Projective and flat covers

We now know that flat covers always exist. We will defer the proof of this fact to Chapter 7 (Theorem 7.4.4) after we establish necessary tools. So in this section, we will only consider the existence of flat precovers and covers for special rings and classes of modules.

Since, for any ring R, any inductive limit of flat R-modules is flat, we have the following as a consequence of Corollary 5.2.7.

Theorem 5.3.1. *For any ring R and any R-module M, if M has a flat precover, then it has a flat cover.*

If R is a Prüfer domain, then the class of flat R-modules is covering by Theorem 4.2.1.

It is easy to see that projective precovers always exist. The following result of Bass [23] tells us when projective modules are covering.

Theorem 5.3.2. *Let $\mathcal{P}roj$ be the class of projective R-modules. Then the following are equivalent*

(1) *Every flat R-module is projective.*

(2) *Every projective precover is a flat precover.*

(3) *$\mathcal{P}roj$ is covering (that is, R is left perfect).*

Proof. (1) ⇒ (2) is trivial.

(2) ⇒ (1). Let F be a flat R-module and $P \to F$ be its projective precover. Then $P \to F$ is also a flat precover by assumption. But projective precovers are surjective. So the diagram

can be completed to a commutative diagram. Hence F is a summand of P.

(1) ⇒ (3). Every R-module has a flat precover since flat means projective. So every R-module has a projective cover by Theorem 5.3.1.

(3) ⇒ (1). Let F be a flat R-module. We consider the minimal projective resolution

$$\cdots \to P_2 \xrightarrow{d_2} P_1 \xrightarrow{d_1} P_0 \to F \to 0$$

If J is the Jacobson radical of R, then $d_n(P_n) \subseteq JP_{n-1}$ since $\mathrm{Ker}(d_{n-1})$ is superfluous in P_{n-1} and so $\mathrm{Ker}(d_{n-1}) \subseteq JP_{n-1}$. So the deleted complex

$$\cdots \to R/J \otimes P_2 \to R/J \otimes P_1 \to R/J \otimes P_0 \to 0$$

has zero differentiation. Hence $\mathrm{Tor}_1(R/J, F) = R/J \otimes P_1 \cong P_1/JP_1$. But since F is flat, $\mathrm{Tor}_1(R/J, F) = 0$. So $P_1 = JP_1$. But then $P_1 = 0$ by Exercise 9. Thus F is a projective left R-module. □

We now argue that finitely generated modules over local rings have projective covers.

Theorem 5.3.3. *If R is a local ring, then every finitely generated R-module has a projective cover.*

Proof. Let M be a finitely generated R-module and $\mathfrak{m}$ be the maximal left ideal of R. Then $R/\mathfrak{m}$ is a division ring and so $M/\mathfrak{m}M$ is a finite-dimensional vector space over $R/\mathfrak{m}$. Thus $M/\mathfrak{m}M \cong (R/\mathfrak{m})^n$ for some integer $n \geq 1$. So we have an epimorphism $\sigma : R^n \to M/\mathfrak{m}M$ which factors through the natural map $\tau : M \to M/\mathfrak{m}M$, that is, there is a map $\varphi : R^n \to M$ such that $\tau \circ \varphi = \sigma$.

If $y \in M$, then there is an $x \in R^n$ such that $\tau(y) = \sigma(x) = \tau \circ \varphi(x)$ since σ is surjective. But then $\varphi(x) - y \in \mathrm{Ker}\,\tau = \mathfrak{m}M$. So $M = \mathrm{Im}\,\varphi + \mathfrak{m}M$. Thus $M = \mathrm{Im}\,\varphi$ by Nakayama lemma. So φ is surjective. Hence $\varphi : R^n \to M$ is a projective precover. Note that $\mathrm{Ker}\,\varphi \subseteq \mathfrak{m}R^n = \mathfrak{m}^n$. We now argue that it is also a cover.

Suppose $f : R^n \to R^n$ is a map such that $\varphi \circ f = \varphi$. Then as in the above, $R^n = \mathrm{Im}\,f + \mathrm{Ker}\,\varphi$. So $\mathrm{Im}\,f = R^n$ since $\mathrm{Ker}\,\varphi \subset \mathfrak{m}R^n$. Therefore f is surjective. But then f splits and so there is a map $f' : R^n \to R^n$ such that $f \circ f' = \mathrm{id}_{R^n}$. Hence $\mathrm{Im}\,f' \oplus \mathrm{Ker}\,f = R^n$. But $\mathrm{Ker}\,f \subset \mathrm{Ker}\,\varphi$ and so $\mathrm{Im}\,f' + \mathrm{Ker}\,\varphi = R^n$. Thus

Im $f' = R^n$ again by Nakayama lemma. Therefore Ker $f = 0$ and so f is one-to-one. Hence the diagram

can be completed only by automorphisms of R^n. That is, $\varphi : R^n \to M$ is a projective cover.   □

Remark 5.3.4. Rings for which every finitely generated module has a projective cover are said to be *semiperfect*. So we see that locals rings are semiperfect and every perfect ring is semiperfect. Further characterizations and examples of perfect and semiperfect rings can be found in Anderson–Fuller [6] or in the original paper of Bass [23].

Recall that given an R-module M, the character module $\mathrm{Hom}_{\mathbb{Z}}(M, \mathbb{Q}/\mathbb{Z})$ is denoted by M^+. With this notation, we have the following result.

Proposition 5.3.5. *Let R be a right coherent ring, M be a right R-module, and E be an injective right R-module containing M. Then $E^+ \to M^+$ is a flat precover.*

Proof. E^+ is a flat left R-module since R is right coherent. We want to show that if F is a flat left R-module, $\mathrm{Hom}_R(F, E^+) \to \mathrm{Hom}_R(F, M^+) \to 0$ is exact, or equivalently, $\mathrm{Hom}_{\mathbb{Z}}(E \otimes_R F, \mathbb{Q}/\mathbb{Z}) \to \mathrm{Hom}_{\mathbb{Z}}(M \otimes_R F, \mathbb{Q}/\mathbb{Z}) \to 0$ is exact which is obvious since F is flat and $\mathbb{Q}/\mathbb{Z}$ is injective. So $E^+ \to M^+$ is a flat precover.   □

Definition 5.3.6. A submodule T of an R-module N is said to be a *pure submodule* if $0 \to A \otimes T \to A \otimes N$ is exact for all right R-modules A, or equivalently, if $\mathrm{Hom}(A, N) \to \mathrm{Hom}(A, N/T) \to 0$ is exact for all finitely presented R-modules A. An exact sequence $0 \to T \to N \to N/T \to 0$ (or $0 \to T \to N$) is said to be *pure exact* if T is a pure submodule of N.

An R-module M is said to be *pure injective* if for every pure exact sequence $0 \to T \to N$ of R-modules, $\mathrm{Hom}(N, M) \to \mathrm{Hom}(T, M) \to 0$ is exact. Clearly, every injective module is pure injective.

Proposition 5.3.7. *For any R-module M, the character module M^+ is a pure injective right R-module.*

Proof. Let $T \subset N$ be a pure submodule of the right R-module N. Then $\mathrm{Hom}_R(N, M^+) \to \mathrm{Hom}_R(T, M^+) \to 0$ is exact if and only if $\mathrm{Hom}_{\mathbb{Z}}(N \otimes_R M, \mathbb{Q}/\mathbb{Z}) \to \mathrm{Hom}_{\mathbb{Z}}(T \otimes_R M, \mathbb{Q}/\mathbb{Z}) \to 0$ is exact. But the latter is exact since $0 \to T \otimes_R M \to N \otimes_R M$ is exact.   □

We note that a similar result holds for right R-modules.

Proposition 5.3.8. *The sequence* $0 \to T \to N$ *of R-modules is pure exact if and only if* $N^+ \to T^+$ *admits a section.*

Proof. If $0 \to T \to N$ is pure exact, then $0 \to M \otimes T \to M \otimes N$ is exact for any right R-module M. Hence $(M \otimes N)^+ \to (M \otimes T)^+ \to 0$ is exact and so $\mathrm{Hom}(M, N^+) \to \mathrm{Hom}(M, T^+) \to 0$ is exact. But if $M = T^+$ then this implies that $N^+ \to T^+$ has a section.

Conversely if $N^+ \to T^+$ has a section, then $\mathrm{Hom}(M, N^+) \to \mathrm{Hom}(M, T^+) \to 0$ is exact for all M, and so $(M \otimes N)^+ \to (M \otimes T)^+ \to 0$ is exact too implying that $0 \to M \otimes T \to M \otimes N$ is exact for all M. This means $0 \to T \to N$ is pure exact. □

Proposition 5.3.9. *Every R-module is a pure submodule of a pure injective R-module.*

Proof. By Proposition 5.3.7, M^{++} is pure injective. To show that the canonical map $M \to M^{++}$ is a pure injection we need to show $M^{+++} \to M^+$ admits a section. But the canonical map $M^+ \to (M^+)^{++} = M^{+++}$ is such a section. □

Remark 5.3.10. If R is commutative, the arguments above hold if we replace M^+ by $\mathrm{Hom}_R(M, E)$ where E is an injective cogenerator of R.

Theorem 5.3.11. *Let R be a right coherent ring. Then every pure injective left R-module has a flat cover* $\varphi : F \to M$ *with F and* $\mathrm{Ker}\, \varphi$ *pure injective.*

Proof. Let M be a pure injective left R-module. Then M is a direct summand of M^{++} since M is a pure submodule of M^{++} by Proposition 5.3.9. But M^+ is a right R-module and so M^{++} has a flat precover $F \to M^{++}$ by Proposition 5.3.5. But then the composition $F \to M^{++} \xrightarrow{\pi} M$ where π is the projection map is a flat precover. Hence M has a flat cover by Theorem 5.3.1.

Now let $0 \to M^+ \to E \to C \to 0$ be exact with E injective. Then since M is a direct summand of M^{++}, F and $\mathrm{Ker}\, \varphi$ are direct summands of E^+ and C^+ respectively whenever $F \to M$ is a flat cover of M. Thus F and $\mathrm{Ker}\, \varphi$ are pure injective. □

We now use Theorem 5.3.11 above to show that the class of submodules of flat left R-modules is covering over a right coherent ring. But first we need the following result from set theory.

Lemma 5.3.12. *Let M and N be R-modules. Then there is a cardinal number* $\aleph_\alpha$ *dependent on* $\mathrm{Card}\, N$ *and* $\mathrm{Card}\, R$ *such that for any morphism* $f : N \to M$, *there is a pure submodule S of M such that* $f(N) \subset S$ *and* $\mathrm{Card}\, S \le \aleph_\alpha$.

Proof. A submodule S is pure in M if for every $m \ge 1$ and every finitely generated $T \subset R^m$ any linear map $T \to S$ has an extension $R^m \to M$ if and only if it has an extension $R^m \to S$ (see Exercise 2). If we consider the direct sum $P = \bigoplus_{(T, R^m)} R^m$

summed over the set of pairs (T, R^m) where $m \geq 1$ and where $T \subset R^m$ is finitely generated we have the submodule $U = \bigoplus_{(T,R^m)} T$ of P. We then see that $S \subset M$ is pure if and only if any linear $U \to S$ has an extension $P \to M$ if and only if it has an extension $P \to S$.

We construct a sequence $S_0 \subset S_1 \subset S_2 \subset \cdots$ of submodules of M as follows. We let $S_0 = f(N)$. Having constructed S_n for $n \geq 0$ we consider the set X of all linear $f : U \to S_n$ such that there is a linear $\bar{f} : P \to M$ agreeing with f. For each such f we choose one such $\bar{f}$ and let $\bar{X}$ be the set of these $\bar{f}$. Then let

$$S_{n+1} = S_n + \sum_{\bar{f} \in \bar{X}} \bar{f}(P).$$

We have $S_0 \subset S_1 \subset S_2 \subset \cdots$ and we let $S = \bigcup_{n=0}^{\infty} S_n$.

Each such S_n has the property that if $U \to S_n$ is linear and can be extended to $P \to M$ then it can be extended to $P \to S_{n+1}$. Then we note that if (T, R^m) is one of the original pairs, any linear $T \to S$ can be factored $T \to S_n \to S$ for some $n \geq 0$ since T is finitely generated. Hence by the criterion mentioned above, $S \subset M$ is pure.

But Card $\bar{X} \leq$ Card R and Card $S_0 \leq$ Card N and so $S_1 = S_0 + \sum_{\bar{f} \in \bar{X}} \bar{f}(P)$ has

$$\text{Card } S_1 \leq \text{Card}(S_0 \bigoplus (\bigoplus_{f \in X} \bar{f}(P)))$$

and so Card $S_1 \leq$ Card $N + ($Card $R)($Card $R)$. But then Card $S_1 \leq \aleph_\alpha$ where $\aleph_\alpha = \max($Card N, Card $R)$. In a similar manner, Card $S_2 \leq \aleph_\alpha$, Card $S_3 \leq \aleph_\alpha, \ldots$. But then Card $S =$ Card $\bigcup_{i=0}^{\infty} S_i \leq$ Card $S_0 +$ Card $S_1 + \cdots \leq \aleph_\alpha + \aleph_\alpha + \cdots = \aleph_0 \cdot \aleph_\alpha = \aleph_\alpha$.   □

Lemma 5.3.13. *Let R be a right coherent ring. Then the class of submodules of flat left R-modules is closed under inductive limits.*

Proof. Let $((S_i), (\varphi_{ji}))$ be a well ordered inductive system with each S_i a submodule of a flat left R-module. We need to show that $\lim_{\to} S_i$ is also a submodule of a flat module. For each i, we have $S_i \subset F_i$ for some flat module F_i. But in general, there is no reason why a morphism $S_i \to S_j$ can be extended to a morphism $F_i \to F_j$ and even so, the different morphisms might not be compatible. Hence we need to choose the embeddings $S_i \subset F_i$ in a functorial manner.

But by the lemma above, there is a cardinal number $\aleph_\alpha$ (dependent on Card S_i and on Card R) such that if $f : S_i \to G$ is any morphism with G flat, then there is a pure and hence flat submodule $F \subset G$ with $f(S_i) \subset F$ and where Card $F \leq \aleph_\alpha$. So let X be a set with Card $X = \aleph_\alpha$. For each i, we consider all morphisms $f : S_i \to F$ where F is a flat left R-module with $F \subset X$ (as sets). Let $F_f = F$, $F_i = \prod F_f$ over all such f, and $S_i \to F_i$ be the morphism $x \mapsto (f(x))$. Then F_i is flat since R is right coherent. Furthermore, if $\varphi : S_i \to S_j$ is a morphism, we define a morphism $F_i \to F_j$ as follows: Let $F_j = \prod G_g$ (over morphisms $g : S_j \to G_g$ described above). To define $F_i \to F_j$, we only need to define $\prod F_f \to G_{g'}$ for each g'. But the composition $S_i \to S_j \to \prod G_g \to G_{g'}$ (the last map being the projection map) is one of the morphisms f', that is, $G_g = F_{f'}$ and $S_i \to G_{g'}$ is the morphism f'. So let $\prod F_f \to G_{g'}$ be the projection map corresponding to f'. Then we see that

is commutative and the morphisms $F_i \to F_j$ are functorial in the obvious sense. So we can define an inductive limit $\varinjlim F_i$. But $S_i \to F_i$ is an injection. So $\varinjlim S_i \to \varinjlim F_i$ is also an injection. Thus we are done since $\varinjlim F_i$ is flat.   □

Theorem 5.3.14. *If R is a right coherent ring, then the class of submodules of flat left R-modules is covering.*

Proof. Let M be a left R-module and $\mathcal{F}$ be the class of submodules of flat left R-modules. Let $M \subset E$ with E injective. Then E has a flat cover $\varphi : F \to E$ by Theorem 5.3.11. Now let $S = \varphi^{-1}(E)$. Then by chasing an obvious diagram, we see that $S \to M$ is an $\mathcal{F}$-precover. Hence M has an $\mathcal{F}$-cover by Corollary 5.2.7 and Lemma 5.3.13 above.   □

Remark 5.3.15. If R is a commutative ring and M is an R-module, then $\mathrm{Hom}_R(M, E)$ is pure injective for each injective R-module E as in Proposition 5.3.7 above. Thus $\mathrm{Hom}_R(M, E)$ has a flat cover by Theorem 5.3.11. Furthermore, just like in the proof of Proposition 5.3.5, if R is noetherian, then $\mathrm{Hom}(E(M), E) \to \mathrm{Hom}(M, E)$ is a flat precover of $\mathrm{Hom}(M, E)$ noting that $\mathrm{Hom}(E(M), E)$ is flat by Theorem 3.2.16. In order to apply this to a specific example, we need the following easy results.

Proposition 5.3.16. *Let R be a local ring and $\mathfrak{p} \in \mathrm{Spec}\, R$. If M is an $\hat{R}_\mathfrak{p}$-module and Matlis reflexive, then M has a flat cover as an R-module.*

Proof. Since $M \cong \mathrm{Hom}(M^v, E(k(\mathfrak{p})))$, M has a flat $\hat{R}_\mathfrak{p}$-module cover $F \to M$ by Remark 5.3.15 above. If G is a flat R-module and $G \to M$ is a morphism, then we have a factorization $G \to G \otimes \hat{R}_\mathfrak{p} \to M$. But $G \otimes \hat{R}_\mathfrak{p}$ is flat as an $\hat{R}_\mathfrak{p}$-module and so $G \otimes \hat{R}_\mathfrak{p} \to M$ can be lifted to F. But then we have a factorization $G \to F \to M$. So $F \to M$ is a flat precover as R-modules. Hence M has a flat cover.   □

Lemma 5.3.17. *A flat precover $\varphi : F \to M$ of an R-module M is a cover if and only if $\mathrm{Ker}\,\varphi$ contains no nonzero direct summand of F.*

Proof. We simply note that if $\varphi : F \to M$ is a cover and $\varphi' : F' \to M$ is a flat precover and $\varphi \circ f = \varphi'$, then f is surjective and $\mathrm{Ker}\,f$ is a summand of F' (see Proposition 5.1.2).   □

Lemma 5.3.18. *If R is right coherent, $\prod M_i$ has a flat cover if and only if each left R-module M_i does.*

Proof. If $F_i \to M_i$ is a flat cover for each i, then $\prod F_i$ is flat and $\prod F_i \to \prod M_i$ is a flat precover and so $\prod M_i$ has a flat cover by Theorem 5.3.1. Conversely, if $F \to \prod M_i$ is a flat cover, then for any j, $F \to \prod M_i \to M_j$ is a flat precover and so M_j has a cover. $\qquad\square$

We note that if $F_i \to M_i$ is a flat cover for each i and the index set I is infinite, $\prod F_i \to \prod M_i$ may fail to be a cover and $\oplus F_i \to \oplus M_i$ may fail to be a precover. If I is finite, $\oplus F_i \to \oplus M_i$ is a cover (see Section 5.5).

Example 5.3.19. For a local ring R, we will let $m(R)$ denote its maximal ideal.

If $\mathfrak{p} \in \operatorname{Spec} R$, then the residue field $k(\mathfrak{p})$ of $R_\mathfrak{p}$ is a reflexive $\hat{R}_\mathfrak{p}$-module. Hence it has a flat cover over R by Proposition 5.3.16. Thus for any set X, $k(\mathfrak{p})^X$, and hence its direct summand $k(\mathfrak{p})^{(X)}$, has a flat cover over R by Lemma 5.3.17. But $k(\mathfrak{p})^{(X)} \cong \operatorname{Hom}(k(\mathfrak{p}), E(k(\mathfrak{p})^{(X)}))$ and so by Remark 5.3.15 above a flat precover of $k(\mathfrak{p})^{(X)}$ is $T = \operatorname{Hom}(E(k(\mathfrak{p})), E(k(\mathfrak{p}))^{(X)})$. The R-module T is the completion of a free $R_\mathfrak{p}$-module with base indexed by X by Theorem 3.4.1. The map $T \to k(\mathfrak{p})^{(X)}$ induces an isomorphism $T/m(\hat{R}_\mathfrak{p})T \to k(\mathfrak{p})^{(X)}$. If S is a summand of T that is in $m(\hat{R}_\mathfrak{p})T$, then $S = m(\hat{R}_\mathfrak{p})S$. But this is impossible unless $S = 0$ since T is Hausdorff in the $m(\hat{R}_\mathfrak{p})$-adic topology. So $T \to k(\mathfrak{p})^{(X)}$ is a flat cover by Lemma 5.3.17.

In particular, if k is a field and Card $X = m < \infty$, then $k[[x_1, \ldots, x_n]]^X \to k^X$ is a flat cover over the local ring $k[[x_1, \ldots, x_n]]$. If X is infinite, $k[[x_1, \ldots, x_n]]^{(X)} \to k^{(X)}$ is not even a precover.

Proposition 5.3.20. *If $\varphi : F \to M$ is a flat cover of an R-module M and $F = F_1 \oplus F_2$, $M = M_1 \oplus M_2$ are decompositions compatible with φ (that is, $\varphi(F_i) \subset M_i$), then $F_i \to M_i$ is a flat cover for $i = 1, 2$.*

Proof. We easily see that $F_i \to M_i$ is a flat precover. But Ker f contains no direct summand of F_i by Lemma 5.3.17 since $\varphi : F \to M$ is a flat cover. So $F_i \to M_i$ is a cover, again by the same lemma. $\qquad\square$

Remark 5.3.21. Let T be as in Example 5.3.19 above. Then any decomposition $T = T_1 \oplus T_2$ gives one of $T/m(\hat{R}_\mathfrak{p})T$. So $T_1 \to T_1/m(\hat{R}_\mathfrak{p})T_1$ is a flat cover by the proposition above. But by Example 5.3.19, $T_1/m(\hat{R}_\mathfrak{p})T_1 \cong k(\mathfrak{p})^{(Y)}$ for some subset Y of X. Thus, since covers are unique, T_1 is also the completion of a free $R_\mathfrak{p}$-module whose dimension is the same as that of $T_1/m(\hat{R}_\mathfrak{p})T_1$ over $k(\mathfrak{p})$. In fact, $T_1 \cong \operatorname{Hom}(E(k(\mathfrak{p})), E(k(\mathfrak{p}))^{(Y)})$. This means that a direct summand of the completion of a free module is again such.

Definition 5.3.22. An R-module M is said to be *cotorsion* if $\mathrm{Ext}^1(F, M) = 0$ for all flat R-modules F. This generalizes the definitions of Harrison [107] and Warfield [177] and agrees with that of Fuchs [97] but differs from that of Matlis [144].

    We note that if M is cotorsion, then $\mathrm{Ext}^i(F, M) = 0$ for all flat R-modules F and all $i \geq 1$. For consider an exact sequence $0 \to K \to P_{i-2} \to \cdots \to P_0 \to F \to 0$ with each P_i projective where $i \geq 2$. Then K is flat and so $\mathrm{Ext}^i(F, M) \cong \mathrm{Ext}^1(K, M) = 0$.

    Our aim now is to study flat covers of cotorsion modules. We need the following preliminary results.

Lemma 5.3.23. *Every pure injective R-module is cotorsion.*

Proof. Let M be a pure injective R-module and $0 \to K \to P \to F \to 0$ be a short projective resolution of a flat R-module F. Then $K \subset P$ is pure (see Exercise 1) and so $\mathrm{Ext}^1(F, M) = 0$.   □

Remark 5.3.24. Let R be commutative and noetherian. If M is any R-module and E is an injective R-module, then $\mathrm{Hom}(M, E)$ is cotorsion since it is pure injective (see Remark 5.3.15).

Lemma 5.3.25. *If $\varphi : F \to M$ is a flat cover, then $\mathrm{Ker}\, \varphi$ is cotorsion.*

Proof. Let F' be a flat module and $0 \to K \to P \to F' \to 0$ be a short projective resolution. Let $f : K \to \mathrm{Ker}\, \varphi$ be any morphism. Then let $P \oplus_f F$ be the amalgamated sum of P and F along K (see Section 4.3). Then $F \subset P \oplus_f F$ and $P \oplus_f F$ is flat. Furthermore, φ can be extended to a morphism $\tau : P \oplus_f F \to M$ which maps P to zero. If we complete

we can assume τ induced the identity on F. But then τ gives a map $P \to \mathrm{Ker}\, \varphi$ which extends f. Thus $0 \to \mathrm{Hom}(F', \mathrm{Ker}\, \varphi) \to \mathrm{Hom}(P, \mathrm{Ker}\, \varphi) \to \mathrm{Hom}(K, \mathrm{Ker}\, \varphi) \to 0$ is exact and so $\mathrm{Ext}^1(F', \mathrm{Ker}\, \varphi) = 0$. Hence $\mathrm{Ker}\, \varphi$ is cotorsion.   □

Corollary 5.3.26. *Let $\varphi : F \to M$ be a flat cover. Then F is cotorsion whenever M is.*

Proof. Since $F \to M$ is surjective, we have the exact sequence $\mathrm{Ext}^1(G, \mathrm{Ker}\, \varphi) \to \mathrm{Ext}^1(G, F) \to \mathrm{Ext}^1(G, M)$. So $\mathrm{Ext}^1(G, F) = 0$ whenever G is flat and M is cotorsion.   □

Lemma 5.3.27. *Let R be a commutative noetherian ring. Then an R-module F is flat and cotorsion if and only if it is a direct summand of a module $\mathrm{Hom}(E, E')$ where E, E' are injective R-modules.*

Proof. $\mathrm{Hom}(E, E')$ is cotorsion by Remark 5.3.24 above and flat by Theorem 3.2.16. Thus a direct summand of $\mathrm{Hom}(E, E')$ is such. Conversely, let F be flat and cotorsion. If E is an injective cogenerator, then $F \hookrightarrow F' = \mathrm{Hom}(\mathrm{Hom}(F, E), E)$ is a pure injection by Remark 5.3.10. But $\mathrm{Hom}(F, E)$ is injective by Theorem 3.2.9 and so F' is flat. But then F'/F is flat and thus $\mathrm{Hom}(F', F) \to \mathrm{Hom}(F, F) \to \mathrm{Ext}^1(F'/F, F) = 0$ is exact. Hence F is a direct summand of F'. $\qquad\square$

We are now in a position to characterize flat covers of cotorsion modules.

Theorem 5.3.28. *Let R be a commutative noetherian ring. Then the following are equivalent for an R-module F.*

(1) *F is a flat cover of some cotorsion module.*

(2) *F is flat and cotorsion.*

(3) *$F \cong \prod T_{\mathfrak{p}}$ (over $\mathfrak{p} \in \mathrm{Spec}\, R$) where $T_{\mathfrak{p}}$ is the completion of a free $R_{\mathfrak{p}}$-module.*

Furthermore, the decomposition in (3) is uniquely determined by the dimension of the free modules.

Proof. (1) $\Rightarrow$ (2) follows from Corollary 5.3.26.

(2) $\Rightarrow$ (3). Suppose F is flat and cotorsion. Then by Lemma 5.3.27, F is a direct summand of $\mathrm{Hom}(E, E')$ for some injective modules E, E'. Thus a flat cotorsion module is a direct summand of a product $\prod T_{\mathfrak{p}}$, over $\mathfrak{p} \in \mathrm{Spec}\, R$ by Theorem 3.3.13. Now let $T = \prod T_{\mathfrak{p}}$. If $\mathfrak{q} \in \mathrm{Spec}\, R$, then $\mathfrak{q}T_{\mathfrak{p}} = T_{\mathfrak{p}}$ for $\mathfrak{q} \not\subset \mathfrak{p}$ and $\cap \mathfrak{q}^n T_{\mathfrak{p}} = 0$ for $\mathfrak{q} \subset \mathfrak{p}$. Hence $T' = \cap \mathfrak{q}^n T_{\mathfrak{p}} = T_{\mathfrak{p}}$ for $\mathfrak{p} \not\subset \mathfrak{q}$. Thus $H = T/T' \cong \prod T_{\mathfrak{p}}$ for $\mathfrak{p} \subset \mathfrak{q}$. But then $\cap_{\mathfrak{p} \subsetneq \mathfrak{q}}(\cap_n \mathfrak{p}^n H) \cong T_{\mathfrak{q}}$. This means that given T, we can "recover" each $T_{\mathfrak{p}}$. The procedure commutes with direct sums, that is, if $T = T_1 \oplus T_2$, then we get an induced decomposed $T_{\mathfrak{p}} = (T_{\mathfrak{p}})_1 \oplus (T_{\mathfrak{p}})_2$ for each $\mathfrak{p} \in \mathrm{Spec}\, R$ so that $T_1 = \prod (T_{\mathfrak{p}})_1$. But as we noted in Remark 5.3.21, $(T_{\mathfrak{p}})_1$ is again the completion of a free module over $R_{\mathfrak{p}}$. Thus F, being a direct summand, is also such a product. This also proves the last statement of the theorem.

(3) $\Rightarrow$ (1). By Example 5.3.19, we have that $T_{\mathfrak{p}} \to T_{\mathfrak{p}}/m(\hat{R}_{\mathfrak{p}})T_{\mathfrak{p}}$ is a flat cover for each $\mathfrak{p} \in \mathrm{Spec}\, R$. Therefore, $\prod T_{\mathfrak{p}} \to \prod T_{\mathfrak{p}}/m(\hat{R}_{\mathfrak{p}})T_{\mathfrak{p}}$ is a flat precover (see proof of Lemma 5.3.18) with kernel $K = \prod m(\hat{R}_{\mathfrak{p}})T_{\mathfrak{p}}$. Let $F = \prod T_{\mathfrak{p}}$ and suppose $S \subset K$ is a direct summand of F. If $\mathfrak{q}$ is a prime ideal such that $S \subset \mathfrak{q}F$, then $S = \mathfrak{q}S$ which implies that the projection of S on $T_{\mathfrak{q}}$ is zero since $T_{\mathfrak{q}}$ is Hausdorff in the $\mathfrak{q}$-adic topology. Thus $S = 0$ if $S \subset \mathfrak{q}F$ for all $\mathfrak{q}$. If not, let $\mathfrak{q}$ be maximal with $S \not\subset \mathfrak{q}F$. But if $\mathfrak{q} \not\subset \mathfrak{p}$, then $\mathfrak{q}T_{\mathfrak{p}} = T_{\mathfrak{p}}$. If $\mathfrak{q} \subsetneq \mathfrak{p}$, then as above, the projection of S on $T_{\mathfrak{p}}$ is zero. But then $S \subset \mathfrak{q}F$ since $\mathfrak{q}T_{\mathfrak{q}} = m(\hat{R}_{\mathfrak{q}})T_{\mathfrak{q}}$, a contradiction. Thus $S = 0$ and K has no nonzero direct summands of F. Hence $\prod T_{\mathfrak{p}} \to \prod T_{\mathfrak{p}}/m(\hat{R}_{\mathfrak{p}})T_{\mathfrak{p}}$ is a flat cover by Lemma 5.3.17. But $T_{\mathfrak{p}}/m(\hat{R}_{\mathfrak{p}})T_{\mathfrak{p}} \cong k(\mathfrak{p})^{(X)}$ for some set X. So $T_{\mathfrak{p}}/m(\hat{R}_{\mathfrak{p}})T_{\mathfrak{p}} \cong \mathrm{Hom}(k(\mathfrak{p}), E(k(\mathfrak{p}))^{(X)})$ and is cotorsion by Remark 5.3.24. Hence $\prod T_{\mathfrak{p}}/m(\hat{R}_{\mathfrak{p}})T_{\mathfrak{p}}$ is cotorsion. $\qquad\square$

Remark 5.3.29. Theorem 5.3.28 above generalizes Harrison's characterization of co-torsion groups in [107] as products $G = \prod T_p$ over primes p where T_p is a direct summand of $\hat{\mathbb{Z}}_p^X$ for some set X, and in which case G is uniquely determined by the dimensions of $T_p/{}_pT_p$ over $\mathbb{Z}/(p)$.

Exercises

1. Let $R = \mathbb{Z}$. Prove that A is a pure subgroup of B if and only if $nA = nB \cap A$ for all $n \in \mathbb{Z}$.

2. Let S be a submodule of an R-module N. Prove that if N/S is a flat R-module, then $S \subset N$ is pure and that the converse holds if N is flat.

3. Let S be a submodule of a flat R-module N. Prove that N/S is flat if and only if $IS = IN \cap S$ for all finitely generated right ideals I of R. Conclude that if N is flat, then S is a pure submodule of N if and only if $IS = IN \cap S$ for all finitely generated right ideals I of R.

4. Let S be a submodule of an R-module N. Prove that the following are equivalent.

   a) S is a pure submodule of N.
   b) $\text{Hom}(A, N) \to \text{Hom}(A, N/S) \to 0$ is exact for all finitely presented R-modules A.
   c) Any linear map $f : T \to S$ where $T \subset R^m$, $m \geq 1$, is finitely generated has an extension $R^m \to N$ if and only if it has an extension $R^m \to S$.

5. Prove the assertion that $k[[x]]^{(\mathbb{N})} \to k^{(\mathbb{N})}$ is not a flat cover over $k[x]$ (here k is a field).

6. Prove that the class $\mathcal{F}$ of submodules of flat R-modules is precovering if and only if every injective R-module has a flat cover.

7. If $S \subset R$ is multiplicative and $F \to M$ is a flat cover of $S^{-1}R$-modules, show that $F \to M$ is also a flat cover of R-modules.

8. If $S \subset R$ is multiplicative and M is an $S^{-1}R$-module, let $F \to M$ be a flat cover of M as an R-module. Prove that $S^{-1}F \cong F$ and that $F \to M$ is a flat cover as $S^{-1}R$-modules.

9. If $S \subset R$ is multiplicative and M is a cotorsion $S^{-1}R$-module, show that M is a cotorsion R-module.
   Hint: If $0 \to M \to U \to F \to 0$ is a short exact sequence of R-modules with F flat, apply the functor $S^{-1}R \otimes_R -$.

10. Let M' be a submodule of an R-module M and suppose that M/M' is flat. Argue that if M has a flat cover, so does M'.

11. (Bass [23, Proposition 2.7]) Let $J = \text{rad}(R)$. Prove that if P is a nonzero projective R-module, then $P \neq JP$.

12. (Bass [23]) An ideal I of R is said to be left (right) T-*nilpotent* if for every sequence $a_1, a_2, \ldots \in I$ there exists an integer n such that $a_1a_2 \ldots a_n = 0$ ($a_n \ldots a_2a_1 = 0$). Prove that if R is left perfect, then $\text{rad}(R)$ is left T-nilpotent.

13. (Nakayama lemma) Prove that if an ideal I of R is left T-nilpotent, then $M \neq IM$ for any nonzero left R-module M.

Hint: If $IM = M \neq 0$, let $a_1 \in I$, $x_1 \in M$ be such that $a_1 x_1 \neq 0$. Use $a_1 x_1$ to construct a sequence $a_1, a_2, \ldots \in I$ for which there is no n such that $a_1 a_2 \ldots a_n = 0$.

14. The *radical* of an R-module M, denoted $\text{rad}(M)$, is the intersection of all maximal submodules of M. Prove that $\text{rad}(M)$ is the sum of all superfluous submodules of M.

15. Let $J = \text{rad}(R)$ and P be a projective R-module. Prove that $\text{rad}(P) = JP$. Conclude that JP contains all superfluous submodules of P.

5.4   Injective covers

We start by showing that the class of injective R-modules is precovering precisely when R is a noetherian ring.

Theorem 5.4.1. *Let $\mathcal{E}$ be the class of injective left R-modules. Then the following are equivalent*

(1) *R is left noetherian.*

(2) *$\mathcal{E}$ is precovering.*

(3) *$\mathcal{E}$ is covering.*

Proof. $(1) \Rightarrow (2)$. If R is left noetherian, then every injective R-module is the direct sum of indecomposable injective left R-modules. Each such module is the injective envelope of a cyclic R-module. Hence, we can find a set of representatives of such (see Gabriel [99]). So there is a family $(E_i)_{i \in I}$ of indecomposable injective left R-modules such that every injective left R-module is the direct sum of copies of the various E_i. If $X_i = \text{Hom}(E_i, M)$ and $E_i^{(X_i)} \to M$ is the evaluation map $(\varphi_f)_{f \in X_i} \mapsto \sum_{f \in X_i} f(\varphi_f)$, then any map $E_i \overset{f}{\to} M$ can be factored through $E_i^{(X_i)} \to M$ by mapping E_i onto the f component of $E_i^{(X_i)}$. Hence $\oplus E_i^{(X_i)} \to M$ is an injective precover of M.

$(2) \Rightarrow (1)$. It suffices to show that every direct sum of injective left R-modules is injective (by Theorem 3.1.17). Let $(E_i)_{i \in I}$ be a family of injective left R-modules, and $E \to \oplus E_i$ be an injective precover. Then there are factorizations $E_j \to E \to \oplus E_i$ where $E_j \to \oplus E_i$ is the canonical injection for each j. These give rise to a map $\oplus E_i \to E$ with the composition $\oplus E_i \to E \to \oplus E_i$ the identity. Hence $\oplus E_i$ is isomorphic to a summand of E and so is injective.

$(2) \Leftrightarrow (3)$. If $\mathcal{E}$ is precovering, then R is left noetherian by the above and so $\mathcal{E}$ is closed under well ordered inductive limits. Hence $\mathcal{E}$ is covering by Corollary 5.2.7. The converse is trivial. $\square$

The proof of the existence of $\mathcal{F}$-covers does not give us the structure of these $\mathcal{F}$-covers. As we saw in Example 5.3.19, one has to appeal to other results in order to get the structure of covers. We now describe the structure of injective covers of some modules.

Lemma 5.4.2. *Let R be a commutative noetherian ring. If M is a finitely generated R-module and $\mathfrak{p}$ is a prime ideal of R with $\mathrm{Hom}(E(R/\mathfrak{p}), M) \neq 0$, then $\mathfrak{p}$ is a maximal ideal.*

Proof. Let $\varphi \in \mathrm{Hom}(E(R/\mathfrak{p}), M)$, $\varphi \neq 0$. By replacing M with $\mathrm{Im}\,\varphi$, we may assume φ is surjective and by going modulo a maximal submodule, we may assume M is simple. Hence we may assume $M = R/\mathfrak{m}$ for some maximal ideal $\mathfrak{m}$.

If $\mathfrak{p} \not\subset \mathfrak{m}$, then let $r \in \mathfrak{p}$ and $r \notin \mathfrak{m}$. Then for each $z \in R/\mathfrak{m}$, $z \neq 0$, $rz \neq 0$. But for each $x \in E(R/\mathfrak{p})$, $r^n x = 0$ for some $n \geq 1$. So for $\varphi(x) = z$ we would have $r^n z = 0$, a contradiction. Hence $\mathfrak{p} \subset \mathfrak{m}$. If $\mathfrak{p} \neq \mathfrak{m}$, let $r \in \mathfrak{m}$, $r \notin \mathfrak{p}$. Then $rE(R/\mathfrak{p}) = E(R/\mathfrak{p})$ and $r^n E(R/\mathfrak{p}) = 0$ for some n. So there is no surjective map $\varphi : E(R/\mathfrak{p}) \to R/\mathfrak{m}$. Hence $\mathfrak{p} = \mathfrak{m}$. □

Theorem 5.4.3. *Let R be a commutative noetherian ring. Then the injective cover of a finitely generated R-module is a direct sum of finitely many copies of $E(R/\mathfrak{m})$ for finitely many maximal ideals $\mathfrak{m}$.*

Proof. Let M be a finitely generated R-module and $E \to M$ be the injective cover. Then $E \cong \bigoplus_{\mathfrak{p} \in \mathrm{Spec}\,R} E(R/\mathfrak{p})$. So E is the direct sum of copies of $E(R/\mathfrak{m})$ over maximal ideals $\mathfrak{m}$ by Lemma 5.4.2 above. But if $\mathrm{Hom}(E(R/\mathfrak{m}), M) \neq 0$, then $\mathfrak{m} \in \mathrm{Ass}(M)$. Furthermore, $\mathrm{Ass}(M)$ has only finitely many primes since M is finitely generated. So E is a direct sum of copies of $E(R/\mathfrak{m})$ for finitely many maximal ideals $\mathfrak{m}$. We now show that there are only finitely many copies of $E(R/\mathfrak{m})$ for each such maximal ideal.

We first recall that $E(M) \cong \oplus E(R/\mathfrak{p})^{(X_\mathfrak{p})}$ over $\mathfrak{p} \in \mathrm{Spec}\,R$ and so

$$\mathrm{Hom}(E(R/\mathfrak{m}), E(M)) \cong \mathrm{Hom}(E(R/\mathfrak{m}), E(R/\mathfrak{m})^n) \cong (\hat{R}_\mathfrak{m})^n$$

where $\mathrm{Card}\,X_\mathfrak{m} = n < \infty$ by Corollary 3.3.11. Hence $\mathrm{Hom}(E(R/\mathfrak{m}), M)$ is a finitely generated $\hat{R}_\mathfrak{m}$-module. Let $\varphi_1, \ldots, \varphi_2 \in \mathrm{Hom}(E(R/\mathfrak{m}), M)$ be generators as an $\hat{R}_\mathfrak{m}$-module. So if $\varphi \in \mathrm{Hom}(E(R/\mathfrak{m}), M)$, then $\varphi = \sum_{i=1}^s \varphi_i \circ \sigma_i$ for some $\sigma_1, \ldots, \sigma_s \in \mathrm{Hom}(E(R/\mathfrak{m}), E(R/\mathfrak{m})) \cong \hat{R}_\mathfrak{m}$. This means we can complete

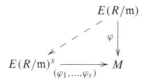

to a commutative diagram.

Now let $E = E_1 \oplus E_2$ where E_1 is the direct sum of all copies of $E(R/\mathfrak{m})$ in some decomposition of E into indecomposable injective R-modules. Then by the above, we can complete

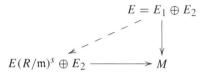

to a commutative diagram. But then

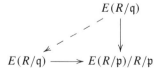

can be completed to a commutative diagram. So $E(R/\mathfrak{m})^s \oplus E_2$ is an injective precover and hence E is a direct summand of $E(R/\mathfrak{m})^s \oplus E_2$. Thus E has finitely many copies of $E(R/\mathfrak{m})$.   □

Similar arguments give the following result.

Theorem 5.4.4. *Let R be a commutative noetherian ring. Then the following statements hold.*

(1) *If $\mathfrak{p} \in \operatorname{Spec} R$, then the injective cover of $E(R/\mathfrak{p})/R/\mathfrak{p}$ is the direct sum of copies of $E(R/\mathfrak{q})$ for prime ideals $\mathfrak{q}$ such that $\mathfrak{q} \supset \mathfrak{p}$.*

(2) *If $\mathfrak{p} \notin \operatorname{Ass}(R)$, then the injective cover of $E(R/\mathfrak{p})/R/\mathfrak{p}$ is the sum of copies of $E(R/\mathfrak{p})$.*

(3) *If $\mathfrak{m}$ is a maximal ideal of R, then the injective cover of $E(R/\mathfrak{m})/R/\mathfrak{m}$ is a direct sum of finitely many copies of $E(R/\mathfrak{m})$.*

Proof. (1) Let $\mathfrak{p}, \mathfrak{q} \in \operatorname{Spec} R$. If $\mathfrak{p} \not\subset \mathfrak{q}$, let $r \in \mathfrak{p}$, $r \notin \mathfrak{q}$. Then multiplication by r on $E(R/\mathfrak{q})$ is an isomorphism and is zero on $R/\mathfrak{p}$. So $\operatorname{Ext}^1(E(R/\mathfrak{q}), R/\mathfrak{p}) = 0$. This means that the diagram

$$
\begin{array}{ccc}
 & & E(R/\mathfrak{q}) \\
 & \nearrow & \downarrow \\
E(R/\mathfrak{q}) & \longrightarrow & E(R/\mathfrak{p})/R/\mathfrak{p}
\end{array}
$$

can be completed when the horizontal map is the natural map and the vertical one is arbitrary. Hence to construct an injective precover of $E(R/\mathfrak{p})/R/\mathfrak{p}$, it suffices to find an injective R-module E and a map $E \to E(R/\mathfrak{p})/R/\mathfrak{p}$ such that

$$
\operatorname{Hom}(E(R/\mathfrak{q}), E) \to \operatorname{Hom}(E(R/\mathfrak{q}), E(R/\mathfrak{p})/R/\mathfrak{p}) \to 0
$$

is exact whenever $\mathfrak{q} \supset \mathfrak{p}$, for then

$$\operatorname{Hom}(E(R/\mathfrak{q}), E \oplus E(R/\mathfrak{p})) \rightarrow \operatorname{Hom}(E(R/\mathfrak{q}), E(R/\mathfrak{q})/R/\mathfrak{q}) \rightarrow 0$$

is exact for any $\mathfrak{q} \in \operatorname{Spec} R$.

$E \oplus E(R/\mathfrak{p}) \rightarrow E(R/\mathfrak{p})/R/\mathfrak{p}$ would be an injective precover since every injective R-module is a direct sum of copies of $E(R/\mathfrak{p})$. So we let E be the direct sum of sufficiently many copies of the R-modules $E(R/\mathfrak{q})$ when $\mathfrak{q} \supset \mathfrak{p}$. Then clearly, there is a map $E \rightarrow E(R/\mathfrak{p})/R/\mathfrak{p}$ satisfying the above. But the injective cover is a direct summand of a precover by Proposition 5.1.2 and thus is also a direct sum of such copies.

(2) If $\mathfrak{p} \notin \operatorname{Ass}(R)$, let $K \subset E(R/\mathfrak{p})$ be the field of fractions of $R/\mathfrak{p}$. If $\mathfrak{p} \subsetneq \mathfrak{q}$ and $E(R/\mathfrak{q}) \rightarrow E(R/\mathfrak{p})/R/\mathfrak{p}$ is a map, consider the composition $E(R/\mathfrak{q}) \rightarrow E(R/\mathfrak{p})/R/\mathfrak{p} \rightarrow E(R/\mathfrak{p})/K$. But $K = (R/\mathfrak{p})_\mathfrak{p}$, $E(R/\mathfrak{p}) = E(R/\mathfrak{p})_\mathfrak{p}$, $E(R/\mathfrak{p})/K = (E(R/\mathfrak{p})/K)_\mathfrak{p}$, $E(R/\mathfrak{q})_\mathfrak{p} = 0$. Thus $E(R/\mathfrak{q}) \rightarrow E(R/\mathfrak{p})/K$ is the zero map. Hence the original map $E(R/\mathfrak{q}) \rightarrow E(R/\mathfrak{p})/R/\mathfrak{p}$ maps $E(R/\mathfrak{q})$ onto $K/(R/\mathfrak{p})$. But $\mathfrak{p} \notin \operatorname{Ass}(R)$ and so there exists an $r \in \mathfrak{p}$ which is not a zero divisor. Thus $E(R/\mathfrak{p})$ is divisible by r. But then any map $E(R/\mathfrak{q}) \rightarrow K/(R/\mathfrak{p})$ is zero since multiplication by r on $K/(R/\mathfrak{p})$ is zero. Consequently, any map $E(R/\mathfrak{q}) \rightarrow E(R/\mathfrak{p})/R/\mathfrak{p}$ is zero for any prime ideal $\mathfrak{q} \supsetneq \mathfrak{p}$. Hence the injective cover of $E(R/\mathfrak{p})/R/\mathfrak{p}$ is a direct sum of copies of $E(R/\mathfrak{p})$.

(3) If $\mathfrak{m}$ is a maximal ideal, then the injective cover of $E(R/\mathfrak{m})/R/\mathfrak{m}$ consists of copies of $E(R/\mathfrak{m})$ by (1) above. We now show that there are finitely many such copies. We recall that $E(R/\mathfrak{p})^v \cong \hat{R}_\mathfrak{p}$, the completion of R at $\mathfrak{p}$, and $(E(R/\mathfrak{p})/R/\mathfrak{p})^v$ is isomorphic to the maximal ideal $m(\hat{R}_\mathfrak{p})$ of $\hat{R}_\mathfrak{p}$. Hence by duality, to find an injective cover of the desired form, we only need to argue that there is an $n \geq 1$ and a map $m(\hat{R}_\mathfrak{p}) \rightarrow \hat{R}_\mathfrak{p}^n$ such that $\operatorname{Hom}_{\hat{R}_\mathfrak{p}}(\hat{R}_\mathfrak{p}^n, \hat{R}_\mathfrak{p}) \rightarrow \operatorname{Hom}_{\hat{R}_\mathfrak{p}}(m(\hat{R}_\mathfrak{p}), \hat{R}_\mathfrak{p}) \rightarrow 0$ is exact. But $\hat{R}_\mathfrak{p}$ is noetherian and so let $\sigma_1, \sigma_2, \ldots, \sigma_n$ be a set of generators of $\operatorname{Hom}_{\hat{R}_\mathfrak{p}}(m(\hat{R}_\mathfrak{p}), \hat{R}_\mathfrak{p})$. Then the map $m(\hat{R}_\mathfrak{p}) \xrightarrow{(\sigma_1, \ldots, \sigma_n)} \hat{R}_\mathfrak{p}^n$ satisfies the requirements. $\qquad\square$

In particular, as we will show below, if $\mathfrak{m}$ is a maximal ideal of a commutative noetherian ring R such that $\operatorname{depth}_\mathfrak{m} R \geq 2$ (see Definition 9.2.5), then the injective cover of $E(R/\mathfrak{m})/R/\mathfrak{m}$ is simply $E(R/\mathfrak{m})$, and the injective cover has at least two copies of $E(R/\mathfrak{m})$ in the case $\operatorname{depth}_\mathfrak{m} R = 1$.

We start with the following general result.

Theorem 5.4.5. *Let R be a commutative noetherian ring, M be a finitely generated R-module with n generators, and G be an injective R-module. If $\operatorname{Ext}^1(M, R) = 0$, then the natural map*

$$G^n \rightarrow G^n/\operatorname{Hom}_R(M, G)$$

is an injective precover. The converse holds if G is an injective cogenerator of R.

Proof. Let E be an injective R-module. Then $\mathrm{Hom}_R(\mathrm{Ext}^1_R(M, R), E) = 0$ implies $\mathrm{Tor}^R_1(\mathrm{Hom}_R(R, E), M) = 0$ and so $\mathrm{Hom}(\mathrm{Tor}_1(E, M), G) = 0$. But then $\mathrm{Ext}^1(E, \mathrm{Hom}(M, G)) = 0$ and so the result follows.

For the converse, note that $G^n \to G^n/\mathrm{Hom}(M, G)$ is an injective precover means that $\mathrm{Ext}^1(E, \mathrm{Hom}(M, G)) = 0$ for all injectives E. If G is an injective cogenerator, this in turn implies $\mathrm{Ext}^1(M, R) = 0$ and so we are done. $\square$

In particular, we have the following result.

Theorem 5.4.6. *Let R be a commutative noetherian ring. Then the following are equivalent for a maximal ideal $\mathfrak{m}$ of R.*

(1) $\mathrm{Ext}^1(R/\mathfrak{m}, R) = 0$.

(2) *The natural map $E(R/\mathfrak{m}) \to E(R/\mathfrak{m})/R/\mathfrak{m}$ is an injective cover.*

(3) $\mathrm{Hom}_{\hat{R}_\mathfrak{m}}(\mathfrak{m}(\hat{R}_\mathfrak{m}), \hat{R}_\mathfrak{m})$ *is cyclic.*

Proof. (1) $\Rightarrow$ (2). $E(R/\mathfrak{m}) \to E(R/\mathfrak{m})/R/\mathfrak{m}$ is an injective precover by the preceding theorem. But the injective cover of $E(R/\mathfrak{m})/R/\mathfrak{m}$ is then a direct summand of $E(R/\mathfrak{m})$ and $E(R/\mathfrak{m})$ is indecomposable. Hence $E(R/\mathfrak{m})$ is the injective cover.

(2) $\Rightarrow$ (1). If $\mathfrak{m}'$ is a maximal ideal distinct from $\mathfrak{m}$, let $r \in \mathfrak{m}$, $r \notin \mathfrak{m}'$. Then r is an isomorphism on $E(R/\mathfrak{m}')$ and is zero on $R/\mathfrak{m}$. So $\mathrm{Hom}(R/\mathfrak{m}, E(R/\mathfrak{m}')) = 0$. Therefore $\mathrm{Ext}^1(E, \mathrm{Hom}(R/\mathfrak{m}, \bigoplus_{\mathfrak{p} \in \mathrm{mSpec}\, R} E(R/\mathfrak{p}))) = \mathrm{Ext}^1(E, R/\mathfrak{m})$. Hence if $\mathrm{Ext}^1(E, R/\mathfrak{m}) = 0$ for all injective modules E, then $\mathrm{Ext}^1(R/\mathfrak{m}, R) = 0$ since $\bigoplus_{\mathfrak{p} \in \mathrm{mSpec}\, R} E(R/\mathfrak{p})$ is an injective cogenerator of R.

(2) $\Rightarrow$ (3). By Matlis duality, if $E(R/\mathfrak{m})^n \to E(R/\mathfrak{m})/R/\mathfrak{m}$ is an injective cover, then n is the least number of generators of $\mathrm{Hom}_{\hat{R}_\mathfrak{m}}(\mathfrak{m}(\hat{R}_\mathfrak{m}), \hat{R}_\mathfrak{m})$.

(3) $\Rightarrow$ (2). If $\mathrm{Hom}_{\hat{R}_\mathfrak{m}}(\mathfrak{m}(\hat{R}_\mathfrak{m}), \hat{R}_\mathfrak{m})$ is cyclic, then as in the proof of Theorem 5.4.4, $E(R/\mathfrak{m}) \to E(R/\mathfrak{m})/R/\mathfrak{m}$ is an injective precover and so is an injective cover. $\square$

Corollary 5.4.7. $\mathrm{depth}_\mathfrak{m} R \geq 2$ *if and only if the natural map $E(R/\mathfrak{m})/R/\mathfrak{m}$ is an injective cover and $\mathrm{depth}_\mathfrak{m} R > 0$.*

Proof. We simply note that $\mathrm{depth}_\mathfrak{m} R = \inf\{i : \mathrm{Ext}^i(R/\mathfrak{m}, R) \neq 0\}$ (see Remark 9.2.9). $\square$

Corollary 5.4.8. $\mathrm{depth}_\mathfrak{m} R = 1$ *if and only if the injective cover of $E(R/\mathfrak{m})/R/\mathfrak{m}$ has at least two copies of $E(R/\mathfrak{m})$ and $\mathrm{depth}_\mathfrak{m} R > 0$.*

Proof. If $\mathrm{depth}_\mathfrak{m} R = 1$, then $E(R/\mathfrak{m}) \to E(R/\mathfrak{m})/R/\mathfrak{m}$ is not an injective cover by Corollary 5.4.7 above. But the injective cover of $E(R/\mathfrak{m})/R/\mathfrak{m}$ is a (finite) sum of copies of $E(R/\mathfrak{m})$ by Theorem 5.4.4. So the cover has at least two copies of $E(R/\mathfrak{m})$. The converse follows from the above corollary. $\square$

Example 5.4.9. Let $R = k[x_1, \ldots, x_n]$ where k is a field. Then $E(k) \to E(k)/k$ is an injective cover if and only if $n \geq 2$ by the above.

Similarly, if R is an n-dimensional local ring with residue field k, then $E(k) \to E(k)/k$ is an injective cover whenever $n \geq 2$.

Exercises

1. Let R be noetherian. Show that every injective cover is an injection if and only if every quotient module of an injective module is injective.

2. Let R be noetherian. Argue that the following are equivalent:

   a) every injective cover is surjective.

   b) every projective module is injective.

   c) R (as a left R-module) is injective.

3. Suppose that $\varphi_i : E_i \to M_i$ is a family of injective covers (for $i \in I$). Show that $\prod \varphi_i : \prod_{i \in I} E_i \to \prod_{i \in I} M_i$ is an injective precover and that this precover is a cover if and only if $\prod \mathrm{Ker}(\varphi_i)$ has no nonzero injective submodules.

4. If $R \to S$ is a ring homomorphism and if $E \to M$ is an injective precover of left R-modules, argue that $\mathrm{Hom}_R(S, E) \to \mathrm{Hom}_R(S, M)$ is an injective precover of left S-modules when S is a flat right R-module.

5. Let C be a left R-module such that $\mathrm{Ext}^1_R(E, C) = 0$ for all injective left R-module E. Prove that $E(C) \to E(C)/C$ is an injective precover and that it is an injective cover if and only if C has no nonzero injective submodules.

5.5 Direct sums and T-nilpotency

A direct sum of covers may fail to be a precover, or it may be a precover and still not be a cover. Namely, if for each $i \in I$, $\psi_i : C_i \to M_i$ is a cover, it may be possible to complete

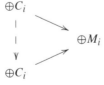

by a map which is not an isomorphism. The next proposition shows when this property of covers is preserved by countable sums. The necessary condition is a sort of T-nilpotency, which when applied to projective covers gives the usual T-nilpotency of the radical of a left perfect ring.

In the situations above, there is no loss in generality in assuming each M_i is a quotient of C_i, say C_i/S_i, and that $\psi_i : C_i \to C_i/S_i$ is the canonical surjection.

Proposition 5.5.1. *If for each* $i = 1, 2, 3, \ldots,$ $S_i \subset C_i$ *is a submodule such that*

can be completed only by automorphisms of C_i, *then the same is true for*

(∗)

if and only if for each sequence $1 \leq k_1 < k_2 < \cdots$ *of positive integers and maps* $f_n : C_{k_n} \to C_{k_{n+1}}$ *with* $\text{Im}(f_n) \subset S_{k_{n+1}}$ *and each* $x \in C_{k_1}$, *there is an* $m \geq 1$ *such that*

$$f_m \circ f_{m-1} \circ \cdots \circ f_1(x) = 0.$$

Proof. We argue the necessity and let the k_n's and f_n's be as stated. Define $\varphi : \oplus C_i \to \oplus C_i$ so that if $i \neq k_n$ for any n then $\varphi_{|C_i}$ is the identity map, and so that $\varphi_{|C_{k_n}}$ agrees with the map $C_{k_n} \to C_{k_n} \oplus C_{k_{n+1}}$ which maps y to $(y, -f_n(y))$. Then φ completes the diagram (∗). Furthermore one checks that if $x \in C_{k_1}$ and if x is in the image of φ, say $\varphi((x_i)) = x$, then $x_i = 0$ for $i \neq k_n$ for all n. Also x_{k_1} must be x and by induction we see that $x_{k_n} = f_{n-1} \circ \cdots \circ f_1(x)$ for $n > 1$. But $(x_i) \in \oplus C_i$ and so $x_{k_n} = 0$ for n sufficiently large.

For the converse, suppose φ completes (∗). Use the matrix notation $\varphi = (\varphi_{ij})$ with $\varphi_{ij} : C_j \to C_i$. Note that for each i, φ_i completes the diagram

and so is an isomorphism, and that for $i \neq j$, φ_{ij} has its image in S_i. Also (φ_{ij}) is locally column finite in the sense that for any j and any $x \in C_j$, $\varphi_{ij}(x) = 0$ except for a finite number of i. Furthermore, any collection of φ_{ij}'s satisfying these conditions gives a φ completing (∗). To argue that φ is an isomorphism we only need to find a ψ which is an isomorphism completing (∗) and such that $\psi \circ \varphi$ or $\varphi \circ \psi$ is an isomorphism. The argument proceeds by showing that φ has a triangular decomposition, that is, it is the product of an upper and lower triangular matrix (corresponding to an automorphism of $\oplus C_i$). If φ is upper triangular, then since its diagonal elements are automorphisms of the C_i, it is a standard argument that it is invertible, and its inverse satisfies the conditions above. This guarantees that it corresponds to an automorphisms of $\oplus C_i$ of the desired type (that is, makes (*) commutative).

So we construct an upper triangular matrix ψ of the desired form so that $\varphi \circ \psi$ is lower triangular. We define ψ as an infinite product $\psi_1 \circ \psi_2 \circ \psi_3 \circ \cdots$. We start by letting

$$
\psi_1 = \begin{bmatrix}
-\varphi_{11}^{-1} & -\varphi_{11}^{-1}\varphi_{12} & \varphi_{11}^{-1}\varphi_{13} & \bullet & \bullet & \bullet \\
0 & \mathrm{id} & 0 & \bullet & \bullet & \bullet \\
0 & 0 & \mathrm{id} & \bullet & \bullet & \bullet \\
\bullet & \bullet & \bullet & \bullet & \bullet & \bullet \\
\bullet & \bullet & \bullet & \bullet & \bullet & \bullet
\end{bmatrix}.
$$

Then $\varphi \circ \psi_1$ has the form

$$
\begin{bmatrix}
\mathrm{id} & 0 & 0 & \bullet & \bullet & \bullet \\
\varphi'_{21} & \varphi'_{22} & \varphi'_{23} & \bullet & \bullet & \bullet \\
\bullet & \bullet & \bullet & \bullet & \bullet & \bullet
\end{bmatrix}.
$$

Hence define ψ_2 as we defined ψ_1 but using the second row of $\varphi \circ \psi_1$, and then similarly defining $\psi_3, \ldots$. It is easy to see that ij entry of $\psi_1 \circ \cdots \circ \psi_n$ is constant for n sufficiently large and so the infinite product converges; and if ψ is this product it gives the desired automorphism of $\oplus C_i$ so that $\varphi \circ \psi$ is lower triangular.

Now assume that φ is lower triangular and that it has the identities id_{C_i} on the diagonal. So $\varphi = \mathrm{id}_{\oplus C_i} - K$ where K is strictly lower triangular and $-K$ has φ_{ij} for its ij entry when $i > j$. Since K is strictly lower triangular, the sum

$$
\varphi' = \mathrm{id} + K + K^2 + \cdots + K^n + \cdots
$$

is well-defined. As a matrix, φ' is φ^{-1}. However we need to argue that it is locally column finite. To argue this, given i, let $x \in C_i$. Then the ji entry of $\varphi'(x)$ for $j > i$ is

$$
\sum \varphi_{jk_s} \circ \varphi_{k_s k_{s-1}} \circ \cdots \circ \varphi_{k_2 k_1} \circ \varphi_{k_1 i}(x)
$$

with the summation taken over all finite sequences $j > k_s > \cdots > k_1 > i$. If for an infinite number of j with $j > i$ the sum is nonzero, an easy application of the König graph theorem (see [129]) allows us to choose $i < k_1 < k_2 < k_3 < \cdots$ with

$$
\varphi_{k_n k_{n-1}} \circ \cdots \circ \varphi_{k_1 i}(x) \neq 0.
$$

Setting $f_n = \varphi_{k_n k_{n-1}}$ for $n \geq 2$ and $f_1 = \varphi_{k_1 i}$ contradicts our hypothesis. □

Corollary 5.5.2. *For a left perfect ring R, if $P_i \to M_i$ are projective covers for each $i = 1, 2, 3, \ldots$, then $\oplus P_i \to \oplus M_i$ is a projective cover.*

Remark 5.5.3. If we apply this result to a countable sum of copies of the projective cover $R \to R/J$ where J is the Jacobson radical of R we get that $\oplus R \to \oplus R/J$ is a projective cover. If $r_1, r_2, \ldots$ is a sequence of elements of J, then using Proposition 5.5.1, let $k_n = n$ for all n and let $f_n : R \to R$ be multiplication by r_n. Then the condition $f_m \circ \cdots \circ f_1(x) = 0$ for $x = 1 \in R$ translates to $r_m \ldots r_1 = 0$. Thus J is right T-nilpotent.

We note that the finite counterpart of Proposition 5.5.1 holds since we can choose $c_k = 0$ for k sufficiently large. So we have the following result.

Proposition 5.5.4. If $C_i \to M_i$, $i = 1, 2, \ldots, n$ are $\mathcal{F}$-covers and $\oplus_{i=1}^n C_i \in \mathcal{F}$, then $\oplus_{i=1}^n C_i \to \oplus_{i=1}^n M_i$ is an $\mathcal{F}$-cover.

Exercises

1. Let $R = \mathbb{Z}_p$ where p is a prime. Then $\mathbb{Z}_p \to \mathbb{Z}/(p)$ is a projective cover. Show that $\mathbb{Z}_p^{(\mathbb{N})} \to \mathbb{Z}/(p)^{(\mathbb{N})}$ is a projective precover but is not a projective cover.

2. Again with $R = \mathbb{Z}_p$ with p a prime, we have that $\hat{\mathbb{Z}}_p \to \mathbb{Z}/(p)$ is a flat cover. Show that $\hat{\mathbb{Z}}_p^{(\mathbb{N})} \to \mathbb{Z}/(p)^{(\mathbb{N})}$ is not a flat precover.

3. Give a direct argument showing that if $\varphi_i : C_i \to M_i$ are $\mathcal{F}$-covers for $i = 1, 2$ and if $C_1 \oplus C_2 \in \mathcal{F}$, then $\varphi = \varphi_1 \oplus \varphi_2 : C_1 \oplus C_2 \to M_1 \oplus M_2$ is an $\mathcal{F}$-cover. Hint: Let $f : C_1 \oplus C_2 \to C_1 \oplus C_2$ be such that $\varphi \circ f = \varphi$. Using matrix notation let $f = \begin{pmatrix} f_{11} & f_{12} \\ f_{21} & f_{22} \end{pmatrix}$ with $f_{ij} : C_j \to C_i$. Then the equality $\varphi \circ f = \varphi$ becomes

$$\begin{pmatrix} \varphi_1 & 0 \\ 0 & \varphi_2 \end{pmatrix} \begin{pmatrix} f_{11} & f_{12} \\ f_{21} & f_{22} \end{pmatrix} = \begin{pmatrix} \varphi_1 & 0 \\ 0 & \varphi_2 \end{pmatrix}.$$

If $g : C_1 \oplus C_2 \to C_1 \oplus C_2$ is an automorphism such that $\varphi \circ g = \varphi$, then it suffices to argue that $f \circ g$ (or $g \circ f$) is an automorphism. Use this observation to show that we can assume $f_{11} = \mathrm{id}_{M_1}$, $f_{22} = \mathrm{id}_{M_2}$ and also that $f_{12} = 0$. So then conclude that $f = \begin{pmatrix} f_{11} & f_{12} \\ f_{21} & f_{22} \end{pmatrix} = \begin{pmatrix} \mathrm{id}_{M_1} & 0 \\ f_{21} & \mathrm{id}_{M_2} \end{pmatrix}$ is an automorphism of $C_1 \oplus C_2$.

Chapter 6

Envelopes

Having introduced covers in the previous two chapters, we now define the dual notion of envelopes. The main aim of this chapter is to prove the existence of envelopes. We first do this by dualizing the results for covers in Sections 5.1 and 5.2. However these results do not prove the existence of injective and pure injective envelopes. In Section 6.6, we will use Maranda's notion of an injective structure to obtain a result that proves the existence of these envelopes.

6.1 $\mathcal{F}$-preenvelopes and envelopes

Definition 6.1.1. Let R be a ring and $\mathcal{F}$ be a class of R-modules, by an $\mathcal{F}$-*preenvelope* of an R-module M we mean a morphism $\varphi : M \to F$ where $F \in \mathcal{F}$ such that for any morphism $f : M \to F'$ with $F' \in \mathcal{F}$, there is a $g : F \to F'$ such that $g \circ \varphi = f$. If furthermore, when $F' = F$ and $f = \varphi$ the only such g are automorphisms of F, then $\varphi : M \to F$ is called an $\mathcal{F}$-*envelope* of M. So if envelopes exist, they are unique up to isomorphism. It is easy to check that if $\mathcal{F}$ is the class of injective modules, then we get the usual injective envelopes. Similarly, we get pure injective envelopes if $\mathcal{F}$ is the class of pure injectives. We note that if the class $\mathcal{F}$ contains injectives, then $\mathcal{F}$-preenvelopes are monomorphisms. If every R-module has an $\mathcal{F}$-(pre)envelope, we say that $\mathcal{F}$ is (*pre*)*enveloping*. For example, we know that the class of injective R-modules is enveloping.

We start with the following result which is dual to Proposition 5.1.2.

Proposition 6.1.2. *Let M be an R-module, then the $\mathcal{F}$-envelope of M, if it is exists, is a direct summand of any $\mathcal{F}$-preenvelope of M.*

Exercises

1. Recall that every subgroup of a free abelian group is free. Deduce that for an abelian group A the following are equivalent:

   a) A has a free preenvelope.
   b) A has a free envelope.

c) A has a direct sum decomposition $A = A_1 \oplus A_2$ with A_2 free and such that $\operatorname{Hom}(A_1, \mathbb{Z}) = 0$.

2.   a) Let $A = \prod_{i=0}^{\infty} Z_i$ with $Z_i = \mathbb{Z}$ for $i \geq 0$. Show that if $B = \bigoplus_{i=0}^{\infty} Z_i$, then A/B has an uncountable divisible subgroup.

   b) Use (a) to deduce that A is not free by arguing that if A were free there would be a decomposition $A = A_1 \oplus A_2$ with $B \subset A_1$ and A_1 countable and so contradict (a).

   c) Argue that A does not have a free preenvelope by noting that if A' is a direct summand of A, then $\operatorname{Hom}(A', \mathbb{Z}) = 0$ implies $A' = 0$.

3. Let $\mathcal{F}$ be a class of modules closed under taking submodules, products and injective envelopes. Then:

   a) Argue that an R-module M has an $\mathcal{F}$-preenvelope if and only if it has an $\mathcal{F}$-envelope and that any $\mathcal{F}$-envelope is surjective.

   b) Show that every M has an $\mathcal{F}$-envelope if and only if $\mathcal{F}$ is closed under products.

   c) If every M has an $\mathcal{F}$-envelope and $K = \operatorname{Ker} \varphi$ for an envelope $\varphi : M \to F$, argue that $\operatorname{Hom}(K, G) = 0$ for all $G \in \mathcal{F}$.

6.2   Existence of preenvelopes

The following is dual to Proposition 5.2.2. We provide a proof for completeness.

Proposition 6.2.1. *If $\mathcal{F}$ is a class of R-modules that is closed under products, then an R-module M has an $\mathcal{F}$-preenvelope if and only if there is a cardinal number $\aleph_\alpha$ such that any morphism $M \to F$ with $F \in \mathcal{F}$ has a factorization $M \to G \to F$ with $G \in \mathcal{F}$ and $\operatorname{Card} G \leq \aleph_\alpha$.*

Proof. If M has an $\mathcal{F}$-preenvelope $M \to F$, then let $\aleph_\alpha = \operatorname{Card} F$. Conversely, there is a set X with cardinality $\aleph_\alpha$ such that any morphism $M \to F$ with $F \in \mathcal{F}$ has a factorization $M \to G \to F$ for some $G \in \mathcal{F}$ with $G \subset X$ (as sets). Now let $(\varphi_i)_{i \in I}$ give all such morphisms $\varphi_i : M \to G_i$ with $G_i \subset X$ (as sets). So any morphism $M \to F$ has a factorization $M \to G_j \to F$ for some j. But then $M \to \prod_I G_i$ is an $\mathcal{F}$-preenvelope. $\square$

Corollary 6.2.2. *Let $\operatorname{Card} M = \aleph_\beta$. Suppose there is an infinite cardinal $\aleph_\alpha$ such that if $F \in \mathcal{F}$ and $S \subset F$ is a submodule with $\operatorname{Card} S \leq \aleph_\beta$, there is a submodule G of F containing S with $G \in \mathcal{F}$ and $\operatorname{Card} G \leq \aleph_\alpha$. Then M has an $\mathcal{F}$-preenvelope.*

Proof. Let $M \xrightarrow{f} F$ be a morphism with $F \in \mathcal{F}$. Then $\operatorname{Card} f(M) \leq \aleph_\beta$. So there is a submodule G of F containing $f(M)$ with $G \in \mathcal{F}$ and $\operatorname{Card} G \leq \aleph_\alpha$. But then we have a factorization $M \to G \to F$. Hence M has an $\mathcal{F}$-preenvelope by Proposition 6.2.1.$\square$

We now provide an example to illustrate how the above arguments apply. But before we do that we need the following.

Definition 6.2.3. An R-module N is said to be *absolutely pure* if it is a pure submodule in every R-module that contains it, or equivalently, if it is pure in every injective R-module that contains N. But then N is absolutely pure if and only if it is pure in $E(N)$ and if and only if $\mathrm{Hom}(M, E(N)) \to \mathrm{Hom}(M, E(N)/N) \to 0$ is exact for all finitely presented R-modules M by Definition 5.3.6. That is, if and only if $\mathrm{Ext}^1(M, N) = 0$ for all finitely presented R-modules M. As a result, absolutely pure modules are also known as *FP-injective* modules.

We now recall from Remark 3.2.3 that an R-module M is finitely presented if and only if there is an exact sequence $0 \to A \to R^n \to M \to 0$ with A finitely generated. Hence $\mathrm{Ext}^1(M, N) = 0$ for all finitely presented R-modules M if and only if $\mathrm{Ext}^1(R^n/A, N) = 0$ for every $n > 0$ and finitely generated $A \subseteq R^n$. So an R-module N is absolutely pure if and only if $\mathrm{Hom}(R^n, N) \to \mathrm{Hom}(A, N) \to 0$ is exact for every $n > 0$ and finitely generated $A \subseteq R^n$.

Proposition 6.2.4. *Let R be a ring. Then the class of absolutely pure R-modules is preenveloping.*

Proof. Let M be an R-module and let $\mathrm{Card}\, M \leq \aleph_\beta$. We need to choose an $\aleph_\alpha$ and construct the G in Corollary 6.2.2. But N is absolutely pure if for each $n \geq 1$ and finitely generated $A \subseteq R^n$ every morphism $A \to N$ has an extension $R^n \to N$. Now let $f : M \to N$ be a morphism. Then setting $B_0 = f(M)$, we consider submodules $B_1 \subset N$ with $B_0 \subset B_1$ such that every morphism $A \to B_0$ has an extension $R^n \to B_1$. We can choose B_1 with a bound on its size depending only on $\aleph_\beta$ and $\mathrm{Card}\, R$. The argument for this claim is similar to that used in the proof of Lemma 5.3.12. Similarly, we can construct $B_1 \subset B_2 \subset \cdots$. Now let $B = \bigcup_{i=1}^\infty B_i$. Then B is absolutely pure and we can find an $\aleph_\alpha$ so that B can always be constructed with $\mathrm{Card}\, B \leq \aleph_\alpha$. Finally, it is easy to check from the definition above that the product of absolutely pure modules is again absolutely pure. $\qquad\square$

Exercises

1. Let R be an integral domain. Modify the proof of Proposition 6.2.4 to show that the class of divisible modules is preenveloping.

2. Let $\mathcal{S}$ be any set of modules. Let $\mathcal{F}$ be the class of modules F such that $\mathrm{Ext}^i(S, F) = 0$ for all $i \geq 1$ and $S \in \mathcal{S}$.

   a) Prove that every module M has an $\mathcal{F}$-preenvelope which is an injection.

   b) If $R = \mathbb{Z}$ and $\mathcal{S}$ consists of all $\mathbb{Z}/(n)$, $n \geq 1$, argue that $\mathcal{F}$ consists of all the divisible $\mathbb{Z}$- modules.

3. Prove that the direct sum of any family of absolutely pure modules is absolutely pure. Deduce that the class of absolutely pure modules coincides with the class of injective modules if and only if the ring is left noetherian.

6.3   Existence of envelopes

The following theorem is dual to Theorem 5.2.3.

Theorem 6.3.1. *Let R be a ring, $\mathcal{F}$ be a class of R-modules that is closed under summands, and M be an R-module. Suppose for any well ordered projective system $((G_\alpha), (\varphi_{\alpha\beta}))$, every map $M \to \varprojlim G_\alpha$ can be factored through some $M \to G$ with $G \in \mathcal{F}$. If M has an $\mathcal{F}$-preenvelope, then it has an $\mathcal{F}$-envelope.*

Proof. We again break up the proof into three lemmas just as in Theorem 5.2.3.   □

Lemma 6.3.2. *For each $F \in \mathcal{F}$ and morphism $M \to F$, there exists an $\mathcal{F}$-preenvelope $M \to G$ and a morphism f in the factorization $M \to G \xrightarrow{f} F$ such that for any morphism g in the factorization $M \to H \xrightarrow{g} G$ where $M \to H$ is also an $\mathcal{F}$-preenvelope, $\operatorname{Im}(f \circ g) = \operatorname{Im} f$.*

Proof. Dual to the proof of Lemma 5.2.4.   □

We now modify the argument of Lemma 5.2.5 in order to prove the following dual result.

Lemma 6.3.3. *There exists an $\mathcal{F}$-preenvelope $M \to G$ such that in any factorization $M \to G \to F$ with $M \to F$ an $\mathcal{F}$-preenvelope, $G \to F$ is surjective.*

Proof. Let α be an infinite ordinal. Then using the lemma above, we can construct a projective system

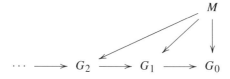

for $\beta < \alpha$ where for each $\beta + 1 < \alpha$, $M \to G_\beta$ is an $\mathcal{F}$-preenvelope and where

$$M \longrightarrow G_{\beta+1}$$
$$\downarrow \quad \swarrow$$
$$G_\beta$$

has the property guaranteed by the lemma. For each $\beta + 1 < \alpha$, let $U_\beta \subset G_\beta$ be this image. Let $M \to F$ be any $\mathcal{F}$-preenvelope of M. Then we can complete

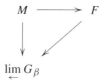

to a commutative diagram by assumption. The map $F \to G_\beta$ can be factored $F \to U_\beta \to G_\beta$ when $\beta + 1 < \alpha$ where $F \to U_\beta$ is surjective by the above. So if $\beta < v < v + 1 < \alpha$ and we consider the commutative diagram

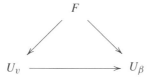

we see that $U_v \to U_\beta$ is surjective. If it is never an isomorphism, then $\mathrm{Ker}(F \to U_v) \subsetneq \mathrm{Ker}(F \to U_\beta)$. But then this implies $\mathrm{Card}\, F \geq \mathrm{Card}(\alpha)$ for any such α, which is impossible. Hence $U_\alpha \to U_\beta$ is an isomorphism for some such $\beta < v < v + 1 \leq \alpha$. This isomorphism is a composition $U_v \to G_{\beta+1} \to U_\beta$ and so it is a retract of $G_{\beta+1}$. Thus $U_v \in \mathcal{F}$ since $\mathcal{F}$ is closed under summands.

We note that $M \to U_v$ is then an $\mathcal{F}$-preenvelope. Furthermore, let $M \to F \to U_v$ be a factorization with $M \to F$ an $\mathcal{F}$-preenvelope. Then consider the diagram

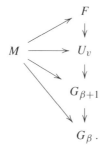

By the assumption on the morphism $G_{\beta+1} \to G_\beta$, the image of $F \to G_\beta$ is U_β. But $U_v \to U_\beta$ is an isomorphism and so $F \to U_v$ is a surjection. Hence if we set $G = U_v$, then $M \to G$ is the desired $\mathcal{F}$-preenvelope.   □

Lemma 6.3.4. *If $\psi : M \to G$ is an $\mathcal{F}$-preenvelope having the property of Lemma 6.3.3, then $\psi : M \to G$ is an $\mathcal{F}$-envelope.*

Proof. Dual to the proof of Lemma 5.2.6.   □

Corollary 6.3.5. *Let $\mathcal{F}$ be a class of R-modules that is closed under summands and well ordered projective limits, and M be an R-module. If M has an $\mathcal{F}$-preenvelope, then it has an $\mathcal{F}$-envelope.*

Proof. Dual to the proof of Corollary 5.2.7.   □

Exercises

1. If M_1 and M_2 satisfy the initial hypothesis of Theorem 6.3.1 and if $\mathcal{F}$ is closed under finite sums, argue that $M_1 \oplus M_2$ also satisfies this hypothesis.

2. Use Theorem 6.3.1 to prove the existence of injective envelopes (note that $M \to E$ with E injective is an injective preenvelope if and only if $M \to E$ is an injection).

3. If $\mathcal{F}$ is enveloping and every R-module M satisfies the initial hypothesis of Theorem 6.3.1, prove that $\mathcal{F}$ is closed under well ordered projective limits.

6.4   Direct sums of envelopes

Using a similar argument to the proof of Proposition 5.5.1, we get the following dual result.

Proposition 6.4.1. *If for each $i = 1, 2, 3, \ldots$, $S_i \subset E_i$ is a submodule such that*

can only be completed by automorphisms of E_i, then the same is true for

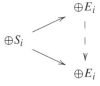

if and only if for each sequence $1 \le k_1 < k_2 < \cdots$ and maps $f_n : E_{k_n} \to E_{k_{n+1}}$ with $f_n(S_{k_n}) = 0$ and for each $x \in E_{k_1}$, there is an $m \ge 1$ such that

$$f_m \circ f_{m-1} \circ \cdots \circ f_1(x) = 0.$$

Corollary 6.4.2. *If $M_i \to E_i$, $i = 1, 2, \ldots, n$ are $\mathcal{F}$-envelopes and $\oplus_{i=1}^{n} E_i \in \mathcal{F}$, then $\oplus_{i=1}^{n} M_i \to \oplus_{i=1}^{n} E_i$ is an $\mathcal{F}$-envelope.*

Proof. We simply set $E_k = 0$ for $k > n$.   □

Corollary 6.4.3. *If for any indexed set I, $S_i \subset E_i$ is a submodule such that*

can be completed only by automorphism of E_i, then

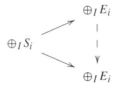

can be completed only by an injection.

Proof. If φ is completes the above, then for any finite subset J of I, consider the commutative diagram

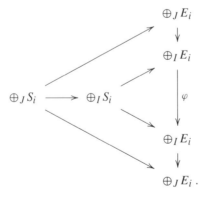

The vertical morphism is an automorphism by Corollary 6.4.2 above. But since this is true for any finite subset J of I, φ is an injection.   □

Corollary 6.4.4. *If for each $i \in I$, $M_i \to E_i$ is an $\mathcal{F}$-envelope and $\oplus M_i$ has an $\mathcal{F}$-envelope, then $\oplus_I M_i \to \oplus_I F_i$ is an $\mathcal{F}$-envelope.*

Proof. Assume $\oplus_I M_i \to G$ is an $\mathcal{F}$-envelope and choose the obvious diagrams.   □

Remark 6.4.5. If $M \subset E$ is an injective envelope of M over a commutative noetherian ring R, then $\oplus M \to \oplus E$ is an injective envelope. Suppose $rM = 0$ for some $r \in R$, then letting $f_n : E \to E$ be multiplication by r for each n, we get that for each $x \in E$, $r^m x = 0$ for some $m \geq 1$ by Proposition 6.4.1. Hence if $I \subset R$ is an ideal and $IM = 0$, then we easily get the familiar result that $I^m x = 0$ for some $m \geq 1$.

Exercises

1. Using Proposition 6.4.1, argue that if $M \subset E$ is an injective envelope over a left noetherian ring and if $f : E \to E$ is linear with $f(M) = 0$, then f is "*locally nilpotent*" on E, that is, for any $x \in E$, $f^n(x) = 0$ for some $n \geq 1$.

2. If f is as in the previous problem, show that we can define a linear map $h :$ $E \to E$ represented by the infinite series $1 - f + f^2 - f^3 + \cdots$. Then if $g = 1 + f$, argue that $h = g^{-1}$.

3. Use the ideas from the previous two problems to argue that if R is left noetherian, M is an essential R-submodule of N, and $g : N \to N$ is such that $g(x) = x$ for all $x \in M$, then g is an automorphism of N.

4. If we take the same hypothesis as in Problem 3 except that we only assume $g(M) = M$ and that $x \mapsto g(x)$ is an automorphism of M, show that g is an automorphism of N.

5. Use the same hypothesis as in Problem 3. Let $f : N \to N$ be linear such that $f(M) \subset M$ and assume that f is locally nilpotent on M. Prove that f is locally nilpotent on N.
   Hint: Define the map $\sigma : N \oplus N \oplus \cdots \to N \oplus N \oplus \ldots$ by $(x_0, x_1, x_2, \ldots) \to$ $(x_0, x_1 - f(x_0), x_2 - f(x_1), \ldots)$. Argue that σ maps $M \oplus M \oplus \ldots$ isomorphically onto itself. Now appeal to Problem 4 to conclude that σ is an automorphism of $N \oplus N \oplus \ldots$. So now use the fact that if $y \in N$, $(y, 0, 0, \ldots)$ is in the image of σ.

6.5   Flat envelopes

In this section, we consider conditions under which flat envelopes and preenvelopes exist. We start with the following result.

Proposition 6.5.1. *A ring R is right coherent if and only if the class of flat left R-modules is preenveloping.*

Proof. Let M be an R-module and let $\text{Card } M \leq \aleph_\beta$. Then by Lemma 5.3.12; there is an infinite cardinal $\aleph_\alpha$ such that if F is a flat module and S is a submodule of F with $\text{Card } S \leq \aleph_\beta$, there is a pure, hence flat, submodule G of F with $S \subset G$ and $\text{Card } G \leq \aleph_\alpha$. Thus M has a flat preenvelope by Corollary 6.2.2 noting that the class of left flat modules is closed under products when the ring is right coherent.

Conversely, let $(F_i)_{i \in I}$ be a family of flat left R-modules. If $\prod F_i$ has a flat preenvelope, then $\prod F_i$ is flat using an argument dual to proof of $2 \Rightarrow 1$ of Theorem 5.4.1.□

Proposition 6.5.2. *If R is right coherent and every projective limit of flat left R-modules is flat, then the class of flat left R-modules is enveloping.*

Proof. By the proposition above we know that the class of flat modules is preenveloping. But then the result follows from Corollary 6.3.5. □

The following results provide examples of modules that do not have flat envelopes.

Proposition 6.5.3. *If $M \to F$ is a flat envelope and M is finitely presented, then F is finitely generated and projective.*

Proof. Since F is flat and M is finitely presented, the map $M \to F$ can be factored through a finitely generated projective P. If $M \to P \to F$ is such a factorization, then we have a commutative diagram

where $F \to P \to F$ is an automorphism, and thus the result follows. □

Corollary 6.5.4. *Let R be a domain, M be a finitely presented R-module and $M \to F$ be a flat envelope. If the sum of countably many copies of M has a flat envelope, then the rank of M equals the rank of F.*

Proof. Let $M \overset{\varphi}{\to} F$ be a flat envelope. Then F is finitely generated and projective by the proposition above. So if rank $M <$ rank F, then $F/\varphi(M)$ has a rank 1 torsion free quotient, say $(F/\varphi(M))/(F'/\varphi(M)) \cong F/F'$. If $x \in F$, $x \neq 0$, there is an injection $F/F' \to Rx$. If $x \in F$, $x \notin F'$, let $f : F \to F$ be the composition $F \to F/F' \to Rx \to F$. Then $f(x) = rx$ with $r \neq 0$ and $f(\varphi(M)) = 0$. Now set $f_i = f$ in Proposition 6.4.1 to get that $f \circ \cdots \circ f(x) = 0$ and so $r^n x = 0$ which is impossible. □

Theorem 6.5.5. *Let R be a domain. Then the class of flat R-modules is enveloping if and only if R is coherent and every projective limit of any projective system of flat R-modules is flat.*

Proof. If the class of flat R-modules is enveloping, then R is coherent by Proposition 6.5.1. Thus the product of flat modules is flat. So it suffices to show that the intersection of flat submodules of a flat module is also flat. But note that if we have an inductive limit $\varinjlim M_i$ of finitely presented modules and if $M_i \to F_i$ is a flat envelope for each i, then by Corollary 6.5.4 above, the rank of M_i is equal to the rank of F_i. Hence for any flat (hence torsion free) module F, the diagram

can be completed uniquely. This implies that we can form $\varinjlim F_i$ and that $\varinjlim M_i \to \varinjlim F_i$ is a flat envelope such that

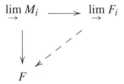

can be completed uniquely for any flat module F.

But R is a coherent domain. So every submodule S of a flat module F is the direct union of finitely generated, hence finitely presented, submodules. This means that the flat envelope of S, say $S \to G$, will have the unique morphism φ in the factorization $S \to G \xrightarrow{\varphi} F$ for each flat module F. Now let $S = \cap F_i$ where (F_i) is some collection of flat submodules of F. Then we can complete the diagram

by a unique injection for each j. So the image of G in any F_j is in $\cap F_i$ and so S is the image of G and hence is flat. The converse follows from Proposition 6.5.2.   □

Exercises

1. Prove that if M is finitely presented and F is flat, then any map $M \to F$ has a factorization $M \to F$ with P finitely generated projective.

2. Let M be a left R-module such that the *algebraic dual* $M^* = \mathrm{Hom}_R(M, R)$ is a finitely generated right R-module. Let $P \to M^* \to 0$ be exact where P is a finitely generated projective module. Argue that $M \to M^{**} \to P^*$ is a flat preenvelope where $M \to M^{**}$ is the canonical linear map.

3. Let $R = k[[x, y]]$ where k is a field. Use Problem 2 to show that $(x, y) \subset k[[x, y]]$ is a flat envelope.

4.   a) Let $M \to F$ be a flat preenvelope over R. Argue that $M[x] \to F[x]$ is a flat preenvelope over $R[x]$.

   b) If $M \to F$ is a flat envelope and if $M[x]$ has a flat envelope over $R[x]$, show that in fact $M[x] \to F[x]$ is such a flat envelope.

5. For a ring R show that every left R-module has a surjective flat envelope if and only if R is right coherent and of weak global dimension at most one.

6.6   Existence of envelopes for injective structures

Theorem 6.3.1 does not prove the existence of injective and pure injective envelopes. In this section, we modify the arguments in Chapter 5 to show that envelopes exist for a wide range of classes that include the classes of injectives and of pure injectives.

We will need the following result.

Lemma 6.6.1. *Let $\mathcal{F}$ be a class of left R-modules that is closed under summands, and M be a left R-module. Suppose M has an $\mathcal{F}$-preenvelope and if $((F_\alpha), (\varphi_{\beta\alpha}))$ is any well ordered inductive system of $\mathcal{F}$-preenvelopes of M, then for some $\mathcal{F}$-preenvelope $M \to F$ there is a factorization $M \to \lim_{\to} F_\alpha \to F$. Then M has an $\mathcal{F}$-envelope.*

Proof. We recall that in the arguments for Lemmas 5.2.4, 5.2.5 and 5.2.6 in Chapter 5, we assumed that if $((C_\alpha), (\varphi_{\beta\alpha}))$ is any inductive system of $\mathcal{F}$-precovers of M, then for some $C \to M$ with $C \in \mathcal{F}$, there is a factorization $\lim_{\to} C_\alpha \to C \to M$ where $\lim_{\to} C_\alpha \to M$ is $\lim_{\to} (C_\alpha \to M)$. Hence the condition on the inductive system $((F_\alpha), (\varphi_{\beta\alpha}))$ of $\mathcal{F}$-preenvelopes is exactly what is needed to carry through all the arguments in these three lemmas to show that M has an $\mathcal{F}$-envelope. $\square$

Definition 6.6.2. A pair $(\mathcal{A}, \mathcal{F})$, where $\mathcal{A}$ is a class of morphisms between R-modules and $\mathcal{F}$ is a class of left R-modules, is called an *injective structure* on the category of left R-modules if

1) $F \in \mathcal{F}$ if and only if $\mathrm{Hom}(N, F) \to \mathrm{Hom}(M, F) \to 0$ is exact for all $M \to N \in \mathcal{A}$.

2) $M \to N \in \mathcal{A}$ if and only if $\mathrm{Hom}(N, F) \to \mathrm{Hom}(M, F) \to 0$ is exact for all $F \in \mathcal{F}$.

3) Every left R-module M has an $\mathcal{F}$-preenvelope $M \to F$.

Definition 6.6.3. If $\mathcal{G}$ is a class of right R-modules, then we say that the pair $(\mathcal{A}, \mathcal{F})$ is *determined* by $\mathcal{G}$ if $M \to N \in \mathcal{A}$ if and only if $0 \to G \otimes M \to G \otimes N$ is exact for all $G \in \mathcal{G}$.

Theorem 6.6.4. *The following statements hold:*

(1) *If an injective structure $(\mathcal{A}, \mathcal{F})$ on the category of left R-modules is determined by a class $\mathcal{G}$ of right R-modules, then every left R-module has an $\mathcal{F}$-envelope.*

(2) *If $\mathcal{G}$ is a set of right R-modules, then there is a unique injective structure $(\mathcal{A}, \mathcal{F})$ determined by $\mathcal{G}$. In this case, $\mathcal{F}$ consists of all F which are isomorphic to a direct summand of products of copies of the left R-modules G^+ for $G \in \mathcal{G}$.*

Proof. (1) Let $((F_\alpha), (\varphi_{\beta\alpha}))$ be an inductive system of $\mathcal{F}$-preenvelopes of M. Then $M \to F_\alpha \in \mathcal{A}$ for each α and so $0 \to G \otimes M \to G \otimes F_\alpha$ is exact for all $G \in \mathcal{G}$. So $M \to \varinjlim F_\alpha \in \mathcal{A}$. But then $\mathrm{Hom}(\varinjlim F_\alpha, F) \to \mathrm{Hom}(M, F) \to 0$ is exact for all $F \in \mathcal{F}$. So if $M \to F$ is an $\mathcal{F}$-preenvelope, we have a factorization $M \to \varinjlim F_\alpha \to F$. Hence M has an $\mathcal{F}$-envelope by Lemma 6.6.1 above.

(2) Now suppose $\mathcal{G}$ is a set of right of R-modules. Let $\mathcal{A}$ be the class of all $M \to N$ such that $0 \to G \otimes M \to G \otimes N$ is exact for all $G \in \mathcal{G}$, and $\mathcal{F}$ consist of all left R-modules F which are isomorphic to direct summands of products of copies of G^+ for G in $\mathcal{G}$. We easily see that $(\mathcal{A}, \mathcal{F})$ is determined by $\mathcal{G}$ and that it is the only one determined by $\mathcal{G}$ if $(\mathcal{A}, \mathcal{F})$ is an injective structure.

We now show that $(\mathcal{A}, \mathcal{F})$ is an injective structure. Let $G \in \mathcal{G}$. Since $0 \to G \otimes M \to G \otimes N$ is exact for $M \to N \in \mathcal{A}$, we get that $\mathrm{Hom}(N, G^+) \to \mathrm{Hom}(M, G^+) \to 0$ is exact. Hence by the choice of $\mathcal{F}$, we see that $\mathrm{Hom}(N, F) \to \mathrm{Hom}(M, F) \to 0$ is exact for all $F \in \mathcal{F}$. Since for $G \in \mathcal{G}, 0 \to G \otimes M \to G \otimes N$ is exact if and only if $\mathrm{Hom}(N, G^+) \to \mathrm{Hom}(M, G^+) \to 0$ is exact, we see that $M \to N \in \mathcal{A}$ if and only if $\mathrm{Hom}(N, F) \to \mathrm{Hom}(M, F) \to 0$ is exact for all $F \in \mathcal{F}$.

Now let M be a left R-module. Since $\mathcal{G}$ is a set, we can easily construct an $F \in \mathcal{F}$ and a morphism $M \to F$ such that for any $G \in \mathcal{G}$ and morphism $M \to G^+$,

can be completed to a commutative diagram. But then for any $F' \in \mathcal{F}, \mathrm{Hom}(F, F') \to \mathrm{Hom}(M, F') \to 0$ is exact. So since $F \in \mathcal{F}, M \to F$ is an $\mathcal{F}$-preenvelope. Thus every left R-module has an $\mathcal{F}$-preenvelope.

Suppose $\mathrm{Hom}(N, L) \to \mathrm{Hom}(M, L) \to 0$ is exact for all $M \to N \in \mathcal{A}$. Let $L \to F$ be an $\mathcal{F}$-preenvelope by the above (so $L \to F \in \mathcal{A}$). Thus $\mathrm{Hom}(F, L) \to \mathrm{Hom}(L, L) \to 0$ is exact. So L is a direct summand of F and so $L \in \mathcal{F}$. Conversely, if $L \in \mathcal{F}$, then L is a direct summand of G^+ and so $\mathrm{Hom}(N, F) \to \mathrm{Hom}(M, F) \to 0$ is exact by the above. Hence $(\mathcal{A}, \mathcal{F})$ is an injective structure. $\square$

Example 6.6.5.

1. If $\mathcal{G} = \{R\}$, then $\mathcal{A}$ is the class of all injections and we get the usual injective envelopes. Alternatively, note that if $\mathcal{F}$ is the class of injective modules, then given an inductive system $((F_\alpha), (\varphi_{\beta\alpha}))$ of $\mathcal{F}$-preenvelopes of a module M, then we have a factorization

$$M \to \varinjlim F_\alpha \to E$$

   for any map $M \to E$ with $E \in \mathcal{F}$ since the direct limit of injections is an injection. Hence $\mathcal{F}$ is enveloping by Lemma 6.6.1.

2. If $\mathcal{G} = \{R/I : I$ is a finitely generated right ideal of $R\}$, then $\mathcal{A}$ consists of all pure injections and we get the pure injective envelopes by Theorem 6.6.4 above noting that every left R-module has a pure injective preenvelope by Proposition 5.3.9.

Proposition 6.6.6. *If R is right coherent, then every left R-module has a pure injective flat envelope.*

Proof. There is a set $\mathcal{G}$ of absolutely pure right R-modules such that every absolutely pure right R-module A is isomorphic to a direct limit of modules G_i with $G_i \in \mathcal{G}$ for each i. For there is a cardinal $\aleph_\alpha$ such that if $S \subset A$ is finitely generated and A is absolutely pure, then there is an absolutely pure submodule B of A with $S \subset B$ and Card $B \le \aleph_\alpha$. Then choosing a set X with Card $X = \aleph_\alpha$, let $\mathcal{G}$ be all absolutely pure right R-modules G with $G \in X$ (as a set).

We now let $(\mathcal{A}, \mathcal{F})$ be the injective structure determined by $\mathcal{G}$. Then $M \to N \in \mathcal{A}$ if and only if $0 \to G \otimes M \to G \otimes N$ is exact for all $G \in \mathcal{G}$. Hence every left R-module has an $\mathcal{F}$-envelope by Theorem 6.6.4. Now, we only need argue that $\mathcal{F}$ consists of all pure injective flat modules. But if $F \in \mathcal{F}$, then F is a direct summand of G^+ for some $G \in \mathcal{G}$ by Theorem 6.6.4. But R is right coherent. So G^+ is flat since G is an absolutely pure right R-module. Hence each $F \in \mathcal{F}$ is flat. But G^+ is also pure injective. So each F in $\mathcal{F}$ is pure injective and flat.

Conversely, suppose F is pure injective and flat. Then F^+ is injective and so absolutely pure. Thus $0 \to F^+ \otimes M \to F^+ \otimes N$ is exact for all $M \to N \in \mathcal{A}$, or equivalently $\operatorname{Hom}(N, F^{++}) \to \operatorname{Hom}(M, F^{++}) \to 0$ is exact for all such $M \to N$. But F is a direct summand of F^{++} since F is pure injective. So $\operatorname{Hom}(N, F) \to \operatorname{Hom}(M, F) \to 0$ is also exact for all $M \to N \in \mathcal{A}$. Hence $F \in \mathcal{F}$. $\square$

Corollary 6.6.7. *If R is right coherent and pure injective as a left R-module, then every finitely presented left R-module has a flat envelope.*

Proof. Let M be a finitely presented left R-module and $\varphi : M \to F$ be a pure injective flat envelope guaranteed by Proposition 6.6.6 above. Then there is a factorization

$M \xrightarrow{f} R^n \xrightarrow{g} F$ of the map φ since M is finitely presented and F is flat. But R^n is flat and pure injective. So $M \xrightarrow{f} R^n$ has a factorization $M \xrightarrow{\varphi} F \xrightarrow{h} R^n$. But then $g \circ h \circ \varphi = \varphi$. So $g \circ h$ is an automorphism since $M \xrightarrow{\varphi} F$ is an envelope. Hence F is a direct summand of R^n, that is, F is finitely generated and projective.

Now suppose $M \to G$ is linear with G flat. Then there is a factorization $M \to R^m \to G$. But R^m is flat and pure injective. So there is a factorization $M \to F \to R^m$ of $M \to R^m$. Hence the map $M \to G$ has a factorization $M \to F \to G$. Thus $M \to F$ is a flat preenvelope and so an envelope.   □

If R is a complete local ring, then R is coherent and pure injective and so every finitely generated R-module has a flat envelope which is finitely generated and free by the above. More generally, we have the following result which we state for completeness.

Proposition 6.6.8. *Let R be right semiperfect and right coherent. Then every finitely presented left R-module has a flat envelope. Such an envelope is finitely generated and projective.*

Proof. Let M be a finitely presented left R-module and let $R^m \to R^n \to M \to 0$ be exact. Then $0 \to M^* \to (R^n)^* \to (R^m)^*$ is exact where M^* denotes the algebraic dual $\operatorname{Hom}(M, R)$. Since R is right coherent, M^* is finitely generated. Since R is right semiperfect, M^* has a projective cover $P \to M^*$ (see Remark 5.3.4). Since M^* is finitely generated, it has a finitely generated projective precover. P is a direct summand of any such precover and so P is finitely generated. Thus P is a reflexive right R-module.

We claim $M \to M^{**} \to P^*$ is the desired envelope. First note that if

is commutative, then the dual diagram

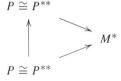

is too. Since $P \to M^*$ is a cover, $P^{**} \to P^{**}$ is an isomorphism. But then $P^* \to P^*$ is also an isomorphism.

To show that $M \to P^*$ is a flat preenvelope, let $M \to F$ be a linear map with F flat. But $M \to F$ can be factored $M \to R^n \to F$ for some $n \geq 1$. So we only need

prove

can be completed to a commutative diagram. But the dual diagram

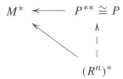

can be completed to a commutative diagram.

Taking the dual of such a completion we get a commutative diagram

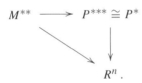

But then the canonical map $M \to M^{**}$ with this diagram gives us the desired commutative diagram

$\square$

Exercises

1. If $\mathcal{G}$ is a class of left R-modules, argue that there is an injective structure $(\mathcal{A}, \mathcal{F})$ on the category of left R-modules with $\mathcal{G} \subset \mathcal{F}$ and such that $\mathcal{F} \subset \mathcal{F}'$ if $(\mathcal{A}', \mathcal{F}')$ is any other injective structure with $\mathcal{G} \subset \mathcal{F}'$.

2. If $\mathcal{G}$ is as in Problem 1, show that $\mathcal{F}$ contains all direct summands of products of modules in $\mathcal{G}$.

3. Let $R = \mathbb{Z}$ and $\mathcal{G} = \{\mathbb{Z}/(p)\}$ where p is some fixed prime. Find the pair $(\mathcal{A}, \mathcal{F})$ determined by $\mathcal{G}$ and then give a description of the $\mathcal{F}$-envelopes.

4. If $(\mathcal{A}, \mathcal{F})$ is the class in Example 6.6.5, argue that there is a largest class $\mathcal{G}$ that determines $(\mathcal{A}, \mathcal{F})$ and then describe $\mathcal{G}$.

6.7 Pure injective envelopes

We saw in Example 6.6.5 that the class of pure injective modules is enveloping. Pure injective envelopes may also be shown to exist by a standard argument using the notion of pure essential extension (see Warfield [177] or Fuchs [95]). We will let $PE(M)$ denote the pure injective envelope of an R-module M.

The following proposition is useful.

Proposition 6.7.1. *Let R be a left coherent ring. If F is a flat right R-module, then $PE(F)$ is also a flat right R-module.*

Proof. By Proposition 5.3.9, the canonical map $F \to F^{++}$ is a pure injection. But F^{++} is pure injective, so $F \to F^{++}$ is a pure injective preenvelope. Hence $PE(F)$ is a direct summand of F^{++}. But F^{1+} is flat since R is coherent and so $PE(F)$ is flat.   □

Corollary 6.7.2. *Let R be a commutative noetherian ring. If F is a flat R- module, then $PE(F) \cong \prod T_{\mathfrak{p}}$, over $\mathfrak{p} \in \mathrm{Spec}\, R$.*

Proof. This follows from Lemma 5.3.23, Theorem 5.3.28 and Proposition 6.7.1 above.   □

For the rest of this section, R will denote a commutative noetherian ring. Our aim in this section is to study pure injective envelopes of flat modules over such rings.

We start with the following result.

Proposition 6.7.3. *Let M be a finitely generated R-module. Then*

$$PE(M) \cong \prod_{\mathfrak{m} \in \mathrm{mSpec}\, R} \hat{M}_{\mathfrak{m}}.$$

Proof. Let $\hat{M}$ be the completion of M with respect to the mSpec R-adic topology on M defined by taking the neighborhoods of 0 to be submodules of IM where I is the intersection of powers of the maximal ideals (see Warfield [177]). Then $\hat{M}$ is algebraically compact and $M \subset \hat{M}$ is a pure injection since M is finitely generated. Thus $\hat{M} \cong PE(M)$. But $\hat{M} \cong \prod_{\mathfrak{m} \in \mathrm{mSpec}\, R} \hat{M}_{\mathfrak{m}}$ and so we are done.   □

Lemma 6.7.4. *If F is a flat R-module, then for each $\mathfrak{p} \in \mathrm{Spec}\, R$, $\hat{F}_{\mathfrak{p}}$ is the completion of a free $R_{\mathfrak{p}}$-module.*

Proof. We can assume R is local and so that $\mathfrak{p} = \mathfrak{m}$ is its maximal ideal. Then $\hat{F}_{\mathfrak{p}} = \varprojlim F/\mathfrak{m}^n F$. For each $n \geq 1$, $F/\mathfrak{m}^n F$ is a flat $R/\mathfrak{m}^n$-module. But $R/\mathfrak{m}^n$ is a local ring of dimension zero. So $F/\mathfrak{m}^n F$ is a projective and hence a free $R/\mathfrak{m}^n$-module. But then $F/\mathfrak{m}^n F \cong F/\mathfrak{m}^{n+1} F \otimes_R R/\mathfrak{m}^n$. Thus the base of $F/\mathfrak{m}^{n+1} F$ over $R/\mathfrak{m}^{n+1}$ is mapped onto the base of $F/\mathfrak{m}^n F$ over $R/\mathfrak{m}^n$ by the canonical map

$F/\mathfrak{m}^{n+1}F \to F/\mathfrak{m}^n F$. So if the base of $F/\mathfrak{m}^n F$ is indexed by X, then we can find maps $R^{(X)} \to F/\mathfrak{m}^n F$ so that the diagram

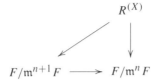

is commutative for each n. Then going modulo $\mathfrak{m}^n$ for various n and chasing an obvious diagram, we get $\varprojlim F/\mathfrak{m}^n F \cong \varprojlim R^{(X)}/\mathfrak{m}^n R^{(X)}$. But the latter is $\widehat{R^{(X)}}$. □

Remark 6.7.5. We note that $F \to \hat{F}_\mathfrak{p}$ induces an isomorphism $F \otimes k(\mathfrak{p}) \to \hat{F}_\mathfrak{p} \otimes k(\mathfrak{p})$. Furthermore, $F \to \hat{F}_\mathfrak{p}$ is a universal map into the completion of a free $R_\mathfrak{p}$-module, that is, a $T_\mathfrak{p}$. Hence any map $F \to T_\mathfrak{p}$ has a unique factorization $F \to \hat{F}_\mathfrak{p} \to T_\mathfrak{p}$.

Proposition 6.7.6. *If F is a flat R-module, then the natural map $F \to \prod \hat{F}_\mathfrak{p}$ is a pure injection for each $\mathfrak{p} \in \operatorname{Spec} R$.*

Proof. By Corollary 6.7.2, $PE(F) \cong \prod T_\mathfrak{p}$ over $\mathfrak{p} \in \operatorname{Spec} R$. Furthermore, the obvious map $F \to T_\mathfrak{p}$ has a factorization $F \to \hat{F}_\mathfrak{p} \to T_\mathfrak{p}$ by Remark 6.7.5 above. This gives a map $f : \prod \hat{F}_\mathfrak{p} \to \prod T_\mathfrak{p}$ such that $F \to \prod T_\mathfrak{p}$ has a factorization $F \to \prod \hat{F}_\mathfrak{p} \to \prod T_\mathfrak{p}$. But then $F \to \prod \hat{F}_\mathfrak{p}$ is a pure injection since $F \to \prod T_\mathfrak{p}$ is. □

Lemma 6.7.7. *Let $\mathfrak{p}, \mathfrak{q} \in \operatorname{Spec} R$. If $T_\mathfrak{q} \neq 0$, then*

$$T_\mathfrak{q} \otimes E(k(\mathfrak{p})) \cong \begin{cases} 0 & \text{if } \mathfrak{p} \not\subset \mathfrak{q} \\ E(k(\mathfrak{p}))^{(X)} & \text{for some set } X \neq \emptyset, \text{ if } \mathfrak{p} \subset \mathfrak{q}. \end{cases}$$

Proof. By Theorem 3.3.12, $T_\mathfrak{q} \otimes E(k(\mathfrak{p})) \cong E(k(\mathfrak{p}))^{(X)}$ for some set X since $T_\mathfrak{q}$ is a flat R-module. If $\mathfrak{p} \subset \mathfrak{q}$, let $T_\mathfrak{q} = \widehat{R_\mathfrak{q}^{(Y)}}$ for some set $Y \neq \emptyset$. Then $R_\mathfrak{q}^{(Y)} \otimes E(k(\mathfrak{p})) \cong (R_\mathfrak{q} \otimes E(k(\mathfrak{p})))^{(Y)} \cong E(k(\mathfrak{p}))^{(Y)}$. But $R_\mathfrak{q}^{(Y)} \to \widehat{R_\mathfrak{q}^{(Y)}}$ is a pure injection. Therefore $R_\mathfrak{q}^{(Y)} \otimes E(k(\mathfrak{p})) \to \widehat{R_\mathfrak{q}^{(Y)}} \otimes E(k(\mathfrak{p}))$ is an injection. So $E(k(\mathfrak{p}))^{(Y)}$ is a direct summand of $T_\mathfrak{q} \otimes E(k(\mathfrak{p}))$ and thus $X \neq \emptyset$.

If $\mathfrak{p} \not\subset \mathfrak{q}$, choose $r \in \mathfrak{p}, r \notin \mathfrak{q}$. Let $S \subset E(k(\mathfrak{p}))$ be a finitely generated submodule. Then there exists an integer $n \geq 1$ such that $r^n S = 0$. Then multiplication by r^n is an automorphism of $T_\mathfrak{q}$ and so $T_\mathfrak{q} \otimes S = 0$. Taking inductive limits, we have $T_\mathfrak{q} \otimes E(k(\mathfrak{p})) = 0$. □

Proposition 6.7.8. $\operatorname{Hom}(\prod_{\mathfrak{p} \not\subset \mathfrak{q}} T_\mathfrak{q}, T_\mathfrak{p}) = 0$.

Proof. We may assume $T_p = \mathrm{Hom}(E(k(p)), E(k(p))^{(X)})$ for some set X. So we have

$$\mathrm{Hom}(\prod_{p \not\subseteq q} T_q, T_p) \cong \mathrm{Hom}((\prod_{p \not\subseteq q} T_q) \otimes E(k(p)), E(k(p))^{(X)}).$$

But if $S \subset E(k(p))$ is a finitely generated R-module, then $(\prod T_q) \otimes S \cong \prod(T_q \otimes S)$. So taking inductive limits, we have $(\prod T_q) \otimes E(k(p)) \cong \prod(T_q \otimes E(k(p)))$ and so the result follows from Lemma 6.7.7 above.   □

Theorem 6.7.9. *If F is a flat R-module, then for each $p \in \mathrm{Spec}\, R$, T_p in $PE(F)$ is isomorphic to a summand of $\hat{F}_p$. If p is maximal such that $F \otimes k(p) \neq 0$, then the map $\hat{F}_p \to T_p$ in the factorization $F \to \hat{F}_p \to T_p$ is an isomorphism.*

Proof. By Proposition 6.7.6, $F \to \prod \hat{F}_p$ is a pure injection. So $f : \prod \hat{F}_p \to \prod T_p$ has a section s making the diagram

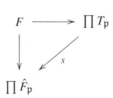

commutative. This guarantees that each $\hat{F}_p \to T_p$ has a section. Thus T_p is isomorphic to a direct summand of $\hat{F}_p$. Note that if $F_p \otimes k(p) = 0$, then $\hat{F}_p = 0$ and so $T_p = 0$.

Now let p be maximal such that $F \otimes k(p) \neq 0$. Then $\hat{F}_p \neq 0$ and $\hat{F}_q = 0$ if $p \subsetneq q$. Furthermore, $\hat{F}_q$ is the completion of a free R_q-module by Lemma 6.7.4, that is, $\hat{F}_q$ is also a T_q (not necessarily the one in $PE(F)$). So $\mathrm{Hom}(\prod_{q \neq p} \hat{F}_q, T_p) = 0$ by Proposition 6.7.8. Similarly $\mathrm{Hom}(\prod_{q \neq p} T_q, \hat{F}_p) = 0$ since $T_q = 0$ whenever $\hat{F}_q = 0$ by the above.

We can now therefore construct the following commutative diagram

$$\prod \hat{F}_q = (\prod_{q \neq p} \hat{F}_q) \oplus \hat{F}_p$$

$$f \downarrow$$

$$F \longrightarrow \prod T_q = (\prod_{q \neq p} T_q) \oplus T_p$$

$$s \downarrow$$

$$\prod \hat{F}_q = (\prod_{q \neq p} \hat{F}_q) \oplus \hat{F}_p$$

and pass to the quotient to get a commutative diagram

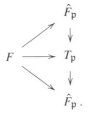

The composition $\hat{F}_{\mathfrak{p}} \to T_{\mathfrak{p}} \to \hat{F}_{\mathfrak{p}}$ is the identity on $\hat{F}_{\mathfrak{p}}$. Since $\hat{F}_{\mathfrak{p}} \to T_{\mathfrak{p}}$ has a section, it must be an isomorphism. $\qquad\square$

Corollary 6.7.10. *Let F be a flat R-module and $\mathfrak{m}$ be a maximal ideal such that $\hat{F}_{\mathfrak{m}} \neq 0$. Then $T_{\mathfrak{m}}$ in $PE(F)$ is isomorphic to $\hat{F}_{\mathfrak{m}}$.*

Proof. If $\hat{F}_{\mathfrak{m}} \neq 0$, then $F_{\mathfrak{m}} \otimes k(\mathfrak{m}) \neq 0$. $\qquad\square$

Proposition 6.7.11. *For any set X, $\hat{R}_{\mathfrak{p}}^X$ is the completion of a free $R_{\mathfrak{p}}$-module and every such completion is a direct summand of $\hat{R}_{\mathfrak{p}}^X$ for some X.*

Proof. $\hat{R}_{\mathfrak{p}}^X \cong \prod T_{\mathfrak{q}}$ over $\mathfrak{q} \in \operatorname{Spec} R$ by Theorem 5.3.28. So if $T_{\mathfrak{q}} \neq 0$, then easily $\operatorname{Hom}(T_{\mathfrak{q}}, \hat{R}_{\mathfrak{p}}) \neq 0$. But then $\operatorname{Hom}(T_{\mathfrak{q}} \otimes E(k(\mathfrak{p})), E(k(\mathfrak{p}))) \neq 0$. Thus $T_{\mathfrak{q}} \otimes E(k(\mathfrak{p})) \neq 0$ and so $\mathfrak{p} \subset \mathfrak{q}$ by Lemma 6.7.7. If $r \notin \mathfrak{p}$, then r is an isomorphism on $\hat{R}_{\mathfrak{p}}^X$ and thus on $T_{\mathfrak{q}}$. But $T_{\mathfrak{q}} \neq 0$ implies $r \notin \mathfrak{q}$. Hence $\mathfrak{q} = \mathfrak{p}$. That is, $\hat{R}_{\mathfrak{p}}^X \cong T_{\mathfrak{p}}$. Now let $T_{\mathfrak{p}}$ be the completion of a free $R_{\mathfrak{p}}$- module. Let $T_{\mathfrak{p}} = \operatorname{Hom}(E(k(\mathfrak{p})), E(k(\mathfrak{p}))^{(X)})$ for some set X. Then $T_{\mathfrak{p}}$ is isomorphic to a direct summand of $\operatorname{Hom}(E(k(\mathfrak{p})), E(k(\mathfrak{p}))^X) \cong \hat{R}_{\mathfrak{p}}^X$ since $E(k(\mathfrak{p}))^{(X)}$ is a direct summand of $E(k(\mathfrak{p}))^X$. $\qquad\square$

Remark 6.7.12. $PE(R) \cong \prod \hat{R}_{\mathfrak{m}}$ over all maximal ideals $\mathfrak{m}$ of R by Proposition 6.7.3. So with our notation, $T_{\mathfrak{p}} = 0$ for every prime ideal $\mathfrak{p}$ that is not maximal. If F is free, say $F = R^{(X)}$, then $R^{(X)} \subset R^X$ and $R^X \hookrightarrow (\prod \hat{R}_{\mathfrak{m}})^X \cong \prod \hat{R}_{\mathfrak{m}}^X$ are pure injections. So $R^{(X)} \hookrightarrow \prod \hat{R}_{\mathfrak{m}}^X$ is also pure. But $\hat{R}_{\mathfrak{m}}^X \cong T_{\mathfrak{m}}$ by Proposition 6.7.11 above. So if F is a free or projective R-module, then $PE(F) \cong \prod \hat{F}_{\mathfrak{m}}$ by Corollary 6.7.10.

We now want to give a necessary and sufficient condition for a map $F \to \prod T_{\mathfrak{p}}$ to be a pure injective envelope of the flat R-module F. We start with the following result.

Lemma 6.7.13. *Let I be an ideal of R and F be a flat pure injective R-module. Then IF is a pure injective R-module.*

Proof. F is flat and cotorsion and so $F \cong \prod T_{\mathfrak{p}}$ over $\mathfrak{p} \in \operatorname{Spec} R$. Therefore, $IF \cong \prod IT_{\mathfrak{p}}$. By Proposition 6.7.11 above, it suffices to show that $I\hat{R}_{\mathfrak{p}}$ is a pure injective $\hat{R}_{\mathfrak{p}}$-module. But $I\hat{R}_{\mathfrak{p}}$ is reflexive and so is pure injective by Remark 5.3.15.   □

Remark 6.7.14. Note that for any $T_{\mathfrak{p}}$, the natural map $\varphi \colon T_{\mathfrak{p}} \to T_{\mathfrak{p}} \otimes k(\mathfrak{p})$ is surjective with kernel $\mathfrak{p}T_{\mathfrak{p}} = m(\hat{R}_{\mathfrak{p}})T_{\mathfrak{p}}$. The lemma above implies that if F is a flat module, then any map $F \to T_{\mathfrak{p}} \otimes k(\mathfrak{p})$ can be lifted to a map $F \to T_{\mathfrak{p}}$. In fact, we showed in Example 5.3.19 that $\varphi \colon T_{\mathfrak{p}} \to T_{\mathfrak{p}} \otimes k(\mathfrak{p})$ is a flat cover. In particular, if $f \colon T_{\mathfrak{p}} \to T_{\mathfrak{p}}$ is such that $\varphi \circ f = \varphi$, then f is an automorphism of $T_{\mathfrak{p}}$. Furthermore, we see that any map $g \colon T_{\mathfrak{p}} \to T_{\mathfrak{p}}$ is an automorphism of $T_{\mathfrak{p}}$ if and only if the induced map $T_{\mathfrak{p}} \otimes k(\mathfrak{p}) \to T_{\mathfrak{p}} \otimes k(\mathfrak{p})$ is an isomorphism. For a vector space V over $k(\mathfrak{p})$, $V \cong T_{\mathfrak{p}} \otimes k(\mathfrak{p})$ for some $T_{\mathfrak{p}}$. Hence we have a map $T_{\mathfrak{p}} \to V$ which can be lifted to a map $T_{\mathfrak{p}} \to T_{\mathfrak{p}}$ inducing an isomorphism $V \cong T_{\mathfrak{p}} \otimes k(\mathfrak{p})$.

Lemma 6.7.15. *For flat modules F and G, $\sigma \colon F \to G$ is a pure injection if and only if $F \otimes k(\mathfrak{p}) \to G \otimes k(\mathfrak{p})$ is an injection for every prime $\mathfrak{p}$.*

Proof. The condition is necessary and so suppose the condition holds. It suffices to prove that $F \to PE(F)$ has a factorization $F \to G \to PE(F)$ since $F \to PE(F)$ is a pure injection. But $PE(F) \cong \prod T_{\mathfrak{p}}$. Thus to show that there is a factorization $F \to G \to PE(F)$ it suffices to show that there is a factorization $F \xrightarrow{\sigma} G \xrightarrow{\varphi} \hat{R}_{\mathfrak{p}}$ for any prime ideal $\mathfrak{p}$ by Proposition 6.7.11. But the diagram

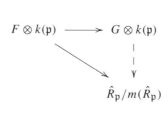

can be completed to a commutative diagram since $F \otimes k(\mathfrak{p}) \to G \otimes k(\mathfrak{p})$ is injective and we have vector spaces. But by Remark 6.7.14 above, the map $G \to \hat{R}_{\mathfrak{p}}/m(\hat{R}_{\mathfrak{p}})$ can be lifted to a map $f_0 \colon G \to \hat{R}_{\mathfrak{p}}$ such that f_0 completes the diagram

to a commutative diagram modulo $m(\hat{R}_{\mathfrak{p}})$. That is, $(\varphi - f_0 \circ \sigma)(F) \subset m(\hat{R}_{\mathfrak{p}})$.

We repeat the procedure using Lemma 6.7.13 where $I = m(\hat{R}_\mathfrak{p})^2$ with the diagram

$$
\begin{array}{ccc}
F & \xrightarrow{\;\sigma\;} & G \\
 & \searrow{\scriptstyle \varphi - f_0 \circ \sigma} & \downarrow{\scriptstyle f_1} \\
 & & m(\hat{R}_\mathfrak{p})
\end{array}
$$

finding f_1 so that $(\varphi - \sigma \circ f_0 - \sigma \circ f_1)(F) \subset m(\hat{R}_\mathfrak{p})^2$. If by induction we have $f_0, f_1, \ldots, f_n$ so that

$$
\begin{array}{ccc}
F & \longrightarrow & G \\
 & \searrow{\scriptstyle \varphi - \sum_{i=0}^{n-1} f_i \circ \sigma} & \downarrow{\scriptstyle f_n} \\
 & & m(\hat{R}_\mathfrak{p})^n
\end{array}
$$

is commutative modulo $m(\hat{R}_\mathfrak{p})^{n+1}$ for each n, let $f = \sum_{i=0}^{\infty} f_k : G \to \hat{R}_\mathfrak{p}$. Then $f \circ \sigma = \varphi$. □

Remark 6.7.16. Since F and G are flat, $F/\mathfrak{p}F$ and $G/\mathfrak{p}G$ are also flat. Thus $F/\mathfrak{p}F$, $G/\mathfrak{p}G$ are torsion free $R/\mathfrak{p}$-modules. That is, $F \otimes R/\mathfrak{p} \to G \otimes R/\mathfrak{p}$ is an injection if and only if $F \otimes k(\mathfrak{p}) \to G \otimes k(\mathfrak{p})$ is an injection.

Theorem 6.7.17. *Let F be a flat R-module. Then $F \to \prod T_\mathfrak{p}$, over $\mathfrak{p} \in \operatorname{Spec} R$, is a pure injective envelope if and only if for each $\mathfrak{q} \in \operatorname{Spec} R$*

(a) *$F \otimes k(\mathfrak{q}) \to (\prod T_\mathfrak{p}) \otimes k(\mathfrak{q})$ is an injection;*

(b) *the image of $F \otimes k(\mathfrak{q})$ in $(\prod T_\mathfrak{p}) \otimes k(\mathfrak{q}) = (T_\mathfrak{q} \otimes k(\mathfrak{q})) \oplus (\prod_{\mathfrak{p} \neq \mathfrak{q}} T_\mathfrak{p}) \otimes k(\mathfrak{q})$ contains $(T_\mathfrak{q} \otimes k(\mathfrak{q})) \oplus 0$.*

Proof. We first construct $\prod T_\mathfrak{p}$ and a map $F \to \prod T_\mathfrak{p}$ satisfying (a) and (b). We then show that any pure injective envelope of F satisfies (a) and (b). Then we argue that if $f : F \to \prod T_\mathfrak{p}$ satisfies (a) and (b), it is in fact a pure injective envelope. This completes the proof the theorem.

Let $X = \operatorname{Spec} R$, and $X_0 = \operatorname{mSpec} R$. For any ordinal $\alpha > 1$, define X_α to be the set of maximal elements of $X - \bigcup_{\beta < \alpha} X_\beta$. Then well order each X_α. Using these orders, well order X so that if $\mathfrak{p} \in X_\alpha$ and $\mathfrak{q} \in X_\beta$ and $\alpha < \beta$, then $\mathfrak{p} \subset \mathfrak{q}$. Thus the $\mathfrak{p} \in X$ can be indexed by $\alpha < \lambda$ for some ordinal λ so that if $\beta < \alpha < \lambda$, then $\mathfrak{p}_\alpha \not\supset \mathfrak{p}_\beta$.

We now construct $T_{\mathfrak{p}_\beta}$ and the map $F \to T_{\mathfrak{p}_\beta}$ by transfinite induction. By Lemma 6.7.4, let $T_{\mathfrak{p}_0} \cong \hat{F}_{\mathfrak{p}_0}$ and let $F \to T_{\mathfrak{p}_0} \cong \hat{F}_{\mathfrak{p}_0}$ be the natural map. Having constructed $T_{\mathfrak{p}_\beta}$ and $F \to T_{\mathfrak{p}_\beta}$ for all $\beta < \alpha < \lambda$, we consider $F \otimes k(\mathfrak{p}_\alpha) \to \prod_{\beta < \alpha} T_{\mathfrak{p}_\beta} \otimes k(\mathfrak{p}_\alpha)$. Let V be its kernel and let $T_{\mathfrak{p}_\alpha}$ be such that there is a surjection $T_{\mathfrak{p}_\alpha} \to V$ inducing an isomorphism $T_{\mathfrak{p}_\alpha} \otimes k(\mathfrak{p}_\alpha) \to V$. Composing $F \to F \otimes k(\mathfrak{p}_\alpha)$ with a projection

$F \otimes k(\mathfrak{p}_\alpha) \to V \cong T_{\mathfrak{p}_\alpha} \otimes k(\mathfrak{p}_\alpha)$, we get a map $F \to V$ which can be lifted to a map $F \to T_{\mathfrak{p}_\alpha}$ as noted in Remark 6.7.14. Then the construction $F \to \prod_{\beta \le \alpha} T_{\mathfrak{p}_\beta}$ is such that $F \otimes k(\mathfrak{p}_\alpha) \to (\prod_{\beta \le \alpha} T_{\mathfrak{p}_\beta}) \otimes k(\mathfrak{p}_\alpha)$ is an injection and its image contains $(T_{\mathfrak{p}_\alpha} \otimes k(\mathfrak{p}_\alpha)) \oplus 0$.

$F \to \prod_{\alpha < \lambda} T_{\mathfrak{p}_\alpha}$ clearly satisfies (a). For $\beta < \lambda$, $(\prod_{\alpha > \beta} T_{\mathfrak{p}_\alpha}) \otimes k(\mathfrak{p}_\beta) = 0$ (by Lemma 6.7.7 and the proof of Proposition 6.7.8) since $\mathfrak{p}_\alpha \not\supseteq \mathfrak{p}_\beta$ for $\beta < \alpha$. Thus $(\prod_{\alpha < \lambda} T_{\mathfrak{p}_\alpha}) \otimes k(\mathfrak{p}_\beta) \cong (\prod_{\alpha \le \beta} T_{\mathfrak{p}_\alpha}) \otimes k(\mathfrak{p}_\beta)$ and so (b) is satisfied by the above.

Now let $F \to \prod U_{\mathfrak{p}_\alpha}$ be a pure injective envelope where each $U_{\mathfrak{p}_\alpha}$ is the completion of a free $R_{\mathfrak{p}_\alpha}$-module. Since $F \to \prod T_{\mathfrak{p}_\alpha}$ satisfies (a) by Lemma 6.7.15 we get a map $f : \prod T_{\mathfrak{p}_\alpha} \to \prod U_{\mathfrak{p}_\alpha}$ making the diagram

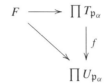

commutative. Similarly we get a map $g : \prod U_\alpha \to \prod T_{\mathfrak{p}_\alpha}$ making the obvious diagram commutative. For a fixed β, we have $\prod_{\alpha < \lambda} T_{\mathfrak{p}_\alpha} \otimes k(\mathfrak{p}_\beta) = \prod_{\alpha \le \beta} T_{\mathfrak{p}_\alpha} \otimes k(\mathfrak{p}_\beta)$. Hence we have the commutative diagram

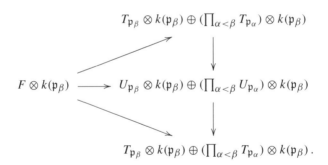

Since $\operatorname{Hom}(T_{\mathfrak{p}_\beta}, \prod_{\alpha < \beta} U_{\mathfrak{p}_\alpha}) = 0$ and $\operatorname{Hom}(U_{\mathfrak{p}_\beta}, \prod_{\alpha < \beta} T_{\mathfrak{p}_\alpha}) = 0$ by Proposition 6.7.8 the vertical maps above map $T_{\mathfrak{p}_\beta} \otimes k(\mathfrak{p}_\beta)$ into $U_{\mathfrak{p}_\beta} \otimes k(\mathfrak{p}_\beta)$ and $U_{\mathfrak{p}_\beta} \otimes k(\mathfrak{p}_\beta)$ into $T_{\mathfrak{p}_\beta} \otimes k(\mathfrak{p}_\beta)$. Since by (b), there is a subspace V of $F \otimes k(\mathfrak{p}_\beta)$ mapped isomorphically onto $T_{\mathfrak{p}_\beta} \otimes k(\mathfrak{p}_\beta)$, we see that the composition $T_{\mathfrak{p}_\beta} \otimes k(\mathfrak{p}_\beta) \to U_{\mathfrak{p}_\beta} \otimes k(\mathfrak{p}_\beta) \to T_{\mathfrak{p}_\beta} \otimes k(\mathfrak{p}_\beta)$ is the identity map.

Now reversing the roles of T and U in the diagram above and using the fact that $F \to \prod U_{\mathfrak{p}_\alpha}$ is a pure injective envelope, a similar argument gives that $U_{\mathfrak{p}_\beta} \otimes k(\mathfrak{p}_\beta) \to T_{\mathfrak{p}_\beta} \otimes k(\mathfrak{p}_\beta) \to U_{\mathfrak{p}_\beta} \otimes k(\mathfrak{p}_\beta)$ is an automorphism of $U_{\mathfrak{p}_\beta} \otimes k(\mathfrak{p}_\beta)$. Thus $T_{\mathfrak{p}_\beta} \otimes k(\mathfrak{p}_\beta) \to U_{\mathfrak{p}_\beta} \otimes k(\mathfrak{p}_\beta)$ is an isomorphism and so $V \subset F \otimes k(\mathfrak{p}_\beta)$ is mapped isomorphically onto $U_{\mathfrak{p}_\beta} \otimes k(\mathfrak{p}_\beta)$ in $(\prod_{\alpha < \lambda} U_{\mathfrak{p}_\alpha}) \otimes k(\mathfrak{p}_\beta)$. Hence (b) is satisfied. Clearly, $F \to \prod U_{\mathfrak{p}_\alpha}$ satisfies (a).

We finally argue that if $F \to \prod T_{\mathfrak{p}_\alpha}$ satisfies (a) and (b), then it is an injective envelope. By (a) and Lemma 6.7.15, $F \to \prod T_{\mathfrak{p}_\alpha}$ is a pure injection. Thus, it suffices to show that if $f : \prod T_{\mathfrak{p}_\alpha} \to \prod T_{\mathfrak{p}_\alpha}$ makes the diagram

commutative, then it is an automorphism of $\prod T_{\mathfrak{p}_\alpha}$.

But since $\prod_{\alpha < \lambda} T_{\mathfrak{p}_\alpha} = \varprojlim(\prod_{\alpha \le \beta} T_{\mathfrak{p}_\alpha})$, it suffices to show inductively that when we pass to the quotients $\prod_{\alpha < \beta} T_{\mathfrak{p}_\alpha}$ for each β, we get an automorphism of $\prod_{\alpha < \beta} T_{\mathfrak{p}_\alpha}$. But $T_{\mathfrak{p}_0} = \hat{F}_{\mathfrak{p}_0}$. Hence f induces the identity map on the quotient $T_{\mathfrak{p}_0}$. Now suppose $\beta < \lambda$ is a limit ordinal and that f induces an automorphism of $\prod_{\alpha \le \gamma} T_{\mathfrak{p}_\gamma}$ for all $\gamma < \beta$. Taking a projective limit, we get that f induces an isomorphism $\prod_{\alpha < \beta} T_{\mathfrak{p}_\alpha} \to \prod_{\alpha < \beta} T_{\mathfrak{p}_\alpha}$. But $\prod_{\alpha < \beta} T_{\mathfrak{p}_\alpha} \to \prod_{\alpha < \beta} T_{\mathfrak{p}_\alpha}$ maps $T_{\mathfrak{p}_\beta}$ into $T_{\mathfrak{p}_\beta}$. So to show it is an isomorphism, we only need to argue that it maps $T_{\mathfrak{p}_\beta}$ onto $T_{\mathfrak{p}_\beta}$. But by (b), $T_{\mathfrak{p}_\beta} \otimes k(\mathfrak{p}_\beta) \cong T_{\mathfrak{p}_\beta} \otimes k(\mathfrak{p}_\beta)$. Thus $T_{\mathfrak{p}_\beta} \to T_{\mathfrak{p}_\beta}$ is an isomorphism.

If β is not an ordinal, say $\beta = \gamma + 1$, then a similar argument shows that $\prod_{\alpha < \gamma} T_{\mathfrak{p}_\alpha} \to \prod_{\alpha \le \gamma} T_{\mathfrak{p}_\alpha}$ is an isomorphism and so is $\prod_{\alpha \le \beta} T_{\mathfrak{p}_\alpha} \to \prod_{\alpha \le \beta} T_{\mathfrak{p}_\alpha}$. This completes the proof. □

Exercises

1. Let $M = R$ in Proposition 6.7.3. Show that $\hat{R}$ is isomorphic to $\mathrm{End}_R(\hat{R})$.
   Hint: use Proposition 6.7.8.

2. If $0 \to F' \to F \to F'' \to 0$ is a short exact sequence of flat modules, show that $PE(F)$ is isomorphic to a direct summand of $PE(F') \oplus PE(F'')$.

3. Find an example as in Problem 2 where $PE(F)$ is not isomorphic to $PE(F') \oplus PE(F'')$.
   Hint: Let $R = \mathbb{Z}$, $F'' = \mathbb{Q}$ and let F be free.

4. Use Theorem 6.7.17 to show that if $R = \mathbb{Z}$ and if $F = \prod_{i=0}^{\infty} Z_i$ with $Z_i = \mathbb{Z}$ for all $i \ge 0$, then $\prod_{i=0}^{\infty} Z_i \to \prod_{i=0}^{\infty} PE(Z_i)$ is a pure injective envelope.

5. Let $S \subset R$ be a multiplicative set. Let $M \subset PE(M)$ be a pure injective envelope of the $S^{-1}R$-module M (as an $S^{-1}R$-module). Explain why $M \subset PE(M)$ is also a pure injective envelope of M as an R-module.

Chapter 7

Covers, Envelopes, and Cotorsion Theories

In this chapter we introduce the notion of cotorsion theory. We prove Eklof and Trlifaj's theorem that under certain conditions cotorsion theories have enough injectives and projectives in which case precovers and preenvelopes exist. As an application of this theory, we prove that every module has a flat cover.

7.1 Definitions and basic results

Definition 7.1.1. Given a class $\mathcal{C}$ of R-modules, we let $^{\perp}\mathcal{C}$ be the class of R-modules F such that $\text{Ext}^1_R(F, C) = 0$ for all $C \in \mathcal{C}$. We let $\mathcal{C}^{\perp}$ be the class of modules G such that $\text{Ext}^1_R(C, G) = 0$ for all $C \in \mathcal{C}$. $^{\perp}\mathcal{C}$ and $\mathcal{C}^{\perp}$ are called *orthogonal classes* of $\mathcal{C}$.

We note that for any $\mathcal{C}, \mathcal{C} \subset {}^{\perp}(\mathcal{C}^{\perp})$ and $\mathcal{C} \subset ({}^{\perp}\mathcal{C})^{\perp}$. Also $\mathcal{C}_1 \subset \mathcal{C}_2$ implies $^{\perp}\mathcal{C}_2 \subset {}^{\perp}\mathcal{C}_1$ and $\mathcal{C}_2^{\perp} \subset \mathcal{C}_1^{\perp}$. From this it follows that $({}^{\perp}(\mathcal{C}^{\perp}))^{\perp} = \mathcal{C}^{\perp}$ and $^{\perp}(({}^{\perp}\mathcal{C})^{\perp}) = {}^{\perp}\mathcal{C}$ for all $\mathcal{C}$.

Definition 7.1.2. A pair $(\mathcal{F}, \mathcal{C})$ of classes of R-modules is called a *cotorsion theory* (for the category of R-modules) if $\mathcal{F}^{\perp} = \mathcal{C}$ and $^{\perp}\mathcal{C} = \mathcal{F}$.

A class $\mathcal{D}$ is said to *generate* the cotorsion theory if $^{\perp}\mathcal{D} = \mathcal{F}$ (and so $\mathcal{D} \subset \mathcal{C}$) and a class $\mathcal{G}$ is said to *cogenerate* $(\mathcal{F}, \mathcal{C})$ if $\mathcal{G}^{\perp} = \mathcal{C}$ (and so $\mathcal{G} \subset \mathcal{F}$).

Example 7.1.3. $(\mathcal{M}, \mathcal{I}nj)$ and $(\mathcal{P}roj, \mathcal{M})$ are cotorsion theories where $\mathcal{M}$ denotes the class of left R-modules and $\mathcal{I}nj$ and $\mathcal{P}roj$ denote the classes of injective and projective modules respectively. The cotorsion theory $(\mathcal{M}, \mathcal{I}nj)$ is cogenerated by the set of modules R/I where I is a left ideal, and is generated by the class of injective modules.

We note that if $(\mathcal{F}, \mathcal{C})$ is a cotorsion theory, then $\mathcal{F}$ and $\mathcal{C}$ are both closed under extensions and summands, and $\mathcal{F}$ contains all the projective modules while $\mathcal{C}$ contains all the injective modules. Also, $\mathcal{F}$ is closed under arbitrary direct sums and $\mathcal{C}$ is closed under arbitrary direct products. If $(\mathcal{F}, \mathcal{C})$ is generated (cogenerated) by a set X (so not just a class), then $(\mathcal{F}, \mathcal{C})$ is generated (cogenerated) by the single module $\prod_{M \in X} M$ ($\bigoplus_{M \in X} M$).

Lemma 7.1.4. *If $\mathcal{F}$ is the class of flat R-modules and if $\mathcal{F}^{\perp} = \mathcal{C}$ (so $\mathcal{C}$ is the class of cotorsion modules, see Definition 5.3.22), then $(\mathcal{F}, \mathcal{C})$ is a cotorsion theory.*

Proof. We only need to prove that $F \in {}^{\perp}\mathcal{C}$ is flat. But pure injective modules are cotorsion. And for any right R-module M, M^{+} is pure injective. But $\mathrm{Ext}^{1}_{R}(F, M^{+}) \cong (\mathrm{Tor}^{R}_{1}(M, F))^{+}$. So if $\mathrm{Ext}^{1}_{R}(F, M^{+}) = 0$, then $\mathrm{Tor}^{R}_{1}(M, F) = 0$. Hence if $F \in {}^{\perp}\mathcal{C}$, then F is flat. □

Definition 7.1.5. A cotorsion theory $(\mathcal{F}, \mathcal{C})$ is said to have *enough injectives* if for every module M there is an exact sequence $0 \to M \to C \to F \to 0$ with $C \in \mathcal{C}$ and $F \in \mathcal{F}$. We say that $(\mathcal{F}, \mathcal{C})$ has *enough projectives* if for every M there is an exact sequence $0 \to C \to F \to M \to 0$ with $F \in \mathcal{F}$ and $C \in \mathcal{C}$.

We note that if $0 \to C \to F \to M \to 0$ is as in the definition, then $F \to M$ is an $\mathcal{F}$-precover for every M. For if $G \in \mathcal{F}$ then $\mathrm{Hom}(G, F) \to \mathrm{Hom}(G, M) \to \mathrm{Ext}^{1}(G, C) = 0$ is exact. Similarly, if $0 \to M \to C \to F \to 0$ is as in the definition then $M \to C$ is a $\mathcal{C}$-preenvelope. The cotorsion theories $(\mathcal{M}, \mathcal{Inj})$ and $(\mathcal{Proj}, \mathcal{M})$ as in Example 7.1.3 have enough projectives and injectives. For example, the exact sequence $0 \to 0 \to M \overset{\mathrm{id}}{\to} M \to 0$ for any M shows that $(\mathcal{M}, \mathcal{Inj})$ has enough projectives.

Definition 7.1.6. Given a class $\mathcal{F}$, a module M is said to have a *special $\mathcal{F}$-precover* if there is an exact sequence $0 \to C \to F \to M \to 0$ with $F \in \mathcal{F}$ and $C \in \mathcal{F}^{\perp}$. M is said to have a *special preenvelope* if there is an exact sequence $0 \to M \to F \to D \to 0$ with $F \in \mathcal{F}$ and $D \in {}^{\perp}\mathcal{F}$.

So if a cotorsion theory $(\mathcal{F}, \mathcal{C})$ has enough injectives and projectives, every module M has a special $\mathcal{F}$-precover and a special $\mathcal{C}$-preenvelope.

Proposition 7.1.7. *If $(\mathcal{F}, \mathcal{C})$ is a cotorsion theory on the category of R-modules having enough injectives (projectives), then it also has enough projectives (injectives).*

Proof. Assume that $(\mathcal{F}, \mathcal{C})$ has enough injectives and let M be a module. Let $0 \to S \to P \to M \to 0$ be exact with P projective. Since there are enough injectives let $0 \to S \to C \to F \to 0$ be exact with $C \in \mathcal{C}$ and $F \in \mathcal{F}$. Using a pushout construction for

we have a commutative diagram

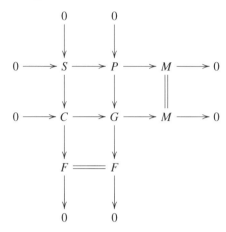

with exact rows and columns. Since $\mathcal{F}$ is closed under extensions, we see by the middle column that $G \in \mathcal{F}$. Hence the middle row gives the desired exact sequence.

A completely dual proof gives that if $(\mathcal{F}, \mathcal{C})$ has enough projectives then it has enough injectives.   □

Exercises

1. Let $(\mathcal{F}, \mathcal{C})$ be a cotorsion theory for the category of R-modules. Argue that the following are equivalent:

   a) $\text{Ext}^i (F, C) = 0$ for all $i \geq 1$, $F \in \mathcal{F}$, $C \in \mathcal{C}$.
   b) If $0 \to F' \to F \to F'' \to 0$ is exact with F, $F'' \in \mathcal{F}$ then $F' \in \mathcal{F}$.
   c) If $0 \to C' \to C \to C'' \to 0$ is exact with C', $C \in \mathcal{C}$ then $C'' \in \mathcal{C}$.

2. If $(\mathcal{F}_i, \mathcal{C}_i)$ are cotorsion theories for the category of R-modules for $i \in I$, prove that there is a cotorsion theory $(\mathcal{F}, \mathcal{C})$ with $\mathcal{F} = \bigcap_{i \in I} \mathcal{F}_i$ and that there is one $(\mathcal{F}', \mathcal{C}')$ with $\mathcal{C}' = \bigcap_{i \in I} \mathcal{C}_i$.

3. Let $(\mathcal{F}, \mathcal{C})$ be the cotorsion theory of the category of $\mathbb{Z}$-modules cogenerated by $\mathbb{Z}/(n)$ for $n \geq 2$. Show that $C \in \mathcal{C}$ if and only if $nC = C$.

4. Argue that the cotorsion theory $(\mathcal{M}, \mathcal{D})$ of the category of $\mathbb{Z}$-modules where $\mathcal{D}$ is the class of divisible $\mathbb{Z}$-modules is not cogenerated by any single finitely generated $\mathbb{Z}$-module.

7.2   Fibrations, cofibrations and Wakamatsu lemmas

In the last section, we showed that when a cotorsion theory $(\mathcal{F}, \mathcal{C})$ has enough projectives, then every module has a special $\mathcal{F}$-precover and a special $\mathcal{F}$-preenvelope. In this section we show that if $\mathcal{F}$ is any class closed under extensions, then any surjective $\mathcal{F}$-cover is special and, dually, any injective $\mathcal{F}$-envelope is special.

Definition 7.2.1. Given a class $\mathcal{F}$ of modules, a linear map $f : M \to N$ is said to be an $\mathcal{F}$-*fibration* if given a submodule S of some module P where $P/S \in \mathcal{F}$ any commutative diagram

$$
\begin{array}{ccc}
S & \xrightarrow{\ i\ } & P \\
\downarrow & \swarrow & \downarrow \\
M & \xrightarrow[\ f\ ]{} & N
\end{array}
$$

can be completed to a commutative diagram where i denotes the canonical injection. An $\mathcal{F}$-*cofibration* is defined dually, that is, f is an $\mathcal{F}$-cofibration if any commutative diagram

$$
\begin{array}{ccc}
M & \longrightarrow & N \\
\downarrow & \swarrow & \downarrow \\
P & \longrightarrow & P/S
\end{array}
$$

with $S \in \mathcal{F}$ can be completed to a commutative diagram.

Proposition 7.2.2. *If $\varphi : F \to M$ is an $\mathcal{F}$-cover for some class $\mathcal{F}$ closed under extensions, then φ is an $\mathcal{F}$-fibration.*

Proof. Given the following commutative diagram where $S \subset P$ is such that $P/S \in \mathcal{F}$

$$
\begin{array}{ccc}
S & \xrightarrow{\ i\ } & P \\
\downarrow & & \downarrow \\
F & \xrightarrow[\ \varphi\ ]{} & M
\end{array}
$$

we form the pushout diagram

$$
\begin{array}{ccc}
S & \longrightarrow & P \\
\downarrow & & \downarrow \\
F & \longrightarrow & G\,.
\end{array}
$$

Then $F \to G$ is an injection and $\mathrm{Coker}(F \to G) \cong P/S \in \mathcal{F}$. But $\mathcal{F}$ is closed under extensions. So $G \in \mathcal{F}$. We can suppose $F \subset G$. By the properties of the pushout diagram, we have a linear $h : G \to M$ which agrees with φ on F. Since $\varphi : F \to M$ is a cover, we have a linear $g : G \to F$ so that $\varphi \circ g = h$. Then since $\varphi \circ (g|F) = \varphi$, we have that $g|F$ is an automorphism of F. If we then replace g by $(g|F)^{-1} \circ g$, we see that we can assume $g|F = \mathrm{id}_F$. So g makes the diagram

commutative and hence makes commutative the diagram

□

The following result is usually called Wakamatsu's lemma.

Corollary 7.2.3. *If $\mathcal{F}$ is a class of modules closed under extensions and if $\varphi : F \to M$ is an $\mathcal{F}$-cover, then $\mathrm{Ker}\,\varphi \in \mathcal{F}^{\perp}$.*

Proof. Let $G \in \mathcal{F}$. We want to argue that $\mathrm{Ext}^i(G, \mathrm{Ker}\,\varphi) = 0$. Let $0 \to S \to P \to G \to 0$ be exact with P projective. Then we need that any $f : S \to \mathrm{Ker}\,\varphi$ can be extended to a linear $P \to \mathrm{Ker}\,\varphi$. But if we consider the commutative diagram

where $S \to F$ agrees with f we see that any $g : P \to F$ that makes the diagram commutative has its image in $\mathrm{Ker}\,\varphi$ and so gives the desired extension.   □

Dually, we have the following

Proposition 7.2.4. *If $\mathcal{C}$ is a class of modules closed under extensions and if $\varphi : M \to C$ is a $\mathcal{C}$-envelope, then φ is a $\mathcal{C}$-cofibration and $\mathrm{Coker}\,\varphi \in {}^{\perp}\mathcal{C}$.*

The argument in the following lemma is another Wakamatsu lemma type of result.

Lemma 7.2.5. *Let $(\mathcal{F}, \mathcal{C})$ be a cotorsion theory for the category of R-modules with enough injectives and projectives. Suppose $0 \to M \to D \to F \to 0$ is an exact sequence with $F \in \mathcal{F}$ such that*

(1) *any diagram*

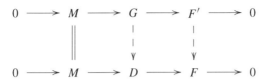

with $F' \in \mathcal{F}$ *can be completed to a commutative diagram, and*

(2) *the diagram*

$$
\begin{array}{ccccccccc}
0 & \longrightarrow & M & \longrightarrow & D & \longrightarrow & F & \longrightarrow & 0 \\
 & & \| & & \downarrow & & \downarrow & & \\
0 & \longrightarrow & M & \longrightarrow & D & \longrightarrow & F & \longrightarrow & 0
\end{array}
$$

can only be completed by automorphisms.

Then $D \in \mathcal{C}$.

Proof. Let $0 \to D \to U \to G \to 0$ be an exact sequence with $U \in \mathcal{C}$ and $G \in \mathcal{F}$. We want to prove that the sequence splits. We use the pushout of the diagram

$$
\begin{array}{ccc}
D & \longrightarrow & F \\
\downarrow & & \\
U & &
\end{array}
$$

to get a commutative diagram

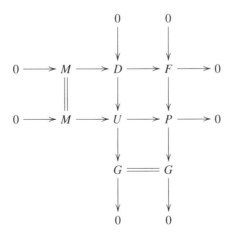

with exact rows and columns. Since $F, G \in \mathcal{F}$, we get that $P \in \mathcal{F}$. But by hypothesis (1), there is a commutative diagram

$$
\begin{array}{ccccccccc}
0 & \longrightarrow & M & \longrightarrow & U & \longrightarrow & P & \longrightarrow & 0 \\
 & & \| & & \downarrow & & \downarrow & & \\
0 & \longrightarrow & M & \longrightarrow & D & \longrightarrow & F & \longrightarrow & 0.
\end{array}
$$

By hypothesis (2), $D \to U \to D$ is an automorphism of D. Hence $0 \to D \to U \to G \to 0$ is split exact and thus $D \in \mathcal{C}$. $\hfill\square$

Theorem 7.2.6. *Let $(\mathcal{F}, \mathcal{C})$ be a cotorsion theory for the category of R-modules with enough injectives and projectives such that $\mathcal{F}$ is closed under well ordered inductive limits. Then every R-module has an $\mathcal{F}$-cover and a $\mathcal{C}$-envelope.*

Proof. The fact that M has an $\mathcal{F}$-cover is a consequence of Theorem 5.2.3.

To prove that M has a $\mathcal{C}$-envelope we first note that we have an exact sequence $0 \to M \to C \to F \to 0$ with $C \in \mathcal{C}$ and $F \in \mathcal{F}$. So we consider the category of such short exact sequences. A morphism in this category will be given by a commutative diagram

$$
\begin{array}{ccccccccc}
0 & \longrightarrow & M & \longrightarrow & C_1 & \longrightarrow & F_1 & \longrightarrow & 0 \\
 & & \| & & \downarrow & & \downarrow & & \\
0 & \longrightarrow & M & \longrightarrow & C_2 & \longrightarrow & F_2 & \longrightarrow & 0.
\end{array}
$$

Given an inductive system in this category, we can take the limit. That is, we get $0 \to M \to \varinjlim C_i \to \varinjlim F_i \to 0$ which will also be an exact sequence. If $0 \to M \to C \to F \to 0$ is in the category, we argue that the diagram

$$
\begin{array}{ccccccccc}
0 & \longrightarrow & M & \longrightarrow & \varinjlim C_i & \longrightarrow & \varinjlim F_i & \longrightarrow & 0 \\
 & & \| & & \big\downarrow & & \big\downarrow & & \\
0 & \longrightarrow & M & \longrightarrow & C & \longrightarrow & F & \longrightarrow & 0
\end{array}
$$

can be completed to a commutative diagram. But we have the exact sequence

$$
\mathrm{Hom}(\varinjlim C_i, C) \to \mathrm{Hom}(M, C) \to \mathrm{Ext}^1(\varinjlim F_i, C) = 0
$$

since $\mathcal{F}$ is closed under direct limits. This gives us our map $\varinjlim C_i \to C$ and then $\varinjlim F_i \to F$ is induced by this map.

So now we get a version of Lemma 5.2.4 which says that given a short exact sequence $0 \to M \to C \to F \to 0$ in our category, there is another exact sequence $0 \to M \to C' \to F' \to 0$ with $F' \in \mathcal{F}$ and a commutative diagram

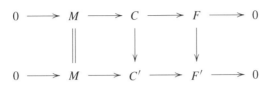

such that if

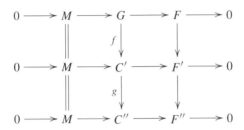

is a commutative diagram with $F'' \in \mathcal{F}$, then $\mathrm{Ker}(g \circ f) = \mathrm{Ker}\, f$. Then modifications of proofs of Lemmas 5.2.5 and 5.2.6 give an exact sequence $0 \to M \to D \to F \to 0$ with $F \in \mathcal{F}$ satisfying hypotheses (1) and (2) of the lemma above. But then $D \in C$ by the lemma and so $M \to D$ is a C-envelope. We note that this envelope is a special C-envelope.   □

Exercises

1. Suppose that for some class $\mathcal{F}$ of modules we have an $\mathcal{F}$-fibration $f_i : M_i \to N_i$ for each $i \in I$. Prove that $\prod_{i \in I} f_i : \prod_{i \in I} M_i \to \prod_{i \in I} N_i$ is an $\mathcal{F}$-fibration.

2. Show that $M \to 0$ is an $\mathcal{F}$-fibration if and only if $\mathrm{Ext}^1(F, M) = 0$ for all $F \in \mathcal{F}$.

3. Argue that if $f : M \to N$ and $g : N \to P$ are $\mathcal{F}$-fibrations, then so is $g \circ f : M \to P$.

4. Argue that $f : M \to N$ is an $\mathcal{F}$-fibration for every class $\mathcal{F}$ if and only if f is surjective with $\mathrm{Ker}\, f$ injective.

5. If $\mathcal{P}roj$ is the class of projective modules, argue that $f : M_1 \to M_2$ is a $\mathcal{P}roj$-fibration if and only if f is surjective.

6. Show that $(\mathcal{I}nj, \mathcal{I}nj^{\perp})$ is a cotorsion theory for the category of R-modules if and only if $R \in \mathcal{I}nj$ and R is left noetherian. In this case show that $(\mathcal{I}nj, \mathcal{I}nj^{\perp})$ has enough injectives and projectives.

7. Let $\mathcal{F}$ be a class of modules closed under extensions, under taking summands and such that every module has a surjective $\mathcal{F}$-cover. Argue that $(\mathcal{F}, \mathcal{F}^{\perp})$ is a cotorsion theory with enough injectives and projectives.

8. Let $(\mathcal{F}, \mathcal{C})$ be a cotorsion theory of R-modules. Let $F \to M$ be an $\mathcal{F}$-cover of the module M and suppose $F \to C$ is a $\mathcal{C}$-envelope of F. Let

$$
\begin{array}{ccc}
F & \longrightarrow & M \\
\downarrow & & \downarrow \\
C & \longrightarrow & D
\end{array}
$$

be a pushout diagram. Prove that $M \to D$ is a $\mathcal{C}$-envelope and that $C \to D$ is an $\mathcal{F}$-cover.

7.3 Set theoretic homological algebra

In this section, we consider ordinal numbers to give results about finding extensions of linear maps and get splitting results.

Lemma 7.3.1. *Given a set X, there exists a limit ordinal λ such that if $(\alpha_x)_{x \in X}$ is a family of ordinals such that $\alpha_x < \lambda$ for all $x \in X$, then there exists an ordinal $\lambda' < \lambda$ such that $\alpha_x \leq \lambda'$ for all $x \in X$.*

Proof. Let Card $X \leq \aleph_\beta$ and let λ be the least ordinal such that $\text{Card}(\lambda) = \aleph_{\beta+1}$. Then λ is a limit ordinal and if $\alpha < \lambda$, $\text{Card}(\alpha) \leq \aleph_\beta$. Now let $(\alpha_x)_{x \in X}$ be any family of ordinals such that $\alpha_x < \lambda$ for all $x \in X$. Well order X and let $\lambda' = \sum_{x \in X} \alpha_x$. Then $\alpha_x \leq \lambda'$ for each $x \in X$ (see Exercises 8 and 9 of Section 1.1) and

$$
\text{Card}(\lambda') = \sum_{x \in X} \text{Card}(\alpha_x) \leq \text{Card } X . \aleph_\beta \leq \aleph_\beta^2 = \aleph_\beta.
$$

So $\lambda' < \lambda$. $\square$

Corollary 7.3.2. *If X and λ are as in the lemma above and if $(Y_\alpha)_{\alpha < \lambda}$ is a family of subsets of a set Y such that $Y_\alpha \subset Y_\beta$ when $\alpha \leq \beta < \lambda$ and such that $Y = \bigcup_{\alpha < \lambda} Y_\alpha$, then for any function $f : X \to Y$ there is an $\alpha < \lambda$ such that $f(X) \subset Y_\alpha$.*

Proof. For any $x \in X$, let $\alpha_x < \lambda$ be any ordinal such that $f(x) \in Y_{\alpha_x}$. Then if $\alpha_x \leq \lambda' < \lambda$ for all x we have $f(X) \subset Y_{\lambda'}$. $\square$

We note that this lemma says that any function $f : X \to Y$ has a decomposition $X \to Y_\alpha \to Y$ with $\alpha < \lambda$ and $Y_\alpha \to Y$ the canonical injection. We will apply this lemma when X and Y are modules, f is linear, and all $Y_\alpha \subset Y$ are submodules.

Definition 7.3.3. *Given an ordinal number λ and a family $(M_\alpha)_{\alpha < \lambda}$ of submodules of a module M, we say that the family is a continuous (well ordered) chain of submodules if $M_\alpha \subset M_\beta$ whenever $\alpha \leq \beta < \lambda$ and if $M_\beta = \bigcup_{\alpha < \beta} M_\alpha$ whenever $\beta < \lambda$ is a limit ordinal. A family $(M_\alpha)_{\alpha \leq \lambda}$ is called a continuous chain if $(M_\alpha)_{\alpha < \lambda+1}$ is such.*

Theorem 7.3.4. *Let M and N be modules and suppose M is the union of a continuous chain of submodules $(M_\alpha)_{\alpha<\lambda}$. Then if $\text{Ext}^1(M_0, N) = 0$ and $\text{Ext}^1(M_{\alpha+1}/M_\alpha, N) = 0$ whenever $\alpha + 1 < \lambda$, then $\text{Ext}^1(M, N) = 0$.*

Proof. We use the principle of transfinite induction in Proposition 1.1.18. So suppose $\beta < \lambda$ and that $\text{Ext}^1(M_\alpha, N) = 0$ for all $\alpha < \beta$. Then we must argue that $\text{Ext}^1(M_\beta, N) = 0$. If β is not a limit ordinal, let $\beta = \alpha + 1$. We have the exact sequence

$$0 \to M_\alpha \to M_{\alpha+1} \to M_{\alpha+1}/M_\alpha \to 0$$

where $\text{Ext}^1(M_\alpha, N) = 0$ by the induction hypothesis and $\text{Ext}^1(M_{\alpha+1}/M_\alpha, N) = 0$ by assumption. So we get that $\text{Ext}^1(M_{\alpha+1}, N) = 0$.

So now assume $\beta < \lambda$ is a limit ordinal such that $\text{Ext}^1(M_\alpha, N) = 0$ for all $\alpha < \beta$. To argue that $\text{Ext}^1(M_\beta, N) = 0$ we let

$$0 \to N \xrightarrow{f} G \xrightarrow{g} M_\beta \to 0$$

be an exact sequence. We must prove that this sequence splits, that is, we should prove that there is a section $s : M_\beta \to G$ for g. For $\alpha < \beta$ we have the exact sequence

$$0 \to N \xrightarrow{f} g^{-1}(M_\alpha) \to M_\alpha \to 0.$$

By hypothesis, each of these sequences splits and so has a section s_α. We use transfinite induction again to argue that we can find compatible such sections, that is, such that if $\alpha \le \alpha' < \beta$ then $s_{\alpha'}|M_\alpha$ agrees with s_α. If we have compatible sections for all $\alpha < \tau \le \beta$ where τ is a limit ordinal, we can let s_τ agree with all s_α for $\alpha < \tau$ and we get compatible sections s_α for all $\alpha \le \tau$.

So the problem reduces to arguing that given a section s_α (for any $\alpha + 1 < \lambda$) there is a section $s_{\alpha+1}$ that agrees with s_α on M_α. Since by hypothesis $\text{Ext}^1(M_{\alpha+1}, N) = 0$ there is a section $t : M_{\alpha+1} \to g^{-1}(M_{\alpha+1})$. But then g is 0 on the image of $s_\alpha - (t|M_\alpha)$ and so $s_\alpha - (t|M_\alpha)$ maps M_α into $f(N) \cong N$. But

$$\text{Ext}^1(M_{\alpha+1}/M_\alpha, f(N)) \cong \text{Ext}^1(M_{\alpha+1}/M_\alpha, N) = 0$$

So $s_\alpha - (t|M_\alpha)$ can be extended to a linear map $u : M_{\alpha+1} \to f(N)$. But then $s_{\alpha+1} = t + u$ gives the desired section. $\square$

Corollary 7.3.5. *Let $(\mathcal{F}, \mathcal{C})$ be any cotorsion theory and suppose a module F is the union of a continuous chain $(F_\alpha)_{\alpha<\lambda}$ of submodules. If $F_0 \in \mathcal{F}$ and $F_{\alpha+1}/F_\alpha \in \mathcal{F}$ whenever $\alpha + 1 < \lambda$, then $F \in \mathcal{F}$.*

Exercises

1. Prove that given any ring R there exists a limit ordinal number λ such that if a module $E = \bigcup_{\alpha < \lambda} E_\alpha$ for injective submodules $E_\alpha \subset E$ where $E_\alpha \subset E_\beta$ whenever $\alpha \leq \beta < \lambda$, then E is injective. Argue that this holds for $\lambda = \omega$ (recall that $\mathrm{Ord}(\mathbb{N}) = \omega$) if and only if R is left noetherian.

2. Let L be the union of a continuous chain of submodules $(L_\alpha)_{\alpha < \lambda}$. Let $n \geq 0$ and suppose $\mathrm{proj\,dim}\, L_0 \leq n$ and $\mathrm{proj\,dim}\, L_{\alpha+1}/L_\alpha \leq n$ when $\alpha + 1 < \lambda$. Then prove that $\mathrm{proj\,dim}\, L \leq n$.
   Hint: If $n = 0$, then $L_{\alpha+1} = L_\alpha \oplus S_\alpha$ for some $S_\alpha \subset L_{\alpha+1}$ when $\alpha + 1 < \lambda$. Let $S_0 = L_0$. Prove that L is the direct sum of all the S_α, $\alpha \leq \lambda$. If $n \geq 1$, use dimension shifting and Theorem 7.3.4.

3. If $(\mathcal{F}, \mathcal{C})$ is the cotorsion theory of the category of $\mathbb{Z}$-modules cogenerated by $\mathbb{Z}/(p)$ for some prime p, argue that if the $\mathbb{Z}$-module F is such that every element has order a power of p then $F \in \mathcal{F}$.

4. Use Problem 3 above to argue that $(\mathcal{F}, \mathcal{C})$ of that problem has enough injectives (and so also enough projectives).

7.4 Cotorsion theories with enough injectives and projectives

In this section, we prove a result guaranteeing enough injectives for any cotorsion theory cogenerated by a set.

We start with a submodule $S \subset P$, a linear $S \to M$, and consider the problem of extending $S \to M$ to P. We form the pushout diagram

$$
\begin{array}{ccc}
S & \longrightarrow & P \\
\downarrow & & \downarrow \\
M & \longrightarrow & G.
\end{array}
$$

From the construction, $M \to G$ is an injection (so we identify M with a submodule of G) and $G/M \cong P/S$. But then also the map $P \to G$ agrees with $S \to M$ on S. So the map $S \to M$ can be extended to a map of P into G.

With a slight generalization of this idea we can construct G with $M \subset G$ so that for every linear $S \to M$ there is a linear $P \to G$ agreeing with $S \to M$ on S. To do so, we consider the evaluation map $S^{(\mathrm{Hom}(S,M))} \to M$ mapping $(x_f)_{f \in \mathrm{Hom}(S,M)}$ to $\sum_{f \in \mathrm{Hom}(S,M)} f(x_f)$, and then the pushout diagram

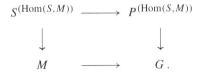

Then we see that we have $M \subset G$ and that for any linear $f : S \to M$ there is a linear $P \to G$ agreeing with f on S. But we also have $G/M \cong (P/S)^{(\mathrm{Hom}(S,M))}$, that is, G/M is a direct sum of copies of P/S.

Theorem 7.4.1. *If a cotorsion theory* $(\mathcal{F}, \mathcal{C})$ *is cogenerated by a set, then it has enough injectives (and so enough projectives).*

Proof. By an earlier comment, we can assume that $(\mathcal{F}, \mathcal{C})$ is cogenerated by a single module A. We let $0 \to S \to P \to A \to 0$ be exact with P projective. Let M be an R-module. For any ordinal λ we use transfinite induction to construct a continuous chain of modules $(M_\alpha)_{\alpha < \lambda}$ so that $M_0 = M$ and so that for any $\alpha + 1 < \lambda$ any linear $S \to M_\alpha$ has an extension $P \to M_{\alpha+1}$, and finally so that $M_{\alpha+1}/M_\alpha$ is a direct sum of copies of $P/S \cong A$ for $\alpha + 1 < \lambda$.

We now use Corollary 7.3.2 with S the set X of that corollary and find a corresponding λ. Let $C = \bigcup_{\alpha < \lambda} M_\alpha$. Then for any linear $S \to C$, there a factorization $S \to M_\alpha \to C$ for some $\alpha < \lambda$. Since λ is a limit ordinal, we have $\alpha + 1 < \lambda$ and so by construction there is a linear $P \to M_{\alpha+1}$ agreeing with $S \to M_\alpha$. This says that any linear $S \to C$ has an extension $P \to C$. But this is equivalent to the fact that $\mathrm{Ext}^1(A, C) = 0$. So $C \in \mathcal{C}$ since A cogenerates $(\mathcal{F}, \mathcal{C})$. Since $A \in \mathcal{F}$, any direct sum U of copies of A is in $\mathcal{F}$. Now let $F = C/M = (\bigcup_{\alpha < \lambda} M_\alpha)/M$. If we let $F_\alpha = M_\alpha/M$ for all $\alpha < \lambda$ and then use Theorem 7.3.4 we get that $\mathrm{Ext}^1(F, D) = 0$ for any $D \in \mathcal{C}$ and so $F \in \mathcal{F}$. Hence we have the desired exact sequence $0 \to M \to C \to F \to 0$. So $(\mathcal{F}, \mathcal{C})$ has enough injectives. $\qquad\square$

Definition 7.4.2. A cotorsion theory $(\mathcal{F}, \mathcal{C})$ with $\mathcal{F}$ the class of flat modules (and so $\mathcal{F}^\perp = \mathcal{C}$ the class of cotorsion modules) is called the *flat cotorsion theory.*

Proposition 7.4.3. *Let R be any ring. The flat cotorsion theory* $(\mathcal{F}, \mathcal{C})$ *of the category of R-modules is cogenerated by a set.*

Proof. Let F be a flat R-module. By Lemma 5.3.12, if Card $R \leq \aleph_\beta$ then for each $x \in F$ there is a pure submodule $S \subset F$ with $x \in S$ such that Card $S \leq \aleph_\beta$ (simply let $N = Rx$ and $f = \mathrm{id}_N$ in the lemma). So we can write F as a union of a continuous chain $(F_\alpha)_{\alpha < \lambda}$ of pure submodules of F such that Card $F_0 \leq \aleph_\beta$ and Card $F_{\alpha+1}/F_\alpha \leq \aleph_\beta$ whenever $\alpha + 1 < \lambda$. But then by Theorem 7.3.4 we see that if C is an R-module, $\mathrm{Ext}^1(F_0, C) = 0$, and $\mathrm{Ext}^1(F_{\alpha+1}/F_\alpha, C) = 0$ whenever $\alpha + 1 < \lambda$, then $\mathrm{Ext}^1(F, C) = 0$. So it follows that if Y is a set of representatives of all flat modules G with Card $G \leq \aleph_\beta$, then C is cotorsion if and only if $\mathrm{Ext}^1(G, C) = 0$ for all $G \in Y$. But then this just says that the given flat cotorsion theory is cogenerated by the set Y.   □

We are now in a position to state the following important result.

Theorem 7.4.4. *The class of flat R-modules is covering for any ring R.*

Proof. This follows immediately from Proposition 7.4.3 above and Theorems 7.4.1 and 5.2.3.   □

Now let $n \geq 1$ be given and for a ring R let $\mathcal{L}$ denote the class of R-modules L such that proj dim $_R L \leq n$.

Proposition 7.4.5. *Let* Card $R \leq \aleph_\beta$ *and* $L \in \mathcal{L}$. *If* $x \in L$, *then there is a submodule* $L' \subset L$ *with* $x \in L'$ *such that* Card $L' \leq \aleph_\beta$ *and* $L', L/L' \in \mathcal{L}$.

Proof. We use the Eilenberg trick, that is, if P is a projective module then there is a free module F such that $P \oplus F$ is free (if $P \oplus P'$ is free just let $F = P' \oplus P \oplus P' \oplus P \oplus \cdots$). Hence if $0 \rightarrow P_n \rightarrow \cdots \rightarrow P_0 \rightarrow L \rightarrow 0$ is a projective resolution of L we can take the direct sum of this complex with complexes of the form $0 \rightarrow \cdots \rightarrow 0 \rightarrow F \xrightarrow{\text{id}} F \rightarrow 0 \rightarrow \cdots \rightarrow 0$ with F free and get a projective resolution $0 \rightarrow F_n \xrightarrow{\partial_n} \cdots \rightarrow F_0 \xrightarrow{\partial_0} L \rightarrow 0$ of L with each of $F_n, \ldots, F_0$ free. If $X_n, \ldots, X_0$ are bases of each of these free modules our objective is to choose subsets $Y_i \subset X_i$ for $i = 0, \ldots, n$ so that if $\langle Y_i \rangle$ is the free submodule of F_i generated by Y_i, then $0 \rightarrow \langle Y_n \rangle \rightarrow \cdots \rightarrow \langle Y_0 \rangle$ is an exact subcomplex of $0 \rightarrow F_n \rightarrow \cdots \rightarrow F_0$ which will give the desired L'. We accomplish this by choosing certain subsets $Z \subset X_i$ and using a zig-zag procedure. At each stage of the procedure we should check that the set Z can be chosen so that Card $Z \leq \aleph_\beta$.

We choose a finite subset $Z_0 \subset X_0$ such that $x \in \partial_0(\langle Z_0 \rangle)$. Then we choose a subset $Z_1 \subset X_1$ so that $\partial_1(\langle Z_1 \rangle) \supset \text{Ker}(\partial_0|\langle Z_0 \rangle)$. We then choose $Z_2 \subset X_2$ so that $\partial_2(\langle Z_2 \rangle) \supset \text{Ker}(\partial_1|\langle Z_1 \rangle)$. We continue this procedure until we have $Z_n \subset X_n$ with $\partial_n(\langle Z_n \rangle) \supset \text{Ker}(\partial_{n-1}|\langle Z_{n-1} \rangle)$. We now enlarge Z_{n-1} to Z'_{n-1} in such a way that $\partial_n(\langle Z_n \rangle) \subset \langle Z'_{n-1} \rangle$. Then we enlarge Z_{n-2} to Z'_{n-2} so that $\partial_{n-1}(\langle Z'_{n-1} \rangle) \subset \langle Z'_{n-2} \rangle$. Continuing in this manner, we construct $Z'_n, Z'_{n-1}, \ldots, Z'_0$ satisfying the obvious conditions. Now we start over and enlarge Z'_1 to Z''_1 in such a way that $\partial_1(\langle Z''_1 \rangle) \supset \text{Ker}(\partial_0|\langle Z'_0 \rangle)$. We then enlarge Z'_2 to Z''_2 and so on. Continuing this zig-zag procedure and eventually letting $Y_i \subset X_i$ be the union of all the subsets of X_i we chose while implementing the procedure we see that the sequence $0 \rightarrow \langle Y_n \rangle \rightarrow \cdots \rightarrow \langle Y_0 \rangle$ is exact. By construction, each Y_i is such that Card $Y_i \leq \aleph_\beta$. Then if we let $L' = \partial_0(\langle Y_0 \rangle)$ we have Card $L' \leq \aleph_\beta$ and $L' \in \mathcal{L}$ since $0 \rightarrow \langle Y_n \rangle \rightarrow \cdots \rightarrow \langle Y_0 \rangle \rightarrow L' \rightarrow 0$ is a projective resolution of L'. The quotient of the exact complex $0 \rightarrow F_n \rightarrow \cdots \rightarrow F_0 \rightarrow L \rightarrow 0$ by the exact subcomplex $0 \rightarrow \langle Y_n \rangle \rightarrow \cdots \rightarrow \langle Y_0 \rangle \rightarrow L' \rightarrow 0$ is an exact complex which gives a projective resolution of L/L'. This then gives that $L/L' \in \mathcal{L}$.   □

Using the notation above we get the following result.

Theorem 7.4.6. $(\mathcal{L}, \mathcal{L}^\perp)$ *is a cotorsion theory of the category of R-modules with enough injectives and projectives.*

Proof. By the previous result we see that any $L \in \mathcal{L}$ can be written $L = \bigcup_{\alpha < \lambda} L_\alpha$ with $(L_\alpha)_{\alpha < \lambda}$ a continuous chain of submodules with $L_0 \in \mathcal{L}$ and $L_{\alpha+1}/L_\alpha \in \mathcal{L}$ when $\alpha + 1 < \lambda$ and with Card L_0, Card $L_{\alpha+1}/L_\alpha \le \aleph_\beta$. Hence if B is the direct sum of a representative set of $L \in \mathcal{L}$ with Card $L \le \aleph_\beta$, we see that $G \in \mathcal{L}^\perp$ if and only if $\operatorname{Ext}^1(B, G) = 0$.

Now let K be any R-module. We use the procedure of Theorem 7.4.1 to get an exact sequence $0 \to K \to A \to L \to 0$ with $A \in \mathcal{L}^\perp$ and $L \in \mathcal{L}$. We note that $L \in \mathcal{L}$ since L can be written as a continuous chain of submodules $L = \bigcup_{\alpha < \lambda} L_\alpha$ with $L_0 \in \mathcal{L}$ and $L_{\alpha+1}/L_\alpha \in \mathcal{L}$ when $\alpha + 1 < \lambda$. Then we apply Problem 2 of Section 3 of this chapter and get that $L \in \mathcal{L}$, that is, proj dim $L \le n$.

Now let M be a module and let $0 \to K \to P \to M \to 0$ be exact with P projective. By the above applied to K we get an exact sequence $0 \to K \to A \to L \to 0$ with $A \in \mathcal{L}^\perp$ and $L \in \mathcal{L}$. Using a pushout of $K \to P$ and $K \to A$, we have a commutative diagram

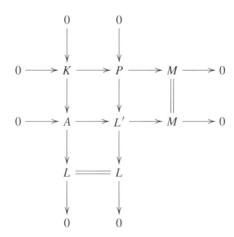

with exact rows and columns. Then we see that since $P, L \in \mathcal{L}$ we have that $L' \in \mathcal{L}$. Now suppose $M \in {}^\perp(\mathcal{L}^\perp)$. Then since $A \in \mathcal{L}^\perp$ the middle row of the diagram splits and so M is a direct summand of L'. Hence $M \in \mathcal{L}$. This shows that $\mathcal{L} = {}^\perp(\mathcal{L}^\perp)$. Hence $(\mathcal{L}, \mathcal{L}^\perp)$ is a cotorsion theory with enough injectives and projectives. $\qquad \square$

Exercises

1. Prove a version of Proposition 7.4.5 but with $n = 0$. So prove that for a ring R there is an $\aleph_\beta$ so that if $x \in P$ where P is a projective R-module there is a direct summand S of P with $x \in S$ and Card $S \le \aleph_\beta$.

Hint: Let $F = P \oplus P'$ be a free module with a base $(x_i)_{i \in I}$. Find finite subsets $I_0 \subset I_1 \subset I_2 \subset \cdots$ of I such that if F_n is the free submodule of F generated by $(x_i)_{i \in I_n}$, then $x \in F_0$ and for $n \geq 0$, F_{n+1} contains the projection of F_n on P and P'. Then if $J = \bigcup_{i=0}^{\infty} I_n$ and if G is generated by the x_j with $j \in I$, deduce that $G = G \cap P + G \cap P'$ and $G \cap P$ is the desired summand. Then note that G and $G \cap P$ are countably generated.

2. Let $\mathcal{A}bs$ be the class of absolutely pure R-modules. Use the methods of this section to argue that $({}^{\perp}\mathcal{A}bs, \mathcal{L})$ is a cotorsion theory with enough injectives and projectives. Argue that $F \in {}^{\perp}\mathcal{A}bs$ if and only if F is a direct summand of a module G which can be written as the union of a continuous chain $(G_\alpha)_{\alpha < \lambda}$ of submodules such that G_0 and $G_{\alpha+1}/G_\alpha$ for $\alpha + 1 < \lambda$ are finitely presented modules. (For the last part of the problem use the material in the proof of Theorem 7.4.1 and Proposition 7.1.7 to argue that for $F \in {}^{\perp}\mathcal{A}bs$, there is an exact sequence $0 \to A \to G \to F \to 0$ with $A \in \mathcal{A}bs$ and with G as described above).

Chapter 8

Relative Homological Algebra and Balance

Covers and envelopes of modules were defined and studied in Chapters 5, 6 and 7. We now redefine these notions for any abelian category.

Let C be an abelian category and $\mathcal{F}$ be a class of objects of C. Then a morphism $\varphi : F \to M$ of C is called an $\mathcal{F}$-*precover* of M if $F \in \mathcal{F}$ and $\mathrm{Hom}(F', F) \to \mathrm{Hom}(F', M) \to 0$ is exact for all $F' \in \mathcal{F}$. If moreover, any morphism $f : F \to F$ such that $\varphi = \varphi \circ f$ is an automorphism of F, then $\varphi : F \to M$ is called an $\mathcal{F}$-*cover* of M (see Definition 5.1.1). An $\mathcal{F}$-*preenvelope* and an $\mathcal{F}$-*envelope* $M \to F$ are defined dually (see Definition 6.1.1). If $\mathcal{F}$-covers and envelopes exist, then they are unique up to isomorphism.

We note that an $\mathcal{F}$-precover $\varphi : F \to M$ in C is not necessarily an epimorphism. But if C has enough projectives and these are in $\mathcal{F}$, then φ is an epimorphism. Similarly, if C contains enough injective objects and these are in $\mathcal{F}$, then an $\mathcal{F}$-preenvelope $M \to F$, if it exists, is a monomorphism.

If every object of C has an $\mathcal{F}$-(pre)cover, $\mathcal{F}$ is said to be (*pre*)*covering*. Similarly, if every object has an $\mathcal{F}$-(pre)envelope, we say that $\mathcal{F}$ is (*pre*)*enveloping*. Eilenberg–Moore [44] consider such classes but with different terminology.

All functors in this chapter will be additive.

8.1 Left and right $\mathcal{F}$-resolutions

Definition 8.1.1. Let C, D, and E be abelian categories and $T : \mathsf{C} \times \mathsf{D} \to \mathsf{E}$ be an additive functor contravariant in the first variable and covariant in the second. If $\mathcal{F}$ is a class of objects of C, we will say that a complex

$$\cdots \to D_1 \to D_0 \to D^0 \to D^1 \to \cdots$$

in D is $T(\mathcal{F}, -)$ *exact* if for every $F \in \mathcal{F}$ the complex

$$\cdots \to T(F, D_1) \to T(F, D_0) \to T(F, D^0) \to T(F, D^1) \to \cdots$$

is an exact sequence in E.

If $\mathcal{F}$ is a class of objects in D, we will say that a complex

$$\cdots \to C_1 \to C_0 \to C^0 \to C^1 \to \cdots$$

in C is $T(-, \mathcal{F})$ *exact* if for each $F \in \mathcal{F}$ the complex

$$\cdots \to T(C^0, F) \to T(C_0, F) \to T(C_1, F) \to \cdots$$

is an exact sequence in E.

Frequently, in place of saying the complex is $T(\mathcal{F}, -)$ exact we say that $T(F, -)$ *makes the complex exact* for all $F \in \mathcal{F}$, and we say that $T(-, F)$ makes the complex exact for all $F \in \mathcal{F}$ when the complex is $T(-, \mathcal{F})$ exact.

We will use the same terminology for finite complexes. The exactness of $T(\mathcal{F}, -)$ and $T(-, \mathcal{F})$ with other choices of variances of T are defined similarly.

Definition 8.1.2. Let C be an abelian category and $\mathcal{F}$ be a class of objects of C. By a *left $\mathcal{F}$-resolution* of an object M of C, we will mean a $\mathrm{Hom}(\mathcal{F}, -)$ exact complex

$$\cdots \to F_1 \to F_0 \to M \to 0$$

(not necessarily exact) with each $F_i \in \mathcal{F}$.

A *right $\mathcal{F}$-resolution* of an object M of C is a $\mathrm{Hom}(-, \mathcal{F})$ exact complex

$$0 \to M \to F^0 \to F^1 \to \cdots$$

(not necessarily exact) with each $F^i \in \mathcal{F}$. Eilenberg–Moore [44] call these resolutions *projective (injective) resolutions of M for the class $\mathcal{F}$*, respectively.

If $\cdots \to F_1 \to F_0 \to M \to 0$ is a left $\mathcal{F}$-resolution, then $K_0 = M$, $K_i = \mathrm{Ker}(F_{i-1} \to F_{i-2})$ for $i \geq 1$, is called an *ith $\mathcal{F}$-syzygy* of M where $F_{-1} = M$. If $0 \to M \to F^0 \to F^1 \to \cdots$ is a right $\mathcal{F}$-resolution of M, then $C^0 = M$, $C^i = \mathrm{Coker}(F^{i-2} \to F^{i-1})$ for $i \geq 1$, is called an *ith $\mathcal{F}$-cosyzygy* of M where $F^{-1} = M$. The complex

$$\cdots \to F_1 \to F_0 \to F^0 \to F^1 \to \cdots$$

(with $F_0 \to F^0$ the composition $F_0 \to M \to F^0$) is called a *complete $\mathcal{F}$-resolution* of M.

Proposition 8.1.3. *Let C be an abelian category and $\mathcal{F}$ be a class of objects of C. If $\mathcal{F}$ is precovering (preenveloping), then every object of C has a left (right) $\mathcal{F}$-resolution. Furthermore, if $\mathcal{F}$ is both precovering and preenveloping, then every object of C has a complete $\mathcal{F}$-resolution.*

Proof. Let M be an object of C. If $F_0 \to M$ is an $\mathcal{F}$-precover, let $F_1 \to \mathrm{Ker}(C_0 \to M)$ also be an $\mathcal{F}$-precover. Proceeding in this manner, we get a complex $\cdots \to F_1 \to F_0 \to M \to 0$ (not necessarily exact) which becomes exact when $\mathrm{Hom}(F, -)$ is applied to it for any $F \in \mathcal{F}$. Similarly for right $\mathcal{F}$-resolutions. $\square$

If $\mathcal{F}$ is precovering and $\cdots \to F_1' \to F_0' \to M \to 0$ is another left $\mathcal{F}$-resolution of M, then there exist morphisms $F_i \to F_i'$ making

$$
\begin{array}{ccccccccc}
\mathbf{F_\bullet} : \cdots & \longrightarrow & F_1 & \longrightarrow & F_0 & \longrightarrow & M & \longrightarrow & 0 \\
& & \downarrow & & \downarrow & & \| & & \\
\mathbf{F_\bullet'} : \cdots & \longrightarrow & F_1' & \longrightarrow & F_0' & \longrightarrow & M & \longrightarrow & 0
\end{array}
$$

into a commutative diagram. Furthermore, any two such collection of morphisms $F_i \to F_i'$ give homotopic morphisms of the deleted complex $\mathbf{F_\bullet} : \cdot \to F_1 \to F_0 \to 0$ to the complex $\mathbf{F_\bullet'} : \cdots \to F_1' \to F_0' \to 0$. Thus we get the usual uniqueness of complexes up to homotopy. Similarly for right $\mathcal{F}$-resolutions.

Definition 8.1.4. If $\mathcal{F}$ is covering, then the left $\mathcal{F}$-resolution of M

$$\cdots \to F_1 \to F_0 \to M \to 0$$

can be constructed so that $F_0 \to M$, $F_1 \to \mathrm{Ker}(F_0 \to M)$, $F_{i+1} \to \mathrm{Ker}(F_i \to F_{i-1})$ for $i \geq 1$, are $\mathcal{F}$-covers. In this case, the complex $\cdots \to F_1 \to F_0 \to M \to 0$ is called a *minimal left $\mathcal{F}$-resolution* of M. Similarly, if $\mathcal{F}$ is enveloping, then a *minimal right $\mathcal{F}$-resolution* of M can be constructed using $\mathcal{F}$-envelopes. So if $\mathcal{F}$ is both covering and enveloping, then M has a *complete minimal $\mathcal{F}$-resolution*. Minimal $\mathcal{F}$-resolutions are unique up to isomorphism.

Exercises

1. If $\mathcal{F}$ is the class of projective R-modules, prove that a complex $\cdots \to F_1 \to F_0 \to M \to 0$ of an R-module M with each $F_i \in \mathcal{F}$ is a left $\mathcal{F}$-resolution if and only if it is exact. State and prove the dual result in the case $\mathcal{F}$ is the class of injective R-modules.

2. (Comparison Theorem) Let $\mathcal{F}$ be precovering and $\mathbf{F} : \cdots \to F_1 \to F_0 \to M \to 0$, $\mathbf{F}' : \cdots \to F_1' \to F_0' \to M' \to 0$ be left $\mathcal{F}$-resolutions of M, M' respectively. Prove that each morphism $\varphi : M \to M'$ induces a chain map $\Phi : \mathbf{F_\bullet} \to \mathbf{F_\bullet'}$ which is unique up to homotopy.

3. Using Problem 2 above, state and prove the Comparison Theorem for a preenveloping class $\mathcal{F}$.

8.2   Derived functors and balance

Let $\mathcal{F}, \mathcal{G}$ be precovering, preenveloping classes of an abelian category C respectively, and T be an additive functor from C to some abelian category. Let $\mathbf{F_\bullet}$ be a deleted complex corresponding to a left $\mathcal{F}$-resolution of an object of C. If T is covariant, then the homology groups of $T(\mathbf{F_\bullet})$ give *left derived functors $L_n T$* of T. Similarly, the *right* *derived functors* $R^n T$ are the nth cohomology groups of $T(\mathbf{G}^\bullet)$ where $\mathbf{G}^\bullet$ corresponds to a deleted right $\mathcal{G}$-resolution. If T is contravariant, then left (right) derived functors can be computed using right $\mathcal{G}$-resolutions (left $\mathcal{F}$-resolutions). Furthermore, for any T, there are natural transformations $L_0 T \to T$ and $T \to R^0 T$. Moreover, derived functors of T for a given object are unique up isomorphism by Proposition 1.4.13 since $\mathcal{F}$-resolutions are unique up to homotopy. We note that $L_0 T \cong T$ if T is right exact and if for each M, $F_1 \to F_0 \to M \to 0$ is exact where $\cdots \to F_1 \to F_0 \to M \to 0$ is a left $\mathcal{F}$-resolution of M. We have $R^0 T \cong T$ if T is left exact and if for each M, $0 \to M \to G^0 \to G^1$ is exact where $0 \to M \to G^0 \to G^1 \to \cdots$ is a right $\mathcal{G}$-resolution of M.

If $F \in \mathcal{F}$, then $L_0 T(F) \cong F$ and $L_n T(F) = 0$ if $n > 0$. Similarly if $G \in \mathcal{G}$, then $R^0 T(G) \cong G$ and $R^n T(G) = 0$ if $n > 0$.

The following result allows us to obtain the familiar connecting homomorphisms.

Lemma 8.2.1 (Horseshoe Lemma). *Let $\mathcal{F}$ be a precovering class closed under finite direct sums of an abelian category* **C**. *Suppose* $0 \to M' \to M \to M'' \to 0$ *is a complex such that* $M', M, M'' \in$ **C** *and such that*

$$0 \to \mathrm{Hom}(F, M') \to \mathrm{Hom}(F, M) \to \mathrm{Hom}(F, M'') \to 0$$

is exact for all $F \in \mathcal{F}$. *If* $\cdots \to F_1' \to F_0' \to M' \to 0$ *and* $\cdots \to F_1'' \to F_0'' \to M'' \to 0$ *are left $\mathcal{F}$-resolutions, then we can construct the following commutative diagram such that the middle column is a left $\mathcal{F}$-resolution of* M.

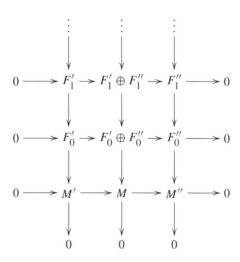

Proof. The map $F_0'' \to M''$ has a factorization $F_0'' \to M \to M''$ since $\mathrm{Hom}(F, M) \to \mathrm{Hom}(F, M'') \to 0$ is exact for all $F \in \mathcal{F}$. So we get a map $F_0' \oplus F_0'' \to M$ and hence

a commutative diagram

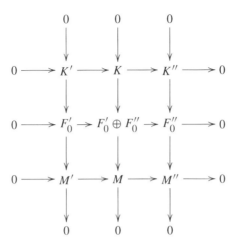

with obvious maps. But then all the rows and columns become exact if we apply $\text{Hom}(F, -)$ to the diagram with $F \in \mathcal{F}$. Thus $F_0' \oplus F_0'' \to M$ is an $\mathcal{F}$-precover. Now we repeat the argument using the complex $0 \to K' \to K \to K'' \to 0$. □

Remark 8.2.2. There is a dual result involving preenveloping classes of C.

Theorem 8.2.3. *Let $\mathcal{F}$ be a precovering class closed under finite direct sums of an abelian category C and $0 \to M' \to M \to M'' \to 0$ be a $\text{Hom}(\mathcal{F}, -)$ exact complex of objects of C. Then*

(1) *If T is a covariant functor, there is a long exact sequence $\cdots \to L_n T(M'') \to L_{n-1}T(M') \to L_{n-1}(M) \to L_{n-1}T(M'') \to \cdots \to L_0 T(M') \to L_0 T(M) \to L_0 T(M'') \to 0$.*

(2) *If T is a contravariant functor, there is a long exact sequence $0 \to R^0 T(M'') \to R^0 T(M) \to R^0 T(M') \to \cdots \to R^{n-1}T(M'') \to R^{n-1}T(M) \to R^{n-1}T(M') \to R^n T(M'') \to \cdots$.*

Proof. If $\mathbf{F}_\bullet'$, $\mathbf{F}_\bullet''$ denote deleted complexes associated with left $\mathcal{F}$-resolutions of M', M'', respectively, then there is a deleted complex $\mathbf{F}_\bullet$ associated with a left $\mathcal{F}$-resolution of M given by the lemma above. So we have an exact sequence of complexes $0 \to \mathbf{F}_\bullet' \to \mathbf{F}_\bullet \to \mathbf{F}_\bullet'' \to 0$. Then one computes the homology of the exact sequences $0 \to T(\mathbf{F}_\bullet') \to T(\mathbf{F}_\bullet) \to T(\mathbf{F}_\bullet'') \to 0$ and $0 \to T(\mathbf{F}_\bullet'') \to T(\mathbf{F}_\bullet) \to T(\mathbf{F}_\bullet') \to 0$, respectively, to get the desired long exact sequences by Theorem 1.4.7. We note that it is not hard to prove that these long exact sequences are functorial (relative to the complexes $0 \to M' \to M \to M'' \to 0$). □

Corollary 8.2.4. *If $M \in \mathcal{F}$, then*

(1) *If T is covariant, then $L_n T(M') \cong L_{n+1}T(M'')$ for all $n \geq 1$.*

(2) *If T is contravariant, then $R^n T(M') \cong R^{n+1} T(M'')$ for all $n \geq 1$.*

Proof. We simply note that if $M \in \mathcal{F}$, then $L_n T(M) = R^n T(M) = 0$ for all $n \geq 1$. $\square$

By Remark 8.2.2, we have the following dual result.

Theorem 8.2.5. *Let $\mathcal{F}$ be a preenveloping class closed under finite direct sums of an abelian category C and $0 \to M' \to M \to M'' \to 0$ be a $\mathrm{Hom}(-, \mathcal{F})$ exact complex of objects of C. Then*

(1) *If T is covariant, then there is a long exact sequence $0 \to R^0 T(M') \to R^0 T(M) \to R^0 T(M'') \to \cdots \to R^{n-1} T(M') \to R^{n-1} T(M) \to R^{n-1} T(M'') \to R^n T(M') \to \cdots$.*

(2) *If T is contravariant, there is a long exact sequence $\cdots \to L_n T(M') \to L_{n-1} T(M'') \to L_{n-1} T(M) \to L_{n-1} T(M') \to \cdots \to L_0 T(M'') \to L_0 T(M) \to L_0 T(M') \to 0$.*

Corollary 8.2.6. *If $M \in \mathcal{F}$, then*

(1) *If T is covariant, then $R^n T(M'') \cong R^{n+1} T(M')$ for all $n \geq 1$.*

(2) *If T is contravariant, then $L_n T(M'') \cong L_{n+1} T(M')$ for all $n \geq 1$.*

Now let $L_0 T \xrightarrow{\sigma} R^0 T$ be the composition of the natural transformations $L_0 T \to T$ and $T \to R^0 T$ where $L_0 T$ is computed relative to some precovering class $\mathcal{F}$ and where $R^0 T$ is computed relative to some preenveloping class $\mathcal{F}'$. Then we will let $\overline{L_0 T}$ and $\overline{R^0 T}$ denote the kernel and cokernel of the morphism σ. We note that if $M \in \mathcal{F}$, then $L_0 T(M) = T(M) = R^0 T(M)$ and so $\overline{L_0 T}(M) = \overline{R^0 T}(M) = 0$. We are now in a position to state the following result.

Theorem 8.2.7. *Let $0 \to M' \to M \to M'' \to 0$ be a complex of objects of an abelian category C. Suppose there is a precovering class $\mathcal{F}$ and preenveloping class $\mathcal{F}'$ of C closed under finite direct sums such that $\mathrm{Hom}(F, -)$, $\mathrm{Hom}(-, F')$ make the complex exact for all $F \in \mathcal{F}$, all $F' \in \mathcal{F}'$ respectively. Then*

(1) *If T is a covariant functor, then there is a long exact sequence $\cdots \to L_1 T(M'') \to \overline{L_0 T}(M') \to \overline{L_0 T}(M) \to \overline{L_0 T}(M'') \to \overline{R^0 T}(M') \to \overline{R^0 T}(M) \to \overline{R^0 T}(M'') \to R^1 T(M') \to \cdots$.*

(2) *If T is a contravariant functor, then there is a long exact sequence $\cdots \to L_1 T(M') \to \overline{L_0 T}(M'') \to \overline{L_0 T}(M) \to \overline{L_0 T}(M') \to \overline{R^0 T}(M'') \to \overline{R^0 T}(M) \to \overline{R^0 T}(M') \to R^1 T(M'') \to \cdots$.*

Proof. By Theorems 8.2.3 and 8.2.5, we have the following diagram

$$\cdots \longrightarrow L_1 T(M'') \longrightarrow L_0 T(M') \longrightarrow L_0 T(M) \longrightarrow L_0 T(M'') \longrightarrow 0$$
$$0 \longrightarrow R^0 T(M') \longrightarrow R^0 T(M) \longrightarrow R^0 T(M'') \longrightarrow R^1 T(M')$$

with exact rows. Chasing this diagram gives part (1) of the theorem. Part (2) follows similarly. □

Definition 8.2.8. The sequences in Theorem 8.2.7 above are called *extended long exact sequences of derived functors.*

Proposition 8.2.9. *Let $\mathcal{F}$ be a class of an abelian category C and*

$$\cdots \to F_1 \to F_0 \to F^0 \to F^1 \to \cdots$$

be a complete $\mathcal{F}$-resolution of an object M of C. Then

(1) *If T is a covariant functor, then the homology groups are $L_i T(M)$, $\overline{L_0 T}(M)$, $\overline{R^0 T}(M)$, and $R^i T(M)$ at $T(F_i)$, $T(F_0)$, $T(F^0)$, and $T(F^i)$ respectively where $i \geq 1$.*

(2) *If T is a contravariant functor, then the homology groups are $L_i T(M)$, $\overline{L_0 T}(M)$, $\overline{R^0 T}(M)$, and $R^i T(M)$ at $T(F^i)$, $T(F^0)$, $T(F_0)$, and $T(F_i)$ respectively where $i \geq 1$.*

Proof. This follows from the definitions. □

Definition 8.2.10. A sequence $\{T^i\}$ of functors is said to be *covariantly right strongly connected* if every exact sequence $0 \to M' \to M \to M'' \to 0$ of R-modules has an associated long exact sequence

$$\cdots \to T^i(M) \to T^i(M'') \to T^{i+1}(M') \to T^{i+1}(M) \to T^{i+1}(M'') \to \cdots$$

which is functorial in such short exact sequences.

We say that it is *contravariantly right strongly connected* if there exists a long exact sequence

$$\cdots \to T^i(M) \to T^i(M') \to T^{i+1}(M'') \to T^{i+1}(M) \to T^{i+1}(M') \to \cdots$$

which is also functorial in the short exact sequences. *Covariantly (contravariantly) left strongly connected* sequences $\{T_i\}$ are similarly defined with connecting homomorphisms $T_i(M'') \to T_{i-1}(M')$ ($T_i(M') \to T_{i-1}(M'')$), respectively.

We note that the sequences $\{R^i T\}$ and $\{L_i T\}$ in Theorems 8.2.3 and 8.2.5 are strongly connected.

Theorem 8.2.11. *Let* $\{F^i\}$ *and* $\{G^i\}$ *be covariantly right strongly connected sequences such that* $F^0 \cong G^0$ *and* $F^i(E) = G^i(E) = 0$ *for all injective R-modules E and for all* $i \geq 1$. *Then* $F^i \cong G^i$ *for each* $i \geq 0$.

Proof. We proceed by induction on i. If $i = 0$, then we are done.

Suppose $i = 1$. Then given an R-module M, we consider a short exact sequence $0 \to M \to E \to L \to 0$ where E is an injective R-module. We then have the following commutative diagram

$$
\begin{array}{ccccccc}
F^0(E) & \longrightarrow & F^0(L) & \longrightarrow & F^1(M) & \longrightarrow & F^1(E) = 0 \\
\downarrow & & \downarrow & & \downarrow{\scriptstyle\varphi} & & \\
G^0(E) & \longrightarrow & G^0(L) & \longrightarrow & G^1(M) & \longrightarrow & G^1(E) = 0
\end{array}
$$

where the first two vertical maps are isomorphisms. So φ is an isomorphism. Thus $F^1(M) \cong G^1(M)$ for all R-modules M. If $i \geq 2$, then $F^{i-1}(L) \cong G^{i-1}(L)$ by the induction hypothesis. Hence $F^i(M) \cong G^i(M)$ since $F^{i-1}(E) = G^{i-1}(E) = F^i(E) = G^i(E) = 0$. □

We also have the following dual result.

Theorem 8.2.12. *Let* $\{F^i\}$ *and* $\{G^i\}$ *be contravariantly right strongly connected sequences such that* $F^0 \cong G^0$ *and* $F^i(P) = G^i(P) = 0$ *for all free R-modules P and for all* $i \geq 1$. *Then* $F^i \cong G^i$ *for each* $i \geq 0$.

Proof. Same proof as in Theorem 8.2.11. One now considers an exact sequence $0 \to L \to P \to M \to 0$ with P free and exact sequences $F^0(P) \to F^0(L) \to F^1(M) \to F^1(P) \to \cdots$, $G^0(P) \to G^0(L) \to G^1(M) \to G^1(P) \to \cdots$. □

There are similar results for left strongly connected sequences.

Most useful applications of derived functors occur when T is a functor of two variables and is balanced as defined below.

Definition 8.2.13. Let C, D, and E be abelian categories and $T : \mathsf{C} \times \mathsf{D} \to \mathsf{E}$ be an additive functor contravariant in the first variable and covariant in the second. Let $\mathcal{F}$ and $\mathcal{G}$ be classes of objects of C and D respectively. Then T is said to be *right balanced* by $\mathcal{F} \times \mathcal{G}$ if for each object M of C, there is a $T(-, \mathcal{G})$ exact complex

$$ \cdots \to F_1 \to F_0 \to M \to 0 $$

with each $F_i \in \mathcal{F}$, and if for every object N of D, there is a $T(\mathcal{F}, -)$ exact complex

$$ 0 \to N \to G^0 \to G^1 \to \cdots $$

with $G^i \in \mathcal{G}$.

If, on the other hand, the complex $\cdots \rightarrow F_1 \rightarrow F_0 \rightarrow M \rightarrow 0$ is $T(\mathcal{G}, -)$ exact and the complex $0 \rightarrow N \rightarrow G^0 \rightarrow G^1 \rightarrow \cdots$ is $T(-, \mathcal{F})$ exact, then T is said to be *left balanced* by $\mathcal{G} \times \mathcal{F}$.

The definitions above are easily modified to give the definitions of a left or right balanced functor relative to $\mathcal{F} \times \mathcal{G}$ with other choices of variances and complexes. For example, if T is covariant in both variables, then we would postulate the existence of complexes $\cdots \rightarrow F_1 \rightarrow F_0 \rightarrow M \rightarrow 0$ and $\cdots \rightarrow G_1 \rightarrow F_0 \rightarrow N \rightarrow 0$ or $0 \rightarrow M \rightarrow F^0 \rightarrow F^1 \rightarrow \cdots$ and $0 \rightarrow N \rightarrow G^0 \rightarrow G^1 \rightarrow \cdots$ with the obvious properties to define left or right balanced functors relative to $\mathcal{F} \times \mathcal{G}$, respectively.

The *double complex* $(T(\mathbf{F}_\bullet, \mathbf{G}^\bullet))$ is defined by

$$(T(\mathbf{F}_\bullet, \mathbf{G}^\bullet))_n = \oplus_{i+j=n} T(F_i, G^j)$$

with differentials

$$d : (T(\mathbf{F}_\bullet, \mathbf{G}^\bullet))_n \rightarrow (T(\mathbf{F}_\bullet, \mathbf{G}^\bullet))_{n-1}$$

defined by $df = d'' f + (-1)^{n+1} d' f$ where d' and d'' are differentials for $\mathbf{F}_\bullet$ and $\mathbf{G}^\bullet$ respectively. We are now in a position to prove the following result.

Theorem 8.2.14. *Let T be contravariant in the first variable and covariant in the second. If T is right balanced by $\mathcal{F} \times \mathcal{G}$, then the double complex $(T(\mathbf{F}_\bullet, \mathbf{G}^\bullet))$ and the complexes $(T(\mathbf{F}_\bullet, N))$ and $(T(M, \mathbf{G}^\bullet))$ have isomorphic homology.*

Proof. Since T is right balanced, we get the following first quadrant commutative diagram with exact rows and columns.

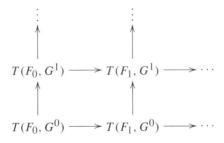

But if $C^i = \mathrm{Ker}(T(F_0, G^i) \rightarrow T(F_0, G^{i+1}))$, $D^i = \mathrm{Ker}(T(F_i, G^0) \rightarrow T(F_{i+1}, G^0))$, and $C^i \rightarrow C^{i+1}$, $D^i \rightarrow D^{i+1}$ are induced by the obvious maps, then the complexes $0 \rightarrow C^0 \rightarrow C^1 \rightarrow \cdots$ and $0 \rightarrow D^0 \rightarrow D^1 \rightarrow \cdots$ have isomorphic homology. But $C^i = T(M, G^i)$ and $D^i = T(F_i, N)$ and so we are done. $\qquad\square$

Remark 8.2.15. The homologies of either complexes in Theorem 8.2.14 above are denoted by $R^n T(M, N)$.

If T is left balanced, we get a corresponding result using a third quadrant diagram. For example, if T is again contravariant in the first variable and covariant in the second, then we may consider complexes

$$0 \to M \to G^0 \to G^1 \to \cdots , \qquad \cdots \to F_1 \to F_0 \to N \to 0$$

with $G^i \in \mathcal{G}$, $F_i \in \mathcal{F}$ subject to required conditions. Then we get a third quadrant diagram from which it follows that the complexes $(T(\mathbf{G}^\bullet, \mathbf{F}_\bullet))$, $(T(\mathbf{G}^\bullet, N))$, $(T(M, \mathbf{F}_\bullet))$ have isomorphic homologies. These homologies are denoted by $L_n T(M, N)$. We get similar results with other choices of variances.

Exercises

1. Prove that for any additive functor T, there are natural transformations $L_0 T \to T$ and $T \to R^0 T$.

2. Suppose $\mathcal{F}$ is precovering and for each M, $F_1 \to F_0 \to M \to 0$ is exact where $\cdots \to F_1 \to F_0 \to M \to 0$ is a left $\mathcal{F}$-resolution. If T is a right exact additive functor, argue that $L_0 T \cong T$.

3. Suppose $\mathcal{G}$ is preenveloping and for each M $0 \to M \to G^0 \to G^1$ is exact where $0 \to M \to G^0 \to G^1 \to \cdots$ is a right $\mathcal{G}$-resolution of M. If T is a left exact additive functor, argue that $R^0 T \cong T$.

4. Let T be an additive functor. Prove that if $\mathcal{F}$ is precovering and $F \in \mathcal{F}$, then $L_0 T(F) \cong F$ and $(L_n T)(F) = 0$ if $n > 0$. Similarly, if $\mathcal{G}$ is preenveloping and $G \in \mathcal{G}$, then $R^0 T(G) \cong G$ and $R^n T(G) = 0$ if $n > 0$.

5. State and prove a result dual to Lemma 8.2.1.

6. Prove that the long exact sequences in Theorem 8.2.3 are functorial relative to the complexes $0 \to M' \to M \to M'' \to 0$.

7. Prove Theorem 8.2.5.

8. Let $\mathcal{F}$ be a precovering class closed under finite direct sums, and K_n for $n \geq 1$ be an nth $\mathcal{F}$-syzygy of M. Prove that

   a) If T is covariant, then $L_{n+1} T(M) \cong L_n T(K_1) \cong \cdots \cong L_1 T(K_n)$.
   b) If T is contravariant, then $R^{n+1} T(M) \cong R^n T(K_1) \cong \cdots \cong R^1 T(K_n)$.

   Conclude that if $K_n \in \mathcal{F}$, then $L_i T(M) = R^i T(M) = 0$ for all $i \geq n + 1$.

9. Let $\mathcal{F}$ be a preenveloping class closed under finite direct sums, and C^n for $n \geq 1$ be an nth $\mathcal{F}$-cosyzygy of M. Prove that

   a) If T is covariant, then $R^{n+1} T(M) \cong R^n T(C^1) \cong \cdots \cong R^1 T(C^n)$.
   b) If T is contravariant, then $L_{n+1} T(M) \cong L_n T(C^1) \cong \cdots \cong L_1 T(C^n)$.

   Conclude that if $C^n \in \mathcal{F}$, then $R^i T(M) = L_i T(M) = 0$ for all $i \geq n + 1$.

10. Under the hypothesis of Theorem 8.2.3, prove that if T is covariant, then $L_0 T$ is right exact and if T is contravariant, then $R^0 T$ is left exact.

11. Prove part (2) of Theorem 8.2.7.

12. Prove Proposition 8.2.9.

13. Complete the proof of Theorem 8.2.12.

14. State and prove results for left strongly connected sequences corresponding to Theorems 8.2.11 and 8.2.12.

15. Prove that the complexes $0 \to C^0 \to C^1 \to \cdots$ and $0 \to D^0 \to D^1 \to \cdots$ in Theorem 8.2.14 have isomorphic homology.

16. Let T be contravariant in the first variable and covariant in the second. Prove that if T is left balanced by $\mathcal{G} \times \mathcal{F}$, then the complexes $(T(\mathbf{G}^\bullet, \mathbf{F}_\bullet))$, $(T(\mathbf{G}^\bullet, N))$, $(T(M, \mathbf{F}_\bullet))$ have isomorphic homology.

17. Let T be covariant in either variable. Prove that if T is left balanced by $\mathcal{F} \times \mathcal{G}$, then the complexes $(T(\mathbf{F}_\bullet, \mathbf{G}_\bullet))$, $(T(\mathbf{F}_\bullet, N))$, $(T(M, \mathbf{G}_\bullet))$ have isomorphic homology.

18. Let T be covariant in either variable. Prove that if T is right balanced by $\mathcal{F} \times \mathcal{G}$, then the complexes $(T(\mathbf{F}^\bullet, \mathbf{G}^\bullet))$, $(T(\mathbf{F}^\bullet, N))$, $(T(M, \mathbf{G}^\bullet))$ have isomorphic homology.

8.3   Applications to modules

Let $T = \text{Hom}(-, -)$ be a functor from a product of abelian categories C and D. Then T is contravariant in the first variable and covariant in the second. If T is right balanced by $\mathcal{F} \times \mathcal{G}$, then the right derived functors $R^n \text{Hom}(M, N)$ can be computed using either a complex $\cdots \to F_1 \to F_0 \to M \to 0$ of M with $F_i \in \mathcal{F}$ or a complex $0 \to N \to G^0 \to G^1 \to \cdots$ of N with $G^i \in \mathcal{G}$ by Theorem 8.2.14. If T is left balanced by $\mathcal{G} \times \mathcal{F}$, then we can compute the left derived functors $L_n \text{Hom}(N, M)$ using either of the complexes by Remark 8.2.15.

We now apply this to categories of modules. But first, we recall and introduce some notation. $_R\mathcal{M}$ ($\mathcal{M}_R$) will denote classes of left (right) R-modules, and given a class $\mathcal{F}$, $\mathcal{F}_{fg}$ and $\mathcal{F}_{fp}$ will denote classes of finitely generated and finitely presented modules in $\mathcal{F}$, respectively. If $\mathcal{F}$ consists of all flat R-modules, then $\mathcal{F}$ will be denoted by $\mathcal{F}lat$. We recall that $\mathcal{A}bs$, $\mathcal{I}nj$, and $\mathcal{P}roj$ denote the classes of absolutely pure, injective, and projective R-modules, and we will let $\mathcal{PP}roj$, and $\mathcal{PI}nj$ denote the classes of pure projective and pure injective modules, respectively. These terms will also be used to denote the corresponding full subcategories. In particular, left $\mathcal{P}roj$-resolution, right $\mathcal{I}nj$-resolution will mean the usual projective and injective resolutions, respectively.

Now let $T = \text{Hom}(-, -)$ be a functor from $_R\mathcal{M} \times {}_R\mathcal{M}$ to the category of abelian groups. Then we have the following:

Example 8.3.1. $\operatorname{Hom}(-,-)$ is right balanced on $_R\mathcal{M} \times {}_R\mathcal{M}$ by $\mathcal{P}roj \times \mathcal{I}nj$. This is standard.

Example 8.3.2. From the definition of a finitely presented module M, it is not hard to see that there is a set $X \subset \mathcal{M}_{fp}$ such that for every $F \in \mathcal{M}_{fp}$, $F \cong G$ for some $G \in X$. For each $F \in X$, set $X_F = \operatorname{Hom}(F, M)$. If $F^{(X_F)} \to M$ is the evaluation map $(\varphi_f)_{X_F} \to \sum f(\varphi_f)$, then any map $F \to M$ factors through $F^{(X_F)} \to M$. So $\operatorname{Hom}(F', \bigoplus_{F \in X} F^{(X_F)}) \to \operatorname{Hom}(F', M) \to 0$ is exact for all $F' \in \mathcal{M}_{fp}$. Thus $\bigoplus_{F \in X} F^{(X_F)} \to M \to 0$ is exact since $R \in \mathcal{M}_{fp}$ and so the sequence $0 \to K \to \bigoplus_{F \in X} F^{(X_F)} \to M \to 0$ is pure exact.

Now recall that a left R-module P is said to be *pure projective* if for every pure exact sequence $0 \to T \to N \to N/T \to 0$, $\operatorname{Hom}(P, N) \to \operatorname{Hom}(P, N/T) \to 0$ is exact. Thus projective and finitely presented modules are pure projective, and in particular $\operatorname{Hom}(P, \oplus F^{(X_F)}) \to \operatorname{Hom}(P, M) \to 0$ is exact whenever P is pure projective.

If $0 \to T \to N \to N/T \to 0$ is pure exact, then $\operatorname{Hom}(\oplus F^{(X_F)}, N) \to \operatorname{Hom}(\oplus F^{(X_F)}, N/T)$ is equivalent to $\prod \operatorname{Hom}(F, N) \to \prod \operatorname{Hom}(F, N/T)$ which is surjective since $\operatorname{Hom}(F, N) \to \operatorname{Hom}(F, N/T)$ is surjective because $F \in \mathcal{PP}roj$. Thus $\oplus F^{(X_F)} \in \mathcal{PP}roj$. Hence since $\oplus F^{(X_F)} \to M \to 0$ is surjective with a pure kernel and $\oplus F^{(X_F)}$ is pure projective, we see that we have a $\mathcal{PP}roj$-precover. Hence we get an exact left $\mathcal{PP}roj$-resolution with pure kernels for each $M \in {}_R\mathcal{M}$. But by Propositions 5.3.9 and 8.1.3, right $\mathcal{PI}nj$-resolutions exist with pure injections for each $N \in {}_R\mathcal{M}$. Hence $\operatorname{Hom}(-,-)$ is right balanced by $\mathcal{PP}roj \times \mathcal{PI}nj$ on $_R\mathcal{M} \times {}_R\mathcal{M}$.

Example 8.3.3. If R is left coherent, then $\operatorname{Hom}(-,-)$ on $\mathcal{M}_{R_{fp}} \times \mathcal{M}_R$ is right balanced by $\mathcal{P}roj_{fg} \times \mathcal{A}bs$. For if R is left coherent, then each object M_R in $\mathcal{M}_{fp}$ has a left $\mathcal{P}roj_{fg}$-resolution (see Remark 3.2.25). Furthermore, any absolutely pure preenvelope is an injection since injectives are absolutely pure. So every module has a right $\mathcal{A}bs$-resolution by Proposition 6.2.4 and Proposition 8.1.3. Hence the result follows as in the above examples.

Example 8.3.4. $\operatorname{Hom}(-,-)$ on $\mathcal{F}lat \times \mathcal{F}lat$ is right balanced by $\mathcal{P}roj \times \mathcal{PI}nj$ since right $\mathcal{PI}nj$-resolutions are exact and left $\mathcal{P}roj$-resolutions of flat modules are pure exact sequences.

Example 8.3.5. If R is left noetherian, then $\operatorname{Hom}(-,-)$ is left balanced on $_R\mathcal{M} \times {}_R\mathcal{M}$ by $\mathcal{I}nj \times \mathcal{I}nj$ since for such an R, every left R-module has a left $\mathcal{I}nj$-resolution by Theorem 5.4.1 and Proposition 8.1.3.

Example 8.3.6. If R is left coherent, then $\operatorname{Hom}(-,-)$ is left balanced on $\mathcal{M}_R \times \mathcal{M}_R$ by $\mathcal{F}lat_R \times \mathcal{F}lat_R$ since every right R-module has a right $\mathcal{F}lat$-resolution by Proposition 6.5.1 and a left $\mathcal{F}lat$-resolution by Theorem 7.4.4.

Notation. When $T = \text{Hom}(-, -)$, right derived functors $R^n \text{Hom}(-, -)$ are denoted by $\text{Ext}^n_R(-, -)$ and left derived functors $L_n \text{Hom}(-, -)$ by $\text{Ext}^R_n(-, -)$ or $\text{Ext}^{-n}_R(-, -)$. In Example 8.3.2, the derived functors are usually denoted by $\text{Pext}^n_R(-, -)$. We note that these functors depend on which classes $\mathcal{F}$ and $\mathcal{G}$ we are using. But this will be clear from the context.

Now let $T = - \otimes -$. Then T is covariant in both variables. If T is left balanced by $\mathcal{F} \times \mathcal{G}$, then we can compute left derived functors $L_n(M \otimes N)$ using either a complex $\cdots \rightarrow F_1 \rightarrow F_0 \rightarrow M \rightarrow 0$ of M with $F_i \in \mathcal{F}$ or a complex $\cdots \rightarrow G_1 \rightarrow G_0 \rightarrow N \rightarrow 0$ of N with $G_i \in \mathcal{G}$. This uses a third quadrant diagram. Similarly, if T is right balanced by $\mathcal{F} \times \mathcal{G}$, then we can compute right derived functors $R^n(M \otimes N)$ using either a complex $0 \rightarrow M \rightarrow F^0 \rightarrow F^1 \rightarrow \cdots$ of M with $F^i \in \mathcal{F}$ or $0 \rightarrow N \rightarrow G^0 \rightarrow G^1 \rightarrow \cdots$ with $G^i \in \mathcal{G}$.

We apply this to categories of modules by again considering $T = - \otimes -$ as a functor from $\mathcal{M}_R \times {}_R\mathcal{M}$ to the category of abelian groups.

Example 8.3.7. $- \otimes -$ is left balanced by $\mathcal{P}roj \times \mathcal{P}roj$ on $\mathcal{M}_R \times {}_R\mathcal{M}$. This is standard.

Example 8.3.8. $- \otimes -$ is left balanced by $\mathcal{F}lat \times \mathcal{F}lat$ on $\mathcal{M}_R \times {}_R\mathcal{M}$. Again this is standard since every module has a flat resolution. The left derived functors of this example coincide with those in Example 8.3.7 above. We note however that a flat resolution is not a left $\mathcal{F}lat$-resolution in general. We will study left $\mathcal{F}lat$-resolutions in Section 8.6.

Example 8.3.9. If R is left coherent, then $- \otimes -$ on $\mathcal{M}_R \times {}_R\mathcal{M}$ is right balanced by $\mathcal{F}lat \times \mathcal{A}bs$. We need to show that if $0 \rightarrow M \rightarrow F^0 \rightarrow F^1 \rightarrow \cdots$ is a right $\mathcal{F}lat$-resolution, which exists by Propositions 6.5.1 and 8.1.3, and G is an absolutely pure left R-module, then $0 \rightarrow M \otimes G \rightarrow F^0 \otimes G \rightarrow F^1 \otimes G \rightarrow \cdots$ is exact. Applying the functor $\text{Hom}_{\mathbb{Z}}(-, \mathbb{Q}/\mathbb{Z})$ and using a standard identity we get the sequence $0 \leftarrow \text{Hom}(M, G^+) \leftarrow \text{Hom}(F^0, G^+) \leftarrow \cdots$. But G^+ is flat and so this sequence is exact. This means the desired sequence is exact. Since right $\mathcal{A}bs$-resolutions are exact, $0 \rightarrow F \otimes N \rightarrow F \otimes G^0 \rightarrow F \otimes G^1 \rightarrow \cdots$ is exact for any right $\mathcal{A}bs$-resolution $0 \rightarrow N \rightarrow G^0 \rightarrow G^1 \rightarrow \cdots$ of N and for any $F \in \mathcal{F}lat$.

Example 8.3.10. If R is again left coherent, then $- \otimes -$ on $\mathcal{M}_{R_{fp}} \times {}_R\mathcal{M}$ is right balanced by $\mathcal{P}roj_{fg} \times \mathcal{A}bs$. For if M is a finitely presented right R-module, then M has a flat preenvelope $M \rightarrow F$ by Proposition 6.5.1. But then $M \rightarrow F$ has a factorization $M \rightarrow P \rightarrow F$ for some finitely generated right R-module P. So $M \rightarrow P$ is a $\mathcal{P}roj_{fg}$-preenvelope of M. Thus M has a right $\mathcal{P}roj_{fg}$-resolution. Hence the example follows as in Example 8.3.9 noting that the right $\mathcal{P}roj_{fg}$-resolution becomes exact if we apply $\text{Hom}(-, P)$ with $P \in \mathcal{P}roj_{fg}$ or $\text{Hom}(-, F)$ with $F \in \mathcal{F}lat$.

Notation. When $T = - \otimes -$, the left derived functors $L_n(- \otimes -)$ are denoted by $\text{Tor}_n^R(-, -)$ and the right derived functors $R^n(- \otimes -)$ are denoted by $\text{Tor}^n_R(-, -)$. Again these depend on the classes $\mathcal{F}$ and $\mathcal{G}$ we are using.

Example 8.3.11. Using the above, we see that if R is left coherent, then $\text{Hom}(-, -)$ is left balanced on $\mathcal{M}_{R_{fp}} \times \mathcal{M}_R$ by $\mathcal{P}roj_{fg} \times \mathcal{F}lat$. Here the derived functors $\text{Ext}_n^R(M, N)$ are obtained by using a right $\mathcal{P}roj_{fg}$-resolution of M or a flat resolution of N.

Exercises

1. Prove Example 8.3.1.

2. Prove Examples 8.3.4 and 8.3.5.

3. Prove Example 8.3.11.

4. Let R be left noetherian, N be an R-module, and C^j a jth cosyzygy of an R-module M. Prove that $\text{Ext}_i^R(C^j, N) \cong \text{Ext}_{i+1}^R(M, N)$ for all $i \geq 1$.

5. Let R be left noetherian and N be an R-module such that its left $\mathcal{I}nj$-resolutions are exact. Let $K^{-i}(N)$ denote the ith syzygy of the minimal left $\mathcal{I}nj$-resolution of N. Prove that $\text{Ext}_i^R(M, N) \cong \text{Ext}_R^1(M, K^{-i-2}(N))$ for all $i \geq 1$ and all R-modules M.

6. Let R be left noetherian and $\mathbf{C}$ denote the full subcategory of all R-modules whose left $\mathcal{I}nj$-resolutions are exact. Prove that the following are equivalent for $N \in Ob(\mathbf{C})$.

   a) $\text{Ext}_i^R(M, N) = 0$ for all $i \geq 1$ and all $M \in Ob(\mathbf{C})$.

   b) $\text{Ext}_1^R(M, N) = 0$ for all $M \in Ob(\mathbf{C})$.

   c) N is an injective R-module.

   Hint: Use Problem 5 above.

7. (Strenström [168, Theorem 3.2] and Würfel [181, Satz 1.6]) Prove that the following are equivalent for a ring R.

   a) R is left coherent.

   b) M^+ is a flat right R-module for each absolutely pure left R-module M.

   c) Every quotient M/S of an absolutely pure left R-module by a pure submodule S is absolutely pure.

   d) Every direct limit of absolutely pure left R-modules is absolutely pure.

8.4 $\mathcal{F}$-dimensions

Definition 8.4.1. If $\mathcal{F}$ is a precovering class of an abelian category $\mathbf{C}$, then an object M of $\mathbf{C}$ is said to have *left $\mathcal{F}$-dimension $\leq n$*, denoted left $\mathcal{F}$- dim $M \leq n$, if there is a left $\mathcal{F}$-resolution of the form $0 \to F_n \to F_{n-1} \to \cdots \to F_1 \to F_0 \to M \to 0$ of M. If n is the least, then we set left $\mathcal{F}$-dim $M = n$ and if there is no such n, we set left $\mathcal{F}$-dim $M = \infty$. In a similar manner, we define the *right $\mathcal{F}$-dimension* of objects of C, denoted right $\mathcal{F}$-dim, if $\mathcal{F}$ is a preenveloping class of C. In particular, *left $\mathcal{P}roj$-dimension, right $\mathcal{I}nj$-dimension* will mean the usual projective and injective dimensions, respectively.

We note that $\mathcal{F}$-dimension depends on both the category and the precovering or preenveloping class. Also, $\mathcal{F}$ might be both precovering and preenveloping in which case left and right $\mathcal{F}$-dimensions of a given object of C may be different.

Definition 8.4.2. Let C be an abelian category and $\mathcal{F}$ be a precovering class of C. Then the *global left $\mathcal{F}$-dimension* of C, denoted gl left $\mathcal{F}$-dim C, is defined by

$$\text{gl left } \mathcal{F}\text{-dim } C = \sup\{\text{left } \mathcal{F}\text{-dim } M : M \in Ob(C)\}$$

and is infinite otherwise. The *global right $\mathcal{F}$-dimension* of C is defined similarly when $\mathcal{F}$ is a preenveloping class.

Again, global dimensions depend on the category and the respective classes. If $\mathcal{F}$ is both precovering and preenveloping, the two global $\mathcal{F}$-dimensions may be different.

Balanced functors are an essential tool in determining the exactness of complexes and comparing different $\mathcal{F}$-dimensions and global $\mathcal{F}$-dimensions. We illustrate this using the category of left R-modules. We start by stating the following standard results where $\text{Ext}^n(M, N)$ are the usual right derived functors obtained by using a right $\mathcal{I}nj$-resolution of N or a left $\mathcal{P}roj$-resolution of M (see Example 8.3.1). We include proofs here for completeness.

Proposition 8.4.3. *The following are equivalent for a ring R, $M \in {}_R\mathcal{M}$, and $n \geq 0$.*

(1) *left $\mathcal{P}roj$-dim $M \leq n$.*

(2) $\text{Ext}_R^{n+k}(M, N) = 0$ *for all $N \in {}_R\mathcal{M}$ and $k \geq 1$.*

(3) $\text{Ext}_R^{n+1}(M, N) = 0$ *for all $N \in {}_R\mathcal{M}$.*

(4) *Every nth $\mathcal{P}roj$-syzygy of M is projective.*

Proof. (1) $\Rightarrow$ (2). Let $0 \to P_n \to P_{n-1} \to \cdots \to P_1 \to P_0 \to M \to 0$ be a left $\mathcal{P}roj$-resolution of M. Then $\text{Hom}(P_{n+k}, N) = 0$ for all $k \geq 1$. So (2) follows.

(2) $\Rightarrow$ (3) and (4) $\Rightarrow$ (1) are trivial.

(3) $\Rightarrow$ (4). Let $K = \text{Ker}(P_n \to P_{n-1})$. Then (3) implies that $\text{Hom}(P_n, N) \to \text{Hom}(K, N) \to 0$ is exact for all N. So by setting $N = K$, we see that $0 \to K \to P_n \to \text{Ker}(P_{n-1} \to P_{n-2}) \to 0$ is split exact and so (4) follows. $\square$

Dually, we have the following

Proposition 8.4.4. *The following are equivalent for a ring R, $N \in {}_R\mathcal{M}$, and $n \geq 0$.*

(1) right $\mathcal{I}nj$- dim $N \leq n$.

(2) $\mathrm{Ext}_R^{n+k}(M, N) = 0$ for all $M \in {}_R\mathcal{M}$, and $k \geq 1$.

(3) $\mathrm{Ext}_R^{n+1}(M, N) = 0$ for all $M \in {}_R\mathcal{M}$.

(4) Every nth $\mathcal{I}nj$-cosyzygy of M is injective.

(5) $\mathrm{Ext}_R^{n+1}(R/I, N) = 0$ for all left ideals I of R.

Proof. The equivalence of (1), (2), (3), (4) follows as in the dual proposition above. (3) $\Rightarrow$ (5) is trivial. We now argue (5) $\Rightarrow$ (4). Let $C = \mathrm{Im}\,(E^{n-1} \to E^n)$. Then it follows from part (1) of Corollary 8.2.6 that $\mathrm{Ext}^1(R/I, C) \cong \mathrm{Ext}^{n+1}(R/I, N) = 0$. So $\mathrm{Hom}(R, C) \to \mathrm{Hom}(I, C) \to 0$ is exact for all left ideals I of R. But then C is injective by Baer's Criterion (Theorem 3.1.3).   □

Now we get the following well known result.

Theorem 8.4.5. *The following are equivalent for a ring R and the category of left R-modules ${}_R\mathcal{M}$.*

(1) gl left $\mathcal{P}roj$- dim ${}_R\mathcal{M} \leq n$.

(2) gl right $\mathcal{I}nj$- dim ${}_R\mathcal{M} \leq n$.

(3) $\mathrm{Ext}_R^{n+1}(M, N) = 0$ for all $M, N \in {}_R\mathcal{M}$.

(4) $\mathrm{Ext}_R^{n+k}(M, N) = 0$ for all $M, N \in {}_R\mathcal{M}$ and $k \geq 1$.

(5) $\sup\{$left $\mathcal{P}roj$- dim $R/I : I$ is a left ideal of $R\} \leq n$.

Proof. This easily follows from Propositions 8.4.3 and 8.4.4 above.   □

This gives the following result concerning the usual left injective and projective global dimensions.

Corollary 8.4.6. gl right $\mathcal{I}nj$- dim ${}_R\mathcal{M}$ = gl left $\mathcal{P}roj$- dim ${}_R\mathcal{M}$ = $\sup\{$left $\mathcal{P}roj$- dim $R/I : I$ is a left ideal of $R\}$.

If R is left coherent, right derived functors $\mathrm{Ext}_R^n(M, N)$ on ${}_R\mathcal{M}_{fp} \times {}_R\mathcal{M}$ can be computed using a left $\mathcal{P}roj_{fg}$-resolution of M or a right $\mathcal{A}bs$-resolution of N by Example 8.3.3. So we have the following results that are analogous to the above.

Proposition 8.4.7. *Let R be left coherent and $n \geq 0$. Then the following are equivalent for $M \in {}_R\mathcal{M}_{fp}$.*

(1) left $\mathcal{P}roj_{fg}$- dim $M \leq n$

(2) $\mathrm{Ext}_R^{n+k}(M, N) = 0$ for all $N \in {}_R\mathcal{M}$ and $k \geq 1$.

(3) $\mathrm{Ext}_R^{n+1}(M, N) = 0$ for all $N \in {}_R\mathcal{M}$.

(4) *Every nth $\mathcal{P}roj_{fg}$-syzygy of M is a finitely generated projective R-module.*

Proof. Similar to Proposition 8.4.3. □

Proposition 8.4.8. *Let R be left coherent and $n \geq 0$. Then the following are equivalent for $N \in {}_R\mathcal{M}$.*

(1) right $\mathcal{A}bs$- dim $N \leq n$.

(2) $\text{Ext}_R^{n+k}(M, N) = 0$ for all $M \in {}_R\mathcal{M}_{fp}$ and $k \geq 1$.

(3) $\text{Ext}_R^{n+1}(M, N) = 0$ for all $M \in {}_R\mathcal{M}_{fp}$.

(4) *Every nth ${}_R\mathcal{A}bs$-cosyzygy of N is absolutely pure.*

(5) $\text{Ext}_R^{n+1}(R/I, N) = 0$ for all finitely generated left ideals I of R.

Proof. We provide a proof here for completeness.

(1) $\Rightarrow$ (2) $\Rightarrow$ (3) $\Rightarrow$ (5) and (4) $\Rightarrow$ (1) are now trivial.

(3) $\Rightarrow$ (4). Let $C = \text{Im}(A^{n-1} \to A^n)$. Then $\text{Ext}^1(F, C) \cong \text{Ext}^{n+1}(F, N) = 0$ for all $F \in \mathcal{M}_{fp}$ again by Corollary 8.2.6. Hence $\text{Hom}(R^m, C) \to \text{Hom}(A, C) \to 0$ is exact for every $n \geq 1$ and finitely generated $A \subseteq R^m$. Thus $C \in \mathcal{A}bs$.

(5) $\Rightarrow$ (3). Let S be the cyclic submodule generated by one of the generators of a finitely presented left R-module F. Then S and F/S are finitely presented since R is left coherent. But both S and F/S have fewer generators than F, and so by induction on the number of generators of F, $\text{Ext}^{n+1}(S, N) = \text{Ext}^{n+1}(F/S, N) = 0$. But then $\text{Ext}^{n+1}(F, N) = 0$. □

Theorem 8.4.9. *Let R be left coherent and $n \geq 0$. Then the following are equivalent.*

(1) gl right $\mathcal{A}bs$- dim ${}_R\mathcal{M} \leq n$.

(2) gl left $\mathcal{P}roj_{fg}$- dim ${}_R\mathcal{M}_{fp} \leq n$.

(3) $\text{Ext}_R^{n+k}(M, N) = 0$ for all $M \in \mathcal{M}_{fp}$, $N \in {}_R\mathcal{M}$, and all $k \geq 1$.

(4) $\text{Ext}_R^{n+1}(M, N) = 0$ for all $M \in \mathcal{M}_{fp}$, $N \in {}_R\mathcal{M}$.

(5) $\text{Ext}_R^{n+1}(R/I, N) = 0$ for all $N \in {}_R\mathcal{M}$ and all finitely generated left ideals I of R.

(6) sup{left $\mathcal{P}roj$- dim R/I : I is a finitely generated left ideal of R} $\leq n$.

Proof. This follows from the preceding two propositions. □

Corollary 8.4.10. gl left $\mathcal{A}bs$- dim ${}_R\mathcal{M} = $ gl left $\mathcal{P}roj_{fg}$- dim ${}_R\mathcal{M}_{fp} = $ sup{left $\mathcal{P}roj$- dim R/I : I is a finitely generated left ideal of R}.

Let $0 \to N \to PE^\circ(N) \to PE^1(N) \to \cdots$ denote the minimal right $\mathcal{P}\mathcal{I}nj$-resolution of an R-module N. Then we have the following result which easily follows from Proposition 6.7.1.

Lemma 8.4.11. *If F is a flat left R-module and R is right coherent, then $PE^k(F)$ and $\mathrm{Im}(PE^{k-1}(F) \to PE^k(F))$ are flat for each $k \geq 0$ where $PE^{-1}(F) = F$.*

Proof. We simply note that $PE^\circ(F)$ is flat by Proposition 6.7.1 and $F \subset PE^\circ(F)$ is pure and so $PE^\circ(F)/F$ is flat. Hence $PE^1(F)$ is flat by Proposition 6.7.1 again and thus we proceed inductively.   □

Now let $\mathrm{Ext}_R^n(M, N)$ denote the right derived functors of $\mathrm{Hom}(M, N)$ on $\mathcal{F}lat \times \mathcal{F}lat$ obtained using a left $\mathcal{P}roj$-resolution of M or a right $\mathcal{P}\mathcal{I}nj$-resolution of N (see Example 8.3.4). Then we get the following result.

Theorem 8.4.12. *The following are equivalent for a right coherent ring R and the full subcategory ${}_R\mathcal{F}lat$.*

(1) gl left $\mathcal{P}roj$-dim ${}_R\mathcal{F}lat \leq n$.

(2) gl right $\mathcal{P}\mathcal{I}nj$-dim ${}_R\mathcal{F}lat \leq n$.

(3) $\mathrm{Ext}_R^{n+1}(M, N) = 0$ *for all* $M, N \in {}_R\mathcal{F}lat$.

(4) $\mathrm{Ext}_R^{n+k}(M, N) = 0$ *for all* $M, N \in {}_R\mathcal{F}lat$ *and* $k \geq 1$.

Proof. (1) $\Rightarrow$ (4), (2) $\Rightarrow$ (4), and (4) $\Rightarrow$ (3) are trivial.

(3) $\Rightarrow$ (1). Let $\cdots \to P_1 \to P_0 \to M \to 0$ be a left $\mathcal{P}roj$-resolution of a flat module M. Then $K = \mathrm{Ker}(P_n \to P_{n-1})$ is flat since the resolution is pure exact. So as in the proof of Proposition 8.4.3, setting $N = K$ yields (1).

(3) $\Rightarrow$ (2). We consider the minimal right $\mathcal{P}\mathcal{I}nj$-resolution $0 \to N \to PE^\circ(N) \to PE^1(N) \to \cdots$. Then $C = \mathrm{Im}(PE^{n-1}(N) \to PE^n(N))$ is flat by Lemma 8.4.11 above. So (3) implies that C is a direct summand of $PE^n(N)$ and thus we are done.   □

Corollary 8.4.13. gl left $\mathcal{P}roj$-dim ${}_R\mathcal{F}lat$ = gl right $\mathcal{P}\mathcal{I}nj$-dim ${}_R\mathcal{F}lat$.

We now use the fact that if R is left noetherian, then $\mathrm{Hom}(-, -)$ is left balanced by $\mathcal{I}nj \times \mathcal{I}nj$ and left derived functors $\mathrm{Ext}_n(-, -)$ can be computed using a right $\mathcal{I}nj$-resolution of M or a left $\mathcal{I}nj$-resolution of N (see Example 8.3.5).

Proposition 8.4.14. *Let R be left noetherian and $n \geq 2$. Then the following are equivalent for $M \in {}_R\mathcal{M}$.*

(1) right $\mathcal{I}nj$-dim $M \leq n$.

(2) $\mathrm{Ext}_{n+k}^R(M, N) = 0$ *for all* $N \in {}_R\mathcal{M}$ *and* $k \geq -1$.

(3) $\mathrm{Ext}_{n-1}^R(M, N) = 0$ *for all* $N \in {}_R\mathcal{M}$.

Proof. (1) $\Rightarrow$ (2). Let $0 \to M \to E^0 \to E^1 \to \cdots \to E^n \to 0$ be a right $\mathcal{I}nj$-resolution of M. Then $0 \to \operatorname{Hom}(E^n, N) \to \operatorname{Hom}(E^{n-1}, N) \to \operatorname{Hom}(E^{n-2}, N)$ is exact and so

$$\operatorname{Ext}_{n-1}(M, N) = \operatorname{Ext}_n(M, N) = 0.$$

But clearly, $\operatorname{Ext}_{n+k}(M, N) = 0$ for all $k \geq 1$. Hence (2) follows.

(2) $\Rightarrow$ (3) is trivial.

(3) $\Rightarrow$ (1). Let $0 \to M \to E^0 \to E^1 \to \cdots$ be a right $\mathcal{I}nj$-resolution of M. Let $C = \operatorname{Im}(E^{n-2} \to E^{n-1})$. Then (3) implies $\operatorname{Ext}_{n-1}(M, E^{n-1}/C) = 0$. But then $E^{n-1}/C \to E^n$ has a retract. Hence $E^{n-1}/C \in \mathcal{I}nj$ and so $0 \to M \to E^0 \to \cdots \to E^{n-1} \to E^{n-1}/C \to 0$ is a right $\mathcal{I}nj$-resolution of M. $\qquad\square$

Proposition 8.4.15. *Let R be left noetherian and $n \geq 2$. Then the following are equivalent for $N \in {}_R\mathcal{M}$.*

(1) left $\mathcal{I}nj$-dim $N \leq n - 2$.

(2) $\operatorname{Ext}^R_{n+k}(M, N) = 0$ for all $M \in {}_R\mathcal{M}$ and $k \geq -1$.

(3) $\operatorname{Ext}^R_{n-1}(M, N) = 0$ for all $M \in {}_R\mathcal{M}$.

Proof. (1) $\Rightarrow$ (2) $\Rightarrow$ (3) are trivial.

(3) $\Rightarrow$ (1). Let $\cdots \to G_1 \to G_0 \to N \to 0$ be a left $\mathcal{I}nj$-resolution and $C = \operatorname{Ker}(G_{n-1} \to G_{n-2})$. Then $\operatorname{Ext}_{n-1}(C, N) = 0$. This implies $C \hookrightarrow G_{n-1}$ can be factored through $G_n \to G_{n-1}$. Hence C is a retract of G_{n-1} and so is injective. But then $G_{n-1}/C \to G_{n-2}$ is an injection with injective image D. An easy check gives that

$$0 \to G_{n-2}/D \to G_{n-3} \to \cdots \to G_0 \to N \to 0$$

is also a left $\mathcal{I}nj$-resolution. So (3) $\Rightarrow$ (1). $\qquad\square$

Theorem 8.4.16. *Let R be left noetherian and $n \geq 2$. Then the following are equivalent.*

(1) gl right $\mathcal{I}nj$-dim ${}_R\mathcal{M} \leq n$.

(2) gl left $\mathcal{I}nj$-dim ${}_R\mathcal{M} \leq n - 2$.

(3) $\operatorname{Ext}^R_{n-1}(M, N) = 0$ for all $M, N \in {}_R\mathcal{M}$.

(4) $\operatorname{Ext}^R_{n+k}(M, N) = 0$ for all $M, N \in {}_R\mathcal{M}$ and all $k \geq -1$.

Proof. This follows from Propositions 8.4.14 and 8.4.15. $\qquad\square$

Corollary 8.4.17. gl left $\mathcal{I}nj$-dim ${}_R\mathcal{M}$ = gl right $\mathcal{I}nj$-dim ${}_R\mathcal{M} - 2$ *and is zero if* gl right $\mathcal{I}nj$-dim ${}_R\mathcal{M} \leq 1$.

Remark 8.4.18. If $n = 2$ in Theorem 8.4.16, then every N has a left $\mathcal{I}nj$-resolution $\cdots \to 0 \to 0 \to E \to N \to 0$. This means that any homomorphism $G \to N$ with G injective can be factored uniquely through E. In this case $\mathcal{I}nj$ is said to be a *coreflective subcategory* of $_R\mathcal{M}$.

We now recall that a left R-module M is said to have *flat dimension* (flat dim M) $\leq n$ if there exists an exact sequence $0 \to F_n \to F_{n-1} \to \cdots \to F_1 \to F_0 \to M \to 0$ with F_i flat. Then we define the *global weak dimension* of a full subcategory $_R\mathsf{C}$, denoted gl w dim $_R\mathsf{C}$, by

$$\text{gl w dim } _R\mathsf{C} = \sup\{\text{flat dim } M : M \in Ob(_R\mathsf{C})\}.$$

We also recall from Example 8.3.8 that flat resolutions can be used to compute the left derived functors $\text{Tor}_n(M, N)$. We thus have the following well known result.

Proposition 8.4.19. *The following are equivalent for a right R-module M and $n \geq 0$.*

(1) flat dim $M \leq n$.

(2) $\text{Tor}^R_{n+k}(M, N) = 0$ *for all $N \in {}_R\mathcal{M}$ and $k \geq 1$.*

(3) $\text{Tor}^R_{n+1}(M, N) = 0$ *for all $N \in {}_R\mathcal{M}$.*

(4) $\text{Tor}^R_{n+1}(M, N) = 0$ *for all $N \in {}_R\mathcal{M}_{fp}$.*

(5) *Every nth $\mathcal{F}lat$-syzygy of M is flat.*

(6) $\text{Tor}^R_{n+1}(M, R/I) = 0$ *for all left ideals I of R.*

Proof. (1) $\Rightarrow$ (2) $\Rightarrow$ (3) $\Rightarrow$ (4) are trivial.

(4) $\Rightarrow$ (5). Let $K = \text{Ker}(F_{n-1} \to F_{n-2})$. Then $\text{Tor}^R_{n+1}(M, N) \cong \text{Tor}^R_1(K, N) = 0$ for all $N \in \mathcal{M}_{fp}$ by part (1) of Corollary 8.2.4. Now let $0 \to L \to F \to K \to 0$ be exact with $F \in \mathcal{F}lat$, then $0 \to L \otimes N \to F \otimes N$ is exact for all $N \in \mathcal{M}_{fp}$. So L is a pure submodule of F and hence K is flat.

(3) $\Rightarrow$ (6), (5) $\Rightarrow$ (1) are trivial.

(6) $\Rightarrow$ (5). Let $K = \text{Ker}(F_{n-1} \to F_{n-2})$. Then $\text{Tor}^R_{n+1}(M, R/I) = 0$ means $\text{Tor}^R_1(K, R/I) = 0$ as in the above. So $0 \to K \otimes I \to K \otimes R \to K \otimes R/I \to 0$ is exact for all left ideals I of R. But then $0 \to (K \otimes R/I)^+ \to (K \otimes R)^+ \to (K \otimes I)^+ \to 0$ is exact and thus $\text{Hom}(R, K^+) \to \text{Hom}(I, K^+) \to 0$ is exact for all left ideals I. Hence K^+ is an injective left R-module by Baer's Criterion and so K is a flat right R-module. $\square$

Theorem 8.4.20. *The following are equivalent for any ring R and $n \geq 0$.*

(1) gl w dim $\mathcal{M}_R \leq n$.

(2) $\text{Tor}^R_{n+1}(M, N) = 0$ *for all $M \in \mathcal{M}_R$, $N \in {}_R\mathcal{M}$.*

(3) gl w dim $_R\mathcal{M} \leq n$.

(4) $\operatorname{Tor}^R_{n+1}(M, N) = 0$ for all $M \in \mathcal{M}_R$, $N \in {}_R\mathcal{M}_{fp}$.

(5) $\operatorname{Tor}^R_{n+1}(M, R/I) = 0$ for all $M \in \mathcal{M}_R$ and all left ideals I of R.

(6) gl w dim ${}_R\mathcal{M}_{fp} \leq n$.

(7) $\sup\{\text{flat dim } R/I : I \text{ is a left ideal of } R\} \leq n$.

Proof. This follows from the preceding proposition. $\qquad\square$

Corollary 8.4.21. gl w dim $\mathcal{M}_R$ = gl w dim ${}_R\mathcal{M}$ = gl w dim ${}_R\mathcal{M}_{fp}$ = sup{flat dim R/I : I is a left ideal of R}.

Remark 8.4.22. If R is left coherent, then every finitely presented flat left R-module is projective. Hence if M is a finitely presented left R-module and flat dim $M \leq n$, then there is an exact sequence $0 \to K \to F_{n-1} \to \cdots \to F_1 \to F_0 \to M \to 0$ with F_i finitely generated and free, and K flat (see Example 8.3.3). But K is finitely presented and so K is projective. Thus left $\mathcal{P}roj_{fg}$-dim $M \leq$ flat dim M. But flat dim $M \leq$ left $\mathcal{P}roj_{fg}$-dim M always. Hence left $\mathcal{P}roj_{fg}$-dim M = flat dim M. Consequently, the integers in Corollaries 8.4.10 and 8.4.21 are equal if R is left coherent.

We will need the following easy result.

Lemma 8.4.23. *If $M_1 \to M_2 \to M_3 \to M_4$ is an exact sequence of left R-modules such that for every finitely presented right R-module P, $P \otimes M_1 \to P \otimes M_2 \to P \otimes M_3 \to P \otimes M_4$ is exact, then $K = \operatorname{Ker}(M_3 \to M_4)$ is a pure submodule of M_3.*

Proof. $P \otimes M_1 \to P \otimes M_2 \to P \otimes M_3 \to P \otimes M_4$ is exact and $P \otimes K \to P \otimes M_3 \to P \otimes M_4$ is a complex. Thus exactness of the first sequence means $0 \to P \otimes K \to P \otimes M_3$ is exact. This means K is a pure submodule of M_3. $\qquad\square$

Let R be left coherent and $\operatorname{Tor}^n(-, -)$ denote the right derived functors of $- \otimes -$ in Example 8.3.10. Then we have the following result.

Proposition 8.4.24. *Let R be left coherent and $n \geq 2$. Then the following are equivalent for $N \in {}_R\mathcal{M}$.*

(1) right $\mathcal{A}bs$-dim $N \leq n$.

(2) $\operatorname{Tor}^{n+k}_R(M, N) = 0$ for all $M \in \mathcal{M}_R$ and all $k \geq -1$.

(3) $\operatorname{Tor}^n_R(M, N) = \operatorname{Tor}^{n-1}_R(M, N) = 0$ for all $M \in \mathcal{M}_R$.

(4) $\operatorname{Tor}^n_R(M, N) = \operatorname{Tor}^{n-1}_R(M, N) = 0$ for all $M \in \mathcal{M}_{R_{fp}}$.

Proof. (1) $\Rightarrow$ (2). Let $0 \to N \to A^0 \to \cdots \to A^n \to 0$ be a right $\mathcal{A}bs$-resolution of N. Then $M \otimes A^{n-2} \to M \otimes A^{n-1} \to M \otimes A^n \to 0$ is exact and so $\operatorname{Tor}^{n-1}(M, N) = \operatorname{Tor}^n(M, N) = 0$. But clearly $\operatorname{Tor}^{n+k}(M, N) = 0$ for $k \geq 1$. Hence (2) holds.

$(2) \Rightarrow (3) \Rightarrow (4)$ is trivial.

$(4) \Rightarrow (1)$. Let $0 \to N \to A^0 \to A^1 \to \cdots$ be a right $\mathcal{A}bs$-resolution of N. Then for any $M \in \mathcal{M}_{R_{fp}}$, $M \otimes A^{n-2} \to M \otimes A^{n-1} \to M \otimes A^n \to M \otimes A^{n+1}$ is exact. So by Lemma 8.4.23, $K = \text{Ker}(A^n \to A^{n+1})$ is pure in A^n. But a pure submodule of an absolutely pure module is absolutely pure and so K is absolutely pure. But then $0 \to N \to A^0 \to \cdots \to A^{n-1} \to K \to 0$ is a right $\mathcal{A}bs$-resolution of N and so (1) holds. $\square$

Proposition 8.4.25. *Let R be left coherent and $n \geq 2$. Then the following are equivalent for $M \in \mathcal{M}_R$.*

(1) right $\mathcal{F}lat$- dim $M \leq n - 2$.

(2) $\text{Tor}_n^{n+k}(M, N) = 0$ for all $N \in {}_R\mathcal{M}$ and all $k \geq -1$.

(3) $\text{Tor}_R^n(M, N) = \text{Tor}_R^{n-1}(M, N)$ for all $N \in {}_R\mathcal{M}$.

Proof. $(1) \Rightarrow (2) \Rightarrow (3)$ is trivial.

$(3) \Rightarrow (1)$. If $0 \to M \to F^0 \to F^1 \to \cdots$ is a right $\mathcal{F}lat$-resolution of M, then $F^{n-2} \otimes N \to F^{n-1} \otimes N \to F^n \otimes N \to F^{n+1} \otimes N$ is exact for any N. By Lemma 8.4.23, $K = \text{Ker}(F^n \to F^{n+1})$ is pure in F^n and so is flat. But $F^{n-2} \to F^{n-1} \to K \to 0$ is exact. Therefore, $L = \text{Ker}(F^{n-2} \to F^{n-1})$ is pure in F^n ^{2} and so is also flat. Thus $0 \to N \to F^0 \to \cdots \to F^{n-3} \to L \to 0$ is a right $\mathcal{F}lat$-resolution of N and so (1) holds. $\square$

Proposition 8.4.26. *Let R be left coherent and $n \geq 2$. Then the following are equivalent for $M \in \mathcal{M}_{R_{fp}}$.*

(1) right $\mathcal{P}roj_{fg}$- dim $M \leq n - 2$.

(2) $\text{Tor}_R^{n+k}(M, N) = 0$ for all $N \in {}_R\mathcal{M}$ and $k \geq -1$.

(3) $\text{Tor}_R^n(M, N) = \text{Tor}_R^{n-1}(M, N) = 0$ for all $N \in {}_R\mathcal{M}$.

Proof. So $(1) \Rightarrow (2) \Rightarrow (3)$ is again trivial.

$(3) \Rightarrow (1)$. Let $0 \to M \to P^0 \to P^1 \to \cdots$ be a right $\mathcal{P}roj_{fg}$-resolution of M. Then by Lemma 8.4.23, $K = \text{Ker}(P^n \to P^{n+1})$ is pure in P^n and so is flat. But $P^{n-2} \to P^{n-1} \to K \to 0$ is exact by assumption since right $\mathcal{P}roj_{fg}$-resolutions are right $\mathcal{F}lat$-resolutions. So if we set $L = \text{Ker}(P^{n-2} \to P^{n-1})$, then $0 \to L \to P^{n-2} \to P^{n-1} \to K \to 0$ is exact and thus P^{n-2}/L is flat. But $P^{n-2}/L \hookrightarrow P^{n-1}$ is a $\mathcal{F}lat$-preenvelope (see Example 8.3.11). So $\text{Hom}(P^{n-1}, P^{n-2}/L) \to \text{Hom}(P^{n-2}/L, P^{n-2}/L) \to 0$ is exact and thus P^{n-2}/L is a direct summand of P^{n-1}. Hence P^{n-2}/L is projective. But then L is a summand of P^{n-2} and so $0 \to M \to P^0 \to P^1 \to \cdots \to P^{n-3} \to L \to 0$ is a right $\mathcal{P}roj_{fg}$-resolution of M.$\square$

Theorem 8.4.27. *Let R be left coherent and $n \geq 2$. Then the following are equivalent.*

(1) gl right $\mathcal{A}bs$-dim $_R\mathcal{M} \leq n$.

(2) gl right $\mathcal{F}lat$-dim $\mathcal{M}_R \leq n - 2$.

(3) gl right $\mathcal{P}roj_{fg}$-dim $\mathcal{M}_{R_{fg}} \leq n - 2$.

(4) $\mathrm{Tor}_R^n(M, N) = \mathrm{Tor}_R^{n-1}(M, N) = 0$ for all $M \in \mathcal{M}_{R_{fp}}$, $N \in {}_R\mathcal{M}$.

(5) $\mathrm{Tor}_R^n(M, N) = \mathrm{Tor}_R^{n-1}(M, N) = 0$ for all $M \in \mathcal{M}_R$, $N \in {}_R\mathcal{M}$.

(6) $\mathrm{Tor}_R^{n+k}(M, N) = 0$ for all $M \in \mathcal{M}_R$, $N \in {}_R\mathcal{M}$ and for all $k \geq -1$.

Proof. The result follows from Propositions 8.4.24, 8.4.25 and 8.4.26 above.   □

Corollary 8.4.28. gl right $\mathcal{F}lat$-dim $\mathcal{M}_R$ = gl right $\mathcal{P}roj_{fg}$-dim $\mathcal{M}_{R_{fp}}$ = gl right $\mathcal{A}bs$-dim $_R\mathcal{M} - 2$ *and are both zero if* gl right $\mathcal{A}bs$-dim $_R\mathcal{M} \leq 1$.

Corollary 8.4.29. gl w dim $\mathcal{M} \leq 2$ *if and only if $\mathcal{F}lat$ is a reflective subcategory of $\mathcal{M}_R$.*

Proof. gl w dim $\mathcal{M}$ = gl right $\mathcal{A}bs$-dim $_R\mathcal{M}$ by Remark 8.4.22. So every object M of $\mathcal{M}_R$ has a right $\mathcal{F}lat$-resolution $0 \to M \to F \to 0$ by Corollary 8.4.28 above. But then every homomorphism $M \to F'$ with $F' \in \mathcal{F}lat$ can be factored uniquely through F. That is, $\mathcal{F}lat$ is a *reflective subcategory* of $\mathcal{M}_R$. For the converse, we simply reverse the steps above.   □

Example 8.4.30. If $R = k[x, y]$, k a field, then the inclusion $(x, y) \to R$ is a rigid flat envelope of (x, y), that is, every homomorphism $M \to F$ with F flat can be factored uniquely through R.

We now characterize left coherent rings with finite self absolutely pure dimension.

Theorem 8.4.31. *If R is left coherent and $n \geq 0$, then the following are equivalent.*

(1) *For every flat left R-module F, there is an exact sequence $0 \to F \to A^0 \to \cdots A^n \to 0$ with each $A^i \in \mathcal{A}bs$.*

(2) *If $0 \to M \to F^0 \to F^1 \to \cdots$ is a right $\mathcal{F}lat$-resolution of M_R, then the sequence is exact at F^k for $k \geq n - 1$ where $F^{-1} = M$.*

(3) *If $0 \to M \to P^0 \to P^1 \to \cdots$ is a right $\mathcal{P}roj_{fg}$-resolution of a finitely presented right R-module M, then the sequence is exact at P^k for $k \geq n - 1$ where $P^{-1} = M$.*

(4) *For every absolutely pure right R-module A, there is an exact sequence $0 \to F_n \to F_{n-1} \to \cdots \to F_1 \to F_0 \to A \to 0$ with each $F_i \in \mathcal{F}lat$.*

(5) *There is an exact sequence $0 \to R \to A^0 \to \cdots \to A^n \to 0$ of left R-modules with each A^i absolutely pure.*

Proof. (1) $\Rightarrow$ (5) is immediate.

(5) $\Rightarrow$ (2). We recall that $- \otimes -$ is right balanced on $\mathcal{M}_R \times {}_R\mathcal{M}$ by $\mathcal{F}lat \times \mathcal{A}bs$ with right derived functors $\text{Tor}^k(-, -)$ (see Example 8.3.9).

If $n \geq 2$, using the exact sequence $0 \to R \to A^0 \to \cdots \to A^n \to 0$, we get $\text{Tor}^k(M, R) = 0$ for $k \geq n - 1$. Computing using $0 \to M \to F^0 \to F^1 \to \cdots$ as in (2), we see that $\text{Tor}^k(M, R)$ is just the kth homology group of this complex, giving the desired result.

For $n = 1$, $0 \to R \to A^0 \to A^1 \to 0$ exact gives $\text{Tor}^1(M, R) = 0$ so that, as above, $F^0 \to F^1 \to F^2$ is exact and $M \otimes R \to \text{Tor}^0(M, R)$ is onto. Computing the latter morphism using $0 \to M \to F^0 \to F^1$ shows that $0 \to M \to F^0 \to F^1$ is exact.

If $n = 0$ then (4) means R is absolutely pure as a left R-module. But the balance of $- \otimes -$ then gives $0 \to M \otimes R \to F^0 \otimes R \to F^1 \otimes R \to \cdots$ is exact. That is, $0 \to M \to F^0 \to F^1 \to \cdots$ is exact.

(2) $\Rightarrow$ (3) is trivial. We remark that (3) for $n = 0$ is equivalent to the requirement that every finitely presented right R-module is a submodule of a free R-module.

(3) $\Rightarrow$ (1). Assume (3) with $n \geq 2$. Let $0 \to F \to A^0 \to A^1 \to \cdots$ be exact with F flat and each A^i absolutely pure. Then by (3), we get $\text{Tor}^k(M, F) = 0$ for $k \geq n - 1$ since F is flat. Computing using $0 \to A^0 \to A^1 \to A^2 \to \cdots$ and using Lemma 8.4.23, we get $K = \text{Ker}(A^n \to A^{n+1})$ is pure in A^n and so K is also absolutely pure. Hence $0 \to F \to A^0 \to \cdots \to A^{n-1} \to K \to 0$ gives the desired exact sequence.

Now let $n = 1$. Then (3) says $M \to P^0 \to P^1 \to \cdots$ is exact. So $\text{Tor}^k(M, F) = 0$ for $k = 0$ and $M \otimes F \to \text{Tor}^0(M, F)$ is onto. Hence if $0 \to F \to A^0 \to A^1 \to \cdots$ is exact, $M \otimes F \to M \otimes A^0 \to M \otimes A^1 \to M \otimes A^2$ is exact for all finitely presented M. By Lemma 8.4.23, we again get the desired exact sequence $0 \to F \to A^0 \to K \to 0$ with $K = \text{Ker}(A^1 \to A^2)$.

If $n = 0$ then $0 \to M \to P^0 \to P^1 \to \cdots$ exact means $\text{Tor}^k(M, F) = 0$ for $k > 0$ and $M \otimes F \to \text{Tor}^0(M, F)$ is an isomorphism. This gives that $0 \to M \otimes F \to M \otimes A^0 \to M \otimes A^1$ is exact for all M which implies F is a pure submodule of A^0 and so is absolutely pure.

The proofs of (3) $\Rightarrow$ (4) and (4) $\Rightarrow$ (3) are similar but use the derived functors $\text{Ext}_n(M, A)$ of Example 8.3.11 and the natural homomorphism $\text{Hom}(M, A) \to \text{Ext}_0(M, A)$. $\square$

Corollary 8.4.32. *If R is left coherent, then*

$$\text{gl right } \mathcal{A}bs\text{- dim } {}_R\mathcal{F}lat = \text{gl w dim } \mathcal{A}bs_R = \text{right } \mathcal{A}bs\text{- dim } {}_R R.$$

If $n = 0$, we get the following.

Corollary 8.4.33. *If R is left coherent, then the following are equivalent.*

(1) *Every flat left R-module is absolutely pure.*

(2) *Every R-module is a submodule of a flat R-module.*

(3) *Every finitely presented right R-module is a submodule of a free module.*

(4) *Every absolutely pure right R-module is flat.*

(5) *R is absolutely pure as a left R-module.*

Lemma 8.4.34. *Let R be left noetherian and G be a left R-module. Then right $\mathcal{I}nj$-$\dim G \leq n$ if and if for any left $\mathcal{I}nj$-resolution $\cdots \to E_1 \to E_0 \to M \to 0$ of each $M \in {}_R\mathcal{M}$, $\mathrm{Hom}(G, E_n) \to \mathrm{Hom}(G, \mathrm{Ker}(E_{n-1} \to E_{n-2})) \to 0$ is exact where $E_{-1} = M$.*

Proof. We proceed by induction on n. For $n \geq 1$, we consider a short exact sequence $0 \to G \to E \to G' \to 0$ with E injective. Then we have the following commutative diagrams

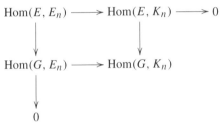

$$\begin{array}{ccccc}
\mathrm{Hom}(E, E_n) & \longrightarrow & \mathrm{Hom}(E, K_n) & \longrightarrow & 0 \\
\downarrow & & \downarrow & & \\
\mathrm{Hom}(G, E_n) & \longrightarrow & \mathrm{Hom}(G, K_n) & & \\
\downarrow & & & & \\
0 & & & &
\end{array}$$

and

$$\begin{array}{ccccccccc}
 & & 0 & & 0 & & 0 & & \\
 & & \downarrow & & \downarrow & & \downarrow & & \\
0 & \longrightarrow & \mathrm{Hom}(G', K_n) & \longrightarrow & \mathrm{Hom}(G', E_{n-1}) & \longrightarrow & \mathrm{Hom}(G', K_{n-1}) & & \\
 & & \downarrow & & \downarrow & & \downarrow & & \\
0 & \longrightarrow & \mathrm{Hom}(E, K_n) & \longrightarrow & \mathrm{Hom}(E, E_{n-1}) & \longrightarrow & \mathrm{Hom}(E, K_{n-1}) & \longrightarrow & 0 \\
 & & \downarrow & & \downarrow & & \downarrow & & \\
0 & \longrightarrow & \mathrm{Hom}(G, K_n) & \longrightarrow & \mathrm{Hom}(G, E_{n-1}) & \longrightarrow & \mathrm{Hom}(G, K_{n-1}) & & \\
 & & & & \downarrow & & & & \\
 & & & & 0. & & & &
\end{array}$$

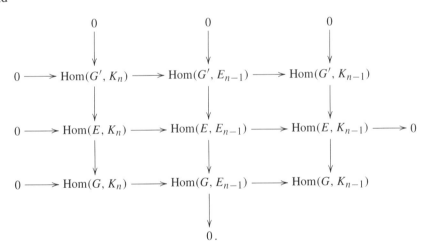

Hence right $\mathcal{I}nj$-$\dim G \leq n$ if and only if right $\mathcal{I}nj$-$\dim G' \leq n - 1$ if and only if $\mathrm{Hom}(G', E_{n-1}) \to \mathrm{Hom}(G', K_{n-1})$ is surjective by induction if and only if $\mathrm{Hom}(E, K_n) \to \mathrm{Hom}(G, K_n)$ is surjective by the second diagram if and only if $\mathrm{Hom}(G, E_n) \to \mathrm{Hom}(G, K_n)$ is surjective by the first diagram.

For $n = 0$, let $K_0 = G$ in the first diagram. Then $\mathrm{Hom}(G, E_0) \to \mathrm{Hom}(G, G)$ is surjective means $\mathrm{Hom}(E, G) \to \mathrm{Hom}(G, G)$ is surjective. Thus $0 \to G \to E$ splits and hence G is injective. The converse is trivial. □

Remark 8.4.35. There is a dual result to Lemma 8.4.34 involving flat dimensions and right $\mathcal{P}roj_{fg}$-resolutions of finitely presented modules over coherent rings.

Theorem 8.4.36. *If R is left noetherian and $n \geq 0$, then the following are equivalent.*

(1) gl right $\mathcal{I}nj$- dim $_R\mathcal{F}lat \leq n$.

(2) *If $0 \to M \to F^0 \to F^1 \to \cdots$ is a right $\mathcal{F}lat$-resolution of a right R-module M, then the sequence is exact at F^k for $k \geq n - 1$ where $F^{-1} = M$.*

(3) *If $0 \to M \to P^0 \to P^1 \to \cdots$ is a right $\mathcal{P}roj_{fg}$-resolution of $M \in \mathcal{M}_{R_{fp}}$, then the sequence is exact at P^k for $k \geq n - 1$ where $P^{-1} = M$.*

(4) gl w dim $\mathcal{I}nj_R \leq n$.

(5) *If $\cdots \to E_1 \to E_0 \to M \to 0$ is a left $\mathcal{I}nj$-resolution of a left R-module M, then the sequence is exact at E_k for $k \geq n - 1$ where $E_{-1} = M$.*

(6) right $\mathcal{I}nj$- dim $_R R \leq n$.

(7) *If $\cdots \to P_1 \to P_0 \to M \to 0$ is a left $\mathcal{P}roj$-resolution of a left R-module M, then the subcomplex $\cdots \to P_{n+1} \to P_n$ is $\mathrm{Hom}(-, \mathcal{F}lat)$ exact.*

Proof. The equivalence of (1), (2), (3), (4) and (6) follows from Theorem 8.4.31 since absolutely pure means injective in this case.

(5) $\Rightarrow$ (6). $E_n \to E_{n-1} \to E_{n-2}$ is exact at E_{n-1} by assumption. Thus $E_n \to \mathrm{Ker}(E_{n-1} \to E_{n-2})$ is surjective. But then right $\mathcal{I}nj$- dim $_R R \leq n$ by Lemma 8.4.34 above.

(6) $\Rightarrow$ (5). Suppose $n \geq 2$ and let $0 \to R \to E^0 \to E^1 \to \cdots \to E^n \to 0$ be a right $\mathcal{I}nj$-resolution of R. Then $\mathrm{Ext}_k(R, M) = 0$ for $k \geq n - 1$. Now computing $\mathrm{Ext}_k(R, M)$ using a left $\mathcal{I}nj$-resolution $\cdots \to E_1 \to E_0 \to M \to 0$, we see that the sequence is exact at $E_k, k \geq n - 1$.

If $n = 1$ and $0 \to R \to E^0 \to E^1 \to 0$ is a right $\mathcal{I}nj$-resolution of R, then $0 \to \mathrm{Hom}(E^1, M) \to \mathrm{Hom}(E^0, M) \to \mathrm{Hom}(R, M)$ is exact. Thus $\mathrm{Ext}_k(R, M) = 0$ for $k \geq 1$ and $\mathrm{Ext}_0(R, M) \to M$ is a monomorphism. But computing $\mathrm{Ext}_0(R, M)$ using a left $\mathcal{I}nj$-resolution, we see that $E_1 \to E_0 \to M$ is exact at E_0. So $\cdots \to E_1 \to E_0 \to M \to 0$ is exact at $E_k, k \geq 0$.

Now let $n = 0$. Then $_R R$ is injective and so every injective precover is surjective and thus $\cdots \to E_1 \to E_0 \to M \to 0$ is exact.

(1) $\Leftrightarrow$ (7). We simply note that gl right $\mathcal{I}nj$- dim $_R\mathcal{F}lat \leq n$ if and only if for each left R-module M, $\mathrm{Ext}^i_R(M, F) = 0$ for all $F \in {}_R\mathcal{F}lat$ and all $i \geq n + 1$. But the latter means that $0 \to \mathrm{Hom}(\mathrm{Ker}(P_{n-1} \to P_{n-2}), F) \to \mathrm{Hom}(P_n, F) \to \mathrm{Hom}(P_{n+1}, F) \to \cdots$ is exact for all $F \in {}_R\mathcal{F}lat$ and so the result follows. $\square$

Corollary 8.4.37. *The following are equivalent for a left noetherian ring R.*

(1) *R is an injective left R-module.*

(2) *Every injective precover of a left R-module is surjective.*

(3) *Every left R-module is a quotient of an injective module.*

(4) *Every flat left R-module is injective.*

(5) *Every projective left R-module is injective.*

(6) *Every $\mathcal{F}$lat-preenvelope of a right R-module is a monomorphism.*

(7) *Every $\mathcal{P}roj_{fg}$-preenvelope of a finitely generated right R-module is a monomorphism.*

(8) *Every finitely generated right R-module is a submodule of a free R-module.*

(9) *Every injective right R-module is flat.*

Proof. (1) $\Leftrightarrow$ (4) $\Leftrightarrow$ (6) $\Leftrightarrow$ (7) $\Leftrightarrow$ (9) follows from Corollary 8.4.33 or Theorem 8.4.36 above.

(1) $\Leftrightarrow$ (2) follows Theorem 8.4.36.

(2) $\Leftrightarrow$ (3), (4) $\Rightarrow$ (5) $\Rightarrow$ (1) and (7) $\Leftrightarrow$ (8) are trivial. $\qquad\square$

Exercises

1. Prove Proposition 8.4.7.

2. Let R be a commutative noetherian ring and M be an R-module. Prove that right $\mathcal{I}nj$- dim $M \leq n$ if and only if $\operatorname{Ext}^{n+1}(R/\mathfrak{p}, M) = 0$ for all $\mathfrak{p} \in \operatorname{Spec} R$. Hint: Use Lemma 2.4.7.

3. Prove that if R is left noetherian then

$$\text{gl right } \mathcal{I}nj\text{- dim } {}_R\mathcal{M} = \text{gl left } \mathcal{P}roj_{fg}\text{- dim } {}_R\mathcal{M}_{fg} = \text{gl w dim } {}_R\mathcal{M}.$$

4. Prove that if R is left noetherian then

$$\begin{aligned}
\text{gl left } \mathcal{I}nj\text{- dim } {}_R\mathcal{M} &= \text{gl right } \mathcal{F}lat\text{- dim } \mathcal{M}_R \\
&= \text{gl right } \mathcal{P}roj_{fg}\text{- dim } \mathcal{M}_{R\,fg} \\
&= \text{gl right } \mathcal{I}nj\text{- dim } {}_R\mathcal{M} - 2
\end{aligned}$$

and zero if gl right $\mathcal{I}nj$- dim ${}_R\mathcal{M} \leq 1$.

5. State and prove results for Example 8.3.6 corresponding to Theorem 8.4.16 and Corollary 8.4.17.

6. Prove the equivalence of parts (3) and (4) of Theorem 8.4.31.

7. A ring R is said to be *left hereditary* if every left ideal of R is projective (and hence hereditary domain means Dedekind domain). Show that the following are equivalent.

   a) *R is left hereditary.*

   b) *Every submodule of a projective R-module is projective.*

   c) Every homomorphic image of an injective R-module is injective.

   d) gl right $\mathcal{I}nj$- dim $_R\mathcal{M} \leq 1$.

   e) $\mathrm{Ext}_R^n(M, N) = 0$ for all R-modules M, N and all $n \geq 2$.

8. Prove that the following are equivalent for a left noetherian ring R.

   a) R is left hereditary.

   b) Every kernel of an $\mathcal{I}nj$-precover is injective.

   c) Every $\mathcal{I}nj$-cover of an R-module is a monomorphism.

   d) Every cokernel of a $\mathcal{P}roj$-preenvelope obtained from a $\mathcal{F}lat$-preenvelope of a finitely presented right R-module is projective.

   e) Every finitely presented right R-module has a surjective $\mathcal{P}roj$-preenvelope.

9. Let R be a commutative noetherian ring and E be an injective R-module. Prove that for each $n \geq 0$, left $\mathcal{F}lat$- dim $\mathrm{Hom}(E, M) \leq n$ for all R-modules M if and only if right $\mathcal{I}nj$- dim $M \otimes E \leq n$ for all R-modules M.

10. Prove that gl left $\mathcal{P}roj$- dim $_R\mathcal{M} = 0$ if and only if R is semisimple.

11. A ring R is said to be *von Neumann regular* if every right R-module is flat. Show that the following are equivalent.

   a) R is von Neumann regular.

   b) gl w dim $_R\mathcal{M} = 0$.

   c) R/I is projective for each finitely generated right ideal I.

12. State and prove the result mentioned in Remark 8.4.35.

8.5 Minimal pure injective resolutions of flat modules

Throughout this section, R will denote a commutative noetherian ring.

If $0 \to F \to PE^0(F) \to PE^1(F) \to \cdots$ is a minimal right $\mathcal{PI}nj$-resolution of a flat R-module F, then for each $n \geq 0$, $PE^n(F)$ is a flat cotorsion module by Lemmas 5.3.23 and 8.4.11. So $PE^0(F) \cong \prod_{\mathfrak{p}\in\mathrm{Spec}\,R} T_\mathfrak{p}$ where $T_\mathfrak{p}$ is the completion of a free $R_\mathfrak{p}$-module by Theorem 5.3.28.

We start with the following result.

Theorem 8.5.1 (Change of rings theorem). *Let F be a flat R-module. If $\sigma : R \to R'$ is a ring homomorphism such that R' is a finitely generated R-module, then $0 \to F \otimes R' \to PE^0(F) \otimes R' \to PE^1(F) \otimes R' \to \cdots$ is a minimal right $\mathcal{PI}nj$-resolution of the R'-module $F \otimes R'$.*

Proof. The sequence is easily pure exact. We now show that $PE^n(F) \otimes R'$ is a pure injective flat R'-module. Clearly it is flat since $PE^n(F)$ is. But $PE^n(F) \cong \prod T_\mathfrak{p}$, over $\mathfrak{p} \in \mathrm{Spec}\,R$. Thus $PE^n(F) \otimes R' \cong (\prod T_\mathfrak{p}) \otimes R'$. But $(\prod T_\mathfrak{p}) \otimes R' \cong \prod(T_\mathfrak{p} \otimes R')$

since R' is finitely generated. Therefore, to show that $PE^n(F) \otimes R'$ is pure injective, it suffices to show that each $T_{\mathfrak{p}} \otimes R'$ is pure injective. By Proposition 6.7.11 this reduces to showing that $\hat{R}_{\mathfrak{p}} \otimes R'$ is pure injective. But $\hat{R}_{\mathfrak{p}} \otimes R' \cong \hat{R}'_{\mathfrak{q}_1} \oplus \cdots \oplus \hat{R}'_{\mathfrak{q}_s}$ where $\mathfrak{q}_1, \ldots, \mathfrak{q}_s$ are the distinct primes lying over $\mathfrak{p}$, and is 0 if there is no such prime.

For minimality, it suffices to show that for any flat R-module F, $F \otimes R' \to PE(F) \otimes R'$ is a pure injective envelope of F. But this will follow once we show that $F \otimes R' \to PE(F) \otimes R'$ satisfies conditions (a) and (b) in Theorem 6.7.17. For any $\mathfrak{p} \in \operatorname{Spec} R$, $F \otimes k(\mathfrak{p}) \to PE(F) \otimes k(\mathfrak{p})$ is an injection and in fact splits. So $F \otimes k(\mathfrak{p}) \otimes R' \to PE(F) \otimes k(\mathfrak{p}) \otimes R'$ also splits. But $k(\mathfrak{p}) \otimes R'$ is the direct sum of a finite number of local rings of dimension 0. If we go modulo the radical of $k(\mathfrak{p}) \otimes R'$, we still get an injection, and in fact, we get a map $F \otimes R' \otimes (k(\mathfrak{q}_1) \oplus \cdots \oplus k(\mathfrak{q}_s)) \to PE(F) \otimes R' \otimes (k(\mathfrak{q}_1) \oplus \cdots \oplus k(\mathfrak{q}_s))$ where $\mathfrak{q}_1, \ldots, \mathfrak{q}_s$ are prime ideals of R' lying over $\mathfrak{p}$ (possibly $s = 0$). Thus $F \otimes R' \to PE(F) \otimes R'$ satisfies (a) of Theorem 6.7.17. The argument that (b) is satisfied is similar. $\square$

Corollary 8.5.2. right $\mathcal{PInj}$- dim $_{R'} R' \leq$ right $\mathcal{PInj}$- dim $_R R$.

Remark 8.5.3. If $R \subset R'$, then for any flat R-module F, $F \otimes R' = 0$ implies $F = 0$ and so we get equality in the corollary. Hence in this case, R is self pure injective if and only if R' is. It is easy to see that for any ring R, R is self pure injective if and only if $\operatorname{Ext}^1_R(F, R) = 0$ for all flat R-modules F. But the latter is equivalent to R being a direct product of a finite number of complete local rings (see Jensen [124, Theorem 8.1]). So if we drop the noetherian condition on R, we get the following.

Corollary 8.5.4. *Let R be a subring of a ring R' which is finitely generated over R. Then R is a direct product of a finite number of complete local rings if and only if R' is.*

Proof. The result follows from the remark above noting that R is noetherian if and only if R' is by Theorem 3.1.18. $\square$

Definition 8.5.5. For a flat R-module F, a prime ideal $\mathfrak{p}$ of R, and $n \geq 0$, $\pi_n(\mathfrak{p}, F)$ is the cardinality of a base of a free $R_{\mathfrak{p}}$-module whose completion is the $T_{\mathfrak{p}}$ in $PE^n(F)$.

Remark 8.5.6. The change of rings theorem (Theorem 8.5.1) above says that if $R \to R'$ is a ring homomorphism such that R' is a finitely generated R-module, then for any flat R-module F, $PE^n(F) \otimes R' \cong PE^n(F \otimes R')$ for all $n \geq 0$. This implies that if $\mathfrak{p}' \subset R'$ is a prime ideal lying over $\mathfrak{p} \subset R$, then $\pi_n(\mathfrak{p}', R') = \pi_n(\mathfrak{p}, R)$. Hence $\mathfrak{p}$ appears in $PE^n(R)$ if and only if $\mathfrak{p}'$ appears in $PE^n(R')$. We note that $\mathfrak{p}$ appears in $PE^n(R)$ if and only if $\hat{R}_{\mathfrak{p}}$ is a summand of $PE^n(R)$.

Theorem 8.5.7. *If F is a flat R-module, $n \geq 0$, and a prime ideal $\mathfrak{p}$ of R is maximal such that $\pi_n(\mathfrak{p}, F) \neq 0$, then $\pi_{n+1}(\mathfrak{q}, F) = 0$ for all primes $\mathfrak{q} \supseteq \mathfrak{p}$.*

Proof. We give the argument for $n = 0$ as it is easy to see how to modify the argument for any $n > 0$. Let $\mathfrak{p}$ be maximal such that $\pi_0(\mathfrak{p}, F) \neq 0$. Let $PE(F) = PE^0(F) = \prod T_\mathfrak{q}$. Then $T_\mathfrak{p} \cong \hat{F}_\mathfrak{p}$ by Theorem 6.7.9. By assumption, $T_\mathfrak{q} = 0$ if $\mathfrak{q} \supsetneq \mathfrak{p}$. If $\mathfrak{p} \not\subset \mathfrak{q}$, let $S \subset k(\mathfrak{p})$ be finitely generated and choose $r \in \mathfrak{p}, r \notin \mathfrak{q}$. Then multiplication by r is zero on S and is an automorphism on $T_\mathfrak{q}$. So $T_\mathfrak{q} \otimes S = 0$. Hence $(\prod_{\mathfrak{p} \not\subset \mathfrak{q}} T_\mathfrak{q}) \otimes S = 0$ and so $(\prod_{\mathfrak{p} \not\subset \mathfrak{q}} T_\mathfrak{q}) \otimes k(\mathfrak{p}) = 0$. Thus $PE(F) \otimes k(\mathfrak{p}) \cong (\prod T_\mathfrak{q}) \otimes k(\mathfrak{p}) \cong T_\mathfrak{p} \otimes k(\mathfrak{p}) \cong \hat{F}_\mathfrak{p} \otimes k(\mathfrak{p})$. This means that $PE(F) \otimes k(\mathfrak{p}) \cong F \otimes k(\mathfrak{p})$.

Now let $0 \to F \to PE(F) \to C \to 0$ be exact. Then $C \otimes k(\mathfrak{p}) = 0$ by the above. If $\mathfrak{q} \supsetneq \mathfrak{p}$, a similar argument as above gives $PE(F) \otimes k(\mathfrak{q}) = 0$ and so $C \otimes k(\mathfrak{q}) = 0$. But $C \otimes k(\mathfrak{q}) = 0$ implies $\hat{C}_\mathfrak{q} = 0$. Hence $T_\mathfrak{q} = 0$ by Theorem 6.7.9. $\square$

As a consequence, we have the following result.

Proposition 8.5.8. *Let F be a flat R-module. If a prime ideal $\mathfrak{p}$ appears in $PE^{n+1}(F)$, then there is a prime ideal $\mathfrak{q} \supsetneq \mathfrak{p}$ which appears in $PE^n(F)$.*

Proof. If $\mathfrak{p}$ appears in $PE^n(F)$, then there is a prime ideal $\mathfrak{q} \supsetneq \mathfrak{p}$ which appears in $PE^n(F)$ by Theorem 8.5.7 above.

Hence suppose no prime ideal $\mathfrak{q} \supseteq \mathfrak{p}$ appears in $PE^n(F) = \prod T_\mathfrak{q}$. If $T_\mathfrak{q} \neq 0$ and $\mathfrak{q} \not\supset \mathfrak{p}$, then $T_\mathfrak{q} \otimes R/\mathfrak{p} = 0$ for if $r \in \mathfrak{p}, r \notin \mathfrak{q}$, then $T_\mathfrak{q} \xrightarrow{r} T_\mathfrak{q}$ is an isomorphism and $R/\mathfrak{p} \xrightarrow{r} R/\mathfrak{p}$ is zero. So $PE^n(F) \otimes R/\mathfrak{p} = 0$. But by the change of rings theorem above, $0 \to F \otimes R/\mathfrak{p} \to PE^0(F) \otimes R/\mathfrak{p} \to \cdots$ is a minimal right $\mathcal{P}Inj$-resolution of $F \otimes R/\mathfrak{p}$ over $R/\mathfrak{p}$. So by minimality, if $PE^n(F) \otimes R/\mathfrak{p} = 0$ then $PE^{n+1}(F) \otimes R/\mathfrak{p} = 0$. But the latter is not possible if $\mathfrak{p}$ appears in $PE^{n+1}(F)$. $\square$

Definition 8.5.9. The *coheight* (coht) of a prime ideal $\mathfrak{p}$ is the supremum of the lengths of strictly increasing chains $\mathfrak{p} = \mathfrak{p}_0 \subset \mathfrak{p}_1 \subset \cdots \subset \mathfrak{p}_s$ of prime ideals. It follows from the definitions that coht $\mathfrak{p} = \dim R/\mathfrak{p}$ and ht $\mathfrak{p}$ + coht $\mathfrak{p} \leq \dim R$ (see Definition 2.4.13).

Corollary 8.5.10. *If $\pi_n(\mathfrak{p}, F) \neq 0$, then* coht $\mathfrak{p} \geq n$.

Proof. $\pi_n(\mathfrak{p}, F) \neq 0$ implies that there exists a prime ideal $\mathfrak{q}_1 \supsetneq \mathfrak{p}$ such that $\pi_{n-1}(\mathfrak{q}_1, F) \neq 0$ by Proposition 8.5.8. One then repeats this argument to get a chain $\mathfrak{p} \subsetneq \mathfrak{q}_1 \subsetneq \mathfrak{q}_2 \subsetneq \cdots \subsetneq \mathfrak{q}_n$ of prime ideals. $\square$

Remark 8.5.11. If $\dim R = n < \infty$, then $\pi_n(\mathfrak{p}, F) \neq 0$ implies that $\mathfrak{p}$ is a minimal prime ideal of R by the corollary above since in this case coht $\mathfrak{p} = \dim R = n$ and so ht $\mathfrak{p} = 0$.

Corollary 8.5.12. *If $\dim R < \infty$, then* gl right $\mathcal{P}Inj$- dim $\mathcal{F}lat \leq \dim R$.

Proof. Suppose $n > \dim R$. Then coht $\mathfrak{p} \leq \dim R < n$ for each $\mathfrak{p} \in \operatorname{Spec} R$. So if $F \in \mathcal{F}lat$, then $\pi_n(\mathfrak{p}, F) = 0$ for all $\mathfrak{p} \in \operatorname{Spec} R$ by Corollary 8.5.10. Thus $PE^n(F) = 0$. Hence right $\mathcal{P}Inj$- dim $F \leq \dim R$. $\square$

Proposition 8.5.13. *Let* $d = $ gl left $\mathcal{P}roj$- dim $\mathcal{F}lat$. *If* $d < \infty$, *then* $d = \sup\{$*left* $\mathcal{P}roj$- dim $R_{\mathfrak{p}} : \mathfrak{p} \in \operatorname{Spec} R\}$.

Proof. Let $G \in \mathcal{F}lat$. Then $\operatorname{Ext}^d(G, F) \neq 0$ for some flat R-module F by Theorem 8.4.12 and so $PE^d(F) \neq 0$. But $PE^d(F) = \prod T_{\mathfrak{p}}$. Now let $\mathfrak{p}$ be the minimal such that $T_{\mathfrak{p}} \neq 0$. Then $\hat{R}_{\mathfrak{p}}$ is a direct summand of $T_{\mathfrak{p}}$ and so the corresponding injection in $\operatorname{Hom}(\hat{R}_{\mathfrak{p}}, PE^d(F))$ gives a nonzero element of $\operatorname{Ext}^d(\hat{R}_{\mathfrak{p}}, F)$. Let the map $\hat{R}_{\mathfrak{p}} \to PE^d(F)$ be restricted to $R_{\mathfrak{p}}$. We claim this gives a nonzero element of $\operatorname{Ext}^d(R_{\mathfrak{p}}, F)$. For if $\operatorname{Ext}^d(R_{\mathfrak{p}}, F) = 0$, then the map $R_{\mathfrak{p}} \to PE^d(F)$ has a factorization $R_{\mathfrak{p}} \to PE^{d-1}(F) \to PE^d(F)$. Now extend the map $R_{\mathfrak{p}} \to PE^{d-1}(F)$ to $\hat{R}_{\mathfrak{p}}$. But by the minimality of $\mathfrak{p}$, this extension is unique by Proposition 6.7.8 and Remark 6.7.5 and so $\operatorname{Ext}^d(\hat{R}_{\mathfrak{p}}, F) = 0$, a contradiction. Hence $\operatorname{Ext}^d(R_{\mathfrak{p}}, F) \neq 0$. But left $\mathcal{P}roj$- dim $R_{\mathfrak{p}} \leq d$ since $R_{\mathfrak{p}}$ is a flat R-module. Therefore it follows that $\sup\{$left $\mathcal{P}roj$- dim $R_{\mathfrak{p}}\} = d$. □

Corollary 8.5.14. *If* $\mathfrak{p}$ *is minimal such that* $d = $ left $\mathcal{P}roj$- dim $R_{\mathfrak{p}}$, *then* coht $\mathfrak{p} \geq d$.

Proof. If $\mathfrak{p}$ is minimal such that $\operatorname{Ext}^d(R_{\mathfrak{p}}, F) \neq 0$ for some flat R-module F, then $\pi_d(\mathfrak{p}, F) \neq 0$ by the proof of Proposition 8.5.13. But then coht $\mathfrak{p} \geq d$ by Corollary 8.5.10. □

Corollary 8.5.15. *Suppose* R *is a domain and* K *is its field of fractions. If left* $\mathcal{P}roj$- dim $K < \dim R < \infty$, *then* gl left $\mathcal{P}roj$- dim $\mathcal{F}lat < \dim R$.

Proof. As noted in Remark 8.5.11, if $n = \dim R$ and F is flat, then $\pi_n(\mathfrak{p}, F) \neq 0$ implies $\mathfrak{p}$ is minimal. Thus $\mathfrak{p} = 0$. But then left $\mathcal{P}roj$- dim $K \leq \operatorname{coht} 0$ by Corollary 8.5.14 and so we are done. □

By Proposition 6.7.3, $PE(R) \cong \prod \hat{R}_{\mathfrak{m}}$ over all maximal ideals $\mathfrak{m}$ of R. So

$$\pi_0(\mathfrak{p}, R) = \begin{cases} 1 & \text{if } \mathfrak{p} = \mathfrak{m} \\ 0 & \text{if } \mathfrak{p} \neq \mathfrak{m}. \end{cases}$$

Therefore, $\pi_1(\mathfrak{m}, R) = 0$ by Theorem 8.5.7 above. So if $\pi_1(\mathfrak{p}, R) \neq 0$, then $\mathfrak{p} \neq \mathfrak{m}$ and coht $\mathfrak{p} \geq 1$ by Corollary 8.5.10.

The following proposition characterizes which prime ideals of height 1 do not appear in $PE^1(R)$.

Proposition 8.5.16. *If* $\mathfrak{p}$ *is a prime ideal of* R *of coheight 1, then* $\pi_1(\mathfrak{p}, R) = 0$ *if and only if* $\mathfrak{p}$ *is contained in a unique maximal ideal* $\mathfrak{m}$ *of* R *such that* $\hat{R}_{\mathfrak{m}} \otimes k(\mathfrak{p}) \cong k(\mathfrak{p})$.

Proof. Let $0 \to R \to PE^0(R) \to C \to 0$ be exact so that $PE^0(C) = PE^1(R)$. Then $\pi_1(\mathfrak{m}, R) = 0$ for all maximal ideals $\mathfrak{m}$ by the above. So if $PE^0(C) = \prod T_{\mathfrak{q}}$, then $T_{\mathfrak{m}} = 0$ for each $\mathfrak{m}$. Hence $T_{\mathfrak{p}} \cong \hat{C}_{\mathfrak{p}}$ for any prime $\mathfrak{p}$ of coheight 1 by Theorem 6.7.9. But $\hat{C}_{\mathfrak{p}} = 0$ if and only if $C \otimes k(\mathfrak{p}) = 0$. So $\pi_1(\mathfrak{p}, R) = 0$ if and only if

$C \otimes k(\mathfrak{p}) = 0$. But $0 \to R \otimes k(\mathfrak{p}) \to (\prod \hat{R}_\mathfrak{m}) \otimes k(\mathfrak{p}) \to C \otimes k(\mathfrak{p}) \to 0$ is exact. So $\dim_{k(\mathfrak{p})}(\prod \hat{R}_\mathfrak{m}) \otimes k(\mathfrak{p}) = 1$. Thus $\hat{R}_\mathfrak{m} \otimes k(\mathfrak{p})$ has dimension 1 for exactly one maximal ideal $\mathfrak{m}$ and so $\mathfrak{p} \subset \mathfrak{m}$ for this $\mathfrak{m}$. For any other maximal ideal $\mathfrak{m}$, $\hat{R}_\mathfrak{m} \otimes k(\mathfrak{p}) = 0$ and so $\mathfrak{p} \not\subset \mathfrak{m}$. This completes the proof. $\square$

Lemma 8.5.17. *Let I be a nilpotent ideal of R and M be an R-module. If F is a flat R-module and $\varphi : N \to F$ is such that the induced map $M/IM \to F/IF$ is an isomorphism, then φ is an isomorphism.*

Proof. Nakayama lemma implies φ is surjective. So $\mathrm{Ker}\,\varphi$ is a pure submodule of M since F is flat. But $\mathrm{Ker}\,\varphi \subset IM$ and so $I\,\mathrm{Ker}\,\varphi = \mathrm{Ker}\,\varphi \cap IM = \mathrm{Ker}\,\varphi$. Hence $\mathrm{Ker}\,\varphi = 0$. $\square$

Remark 8.5.18. Let I be an ideal of R. If $I \subset \mathfrak{p}$ for a prime ideal $\mathfrak{p}$ of R, then the identification $\hat{R}_\mathfrak{p}/I\hat{R}_\mathfrak{p} \cong (\widehat{R/I})_{\mathfrak{p}/I}$ and Proposition 6.7.11 show that $T_\mathfrak{p} \otimes R/I$ is the completion of a free $(R/I)_{\mathfrak{p}/I}$-module. The identification $T_\mathfrak{p} \otimes k(\mathfrak{p}) \cong (T_\mathfrak{p}/IT_\mathfrak{p}) \otimes k(\mathfrak{p}/I)$ shows that the free $(R/I)_{\mathfrak{p}/I}$-module in question has a base having the same cardinality as that of the base of the free module whose completion is $T_\mathfrak{p}$. That is, if F is a flat R-module and $I \subset \mathfrak{p}$, then $\pi_n(\mathfrak{p}/I, F \otimes R/I) = \pi_n(\mathfrak{p}, F)$ for each $n \geq 0$.

We also note that if $I \not\subset \mathfrak{p}$, then $T_\mathfrak{p} \otimes R/I = 0$. But if $I \subset \mathfrak{p}$, then $\hat{R}_\mathfrak{p}$ is complete in the I-adic topology. So a pure injective flat module $F = \prod T_\mathfrak{p}$ will be complete if $T_\mathfrak{p} = 0$ whenever $I \not\subset \mathfrak{p}$.

These remarks give the following result.

Proposition 8.5.19. *If for each $n \geq 1$, F_n is a pure injective flat R/I^n-module and $F_{n+1} \to F_n$ are surjective maps with kernels $I^n F_{n+1}$, then $\varprojlim F_n$ is a pure injective flat R-module.*

Proof. Let $F_1 = \prod T_{\mathfrak{p}/I}$ (over primes ideals $\mathfrak{p}$ of R such that $I \subset \mathfrak{p}$) where each $T_{\mathfrak{p}/I}$ is the completion of a free $(R/I)_{\mathfrak{p}/I}$-module. From the remark above, we can construct a pure injective flat R-module $F = \prod T_\mathfrak{p}$ (for $\mathfrak{p} \supset I$) and an onto map $F \to F_1$, with kernel IF. We now consider

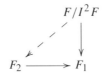

and we apply Lemma 6.7.13 to get a map $F \to F_2$ which can be used to make the diagram above commutative. By Lemma 8.5.17, $F/I^n F \to F_2$ is an isomorphism. We repeat the argument and get a map $F \to \varprojlim F_n$ inducing an isomorphism $F/I^n F \to F_n$ for each n. But then, F is complete with respect to the I-adic topology. Hence $F \cong \varprojlim F/I^n F$ and so $F \cong \varprojlim F/I^n F = \varprojlim F_n$. $\square$

Remark 8.5.20. Note that if $F = \varprojlim F_n$ in the above is written as $\prod T_{\mathfrak{p}}$, then $T_{\mathfrak{p}} = 0$ if $I \not\subset \mathfrak{p}$. Moreover, if for any n, $F_n = \prod S_{\mathfrak{p}/I^n}$ where each $S_{\mathfrak{p}/I^n}$ is the completion of a free $(R/I^n)_{\mathfrak{p}/I^n}$-module, then $T_{\mathfrak{p}} \otimes R/I^n \cong S_{\mathfrak{p}/I^n}$. Hence the bases of the free modules whose completions are $T_{\mathfrak{p}}$ and $S_{\mathfrak{p}/I^n}$ have the same cardinality.

Theorem 8.5.21. *Let $I \subset R$ be an ideal such that R/I has finite Krull dimension. If R is complete with respect to the I-adic topology, then for any prime $\mathfrak{p}$ of R and $n \geq 0$, $\pi_n(\mathfrak{p}, R) \neq 0$ implies $I \subset \mathfrak{p}$ (and so $\pi_n(\mathfrak{p}, R) = \pi_n(\mathfrak{p}/I, R/I)$).*

Proof. We consider the rings R/I^n for $n \geq 1$ and their minimal right $\mathcal{PI}nj$-resolutions over themselves. By the Change of rings theorem 8.5.1, we can identify the sequence $0 \to R/I^{n+1} \otimes R/I^n \to PE^\circ(R/I^{n+1}) \otimes R/I^n \to \cdots$ with the sequence $0 \to R/I^n \to PE^\circ(R/I^n) \to \cdots$. Thus we have surjective maps $PE^i(R/I^{n+1}) \to PE^i(R/I^n)$ with kernels $I^n PE^i(R/I^{n+1})$ making the obvious diagram commutative.

We wish to show that the sequence $0 \to \varprojlim R/I^n (= R) \to \varprojlim PE^\circ(R/I^n) \to \cdots$ is a minimal right $\mathcal{PI}nj$-resolution of R. We first consider the short exact sequence $0 \to R/I^n \to PE^\circ(R/I^n) \to C_n \to 0$. Since $R/I^{n+1} \to R/I^n$ is surjective for each n, the sequence $0 \to R \to \varprojlim PE^\circ(R/I^n) \to \varprojlim C_n \to 0$ is exact by Theorem 1.5.13. But $C_{n+1} \to C_n$ is also surjective for each $n \geq 1$. Hence we repeat the argument to get that the sequence is exact. Furthermore, each $\varprojlim PE^i(R/I^n)$ is flat and pure injective by Proposition 8.5.19.

But if the Krull dimension is finite, say d, then $\pi_i(\mathfrak{p}/I^n, R/I^n) = 0$ for all n and all $i > d$ by Corollary 8.5.12. So the sequence $0 \to R \to \varprojlim PE^\circ(R/I^n) \to \cdots \to \varprojlim PE^d(R/I^n) \to 0$ is exact. Hence it is pure exact since all the modules in the sequence are flat. Finally note that each module in the sequence satisfies (a) and (b) of Theorem 6.7.17. (a) follows from pure exactness, and for $I \subset \mathfrak{p}$, (b) follows from $k(\mathfrak{p}) \cong k(\mathfrak{p}/I)$ and the fact that the resolution of R/I is minimal. If $I \not\subset \mathfrak{p}$, then $T_{\mathfrak{p}} = 0$ and so there is nothing to prove. So each module $\varprojlim PE^i(R/I^n)$ is a pure injective envelope by Theorem 6.7.17. Hence the sequence is minimal. This completes the proof. □

Corollary 8.5.22. right $\mathcal{PI}nj$-$\dim_{R/I} R/I = $ right $\mathcal{PI}nj$-$\dim_R R$.

Proof. Suppose right $\mathcal{PI}nj$-$\dim_R R = n$, then $\pi_n(\mathfrak{p}, R) \neq 0$. So $I \subset \mathfrak{p}$ and $\pi_n(\mathfrak{p}/I, R/I) \neq 0$ by the theorem above. Hence right $\mathcal{PI}nj$-$\dim_R R \leq$ right $\mathcal{PI}nj$-$\dim_{R/I} R/I$. Thus the result follows from Corollary 8.5.2. □

Corollary 8.5.23. *If R is complete with respect to the I-adic topology and $\mathfrak{p} \subset R$ is a prime ideal such that $\mathrm{coht}\,\mathfrak{p} = 1$ and $I \not\subset \mathfrak{p}$, then $\mathfrak{p}$ is contained in a unique maximal ideal $\mathfrak{m}$ and $\hat{R}_{\mathfrak{m}} \otimes k(\mathfrak{p}) \cong k(\mathfrak{p})$.*

Proof. Since $I \not\subset \mathfrak{p}$, $\pi_1(\mathfrak{p}, R) = 0$ by the theorem. Then the result follows from Proposition 8.5.16. □

Remark 8.5.24. The first part of the conclusion of Corollary 8.5.23 holds without assuming that R/I has finite Krull dimension. For I is contained in the radical of R and so R is a Zariski ring with the I-adic topology. Thus $\mathfrak{p}$ is a closed ideal and $D = R/\mathfrak{p}$ is a one-dimensional noetherian domain which is complete in the $\bar{I}$-adic topology where $\bar{I}$ is the image of I. So $\bar{I}$ is in the radical of D. Hence D is semilocal and complete in the topology determined by its Jacobson radical. But then D is the product of local rings by Theorem 2.5.20. So D is local since D is a domain. Thus $\mathfrak{p}$ is contained in a unique maximal ideal.

Lemma 8.5.25. *If F is a flat cotorsion R-module, then for each $\mathfrak{p} \in \operatorname{Spec} R$,*

$$k(\mathfrak{p}) \otimes \operatorname{Hom}(\hat{R}_\mathfrak{p}, F) \cong k(\mathfrak{p}) \otimes \operatorname{Hom}(R_\mathfrak{p}, F) \cong k(\mathfrak{p}) \otimes T_\mathfrak{p}.$$

Proof. By Theorem 5.3.28, $F \cong \prod T_\mathfrak{q}$ over $\mathfrak{q} \in \operatorname{Spec} R$. So

$$\operatorname{Hom}(\hat{R}_\mathfrak{p}, F) \cong \prod \operatorname{Hom}(\hat{R}_\mathfrak{p}, T_\mathfrak{q}).$$

But $T_\mathfrak{q} \cong \operatorname{Hom}(E(k(\mathfrak{q})), E(k(\mathfrak{q}))^{(X)})$ for some set X. So

$$\operatorname{Hom}(\hat{R}_\mathfrak{p}, F) \cong \prod \operatorname{Hom}(\hat{R}_\mathfrak{p} \otimes E(k(\mathfrak{q})), E(k(\mathfrak{q}))^{(X)}).$$

But $\hat{R}_\mathfrak{p} \otimes E(k(\mathfrak{q})) \cong E(k(\mathfrak{q}))^{(X_\mathfrak{q})}$ for some set $X_\mathfrak{q}$ by Lemma 6.7.7. So $\operatorname{Hom}(\hat{R}_\mathfrak{p}, \prod T_\mathfrak{q}) \cong \prod T_\mathfrak{q}^{X_\mathfrak{q}}$. But then $\operatorname{Hom}(\hat{R}_\mathfrak{p}, F) \cong \prod_{\mathfrak{q} \subseteq \mathfrak{p}} T_\mathfrak{q}^{X_\mathfrak{q}}$ by Proposition 6.7.8. Now let $S \subset k(\mathfrak{p})$ be finitely generated. Then $S \otimes \prod_{\mathfrak{q} \subseteq \mathfrak{p}} T_\mathfrak{q}^{X_\mathfrak{q}} \cong \prod (S \otimes T_\mathfrak{q})^{X_\mathfrak{q}}$. But if $\mathfrak{q} \subsetneq \mathfrak{p}$, let $r \in \mathfrak{p}$, $r \notin \mathfrak{q}$. Then multiplication by r is zero on S and is an automorphism on $T_\mathfrak{q}$. Hence $S \otimes T_\mathfrak{q} = 0$ and $S \otimes \prod_{\mathfrak{q} \subseteq \mathfrak{p}} T_\mathfrak{q}^{X_\mathfrak{q}} \cong S \otimes T_\mathfrak{p}^{X_\mathfrak{p}}$. Thus $k(\mathfrak{p}) \otimes \prod_{\mathfrak{q} \subseteq \mathfrak{p}} T_\mathfrak{q}^{X_\mathfrak{q}} \cong k(\mathfrak{p}) \otimes T_\mathfrak{p}^{X_\mathfrak{p}}$. But if $\mathfrak{q} = \mathfrak{p}$, then $\hat{R}_\mathfrak{p} \otimes E(k(\mathfrak{p})) \cong E(k(\mathfrak{p}))$ and so $\operatorname{Card} X_\mathfrak{p} = 1$. Thus half of the result follows.

If $\mathfrak{q} \subsetneq \mathfrak{p}$, then $R_\mathfrak{p} \otimes E(k(\mathfrak{q})) = 0$ as in the proof of Lemma 6.7.7. If $\mathfrak{q} \subseteq \mathfrak{p}$, then $R_\mathfrak{p} \otimes E(k(\mathfrak{q})) \cong E(k(\mathfrak{q}))$. So $k(\mathfrak{p}) \otimes \operatorname{Hom}(R_\mathfrak{p}, F) \cong k(\mathfrak{p}) \otimes \prod_{\mathfrak{q} \subseteq \mathfrak{p}} T_\mathfrak{q} \cong k(\mathfrak{p}) \otimes T_\mathfrak{p}$ as in the above. □

Proposition 8.5.26. *The complexes obtained from the minimal right $\mathcal{PI}nj$-resolution of a flat R-module by applying the functors*

$$k(\mathfrak{p}) \otimes \operatorname{Hom}(\hat{R}_\mathfrak{p}, -) \text{ and } k(\mathfrak{p}) \otimes \operatorname{Hom}(R_\mathfrak{p}, -)$$

are the same and have zero differentiation.

Proof. Let F be a flat R-module. Then $PE^n(F) \cong \prod T_\mathfrak{q}^n$, over $\mathfrak{q} \in \operatorname{Spec} R$, where $T_\mathfrak{q}^n$ denotes the $T_\mathfrak{q}$ at the nth place in the minimal right $\mathcal{PI}nj$-resolution of F. If we apply either of the functors to the resolution, we get the complex $0 \to k(\mathfrak{p}) \otimes T_\mathfrak{p}^0 \to k(\mathfrak{p}) \otimes T_\mathfrak{p}^1 \to \cdots$ by Lemma 8.5.25.

As in the lemma, $\mathrm{Hom}(R_{\mathfrak{p}}, F) \cong \prod_{\mathfrak{q} \subseteq \mathfrak{p}} T_{\mathfrak{q}}^n$. So applying $\mathrm{Hom}(R_{\mathfrak{p}}, -)$ to the minimal right $\mathcal{PInj}$-resolution, we get a map $\prod_{\mathfrak{q} \subseteq \mathfrak{p}} T_{\mathfrak{q}}^n \to \prod_{\mathfrak{q} \subseteq \mathfrak{p}}^{n+1}$. But by Proposition 6.7.8, $\mathrm{Hom}(\prod_{\mathfrak{q} \subsetneq \mathfrak{p}} T_{\mathfrak{q}}^n, T_{\mathfrak{p}}^{n+1}) = 0$. So passing to quotients, that is $\prod_{\mathfrak{q} \subseteq \mathfrak{p}} T_{\mathfrak{q}}^n / \prod_{\mathfrak{q} \subsetneq \mathfrak{p}} T_{\mathfrak{q}}^n \cong T_{\mathfrak{p}}^n$, we get a map $T_{\mathfrak{p}}^n \to T_{\mathfrak{p}}^{n+1}$. Thus we have a commutative diagram

$$
\begin{array}{ccc}
T_{\mathfrak{p}}^n & \longrightarrow & T_{\mathfrak{p}}^{n+1} \\
\downarrow & & \downarrow \\
k(\mathfrak{p}) \otimes T_{\mathfrak{p}}^n & \longrightarrow & k(\mathfrak{p}) \otimes T_{\mathfrak{p}}^{n+1}
\end{array}
$$

with the bottom map the map of our complex.

If this map is not zero, let $x \in T_{\mathfrak{p}}^n$ have a nonzero image in $k(\mathfrak{p}) \otimes T_{\mathfrak{p}}^{n+1}$. Then $x \hat{R}_{\mathfrak{p}}$ will be a direct summand of $T_{\mathfrak{p}}^n$ which is mapped isomorphically onto a direct summand of $T_{\mathfrak{p}}^{n+1}$. So $x \hat{R}_{\mathfrak{p}} \subset \prod_{\mathfrak{q} \subseteq \mathfrak{p}} T_{\mathfrak{q}}^n$ will be mapped isomorphically onto a direct summand of $\prod_{\mathfrak{q} \subseteq \mathfrak{p}} T_{\mathfrak{q}}^{n+1}$ by Proposition 6.7.8. This contradicts the minimality of the right $\mathcal{PInj}$-resolution.   □

We are now in a position to prove the following result.

Theorem 8.5.27. *The following are equivalent for an integer $n \geq 0$.*

(1) gl right $\mathcal{PInj}$-dim $\mathcal{F}lat \leq n$.

(2) gl left $\mathcal{P}roj$-dim $\mathcal{F}lat \leq n$.

(3) left $\mathcal{P}roj$-dim $\hat{R}_{\mathfrak{p}} \leq n$ for all $\mathfrak{p} \in \mathrm{Spec}\, R$ and gl right $\mathcal{PInj}$-dim $\mathcal{F}lat < \infty$.

(4) *The subcomplex* $\mathrm{Hom}(\hat{R}_{\mathfrak{p}}, PE^n(F)) \to \mathrm{Hom}(\hat{R}_{\mathfrak{p}}, PE^{n+1}(F)) \to \cdots$ *is pure exact for all $\mathfrak{p} \in \mathrm{Spec}\, R$ and all $F \in \mathcal{F}lat$.*

(5) left $\mathcal{P}roj$-dim $R_{\mathfrak{p}} \leq n$ for all $\mathfrak{p} \in \mathrm{Spec}\, R$ and gl left $\mathcal{P}roj$-dim $\mathcal{F}lat < \infty$.

(6) *The subcomplex* $\mathrm{Hom}(R_{\mathfrak{p}}, PE^n(F)) \to \mathrm{Hom}(R_{\mathfrak{p}}, PE^{n+1}(F)) \to \cdots$ *is pure exact for all $\mathfrak{p} \in \mathrm{Spec}\, R$ and all $F \in \mathcal{F}lat$.*

Proof. (1) ⇔ (2) is part of Theorem 8.4.12.

(2) ⇒ (3). $\hat{R}_{\mathfrak{p}}$ is flat and (2) ⇒ (1). So (3) follows trivially.

(3) ⇒ (4). Let right $\mathcal{PInj}$-dim $F = m$. If $m \leq n$, then (4) follows trivially. If $m > n$, then

$$
\mathrm{Hom}(\hat{R}_{\mathfrak{p}}, PE^n(F)) \to \mathrm{Hom}(\hat{R}_{\mathfrak{p}}, PE^{n+1}(F)) \to \cdots \xrightarrow{\sigma^{m-1}} \mathrm{Hom}(\hat{R}_{\mathfrak{p}}, PE^m(F)) \to 0
$$

is exact since left $\mathcal{P}roj$-dim $\hat{R}_{\mathfrak{p}} \leq n$ recalling that $\mathrm{Hom}(-, -)$ is right balanced on $\mathcal{F}lat \times \mathcal{F}lat$ by $\mathcal{P}roj \times \mathcal{PInj}$ (see Example 8.3.4). But

$$
\mathrm{Hom}(\hat{R}_{\mathfrak{p}}, PE^{m-1}(F)) / \mathrm{Ker}\, \sigma^{m-1} \cong \mathrm{Hom}(\hat{R}_{\mathfrak{p}}, PE^m(F))
$$

is flat, being a product of T_q's (see proof of Lemma 8.5.25). Hence $\operatorname{Ker} \sigma^{m-1}$ is a pure submodule and so is flat since $\operatorname{Hom}(\hat{R}_\mathfrak{p}, PE^{m-1}(F))$ is likewise flat. We now proceed in this fashion to get the result.

(4) $\Rightarrow$ (1). We apply $k(\mathfrak{p}) \otimes -$ to the pure exact subcomplex in (4). Then by Proposition 8.5.26 above, we get an exact complex $k(\mathfrak{p}) \otimes T_\mathfrak{p}^n \to k(\mathfrak{p}) \otimes T_\mathfrak{p}^{n+1} \to \cdots$ with zero differentiation. Thus $k(\mathfrak{p}) \otimes T_\mathfrak{p}^i = 0$ for $i > n$ and hence $T_\mathfrak{p}^i = 0$ for $i > n$. This is true for each $\mathfrak{p} \in \operatorname{Spec} R$ and thus $PE^i(F) = 0$ for $i > n$.

(2) $\Rightarrow$ (5). Same proof as (2) $\Rightarrow$ (3) since $R_\mathfrak{p}$ is flat.

(5) $\Rightarrow$ (6). Same proof as (3) $\Rightarrow$ (4) since $\operatorname{Hom}(R_\mathfrak{p}, PE^i(F))$ is also flat.

(6) $\Rightarrow$ (1). A similar proof to (4) $\Rightarrow$ (1). $\qquad\square$

Corollary 8.5.28. *If R has finite Krull dimension, then the following integers are equal.*

(1) gl right $\mathcal{PI}nj$- dim $\mathcal{F}lat$.

(2) gl left $\mathcal{P}roj$- dim $\mathcal{F}lat$.

(3) $\sup\{$left $\mathcal{P}roj$- dim $R_\mathfrak{p} : \mathfrak{p} \in \operatorname{Spec} R\}$.

(4) $\sup\{$left $\mathcal{P}roj$- dim $\hat{R}_\mathfrak{p} : \mathfrak{p} \in \operatorname{Spec} R\}$.

Furthermore, this common integer is at most $\dim R$.

Proof. The result follows from Theorem 8.5.27 above and Corollary 8.5.12. It also follows directly from Corollary 8.4.13, Corollary 8.5.12, and Proposition 8.5.13 and its proof. $\qquad\square$

Remark 8.5.29. left $\mathcal{P}roj$- dim $R_\mathfrak{p} = i$ with $T_\mathfrak{p}^i = 0$ for all flat modules can occur. For example, when $R = \mathbb{Z}$ and $p \in \mathbb{Z}$ is prime, left $\mathcal{P}roj$- dim $\mathbb{Z}_{(p)} = 1$ but $T_\mathfrak{p}^1 = 0$ for all torsion free groups G.

We now characterize perfect rings by letting $n = 0$ in Theorem 8.5.27. But first we need the following result.

Lemma 8.5.30. *Let R be local and F be a free R-module with infinite base. If F is complete, then the maximal ideal $\mathfrak{m}$ is nilpotent.*

Proof. Suppose $\mathfrak{m}$ is not nilpotent. Then $\mathfrak{m} \supsetneq \mathfrak{m}^2 \supsetneq \cdots$. Let $r_i \in \mathfrak{m}^i - \mathfrak{m}^{i+1}$ and (x_i), $i \in \mathbb{N}$ be the base of F. Let $y_n = \sum_{i=1}^n r_i x_i$. Then (y_n) is a cauchy sequence which does not converge in F. For if $y = \lim y_n \in F$, let $y = \sum_{i=1}^m a_i x_i$. Let n be so large that $y_n - y \in \mathfrak{m}^k F = \sum_{i=1}^\infty \mathfrak{m}^k x_i$ where $k = \max(m+1, n+1)$. But then $y_n \in \mathfrak{m}^k$ which is impossible since $k > n$. $\qquad\square$

Theorem 8.5.31. *The following are equivalent for a ring R.*

(1) *Every flat R-module is pure injective.*

(2) *Every flat R-module is projective (R is perfect).*

(3) $\hat{R}_P$ *is projective for each* $\mathfrak{p} \in \mathrm{Spec}\, R$ *and* gl right $\mathcal{PInj}$-dim $\mathcal{F}lat < \infty$.

(4) $0 \to \mathrm{Hom}(\hat{R}_\mathfrak{p}, F) \to \mathrm{Hom}(\hat{R}_\mathfrak{p}, PE^0(F)) \to \mathrm{Hom}(\hat{R}_\mathfrak{p}, PE^1(F)) \to \cdots$ *is pure exact for all* $\mathfrak{p} \in \mathrm{Spec}\, R$ *and all* $F \in \mathcal{F}lat$.

(5) $R_\mathfrak{p}$ *is projective for each* $\mathfrak{p} \in \mathrm{Spec}\, R$ *and* gl left $\mathcal{PInj}$-dim $\mathcal{F}lat < \infty$.

(6) $0 \to \mathrm{Hom}(R_\mathfrak{p}, F) \to \mathrm{Hom}(R_\mathfrak{p}, PE^0(F)) \to \cdots$ *is pure exact for all* $\mathfrak{p} \in$ Spec R *and all* $F \in \mathcal{F}lat$.

(7) *The pure injective envelope of every flat module is projective.*

(8) *The pure injective envelope of every free module is projective.*

(9) dim $R = 0$.

Proof. (1) through (6) follow from Theorem 8.5.27.

　　(1) $\Rightarrow$ (7) since $PE(F)$ is flat for any flat F.

　　(7) $\Rightarrow$ (8) is trivial.

　　(8) $\Rightarrow$ (9). Let $F = R^{(X)}$. Then $PE(F) \cong \prod \hat{F}_\mathfrak{m}$, over maximal ideals $\mathfrak{m}$, by Remark 6.7.12. So if $PE(F)$ is projective, then $\hat{R}_\mathfrak{m}$ is projective and thus free for each maximal ideal $\mathfrak{m}$. Now let X be infinite. Then by Lemma 8.5.30, the maximal ideal of $R_\mathfrak{m}$ is nilpotent for each maximal ideal $\mathfrak{m}$ of R. Hence dim $R = 0$.

　　(9) $\Rightarrow$ (1) follows from Corollary 8.5.12.   □

Exercises

1. Argue that a ring R is self pure injective if and only if $\mathrm{Ext}_R^1(F, R) = 0$ for all flat R-module F.

2. Let F be a flat R-module. Prove that the map $F \otimes R' \to PE(F) \otimes R'$ in Theorem 8.5.1 satisfies part (b) of Theorem 6.7.17.

3. Prove that if F is a pure injective flat R-module, then $F \otimes R/I$ is a pure injective flat R/I-module.

4. Prove Corollary 8.5.28.

5. Prove that if F is a flat R-module, then $\pi_n(\mathfrak{p}/I, F/IF) = \pi_n(\mathfrak{p}, F)$ for all prime ideals $\mathfrak{p}$ such that $I \subset \mathfrak{p}$.

6. Prove that if left $\mathcal{P}roj$-dim $R_\mathfrak{p} \leq n$ and dim $R < \infty$, then $T_\mathfrak{p}^i = 0$ for all $i > n$.

8.6　λ and μ-dimensions

In this section $\mathcal{F}$ will be a class of R-modules, and we will assume $\mathcal{F}$ is closed under finite direct sums.

　　We will be concerned with the question of when a module M has an $\mathcal{F}$-precover $\varphi : F \to M$ and also when Ker φ has an $\mathcal{F}$-precover $G \to$ Ker φ. When this is the case, we see that for $H \in \mathcal{F}$, $\text{Hom}(H, G) \to \text{Hom}(H, F) \to \text{Hom}(H, M) \to 0$ is an exact sequence.

The λ-*dimension* of M (relative to $\mathcal{F}$) will tell us how long we can continue this procedure. If M is a left R-module, we will define $\lambda_{\mathcal{F}}(M)$ (or we will write $\lambda(M)$ when $\mathcal{F}$ is understood) to be either an integer $n \geq -1$ or ∞.

Definition 8.6.1. A finite subcomplex $F_n \to \cdots \to F_0 \to M \to 0$ of a left $\mathcal{F}$-resolution of M is called a *partial* left $\mathcal{F}$-resolution of M of length n. Partial right resolutions are defined similarly.

We say $\lambda(M) = -1$ if M does not have an $\mathcal{F}$-precover. If $n \geq 0$ we say that $\lambda(M) = n$ if there is a partial left $\mathcal{F}$-resolution $F_n \to \cdots \to F_1 \to F_0 \to M \to 0$ of M of length n and if there exists no longer such complex. We say $\lambda(M) = \infty$ if there exists a partial left $\mathcal{F}$-resolution for every $n \geq 0$.

If $\lambda(M) = \infty$, it is natural to ask whether there is an infinite left $\mathcal{F}$-resolution $\cdots \to F_2 \to F_1 \to F_0 \to M \to 0$ of M. We will show that this is indeed the case.

Lemma 8.6.2. *If M is a left R-module and $F \in \mathcal{F}$, then $F \oplus M$ has an $\mathcal{F}$-precover if and only if M has an $\mathcal{F}$-precover.*

Proof. If $G \to M$ is an $\mathcal{F}$-precover, then so is $F \oplus G \to F \oplus M$.

Conversely if $\varphi : G \to F \oplus M$ is an $\mathcal{F}$-precover, then so is $\pi_2 \circ \varphi : G \to M$ with $\pi_2 : F \oplus M \to M$ the projection map.   □

In the language above, this result says $\lambda(F \oplus M) \geq 0$ if and only if $\lambda(M) \geq 0$. We will use this to prove that in fact $\lambda(F \oplus M) = \lambda(M)$.

The next result is called Schanuel's lemma when $\mathcal{F}$ is the class of projective modules.

Lemma 8.6.3. *If $F \to M$ and $G \to M$ are $\mathcal{F}$-precovers with kernels K and L respectively, then $K \oplus G \cong L \oplus F$.*

Proof. We consider the pullback diagram

By the definition of a precover, there is a factorization $G \to F \to M$ of the map $G \to M$. This means that $P \to G$ has a section by the property of a pullback and so $P \cong K \oplus G$ since $\text{Ker}(P \to G) \cong \text{Ker}(F \to M) = K$. Similarly $P \cong L \oplus F$ and so $L \oplus F \cong K \oplus G$.   □

Corollary 8.6.4. *Let $n \geq 0$ and $F_n \to \cdots \to F_1 \to F_0 \to M \to 0$ and $G_n \to \cdots \to G_1 \to G_0 \to M \to 0$ be partial left $\mathcal{F}$-resolutions. If $K = \text{Ker}(F_n \to F_{n-1})$ (or $\text{Ker}(F_0 \to M)$ if $n = 0$) and $L = \text{Ker}(G_n \to G_{n-1})$ (or $\text{Ker}(G_0 \to M)$ if $n = 0$), then*

$$K \oplus G_n \oplus F_{n-1} \oplus \cdots \cong L \oplus F_n \oplus G_{n-1} \oplus \cdots.$$

Proof. By induction on n. If $n = 0$, this is the preceding result. So we assume $n > 0$. Then we have the complexes $F_n \to F_{n-1} \to \cdots \to F_2 \to F_1 \oplus G_0 \to \text{Ker}(F_0 \to M) \oplus G_0 \to 0$ and $G_n \to G_{n-1} \to \cdots \to G_2 \to G_1 \oplus F_0 \to \text{Ker}(G_0 \to M) \oplus F_0 \to 0$ which are $\text{Hom}(\mathcal{F}, -)$ exact and which have $F_n, G_n, \ldots, F_2, G_2, F_1 \oplus G_0, G_1 \oplus F_0 \in \mathcal{F}$. Also $\text{Ker}(F_0 \to M) \oplus G_0 \cong \text{Ker}(G_0 \to M) \oplus F_0$ by the previous result. Then an appeal to an induction hypothesis gives the result. $\qquad\square$

Proposition 8.6.5. *If $F \in \mathcal{F}$, then $\lambda(F \oplus M) = \lambda(M)$.*

Proof. We prove that for $n \geq -1$, $\lambda(F \oplus M) \geq n$ if and only if $\lambda(M) \geq n$. This is trivial if $n = -1$ and is true for $n = 0$ by Lemma 8.6.2. So we proceed by induction and let $n > 0$.

Suppose $\lambda(M) \geq n$. If $F_n \to \cdots \to F_0 \to M \to 0$ is a partial left $\mathcal{F}$-resolution, then we see from the complex $F \oplus F_n \to \cdots \to F \oplus F_0 \to F \oplus M \to 0$ that $\lambda(F \oplus M) \geq n$.

Conversely suppose $\lambda(F \oplus M) \geq n$ and $G_n \to \cdots \to G_0 \to F \oplus M \to 0$ is a partial left $\mathcal{F}$-resolution of $F \oplus M$. We know $\lambda(M) \geq 0$ and so let $F_0 \to M$ be an $\mathcal{F}$-precover with kernel K. Let $L = \text{Ker}(G_0 \to F \oplus M)$. Then $F \oplus F_0 \to F \oplus M$ is also an $\mathcal{F}$-precover with kernel K. So $L \oplus F \oplus F_0 \cong K \oplus G_0$ by Lemma 8.6.3. But clearly $\lambda(L) \geq n - 1$ and so $\lambda(L \oplus F \oplus F_0) \geq n - 1$. But then $\lambda(K \oplus G_0) \geq n - 1$ gives $\lambda(K) \geq n - 1$ by induction. Hence $\lambda(M) \geq n$. $\qquad\square$

Proposition 8.6.6. *If $\lambda(M) \geq n > k \geq 0$ and $F_k \to \cdots \to F_0 \to M \to 0$ is a partial left $\mathcal{F}$-resolution of M, then $\lambda(K) \geq n - k - 1$ where $K = \text{Ker}(F_k \to F_{k-1})$ and $F_{-1} = M$. In particular if $\lambda(M) = n$, then $\lambda(K) = n - k - 1$.*

Proof. Since $\lambda(M) \geq n$, there is a partial left $\mathcal{F}$-resolution $G_n \to \cdots \to G_0 \to M \to 0$. Let $L = \text{Ker}(G_k \to G_{k-1})$. Then $\lambda(L) \geq n - k - 1$. By Corollary 8.6.4, we have $L \oplus F_k \oplus G_{k-1} \oplus \cdots \cong K \oplus G_k \oplus F_{k-1} \oplus \cdots$ and so we see that $\lambda(L) = \lambda(K)$ by Proposition 8.6.5. Hence $\lambda(K) \geq n - k - 1$. $\qquad\square$

Corollary 8.6.7. *If $\lambda(M) = \infty$ then there is an infinite left $\mathcal{F}$-resolution*

$$\cdots \to F_2 \to F_1 \to F_0 \to M \to 0$$

of M.

Proof. If $n > 0$ and $F_n \to \cdots \to F_0 \to M \to 0$ satisfies the usual conditions and $K = \mathrm{Ker}(F_n \to F_{n-1})$, then $\lambda(K) = \infty$. So this complex can be extended to a complex $F_{n+1} \to F_n \to \cdots \to F_0 \to M \to 0$ satisfying the conditions. Continuing this way we get the desired complex. $\qquad\square$

Lemma 8.6.8. *If $M_1 \to M_2$ is a linear map such that* $\mathrm{Hom}(F, M_1) \to \mathrm{Hom}(F, M_2)$ *is an isomorphism for all $F \in \mathcal{F}$, then $\lambda(M_1) = \lambda(M_2)$.*

Proof. If $\lambda(M_1) \geq n$ and $F_n \to \cdots \to F_0 \to M_1 \to 0$ is a partial left $\mathcal{F}$-resolution, then so is $F_n \to \cdots \to F_0 \to M_2 \to 0$ with $F_0 \to M_2$ the composition $F_0 \to M_1 \to M_2$. Hence $\lambda(M_2) \geq n$.

If $\lambda(M_2) \geq n$ and $F_n \to \cdots \to F_0 \to M_2 \to 0$ is a partial left resolution, then by hypothesis, $F_0 \to M_2$ has a lifting $F_0 \to M_1$ (so $F_0 \to M_2$ is the composition $F_0 \to M_1 \to M_2$).

Then we see that $F_1 \to F_0 \to M_1$ is a complex since $\mathrm{Hom}(F_1, M_1) \to \mathrm{Hom}(F_1, M_2)$ is an isomorphism and $F_1 \to F_0 \to M_2$ is 0. Hence $F_n \to \cdots \to F_1 \to F_0 \to M_1 \to 0$ is a complex. Our hypotheses guarantee it is $\mathrm{Hom}(\mathcal{F}, -)$ exact. Hence $\lambda(M_1) \geq n$. Hence we can conclude $\lambda(M_1) = \lambda(M_2)$. $\qquad\square$

In the next theorem we will consider complexes $0 \to M' \to M \to M'' \to 0$ of modules which are $\mathrm{Hom}(\mathcal{F}, -)$ exact. We note that in this case, if $K = \mathrm{Ker}(M \to M'')$ then $M' \to K$ is such that $\mathrm{Hom}(F, M') \to \mathrm{Hom}(F, K)$ is an isomorphism for all $F \in \mathcal{F}$ and so $\lambda(M') = \lambda(K)$ by Lemma 8.6.8.

Theorem 8.6.9. *Let $0 \to M' \to M \to M'' \to 0$ be a $\mathrm{Hom}(\mathcal{F}, -)$ exact complex of left R-modules, then*

(1) $\lambda(M'') \geq \min(\lambda(M') + 1, \lambda(M))$

(2) $\lambda(M) \geq \min(\lambda(M'), \lambda(M''))$

(3) $\lambda(M' \geq \min(\lambda(M), \lambda(M'') - 1)$

Proof. We prove (1). We only need to prove that if $n \geq -1$ is an integer and $\min(\lambda(M')+1, \lambda(M)) \geq n$, then $\lambda(M'') \geq n$. If $n = -1$, this is trivially true. If $n = 0$, then $\lambda(M) \geq 0$ means M has an $\mathcal{F}$-precover $F \to M$. By hypothesis, $\mathrm{Hom}(G, M) \to \mathrm{Hom}(G, M'') \to 0$ is exact if $G \in \mathcal{F}$. So $\mathrm{Hom}(G, F) \to \mathrm{Hom}(G, M) \to \mathrm{Hom}(G, M'')$ is surjective. Thus $F \to M''$ is an $\mathcal{F}$-precover and so $\lambda(M'') \geq 0$.

We now suppose $n > 0$. We have $\lambda(M') \geq n - 1 \geq 0$ and $\lambda(M) \geq n$. So we have partial left $\mathcal{F}$-resolutions $F'_{n-1} \to \cdots \to F'_0 \to M' \to 0$ and $F_n \to F_{n-1} \to \cdots \to F_0 \to M \to 0$. Hence we have a commutative diagram

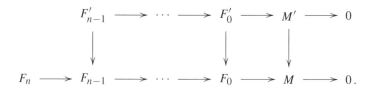

This diagram gives rise to the complex $F_n \oplus F'_{n-1} \to F_{n-1} \oplus F'_{n-2} \to \cdots \to$
$F_1 \oplus F'_0 \to F_0 \oplus M' \to M \to 0$.

But then we have a commutative diagram

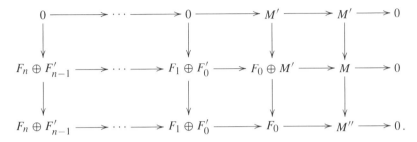

We now apply the additive functor $\mathrm{Hom}(F, -)$ with any $F \in \mathcal{F}$ to all the diagrams
above. Then by Proposition 1.4.14 and our previous remarks, we see that $F_n \oplus F'_{n-1} \to$
$F_n \oplus F'_{n-2} \to \cdots \to F_1 \oplus F'_0 \to F_0 \to M'' \to 0$ is $\mathrm{Hom}(\mathcal{F}, -)$ exact. Hence
$\lambda(M'') \geq n$.

The proof of (3) is similar. We need to argue that if $\min(\lambda(M), \lambda(M'') - 1) \geq n$,
then $\lambda(M') \geq n$. We can assume $n \geq 0$. Then we get a commutative diagram

$$
\begin{array}{ccccccccc}
F_n & \longrightarrow & \cdots & \longrightarrow & F_0 & \longrightarrow & M & \longrightarrow & 0 \\
\downarrow & & & & \downarrow & & \downarrow & & \\
F''_{n+1} & \longrightarrow & F''_n & \longrightarrow \cdots \longrightarrow & F''_0 & \longrightarrow & M'' & \longrightarrow & 0
\end{array}
$$

and the complex $F''_{n+1} \oplus F_n \to \cdots \to F''_1 \oplus F_0 \to F''_0 \oplus M \to M'' \to 0$. But then
we get a commutative diagram

$$
\begin{array}{ccccccccc}
F''_{n+1} \oplus F_n & \longrightarrow & \cdots & \longrightarrow & F''_1 \oplus F_0 & \longrightarrow & F''_0 \oplus M & \longrightarrow & M'' & \longrightarrow & 0 \\
\downarrow & & & & \downarrow & & \downarrow & & \downarrow & & \\
0 & \longrightarrow & \cdots & \longrightarrow & 0 & \longrightarrow & M'' & \longrightarrow & M'' & \longrightarrow & 0
\end{array}
$$

The kernel of the corresponding map of complexes is the complex $F''_{n+1} \oplus F_n \to$
$\cdots \to F''_1 \oplus F_0 \to P \to 0$ where $P = \mathrm{Ker}(F''_0 \oplus M \to M'')$. So

$$
\begin{array}{ccc}
P & \longrightarrow & M \\
\downarrow & & \downarrow \\
F''_0 & \longrightarrow & M''
\end{array}
$$

is a pullback diagram. Hence by our hypothesis on $0 \to M' \to M \to M'' \to 0$, we
see that the map $F''_0 \to M''$ has a lifting $F''_0 \to M$. But by the property of a pullback
this means $P \to F''_0$ has a section. Hence $P \cong F''_0 \oplus K$ where $K = \mathrm{Ker}(M \to M'')$.

But as in the argument for (1), we see that $F''_{n+1} \oplus F_n \to \cdots \to F''_1 \oplus F_0 \to P \to 0$ is $\mathrm{Hom}(\mathcal{F}, -)$ exact. This means $\lambda(P) \geq n$. But since $P \cong F''_0 \oplus K$ we get that $\lambda(K) \geq n$ by Proposition 8.6.5. But then by Lemma 8.6.8, we get $\lambda(M') \geq n$.

We now prove (2). We assume $\lambda(M'), \lambda(M'') \geq n \geq 0$ and argue $\lambda(M) \geq n$. Let $F'_n \to \cdots \to F'_0 \to M' \to 0$ and $F''_n \to \cdots \to F''_0 \to M'' \to 0$ be partial left $\mathcal{F}$-resolutions of M' and M'' respectively. Then by the Horseshoe Lemma 8.2.1, we get a partial $\mathcal{F}$-resolution of M of length n. Hence $\lambda(M) \geq n$.   □

Remark 8.6.10. When $\mathcal{F}$ is the class of finitely generated projective modules, then a short complex $0 \to M' \to M \to M'' \to 0$ is exact if and only if it is $\mathrm{Hom}(\mathcal{F}, -)$ exact. In this case $\lambda(M) \geq 0$ if and only if M is finitely generated and $\lambda(M) \geq 1$ if and only if M is finitely presented. These λ-dimensions are defined in Bourbaki [29, page 41] and Theorem 8.6.9 corresponds to their Exercise 6.

Definition 8.6.11. For a module M, we define $\bar{\lambda}(M)$ (or $\bar{\lambda}_{\mathcal{F}}(M)$) to be -1 if M does not have a special $\mathcal{F}$-precover. If there is an exact sequence $F_n \to \cdots \to F_0 \to M \to 0$ where $F_0 \to M$, $F_i \to K_{i-1}$ (where $K_1 = \mathrm{Ker}(F_0 \to M)$ and $K_i = \mathrm{Ker}(F_{i-1} \to F_{i-2})$ for $i > 1$) are special precovers and if there is no longer such sequence, we write $\bar{\lambda}(M) = n$. We say $\bar{\lambda}(M) - \infty$ if there is such a sequence for each $n \geq 0$.

The proofs of several results concerning $\bar{\lambda}$-dimensions are straightforward modifications of the corresponding results about λ-dimensions. These include Propositions 8.6.5, 8.6.6 and Corollary 8.6.7.

Proposition 8.6.12. *If $\mathcal{F}$ is such that $\lambda(M) \geq 0$ implies $\bar{\lambda}(M) \geq 0$ for all R-modules M, then $\lambda(M) = \bar{\lambda}(M)$ for all M.*

Proof. Clearly $\lambda(M) \geq \bar{\lambda}(M)$. So we argue that $\lambda(M) \geq n$ implies $\bar{\lambda}(M) \geq n$ for $n \geq 0$. By hypothesis, this is true if $n = 0$. So suppose $\lambda(M) \geq n > 0$. Then we have $\bar{\lambda}(M) > 0$ and so let $F \to M$ be a special precover with kernel K. Then by Proposition 8.6.6, $\lambda(K) \geq n - 1$. So $\bar{\lambda}(K) \geq n - 1$ by induction. Hence $\bar{\lambda}(M) \geq n$.□

Theorem 8.6.13. *If $0 \to M' \to M \to M'' \to 0$ is an exact sequence, then*

$$\bar{\lambda}(M'') \geq \min(\bar{\lambda}(M') + 1, \bar{\lambda}(M)).$$

Proof. The argument is a straightforward modification of the proof of (1) of Theorem 8.6.9.   □

Theorem 8.6.14. *If $0 \to M' \to M \to M'' \to 0$ is exact and $\mathrm{Hom}(\mathcal{F}, -)$ exact, then*

$$\bar{\lambda}(M) \geq \min(\bar{\lambda}(M'), \bar{\lambda}(M'')).$$

Proof. This argument is like that for (2) of Theorem 8.6.9.   □

Definition 8.6.15. The class $\mathcal{F}$ is said to be *resolving* if $\mathcal{F}$ contains all the projective modules, $\mathcal{F}$ is closed under extensions and if whenever $0 \rightarrow F' \rightarrow F \rightarrow F'' \rightarrow 0$ is exact with $F, F'' \in \mathcal{F}$, F' is also in $\mathcal{F}$.

Theorem 8.6.16. *If $\mathcal{F}$ is resolving and $0 \rightarrow M' \rightarrow M \rightarrow M'' \rightarrow 0$ is an exact sequence of modules, then*

$$\bar{\lambda}(M') \geq \min(\bar{\lambda}(M), \bar{\lambda}(M'') - 1).$$

Proof. We prove by induction on n that if $\bar{\lambda}(M) \geq n$ and $\bar{\lambda}(M'') \geq n + 1$ then $\bar{\lambda}(M') \geq n$.

Let $n = 0$. So $\bar{\lambda}(M'') \geq 1$ and $\bar{\lambda}(M) \geq 0$. Then let $0 \rightarrow K_0'' \rightarrow F_0'' \rightarrow M'' \rightarrow 0$, $0 \rightarrow K_1'' \rightarrow F_1'' \rightarrow K_0'' \rightarrow 0$, and $0 \rightarrow K_0 \rightarrow F_0 \rightarrow M \rightarrow 0$ be exact sequences with $K_0, K_0'', K_1'' \in \mathcal{F}^\perp$ and $F_0'', F_1'', F_0 \in \mathcal{F}$. Then we can construct a pullback H (with maps $H \rightarrow M$ and $H \rightarrow F_0''$) of $M \rightarrow M''$ and $F_0'' \rightarrow M''$ and get a commutative diagram with exact sequences $0 \rightarrow K_0'' \rightarrow H \rightarrow M \rightarrow 0$ and $0 \rightarrow M' \rightarrow H \rightarrow F_0'' \rightarrow 0$.

Since $K_0'' \in \mathcal{F}^\perp$, the sequence $0 \rightarrow K_0'' \rightarrow H \rightarrow M \rightarrow 0$ is $\mathrm{Hom}(\mathcal{F}, -)$ exact. So by Horseshoe Lemma 8.2.1, we get exact sequences $0 \rightarrow K \rightarrow F_1'' \oplus F_0 \rightarrow H \rightarrow 0$ and $0 \rightarrow K_1'' \rightarrow K \rightarrow K_0 \rightarrow 0$ with $K \in \mathcal{F}^\perp$ since $K_1'', K_0 \in \mathcal{F}^\perp$. Therefore, we can now form the pullback of $M' \rightarrow H$ and $F_1'' \oplus F_0 \rightarrow H$ to get the following commutative diagram

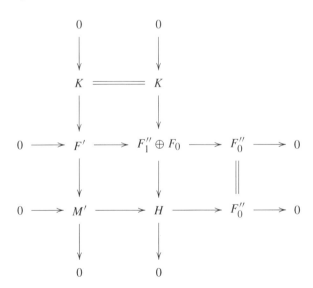

with exact rows and columns. Since $F_1'' \oplus F_0, F_0'' \in \mathcal{F}$ and $\mathcal{F}$ is resolving, $F' \in \mathcal{F}$. As noted above, $K \in \mathcal{F}^\perp$. Hence $F' \rightarrow M'$ is a special $\mathcal{F}$-precover and so $\bar{\lambda}(M') \geq 0$.

Now assume $n > 0$ and use the construction above. Then by the exactness and $\mathrm{Hom}(\mathcal{F}, -)$ exactness of $0 \rightarrow K_1'' \rightarrow K \rightarrow K_0 \rightarrow 0$ ($K_1'' \in \mathcal{F}^\perp$ gives the

$\text{Hom}(\mathcal{F}, -)$ exactness), we get $\bar{\lambda}(K) \geq \min(\bar{\lambda}(K_1''), \bar{\lambda}(K_0))$ by Theorem 8.6.14. But $\min(\bar{\lambda}(K_1''), \bar{\lambda}(K_0)) \geq n - 1$ by the $\bar{\lambda}$-dimension counterpart of Proposition 8.6.6 (or we can assume we chose K_1'' and K_0 so that the inequality holds). But then $\bar{\lambda}(K) \geq n - 1$ implies $\bar{\lambda}(M') \geq n$. □

Remark 8.6.17. If $\mathcal{F}$ is a class of left R-modules we can define the μ-*dimension* of a left R-module relative to $\mathcal{F}$ (denoted $\mu_{\mathcal{F}}(M)$ or $\mu(M)$ if $\mathcal{F}$ is understood) with the definition dual to the definition of the λ-dimensions above. So $\mu_{\mathcal{F}}(M) = -1$ means that M does not have an $\mathcal{F}$-preenvelope and $\mu_{\mathcal{F}}(M) = n$ with $0 \leq n < \infty$ means there is a complex $0 \to M \to F^0 \to \cdots \to F^n$ with each $F^i \in \mathcal{F}$ such that if $G \in \mathcal{F}$, $\text{Hom}(-, G)$ makes the complex exact and that there is no longer such complex. Then $\mu_{\mathcal{F}}(M) = \infty$ will mean there is such a complex for every $n \geq 0$.

We will not state them but only note that all of the definitions and results for λ-dimensions have their counterpart concerning μ-dimensions (or $\bar{\mu}$-dimensions). For each of these the proof is just the dual of the proof of the corresponding result.

Exercises

1. Prove that if $M = M' \oplus M''$, then $\lambda(M) = \min(\lambda(M'), \lambda(M''))$.

2. Let $\mathcal{F} = \mathcal{P}roj_{fg}$ and M_1, M_2 be submodules of an R-module M such that $\lambda(M_i) \geq 0$ for $i = 1, 2$. Prove that $\lambda(M_1 + M_2) \geq 0$ if and only if $\lambda(M_1 \cap M_2) \geq 0$.

3. Prove that if $F \in \mathcal{F}$, then $\bar{\lambda}(F \oplus M) = \bar{\lambda}(M)$.

4. Suppose $\bar{\lambda}(M) \geq n > k \geq 0$ and $F_k \to F_{k-1} \to \cdots \to F_0 \to M \to 0$ is a partial left $\mathcal{F}$-resolution where $F_0 \to M$ and $F_i \to F_{i-1}$ for $i > 0$ are special $\mathcal{F}$-precovers. Prove that if $K = \text{Ker}(F_k \to F_{k-1})$, then $\bar{\lambda}(K) \geq n - k - 1$.

5. Prove that if $\bar{\lambda}(M) = \infty$, then there is an infinite left $\mathcal{F}$-resolution $\cdots \to F_1 \to F_0 \to M \to 0$ of M where $F_0 \to M$ and $F_i \to F_{i-1}$ for $i > 0$ are special $\mathcal{F}$-precovers.

6. Prove Remark 8.6.10.

7. Prove Theorem 8.6.13.

8. Prove Theorem 8.6.14.

Chapter 9

Iwanaga–Gorenstein and Cohen–Macaulay Rings and Their Modules

In this chapter we will show that the property of being Iwanaga–Gorenstein or Cohen–Macaulay imposes nice conditions on the homological properties of modules over such rings.

9.1 Iwanaga–Gorenstein rings

Definition 9.1.1. A ring R is called an *Iwanaga–Gorenstein* ring (or simply a *Gorenstein* ring) if R is both left and right noetherian and if R has finite self-injective dimension on both the left and the right.

We will first consider these conditions on one side only.

Proposition 9.1.2. *If R is left (right) noetherian and the left (right) self-injective dimension of R is $n < \infty$, then* inj dim $F \le n$ *for every flat left (right) R-module. And if* flat dim $M < \infty$ *for a left (right) R-module M, then* proj dim $M \le n$.

Proof. We give the proof on the left. But inj dim $\varinjlim N_i \le \sup\{$inj dim $N_i\}$ for any inductive system of left R-modules since R is left noetherian, and inj dim $P \le n$ for any projective left R-module since inj dim $_R R = n$. So the first claim follows since a flat left R-module is the inductive limit of projective left R-modules.

Now let flat dim $M < \infty$, $m > n$, and $m >$ flat dim M. Then let $0 \to F \to P_{m-1} \to \cdots \to P_0 \to M \to 0$ be exact with $P_0, \ldots, P_{m-1}$ projective. Then F is flat. But by the above, inj dim $F < m$ and so $\text{Ext}^m(M, F) = 0$. This means that id $: F \to F$ can be extended to $P_{m-1} \to F$ and so F is a summand of P_{m-1}. If $P_{m-1} = F \oplus G$, then we have a projective resolution $0 \to G \to P_{m-2} \to \cdots \to P_0 \to M \to 0$. So proj dim $M < \infty$. If $m - 1 > n$, then we repeat the procedure with G replacing F. So we see that proj dim $M \le n$. $\square$

Corollary 9.1.3. *If F is a flat left (right) R-module, then* proj dim $F \le n$.

Lemma 9.1.4. *If $M \subset N$ is a pure submodule of the left R-module N, then*

$$\text{flat dim } M \le \text{flat dim } N.$$

Proof. If flat dim $N = \infty$ the result is trivial. So suppose flat dim $N = n < \infty$. Let G be any right R-module and $0 \to S \to P_n \to \cdots \to P_0 \to G \to 0$ be a partial projective resolution of G.

Consider the commutative diagram

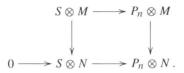

The bottom row is exact since $\text{Tor}_{n+1}(G, N) = 0$. The two vertical arrows are injections since $M \subset N$ is pure. Hence $0 \to S \otimes M \to P_n \otimes M$ is exact and so $\text{Tor}_{n+1}(G, M) = 0$. Hence flat dim $M \leq n$.   □

The lemma above has a right counterpart. When necessary, we will appeal to such counterparts even though the result is stated for only one side.

Lemma 9.1.5. *If R is left noetherian and $M \subset N$ is a pure submodule of the left R-module N, then*
$$\text{inj dim } M \leq \text{inj dim } N.$$

Proof. We suppose that $\text{inj dim } N = n < \infty$. Then $\text{Ext}^{n+1}(R/I, N) = 0$ for all left ideals I of R. Let $0 \to S \to P_n \to \cdots \to P_0 \to R/I \to 0$ be a partial projective resolution of R/I with $P_0, \ldots, P_n$ finitely generated projective modules. Since $\text{Ext}^{n+1}(R/I, N) = 0$, we have $\text{Hom}(P_n, N) \to \text{Hom}(S, N) \to 0$ is exact. But since $M \subset N$ is pure, a linear $S \to M$ has an extension $P_n \to N$ if and only if there is an extension $P_n \to M$. Hence $\text{Hom}(P_n, M) \to \text{Hom}(S, M) \to 0$ is exact. This means that $\text{Ext}^{n+1}(R/I, M) = 0$ for all such I and so inj dim $M \leq n$ by Proposition 8.4.4.   □

Proposition 9.1.6. *Let R be left and right noetherian and $\text{inj dim }_R R = n < \infty$. Then the following are equivalent.*

(1) $\text{inj dim } R_R < \infty$.

(2) flat dim $E < \infty$ for all injective left R-modules E.

(3) flat dim $E \leq n$ for all injective left R-modules E.

Proof. Suppose (1) holds. Let E be an injective left R-module. Then the character module E^+ is a flat right R-module. So by Proposition 9.1.2, E^+ has finite injective dimension. But then E^{++} has finite flat dimension. Since $E \subset E^{++}$ is pure by Proposition 5.3.9, flat dim $E \leq$ flat dim E^{++} by Lemma 9.1.4. So flat dim E is finite and so (2) holds.

Now assume (2). Then R^+ is an injective left R-module and so has finite flat dimension. Hence R^{++} has finite injective dimension and so R_R has finite injective dimension by Lemma 9.1.5.

Thus (1) and (2) are equivalent. Then by Proposition 9.1.2, we see that (1), (2) and (3) are equivalent. □

Proposition 9.1.7. *If R is Gorenstein, then the following are equivalent for a left R-module M.*

(1) inj dim $M < \infty$.

(2) proj dim $M < \infty$.

(3) flat dim $M < \infty$.

Proof. By Proposition 9.1.2 we see that (2) and (3) are equivalent and that (3) implies (1).

To argue (1) $\Rightarrow$ (3), let M have finite injective dimension. Then the right module M^+ has finite flat dimension. So by (2) $\Rightarrow$ (1) (on the right) we get that M^+ has finite injective dimension. So M^{++} has finite flat dimension. Then Lemma 9.1.5 gives that M has finite flat dimension. □

Proposition 9.1.8. *If R is left and right noetherian and* inj dim $_R R = m < \infty$ *and* inj dim $R_R = n < \infty$, *then $m = n$.*

Proof. Since inj dim $R_R = n$, $\mathrm{Ext}^n(M, R) \neq 0$ for some finitely generated right R-module M. Noting that $\mathrm{Ext}^n(M, R)$ is a left R-module, we have $\mathrm{Hom}(\mathrm{Ext}^n(M, R), E) \neq 0$ for some injective left R-module E. But there are natural isomorphisms

$$\mathrm{Hom}(\mathrm{Ext}^n_R(M, R), E) \cong \mathrm{Tor}^R_n(\mathrm{Hom}(R, E), M) \cong \mathrm{Tor}^R_n(M, E)$$

and so $\mathrm{Tor}^R_n(M, E) \neq 0$. But by Proposition 9.1.6, flat dim $_R E \leq m$. Hence $m \geq n$. But the same type argument gives $m \leq n$ and so $m = n$. □

Definition 9.1.9. A Gorenstein ring with inj dim $_R R$ at most n is called n-*Gorenstein*. We note that in this case inj dim R_R is also at most n by the above.

Theorem 9.1.10. *If R is n-Gorenstein, then the following are equivalent for a left R-module M.*

(1) inj dim $M < \infty$.

(2) proj dim $M < \infty$.

(3) flat dim $M < \infty$.

(4) inj dim $M \leq n$.

(5) proj dim $M \leq n$.

(6) flat dim $M \leq n$.

Proof. By Proposition 9.1.7, (1), (2) and (3) are equivalent and by Proposition 9.1.2, (3), (5) and (6) are equivalent. And (4) $\Rightarrow$ (1) trivially.

To prove (1) $\Rightarrow$ (4), let $0 \to M \to E^0 \to \cdots \to E^k \to 0$ be an injective resolution of a left R-module M. Then flat dim $M^+ \le k$ and so flat dim $M^+ \le n$ by the equivalence of (3) and (6) on the right. But then inj dim $M^{++} \le n$ and so inj dim $M \le n$ by Lemma 9.1.5. This completes the proof.   □

Theorem 9.1.11. *The following are equivalent for a left and right noetherian ring R.*

(1) *R is n-Gorenstein.*

(2) *The injective dimension of all left and right flat R-modules is at most n.*

(3) *The injective dimension of all left and right projective R-modules is at most n.*

(4) *The flat dimension of all left and right injective R-modules is at most n.*

(5) *The projective dimension of all left and right injective R-modules is at most n.*

(6) *If $\cdots \to P_1 \to P_0 \to M \to 0$ is a projective resolution of a left (right) R-module M, then the subcomplex $\cdots \to P_{n+1} \to P_n$ is a right $\mathcal{F}$lat-resolution.*

(7) *If $0 \to M \to F^0 \to E^1 \to \cdots$ is an injective resolution of a left (right) R-module M, then the subcomplex $E^n \to E^{n+1} \to \cdots$ is a left $\mathcal{I}$nj-resolution.*

Proof. The equivalence of (1), (2), (4), and (6) follows from Theorem 8.4.36.

(2) $\Rightarrow$ (3) $\Rightarrow$ (1) and (5) $\Rightarrow$ (4) are trivial.

(4) $\Rightarrow$ (5). Let E be an injective left or right R-module. Then there is an exact sequence $0 \to F_n \to P_{n-1} \to \cdots \to P_1 \to P_0 \to E \to 0$ with P_i projective and F_n flat by assumption. But proj dim $F_n < \infty$ by Theorem 9.1.10 since (4) is equivalent to (1). So proj dim $E < \infty$ and thus proj dim $E \le n$ again by Theorem 9.1.10.

(5) $\Leftrightarrow$ (7). We simply note that proj dim $E \le n$ if and only if $\mathrm{Ext}^i(E, M) = 0$ for all $i \ge n + 1$.   □

We may now add the following to Corollary 8.4.37.

Corollary 9.1.12. *Let R be left and right noetherian. Then R is injective as a left and right R-module if and only if every injective left and right R-module is projective.*

Exercises

1. The *copure injective dimension* (cid) of an R-module M is the largest positive integer n such that $\mathrm{Ext}^n(E, M) \ne 0$ for some injective R-module E. If M is strongly copure injective, that is, $\mathrm{Ext}^i(E, M) = 0$ for all injectives E and all $i \ge 0$, we set cid $M = 0$. Show that the following are equivalent for a left and right noetherian ring R.

   a) *R is n-Gorenstein.*

    b) cid $M \leq n$ for all left and right R-modules M.

    c) If $0 \to M \to E^0 \to E^1 \to \cdots \to E^{n-1} \to C \to 0$ is an exact sequence of any left (right) R-module M with each E^i injective, then C is strongly copure injective.

2. The *copure flat dimension* (cfd) of an R-module is the largest positive integer n such that $\mathrm{Tor}_n(E, M) \neq 0$ for some injective R-module E. If there is no such n, we set cfd $M = 0$ and we say M is *strongly copure flat*. State and prove the counterpart of the exercise above for copure flat dimension.

3. An R-module M is said to be *copure flat* if $\mathrm{Tor}_1(E, M) = 0$ for all injective R-modules E, and *copure injective* if $\mathrm{Ext}^1(E, M) = 0$ for all injective R-modules E. Prove that the following are equivalent for a left and right noetherian ring R.

    a) R is 1-Gorenstein.

    b) An R-module (left and right) M is copure injective if and only if it is strongly copure injective.

    c) Every copure injective R-module (left and right) is h-divisible.

    d) Every homomorphic image of a copure injective R-module (left and right) is copure injective.

    e) Every h-divisible R-module (left and right) is copure injective.

    f) An R-module (left and right) M is copure flat if and only if it is strongly copure flat.

    g) Every submodule of a copure flat R-module (left and right) is copure flat.

    h) Every submodule of a flat R-module (left and right) is copure flat.

4. Let R be a noetherian local ring. Prove that if R is Gorenstein, then $R_{\mathfrak{p}}$ is Gorenstein for each $\mathfrak{p} \in \mathrm{Spec}\, R$.

9.2 The minimal injective resolution of R

In this section, R will be a commutative noetherian ring. We will study the injective resolution of R when R is Gorenstein.

    We recall from Section 3.3 that there is a bijective correspondence between the prime ideals $\mathfrak{p}$ of R and the indecomposable injective modules with $\mathfrak{p}$ corresponding to $E(R/\mathfrak{p})$. Also, each injective module E can be written uniquely, up to isomorphism, as a direct sum of such $E(R/\mathfrak{p})$'s. If M is an R-module, then the cardinality of the summands of $E^i(M)$ isomorphic to $E(R/\mathfrak{p})$ in such a decomposition is denoted $\mu_i(\mathfrak{p}, M)$. The invariants $\mu_i(\mathfrak{p}, M)$ are called *Bass invariants*. In several results below we will consider these invariants under various change of rings.

    If $S \subset R$ is a multiplicative set, then $S^{-1}E(R/\mathfrak{p}) = 0$ if $S \cap \mathfrak{p} \neq \emptyset$ and $S^{-1}E(R/\mathfrak{p}) = E_{S^{-1}R}(S^{-1}R/S^{-1}\mathfrak{p})$ if $S \cap \mathfrak{p} = \emptyset$ by Theorems 3.3.3 and 3.3.8.

Lemma 9.2.1. *If $S \cap \mathfrak{p} = \emptyset$ for a prime ideal $\mathfrak{p}$ of R and a multiplicative set $S \subset R$, then for every R-module M, $\mu_i(\mathfrak{p}, M) = \mu_i(S^{-1}\mathfrak{p}, S^{-1}M)$.*

Proof. This follows from the above and the fact that $0 \to S^{-1}M \to S^{-1}E^0(M) \to S^{-1}E^1(M) \to \cdots$ is a minimal injective resolution of the $S^{-1}R$-module $S^{-1}M$ by Remark 3.3.4. □

Lemma 9.2.2. *Let M be an R-module and $r \in R$ be such that $R \xrightarrow{r} R$ and $M \xrightarrow{r} M$ are both injections, and set $\bar{R} = R/(r)$. If $0 \to M \to E^0(M) \to E^1(M) \to \cdots$ is a minimal injective resolution of M and $0 \to M \to E^0(M) \to C \to 0$ is exact, then*

$$0 \to \mathrm{Hom}(\bar{R}, C) \to \mathrm{Hom}(\bar{R}, E^1(M)) \to \mathrm{Hom}(\bar{R}, E^2(M)) \to \cdots$$

is a minimal injective resolution of $\mathrm{Hom}(\bar{R}, C)$ as an $\bar{R}$-module. Furthermore

$$\mathrm{Hom}(\bar{R}, C) \cong M/rM.$$

Proof. We note that if r is a unit of R then $\bar{R} = 0$ and $M/rM = 0$, and so the result is trivial. So we assume r is not a unit of R. Then by the exactness of $0 \to R \xrightarrow{r} R \to \bar{R} \to 0$, we see that $\mathrm{proj\,dim}\,\bar{R} = 1$. Hence $\mathrm{Ext}^i(\bar{R}, M) = 0$ for $i \geq 2$. This implies the exactness of the sequence $0 \to \mathrm{Hom}(\bar{R}, C) \to \mathrm{Hom}(\bar{R}, E^1(M)) \to \mathrm{Hom}(\bar{R}, E^2(M)) \to \cdots$ Also, each $\mathrm{Hom}(\bar{R}, E^2(M))$ is an injective $\bar{R}$-module for $i \geq 1$. Since C is essential in $E^1(M)$, we see that $\mathrm{Hom}(\bar{R}, C) \subset C$ is essential in $\mathrm{Hom}(\bar{R}, E^1(C)) \subset E^1(C)$. Proceeding in this manner we see that in fact this sequence is a minimal injective resolution of $\mathrm{Hom}(\bar{R}, C)$ over $\bar{R}$.

For the last claim we consider the commutative diagram with exact rows

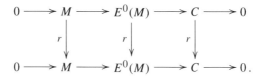

Since $M \xrightarrow{r} M$ is an injection and $M \subset E^0(M)$ is essential, $E^0(M) \xrightarrow{r} E^0(M)$ is an injection. But then the Snake lemma gives the exact sequence $0 \to \mathrm{Ker}(C \xrightarrow{r} C) \to \mathrm{Coker}\,(M \xrightarrow{r} M) \to 0$, that is, $\mathrm{Hom}(\bar{R}, C) \cong M/rM$. □

Corollary 9.2.3. *If $r \in \mathfrak{p}$ is such that $R \xrightarrow{r} R$ and $M \xrightarrow{r} M$ are both injections, then $\mu_i(\mathfrak{p}/(r), M/rM) = \mu_{i+1}(\mathfrak{p}, M)$.*

Proof. This follows from the fact that when $\bar{R} = R/(r)$, $\mathrm{Hom}_R(\bar{R}, E(R/\mathfrak{p}))$ is the injective envelope of the $\bar{R}$-module $\bar{R}/(\mathfrak{p}/(r)) \cong R/\mathfrak{p}$. □

Theorem 9.2.4. *If M is an R-module and $\mathfrak{p} \subset R$ is a prime ideal, then $\mu_i(\mathfrak{p}, M) = \dim_{k(\mathfrak{p})} \mathrm{Ext}^i_{R_\mathfrak{p}}(k(\mathfrak{p}), M_\mathfrak{p}) = \dim_{k(\mathfrak{p})} \mathrm{Ext}^i_R(R/\mathfrak{p}, M)_\mathfrak{p}$. In particular, if M is finitely generated, $\mu_i(\mathfrak{p}, M) < \infty$ for all $i \geq 0$.*

Proof. By Lemma 9.2.1, we may assume $R = R_{\mathfrak{p}}$. Then given $0 \to M \to E^0(M) \to E^1(M) \to \cdots$, let T_i be the submodule of $E^i(M)$ annihilated by $\mathfrak{p}$. Then $\mu_i(\mathfrak{p}, M) = \dim_k T_i$ with $k = R/\mathfrak{p}$. Minimality guarantees that $E^i(M) \to E^{i+1}(M)$ maps T_i to 0. Hence the complex

$$0 \to \mathrm{Hom}(k, E^0(M)) \to \mathrm{Hom}(k, E^1(M)) \to \cdots$$

has 0 differentiation. Hence $\mathrm{Ext}^i_R(k, M) \cong \mathrm{Hom}(k, E^i(M)) \cong T_i$.   □

Definition 9.2.5. Let M be an R-module. Then a sequence $r_1, r_2, \ldots, r_s$ of elements in an ideal I of R is called an *M-sequence* in I if $(r_1, \ldots, r_s)M \neq M$ and x_i is not a zero divisor on $M/(r_1, \ldots, r_{i-1})M$ for $1 \leq i \leq s$. If $r_1, r_2, \ldots$ is an M-sequence, then $(r_1) \subset (r_1, r_2) \subset \cdots$ is a strictly increasing ascending chain of ideals in I and thus must stop since R is noetherian. So each M-sequence in I can be extended to a maximal M-sequence in I. If M is finitely generated, then all maximal M-sequences in I have the same length by Corollary 9.2.8 below. This length is called the *depth of M in I* and is denoted by $\mathrm{depth}_I M$.

If R is a local ring with maximal ideal $\mathfrak{m}$, then the depth of M in $\mathfrak{m}$ is called the *depth of M* and is denoted by $\mathrm{depth}_R M$.

Proposition 9.2.6. *Let R be a local ring and let $r_1, \ldots, r_s$ be an R-sequence. Then $\mathrm{proj\,dim}\, R/(r_1, \ldots, r_s) = s$ and $\mathrm{Ext}^s(R/(r_1, \ldots, r_s), M) \neq 0$ for all finitely generated R-modules $M \neq 0$.*

Proof. By induction on $s \geq 1$. If $s = 1$, then the exactness of $0 \to R \xrightarrow{r} R \to R/(r_1) \to 0$ gives $\mathrm{proj\,dim}\, R/(r_1) \leq 1$. But if $M \neq 0$, the exactness of

$$\mathrm{Hom}(R, M) \xrightarrow{r_1} \mathrm{Hom}(R, M) \to \mathrm{Ext}^1(R/(r_1), M) \to 0$$

and Nakayama lemma give that $\mathrm{Ext}^1(R/(r_1), M) \neq 0$. Hence in general

$$0 \to R/(r_1, \ldots, r_{s-1}) \xrightarrow{r_s} R/(r_1, \ldots, r_{s-1}) \to R/(r_1, \ldots, r_{s-1}, r_s) \to 0$$

is exact and so by induction we see that $\mathrm{proj\,dim}\, R/(r_1, \ldots, r_s) \leq s$. But then for a finitely generated $M \neq 0$ we have that

$$\mathrm{Ext}^{s-1}(R/(r_1, \ldots, r_{s-1}), M) \xrightarrow{r_s} \mathrm{Ext}^{s-1}(R/(r_1, \ldots, r_{s-1}), M) \to$$
$$\mathrm{Ext}^s(R/(r_1, \ldots, r_s), M) \to 0$$

is exact. So by the induction hypothesis, $\mathrm{Ext}^{s-1}(R/(r_1, \ldots, r_{s-1}), M) \neq 0$ and then by Nakayama lemma $\mathrm{Ext}^s(R/(r_1, \ldots, r_s), M) \neq 0$ giving

$$\mathrm{proj\,dim}\, R/(r_1, \ldots, r_s) = s$$

and thus completing the proof.   □

Proposition 9.2.7. *If M and N are R-modules and $r_1, r_2, \ldots, r_s$ is an N-sequence contained in $\mathrm{Ann}(M)$, then*

$$\mathrm{Ext}^i(M, N) \cong \begin{cases} 0 & \text{for } 0 \leq i < s \\ \mathrm{Hom}(M, N/(r_1, \ldots, r_s)N) & \text{for } i = s. \end{cases}$$

In particular, $\mathrm{Hom}(M, N/(r_1, \ldots, r_s)N)$ depends only on s and not on the sequence $r_1, \ldots, r_s$.

Proof. By induction on $s \geq 0$. If $s = 0$, the result is trivial. If $s = 1$, then $\mathrm{Ext}^0(M, N) = \mathrm{Hom}(M, N) = 0$ since $r_1 M = 0$ and $N \xrightarrow{r_1} N$ is an injection. The exact sequence $0 \to N \xrightarrow{r_1} N \to N/r_1 N \to 0$ gives the exact sequence

$$\mathrm{Hom}(M, N) = 0 \to \mathrm{Hom}(M, N/r_1 N) \to \mathrm{Ext}^1(M, N) \xrightarrow{r_1} \mathrm{Ext}^1(M, N).$$

But $\mathrm{Ext}^1(M, N) \xrightarrow{r_1} \mathrm{Ext}^1(M, N)$ is 0 since $r_1 M = 0$. Then the proof can be completed by an induction on s. □

Applying this to M and $N/(r_1, \ldots, r_s)N$ when $r_1, \ldots, r_s$ is a maximal N-sequence in $\mathrm{Ann}(M)$, we get

Corollary 9.2.8. *If N is finitely generated and $IN \neq N$ for an ideal I, then every maximal N-sequence in I has length equal to the least i such that $\mathrm{Ext}^i(R/I, N) \neq 0$.*

Remark 9.2.9. We note that if N is finitely generated and $IN \neq N$ for an ideal I, then $\mathrm{depth}_I N = \inf\{i : \mathrm{Ext}^i(R/I, N) \neq 0\}$ by the above. So if R is local with maximal ideal $\mathfrak{m}$ and residue field k, then $\mathrm{depth}\, N = \inf\{i : \mathrm{Ext}^i(k, N) \neq 0\} = \inf\{i : \mu_i(\mathfrak{m}, N) \neq 0\}$.

Corollary 9.2.10. *If M is finitely generated, then the least i such that $\mathrm{Ext}^i(M, R) \neq 0$ is the maximum length of an R-sequence contained in $\mathrm{Ann}(M)$.*

Remark 9.2.11. If M is a finitely generated R-module, then the least i such that $\mathrm{Ext}^i(M, R) \neq 0$ is called the *grade* of M and is denoted grade M. By Corollary 9.2.10 above, we see that grade $M = \mathrm{depth}_I M$ where $I = \mathrm{Ann}(M)$.

Corollary 9.2.12. *For a prime ideal $\mathfrak{p} \subset R$, $\mathrm{depth}_{R_\mathfrak{p}} R_\mathfrak{p}$ is the least i such that $\mu_i(\mathfrak{p}, R) > 0$.*

Proof. Apply the previous corollary when $M = k(\mathfrak{p})$. □

Proposition 9.2.13. *If $\mathfrak{p} \subset \mathfrak{q}$ are distinct prime ideals of R with no prime ideal between them and M is finitely generated, then*

$$\mu_i(\mathfrak{p}, M) \neq 0 \quad \text{implies} \quad \mu_{i+1}(\mathfrak{q}, M) \neq 0.$$

Proof. We assume $R = R_\mathfrak{p}$. Let $r \in \mathfrak{q}$, $r \notin \mathfrak{p}$, $B = R/\mathfrak{p}$ and $C = B/rB$. Then C has finite length and $B \xrightarrow{r} B$ is an injection. So the exact sequence $0 \to B \xrightarrow{r} B \to C \to 0$ gives rise to the exact sequence

$$\operatorname{Ext}_R^i(B, M) \xrightarrow{r} \operatorname{Ext}_R^i(B, M) \to \operatorname{Ext}^{i+1}(C, M).$$

But $\operatorname{Ext}^i(B, M) \neq 0$ by Theorem 9.2.4. Since R is local and $r \in \mathfrak{p}$, we get $\operatorname{Ext}^{i+1}(C, M) \neq 0$. But then since C has finite length, it is easy to argue that $\operatorname{Ext}^{i+1}(k, M) \neq 0$ where $k = R/\mathfrak{p}$. So the result follows by Theorem 9.2.4. $\square$

Corollary 9.2.14. *If R is local with residue field k and M is a finitely generated R-module, then*
$$\operatorname{inj} \dim M = \sup\{i : \operatorname{Ext}^i(k, M) \neq 0\}.$$

Proof. This follows from Theorem 9.2.4 and Proposition 9.2.13. $\square$

Corollary 9.2.15. *If M is finitely generated and $\operatorname{inj} \dim {}_R M = r < \infty$, then $\dim M \leq r$ and $\mu_r(\mathfrak{p}, M) > 0$ implies $\mathfrak{p}$ is maximal.*

Proof. If $\mathfrak{p} \in \operatorname{Supp}(M)$ is a minimal prime ideal, then $R/\mathfrak{p}$ is isomorphic to a submodule of M by Theorem 2.4.12. So $\mu_0(\mathfrak{p}, M) > 0$. Then if $\mathfrak{p} = \mathfrak{p}_0 \subsetneq \mathfrak{p}_1 \subsetneq \mathfrak{p}_2 \subsetneq \cdots \subsetneq \mathfrak{p}_s$ is a chain of prime ideals of $\operatorname{Supp}(M)$ with no prime ideals between successive terms, then by Proposition 9.2.13 $\mu_s(\mathfrak{p}_s, M) > 0$ and so $\operatorname{inj} \dim M \geq s$. That is, $r \geq s$. This gives the first conclusion. For the second, if $\mu_r(\mathfrak{p}, M) > 0$ and $\mathfrak{p}$ is not maximal, by Proposition 9.2.13 we get a prime ideal $\mathfrak{p}'$ with $\mathfrak{p} \subset \mathfrak{p}'$ and $\mu_{r+1}(\mathfrak{p}', M) > 0$, contradicting our assumption $\operatorname{inj} \dim M = r$. $\square$

Theorem 9.2.16. *Let $(R, \mathfrak{m}, k)$ be local and M, N be nonzero finitely generated R-modules. If $\operatorname{inj} \dim N < \infty$, then*
$$\operatorname{depth} M + \sup\{i : \operatorname{Ext}_R^i(M, N) \neq 0\} = \operatorname{inj} \dim N.$$

Proof. If $M = k$, then the result follows by Corollary 9.2.14. If $\operatorname{depth} M = 0$, then there is an embedding $k \subset M$ and thus we have an exact sequence $\operatorname{Ext}^n(M/k, N) \to \operatorname{Ext}^n(M, N) \to \operatorname{Ext}^n(k, N) \to \operatorname{Ext}^{n+1}(M/k, N) = 0$. But if $\operatorname{inj} \dim N = n$, then $\operatorname{Ext}^n(k, N) \neq 0$ and so $\operatorname{Ext}^n(M, N) \neq 0$. Thus the result follows in this case.

We now proceed by induction on $\operatorname{depth} M$. If $\operatorname{depth} M > 0$, let $r \in \mathfrak{m}$ be a nonzero divisor on M. Then it follows from the exact sequence $\operatorname{Ext}^i(M, N) \xrightarrow{r} \operatorname{Ext}^i(M, N) \to \operatorname{Ext}^{i+1}(M/rM, N) \to \operatorname{Ext}^{i+1}(M, N)$ and Nakayama lemma that if $\operatorname{Ext}^i(M, N) \neq 0$ then $\operatorname{Ext}^{i+1}(M/rM, N) \neq 0$. It also trivially follows that if $\operatorname{Ext}^i(M, N) = \operatorname{Ext}^{i+1}(M, N) = 0$ then $\operatorname{Ext}^{i+1}(M/rM, N) \neq 0$. Hence

$$\sup\{i : \operatorname{Ext}^i(M/rM, N) \neq 0\} = \sup\{i : \operatorname{Ext}^i(M, N) \neq 0\} - 1.$$

Using a similar argument and Remark 9.2.9 we have $\operatorname{depth} M = \operatorname{depth} M/rM + 1$ and thus the result follows by induction. $\square$

Corollary 9.2.17. *If N is a nonzero finitely generated R-module of finite injective dimension, then* inj dim $N = $ depth R.

Proof. We simply set $M = R$ in the theorem.   □

Corollary 9.2.18. *If R has a nonzero finitely generated R-module of finite injective dimension and M is a finitely generated R-module, then* depth $M \leq$ depth R.

Proof. We simply note that $\sup\{i : \mathrm{Ext}^i(M, N) \neq 0\} \geq 0$ where N is as in the theorem above.   □

Remark 9.2.19. For completeness, we note that dually, if $\cdots \rightarrow F_1 \rightarrow F_0 \rightarrow M \rightarrow 0$ is a minimal free resolution of a finitely generated R-module, then the minimal number of generators of F_i, called the *rank* of F_i, is given by $\beta_i(M) = \dim_k \mathrm{Tor}_i(M, k)$ and one easily gets proj dim $M = \sup\{i : \mathrm{Tor}_i(M, k) \neq 0\}$. If depth $R > 0$, depth $M > 0$ and $x \in \mathfrak{m}$ is a nonzero divisor on R and M, then we have depth$_{R/rR} M/rM = $ depth $M - 1$, depth$_{R/rR} R/rR = $ depth $R - 1$, and proj dim$_{R/rR} M/rM = $ proj dim M. So an induction on depth M gives the following result which is dual to Theorem 9.2.16.

Theorem 9.2.20 (Auslander and Buchsbaum Theorem). *Let $(R, \mathfrak{m}, k)$ be local and M be a finitely generated R-module. If* proj dim $M < \infty$, *then*

$$\text{proj dim } M + \text{depth } M = \text{depth } R.$$

Proposition 9.2.21. *Let R be a local ring, $\mathfrak{p} \in \mathrm{Spec}\, R$, and $M \neq 0$ be a finitely generated R-module. If* dim $R/\mathfrak{p} < $ depth M, *then* $\mathrm{Ext}_R^i(R/\mathfrak{p}, M) = 0$ *for* $i < $ depth $M - $ dim $R/\mathfrak{p}$.

Proof. By induction on dim $R/\mathfrak{p}$. Let $\mathfrak{m}$ be the maximal ideal of R. If dim $R/\mathfrak{p} = 0$, then $\mathfrak{p} = \mathfrak{m}$ and so the result follows (see Remark 9.2.9). Now suppose dim $R/\mathfrak{p} > 0$. Then $\mathfrak{p} \neq \mathfrak{m}$ and so choose an element r in $\mathfrak{m} - \mathfrak{p}$. Then we consider the exact sequence $0 \rightarrow R/\mathfrak{p} \xrightarrow{r} R/\mathfrak{p} \rightarrow R/(rR + \mathfrak{p}) \rightarrow 0$. But by Lemma 2.4.7, there is a chain $0 = M_0 \subset M_1 \subset \cdots \subset M_{n-1} \subset M_n = R/(rR + \mathfrak{p})$ of submodules of $R/(rR + \mathfrak{p})$ such that $M_i/M_{j-1} \cong R/\mathfrak{p}_j$ for some $\mathfrak{p}_j \in \mathrm{Spec}\, R$. Moreover, $\mathfrak{p} \subset \mathfrak{p}_j$ and so dim $R/\mathfrak{p}_j < $ dim $R/\mathfrak{p}$ for each j. So by induction, for each j, $\mathrm{Ext}^i(R/\mathfrak{p}_j, M) = 0$ for $i < $ depth $M - $ dim $R/\mathfrak{p}_j \leq $ depth $M - $ dim $R/\mathfrak{p} + 1$. Hence $\mathrm{Ext}^i(R/(rR + \mathfrak{p}), M) = 0$ for $i < $ depth $M - $ dim $R/\mathfrak{p} + 1$. But then for each $i < $ depth $M - $ dim $R/\mathfrak{p}$, we have an exact sequence

$$0 \rightarrow \mathrm{Ext}^i(R/\mathfrak{p}, M) \xrightarrow{r} \mathrm{Ext}^i(R/\mathfrak{p}, M) \rightarrow \mathrm{Ext}^{i+1}(R/(rR + \mathfrak{p}), M) = 0.$$

So the result follows from Nakayama lemma.   □

Theorem 9.2.22. *If R is local and $M \neq 0$ is finitely generated, then*

$$\text{depth } M \leq \dim R/\mathfrak{p} \text{ for all } \mathfrak{p} \in \text{Ass } M.$$

Proof. If $\mathfrak{p} \in \text{Ass } M$, then $\text{Hom}(R/\mathfrak{p}, M) \neq 0$ and so $\text{depth } M \leq \dim R/\mathfrak{p}$ by the proposition above.   □

Corollary 9.2.23. *If R is local, then $\text{depth } M \leq \dim M$ for all nonzero finitely generated R-modules M.*

Proof. If $\mathfrak{p} \in \text{Ass } M$, then $\text{Ann}(M) \subseteq \mathfrak{p}$. So $\dim R/\mathfrak{p} \leq \dim R/\text{Ann}(M) = \dim M$ for each $\mathfrak{p} \in \text{Ass}(M)$. Hence $\text{depth } M \leq \dim M$ by the theorem.   □

Definition 9.2.24. Suppose R is local and M is a finitely generated R-module. Then $\text{depth } M \leq \dim M \leq \dim R$ from the above. If $\text{depth } M = \dim M$, then we say M is *Cohen–Macaulay*. If R is Cohen–Macaulay as an R-module, then R is said to be a *Cohen–Macaulay ring*. If $\text{depth } M = \dim R$, then M is said to be *maximal Cohen–Macaulay*.

Corollary 9.2.25. *If $\text{inj dim }_R R < \infty$, then $\text{inj dim }_R R = \dim R$ and R is a Cohen–Macaulay ring.*

Proof. $\text{inj dim } R = \text{depth } R$ by Corollary 9.2.17 and $\text{inj dim } R \geq \dim R$ by Corollary 9.2.15. But $\text{depth } R \leq \dim R$ by the above and so $\text{depth } R = \dim R$ and $\text{inj dim } R = \dim R$.   □

Corollary 9.2.26. *If for a prime ideal $\mathfrak{p}$ $\text{inj dim } R_\mathfrak{p} = \infty$, then $\mu_i(\mathfrak{p}, R) > 0$ for all $i \geq \text{ht } \mathfrak{p}$.*

Proof. We can again assume $R = R_\mathfrak{p}$ and use induction on $\text{ht } \mathfrak{p}$. If $\text{ht } \mathfrak{p} = 0$, then $\mathfrak{p}$ is the only prime ideal and so $\mu_i(\mathfrak{p}, R) = 0$ implies $E^i(R) = 0$ which contradicts our hypothesis.

Assume $\text{ht } \mathfrak{p} = h > 0$ and let $\mathfrak{q} \neq \mathfrak{p}$ be a prime ideal. If $s = \dim R/\mathfrak{q}$, then $\text{ht } \mathfrak{p} + s \leq h$. If $\text{inj dim }_{R_\mathfrak{q}} R_\mathfrak{q} = \infty$, then by the induction hypothesis $\mu_i(\mathfrak{q}, R) > 0$ for $i \geq \text{ht } \mathfrak{q}$. So $\mu_{i+s}(\mathfrak{p}, R) > 0$ for $i \geq \text{ht } \mathfrak{q}$ and then by the above $\mu_j(\mathfrak{p}, R) > 0$ for $j \geq h = \text{ht } \mathfrak{p}$.

So now assume $\text{inj dim }_{R_\mathfrak{q}} R_\mathfrak{q} < \infty$ for all $\mathfrak{q} \neq \mathfrak{p}$. Then $E^i(R)_\mathfrak{q} = 0$ for $i > \text{ht } \mathfrak{q}$ and so for $i \geq \text{ht } \mathfrak{p}$. Therefore $E^i(R)$ must be the direct sum of copies of $E(R/\mathfrak{p})$'s only. Since $E^i(R) \neq 0$ for $i \geq \text{ht } \mathfrak{p}$, we have $\mu_i(\mathfrak{p}, R) > 0$ for $i \geq \text{ht } \mathfrak{p}$.   □

Theorem 9.2.27. *If $(R, \mathfrak{m}, k)$ is a local ring, then the following are equivalent.*

(1) $\text{inj dim } R < \infty$.

(2) $\mu_i(\mathfrak{m}, R) = 0$ for all $i > \dim R$.

(3) $\mu_i(\mathfrak{m}, R) = 0$ *for some* $i > \dim R$.

(4) $\mu_i(\mathfrak{m}, R) = \begin{cases} 0 & \text{for } i < \dim R \\ 1 & \text{for } i = \dim R. \end{cases}$

(5) *R is Cohen–Macaulay and* $\mu_{\dim R}(\mathfrak{m}, R) = 1$.

(6) $\mu_i(\mathfrak{p}, R) = \delta_{i\,\mathrm{ht}(\mathfrak{p})}$ *for all prime ideals* $\mathfrak{p}$ *where* $\delta_{i\,\mathrm{ht}(\mathfrak{p})}$ *is the Kronecker delta.*

Proof. (1) $\Rightarrow$ (2) by Corollary 9.2.25, (2) $\Rightarrow$ (3) is trivial, and (3) $\Rightarrow$ (1) by Corollary 9.2.26.

(1) $\Rightarrow$ (4). By Corollary 9.2.25, R is Cohen–Macaulay and by Corollary 9.2.12 this is equivalent to $\mu_i(\mathfrak{m}, R) = 0$ for $i < \mathrm{ht}\,\mathfrak{m} = \dim R$.

Now let I be generated by a maximal R-sequence $r_1, r_2, \ldots, r_d$ where $d = \dim R$. Then by several applications of Corollary 9.2.25, R/I is self-injective. But by Proposition 3.4.2 this means $\mu_d(\mathfrak{m}, R) = \mu_0(R/I) = 1$.

(4) $\Rightarrow$ (2). With the same notation, $\mu_0(m(R/I), R/I) = 1$ and so R/I is self injective by Proposition 3.4.2. Hence $\mu_{i+d}(\mathfrak{m}, R) = \mu_i(\mathfrak{m}/I, R/I) = 0$ for all $i > 0$.

(4) $\Leftrightarrow$ (5) is trivial.

(2) and (4) give $\mu_i(\mathfrak{m}, R) = \delta_{\mathrm{id}}$ where $d = \mathrm{ht}\,\mathfrak{m}$. Since $\mathrm{inj\,dim}\,R < \infty$ gives $\mathrm{inj\,dim}\,R_{\mathfrak{p}} < \infty$ for all prime ideals $\mathfrak{p} \subset R$, we get (1) $\Rightarrow$ (6). (6) $\Rightarrow$ (1) is easy. $\square$

Corollary 9.2.28. *If R is a Gorenstein local ring and $I \subset R$ is generated by an R-sequence, then R/I is Gorenstein.*

Proof. This follows from the theorem and applications of Corollary 9.2.3. $\square$

Corollary 9.2.29. *If R is local, then R is Gorenstein if and only if $\hat{R}$ is.*

Proof. If R is Gorenstein and $0 \to R \to E^0(R) \to \cdots \to E^n(R) \to 0$ is an injective resolution of R, then $0 \to \hat{R} = \hat{R} \otimes_R R \to \hat{R} \otimes_R E^0(R) \to \cdots \to \hat{R} \otimes_R E^n(R) \to 0$ is an injective resolutions of $\hat{R}$ as an $\hat{R}$-module. So $\hat{R}$ is Gorenstein.

If $\hat{R}$ is Gorenstein, then $\mathrm{inj\,dim}_{\hat{R}} \hat{R} < \infty$ and so $\mathrm{inj\,dim}_R \hat{R} < \infty$ since $\hat{R}$ is a flat R-module. But $R \subset \hat{R}$ is pure. Thus $\mathrm{inj\,dim}_R R \leq \mathrm{inj\,dim}_R \hat{R} < \infty$ by Lemma 9.1.5. $\square$

Exercises

1. Let M be an R-module and $r \in R$ be such that $R \overset{r}{\to} R$ and $M \overset{r}{\to} M$ are both injections, and set $\bar{R} = R/(r)$. If N is an R-module such that $rN = 0$, prove that

$$\mathrm{Ext}_R^{i+1}(N, M) \cong \mathrm{Ext}_{\bar{R}}^i(N, M/rM)$$

for all $i \geq 0$. Conclude that $\mathrm{inj\,dim}_R M \geq \mathrm{inj\,dim}_R M/rM + 1$.

2. Let R be a noetherian local ring with maximal ideal $\mathfrak{m}$ and M be a finitely generated R-module. If $r \in \mathfrak{m}$ is as in Problem 1 above, argue that

a) $\operatorname{inj\,dim}_R M = \operatorname{inj\,dim}_{\bar{R}} M/rM + 1$.

b) $\operatorname{depth} M = \operatorname{depth}_{\bar{R}} M/rM + 1$.

c) $\operatorname{depth} M/rM = \operatorname{depth}_{\bar{R}} M/rM$.

3. Let R be a local ring and M a finitely generated R-module. Argue that

   a) $\operatorname{depth} M = \operatorname{depth}_{\hat{R}} \hat{M}$.

   b) M is Cohen–Macaulay if and only if $\hat{M}$ is Cohen–Macaulay.

4. Let $(R, \mathfrak{m}, k)$ be a local ring. Prove that if M is a finitely generated R-module, then $\operatorname{proj\,dim} M = \sup\{i : \operatorname{Tor}_i(M, k) \neq 0\} \leq \operatorname{proj\,dim} k$.

5. Let $(R, \mathfrak{m}, k)$ be a local ring and M be a finitely generated R-module. Prove that if $r \in \mathfrak{m}$ is such that $R \xrightarrow{r} R$ and $M \xrightarrow{r} M$ are both injections, then $\operatorname{proj\,dim} M = \operatorname{proj\,dim}_{R/rR} M/rM$.

6. Prove Theorem 9.2.20.

7. Let $(R, \mathfrak{m}, k)$ be a local ring, M a finitely generated R-module, $r_1, \ldots, r_d$ an M-regular sequence in $\mathfrak{m}$, and $\bar{M} = M/(r_1, \ldots, r_d)M$. Prove that

   a) $\dim \bar{M} = \dim M - d$.

   b) $\operatorname{depth} \bar{M} = \operatorname{depth} M - d$.

   c) M is Cohen–Macaulay if and only if $\bar{M}$ is.

   d) $\operatorname{proj\,dim} \bar{M} = \operatorname{proj\,dim} M + d$.

9.3 More on flat and injective modules

We now show that when a commutative noetherian ring R is *generically Gorenstein* (that is, $R_{\mathfrak{p}}$ is Gorenstein for all minimal prime ideals $\mathfrak{p}$ of R), we get nice properties holding on modules over R.

In this section, R will again denote a commutative noetherian ring.

Lemma 9.3.1. *For a prime ideal* $\mathfrak{p} \subset R$, $E(R/\mathfrak{p}) \otimes E(R/\mathfrak{p}) \neq 0$ *if and only if* $\operatorname{depth} R_{\mathfrak{p}} = 0$.

Proof. Suppose $E(R/\mathfrak{p}) \otimes E(R/\mathfrak{p}) \neq 0$. Then since $E(R/\mathfrak{p})$ is an injective cogenerator for $R_{\mathfrak{p}}$-modules,

$$(E(R/\mathfrak{p}) \otimes E(R/\mathfrak{p}))^v \cong \operatorname{Hom}(E(R/\mathfrak{p}), E(R/\mathfrak{p})^v) \cong \operatorname{Hom}(E(R/\mathfrak{p}), \widehat{R_{\mathfrak{p}}}) \neq 0.$$

But $E(R/\mathfrak{p})$ is an artinian $R_{\mathfrak{p}}$-module. So $k(\mathfrak{p}) \subset \widehat{R_{\mathfrak{p}}}$. But then depth $\widehat{R_{\mathfrak{p}}} = 0$ (over $\widehat{R_{\mathfrak{p}}}$) and thus depth $R_{\mathfrak{p}} = 0$ (over $R_{\mathfrak{p}}$).

Conversely, if depth $R_{\mathfrak{p}} = 0$ then, $k(\mathfrak{p}) \subset R_{\mathfrak{p}}$ and so $k(\mathfrak{p}) \subset \widehat{R_{\mathfrak{p}}}$. So by Matlis duality, we get a surjection $\widehat{R_{\mathfrak{p}}}^v = E(R/\mathfrak{p}) \to k(\mathfrak{p})^v = k(\mathfrak{p})$. But this implies $\operatorname{Hom}(E(R/\mathfrak{p}), \widehat{R_{\mathfrak{p}}}) \neq 0$ and so, as above, $E(R/\mathfrak{p}) \otimes E(R/\mathfrak{p}) \neq 0$. $\square$

Lemma 9.3.2. *For a prime ideal* $\mathfrak{p} \subset R$, $E(R/\mathfrak{p}) \otimes E(R/\mathfrak{p})$ *is a nonzero injective module if and only if* $R_\mathfrak{p}$ *is Gorenstein of Krull dimension* 0.

Proof. The Matlis dual of $E(R/\mathfrak{p}) \otimes E(R/\mathfrak{p})$ is $\operatorname{Hom}(E(R/\mathfrak{p}), \widehat{R_\mathfrak{p}})$. If $I \subset \widehat{R_\mathfrak{p}}$ is the largest ideal in $\widehat{R_\mathfrak{p}}$ of finite length, then I^v is the largest quotient of $E(R/\mathfrak{p})$ of finite length (over $\widehat{R_\mathfrak{p}}$). Then since $\sigma(E(R/\mathfrak{p}))$ has finite length for any $\sigma \in \operatorname{Hom}(E(R/\mathfrak{p}), \widehat{R_\mathfrak{p}})$, we get $\operatorname{Hom}(E(R/\mathfrak{p}), \widehat{R_\mathfrak{p}}) = \operatorname{Hom}(I^v, I)$. This shows that $\operatorname{Hom}(E(R/\mathfrak{p}), \widehat{R_\mathfrak{p}})$ has finite length and so it is Matlis reflexive. But it is the dual of $E(R/\mathfrak{p}) \otimes E(R/\mathfrak{p})$. So $E(R/\mathfrak{p}) \otimes E(R/\mathfrak{p})$ has finite length (and is also reflexive).

So now assume $E(R/\mathfrak{p}) \otimes E(R/\mathfrak{p})$ is nonzero and injective. Then its Matlis dual $\operatorname{Hom}(E(R/\mathfrak{p}), \widehat{R_\mathfrak{p}})$ is flat. But by the preceding, it is also a finitely generated $\widehat{R_\mathfrak{p}}$-module and so it is a nonzero free $\widehat{R_\mathfrak{p}}$-module. Thus since $\operatorname{Hom}(E(R/\mathfrak{p}), \widehat{R_\mathfrak{p}})$ has finite length, so does $\widehat{R_\mathfrak{p}}$. Thus $\widehat{R_\mathfrak{p}}$ has Krull dimension 0. Hence $\widehat{R_\mathfrak{p}} = R_\mathfrak{p}$. Let $\bar{\mathfrak{p}} = \mathfrak{p} R_\mathfrak{p}$ and $\bar{\mathfrak{p}}^n = 0$ and $\bar{\mathfrak{p}}^{n-1} \neq 0$ for some $n \geq 1$. Since $\operatorname{Hom}(E(R/\mathfrak{p}), \widehat{R_\mathfrak{p}}) \neq 0$ is free, $\bar{\mathfrak{p}}^{n-1} \operatorname{Hom}(E(R/\mathfrak{p}), \widehat{R_\mathfrak{p}}) \neq 0$. If for every $\sigma \in \operatorname{Hom}(E(R/\mathfrak{p}), \widehat{R_\mathfrak{p}})$, $\sigma(E(R/\mathfrak{p})) \subset \bar{\mathfrak{p}}$, then $\bar{\mathfrak{p}}^{n-1}\sigma = 0$ and so $\bar{\mathfrak{p}}^{n-1} \operatorname{Hom}(E(R/\mathfrak{p}), \widehat{R_\mathfrak{p}}) = 0$, a contradiction. This means $\sigma(E(R/\mathfrak{p})) = \widehat{R_\mathfrak{p}}$ for some σ. So $\widehat{R_\mathfrak{p}} = R_\mathfrak{p}$ is a direct summand of $E(R/\mathfrak{p})$ and so is injective.

Conversely, if $R_\mathfrak{p}$ is Gorenstein of dimension 0, then $R_\mathfrak{p}$ is self-injective and artinian and so $R_\mathfrak{p} = E(R_\mathfrak{p}) \cong E(R/\mathfrak{p})$. Thus $E(R/\mathfrak{p}) \otimes E(R/\mathfrak{p}) \cong R_\mathfrak{p} = E(R/\mathfrak{p})$ is a nonzero injective module.   □

In the next theorem, we will use the notation $F(M) \to M$ for the flat cover of an R-module M which exists by Theorem 7.4.4.

Theorem 9.3.3. *The following are equivalent for a commutative noetherian ring* R.

(1) $E(R)$ *is flat.*

(2) $R_\mathfrak{p}$ *is a Gorenstein ring of Krull dimension* 0 *for all* $\mathfrak{p} \in \operatorname{Ass}(R)$.

(3) $E(F)$ *is flat for all flat* R-modules F.

(4) $F(E)$ *is injective for all injective* R-modules E.

(5) $E \otimes E'$ *is an injective module for all injective* R-modules E *and* E'.

(6) $S^{-1}R$ *is an injective* R-module where S *is the set of nonzero divisors of* R.

Proof. (1) $\Rightarrow$ (2). Since $R/\mathfrak{p}$ is isomorphic to a submodule of R, $E(R/\mathfrak{p})$ is a summand of $E(R)$ and so is flat. But then $\widehat{R_\mathfrak{p}} = \operatorname{Hom}_R(E(R/\mathfrak{p}), E(R/\mathfrak{p}))$ is injective as an R-module (and so as an $\widehat{R_\mathfrak{p}}$-module) since the first $E(R/\mathfrak{p})$ is flat and the second is injective. But if $\widehat{R_\mathfrak{p}}$ is self-injective, it has Krull dimension 0 and so is artinian. But then $R_\mathfrak{p}$ is artinian and $R_\mathfrak{p} = \widehat{R_\mathfrak{p}}$.

(2) $\Rightarrow$ (1). If $\mathfrak{p} \in \operatorname{Ass}(R)$, then depth $R_\mathfrak{p} = 0$. So if $R_\mathfrak{p}$ is Gorenstein, $R_\mathfrak{p}$ has Krull dimension 0. Thus $R_\mathfrak{p}$ is self-injective and so $E(R/\mathfrak{p}) = R_\mathfrak{p}$. So $E(R/\mathfrak{p})$ is a flat R-module. Since $E(R)$ is the direct sum of $E(R/\mathfrak{p})$'s with $\mathfrak{p} \in \operatorname{Ass}(R)$, $E(R)$ is also flat.

(3) $\Rightarrow$ (1) trivially.

(2) $\Rightarrow$ (6). If $\mathrm{Ass}(R) = \{\mathfrak{p}_1, \ldots, \mathfrak{p}_t\}$, then $S = R - (\mathfrak{p}_1 \cup \cdots \cup \mathfrak{p}_t)$ is the set of nonzero divisors of R. So $S^{-1}R$ is a semi local ring of Krull dimension 0 with prime ideals $S^{-1}\mathfrak{p}_1, \ldots, S^{-1}\mathfrak{p}_t$. Hence $S^{-1}R$ is isomorphic to $R_{\mathfrak{p}_1} \times \cdots \times R_{\mathfrak{p}_t}$. But each $R_{\mathfrak{p}_i}$ is an injective R-module. So $S^{-1}R$ is injective.

(6) $\Rightarrow$ (1). This is true since $R_{\mathfrak{p}}$ with $\mathfrak{p} \in \mathrm{Ass}(R)$ is a localization of $S^{-1}R$.

(6) $\Rightarrow$ (3). By (6), $E(R) = S^{-1}R$ and so $E(R^n) = S^{-1}R^n$ for $n \geq 1$. But then since any flat R-module F is the inductive limit of finitely generated free modules, it is easy to see that $S^{-1}F = E(F)$.

(1) $\Rightarrow$ (5). We only need show that $E(R/\mathfrak{p}) \otimes E(R/\mathfrak{q})$ is injective for any prime ideals $\mathfrak{p}$ and $\mathfrak{q}$ of R. We note that if $r \in R$ is not a zero divisor on R, then $E \overset{r}{\to} E$ is surjective for any injective R-module E. And if $r \in \mathfrak{p}$ for a prime ideal $\mathfrak{p}$, then $E(R/\mathfrak{p}) \overset{r}{\to} E(R/\mathfrak{p})$ is locally nilpotent. Hence if $r \in \mathfrak{p}$ and r is not a zero divisor on R, then $E(R/\mathfrak{p}) \otimes E(R/\mathfrak{q}) = 0$. For if $x \in E(R/\mathfrak{p})$, we have $r^n x = 0$ for some $n \geq 1$, and if $y \in E(R/\mathfrak{q})$, $y = r^n z$ for some $z \in E(R/\mathfrak{q})$. So $x \otimes y = x \otimes r^n z = xr^n \otimes z = 0 \otimes z = 0$. Hence $E(R/\mathfrak{p}) \otimes E(R/\mathfrak{q}) = 0$ if $\mathfrak{p} \notin \mathrm{Ass}(R)$ or $\mathfrak{q} \notin \mathrm{Ass}(R)$. In a similar manner we see that $E(R/\mathfrak{p}) \otimes E(R/\mathfrak{q}) = 0$ when $\mathfrak{p} \not\subset \mathfrak{q}$. For then if $r \in \mathfrak{p}$, $E(R/\mathfrak{p}) \overset{r}{\to} E(R/\mathfrak{p})$ is locally nilpotent and $E(R/\mathfrak{q}) \overset{r}{\to} E(R/\mathfrak{q})$ is an isomorphism.

So we are reduced to showing that $E(R/\mathfrak{p}) \otimes E(R/\mathfrak{p})$ is injective if $\mathfrak{p} \in \mathrm{Ass}(R)$. But this follows from the fact that $E(R/\mathfrak{p}) = R_{\mathfrak{p}}$ (see (1) $\Rightarrow$ (2)) and the fact that $R_{\mathfrak{p}} \otimes_R R_{\mathfrak{p}} = R_{\mathfrak{p}}$.

(5) $\Rightarrow$ (2). Let $\mathfrak{p} \in \mathrm{Ass}(R)$. Then $R/\mathfrak{p} \subset R$ and so depth $R_{\mathfrak{p}}$ is 0. So by Lemma 9.3.1, $E(R/\mathfrak{p}) \otimes E(R/\mathfrak{p}) \neq 0$. But then by Lemma 9.3.2 and the assumption (5), we get (2).

(3) $\Rightarrow$ (4). Let G be an injective R-module and let $F(G) \to G$ be a flat cover. Then $F(G) \to G$ can be extended to $E(F(G)) \to G$. But $E(F(G))$ is flat by assumption. So

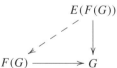

can be completed to a commutative diagram. But then the restriction of $E(F(G)) \to G$ to $F(G)$ is an automorphism of $F(G)$ (by the definition of a flat cover). Thus $F(G)$ is a direct summand of $E(F(G))$ and so is injective.

(4) $\Rightarrow$ (3). Let L be flat. Then the diagram

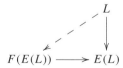

can be completed to a commutative diagram. By (4), $F(E(L))$ is injective and so $L \to F(E(L))$ (which is an injection) can be extended to an injection $E(L) \to F(E(L))$.

Therefore we can assume $E(L) \subset F(E(L))$. Then $E(L)$ is flat since it is a direct summand of a flat module. □

We note that when $R_{\mathfrak{p}}$ is Gorenstein for $\mathfrak{p} \in \mathrm{Ass}(R)$, then $\mathrm{ht}\, \mathfrak{p} = 0$, that is, $\mathfrak{p}$ is a minimal prime ideal. Since any minimal prime $\mathfrak{p}$ is in $\mathrm{Ass}(R)$, (2) of the theorem simply says R is generically Gorenstein.

Exercises

1. Prove that if $\mathfrak{p} \subseteq \mathfrak{q}$, then $E(R/\mathfrak{p}) \otimes E(R/\mathfrak{q}) = 0$.

2. Prove that if R is Gorenstein, then $E(R)$ is a flat R-module.

3. Prove that if $E(R)$ is flat, then $R_{\mathfrak{p}} \otimes_R R_{\mathfrak{p}} \cong R_{\mathfrak{p}}$.

9.4   Torsion products of injective modules

R will again always be commutative and noetherian. In the last section it was shown that R is generically Gorenstein if and only if $E \otimes E'$ is injective whenever E and E' are injective R-modules. In this section, we will show that $\mathrm{Tor}_i(E, E')$ is injective for all injective E and E' and all $i \geq 0$ if and only if $R_{\mathfrak{p}}$ is Gorenstein for all prime ideals $\mathfrak{p}$. Since these two properties are local we will also assume R is a local ring with maximal ideal $\mathfrak{m}$ and residue field k.

Lemma 9.4.1. *We have*

$$\mathrm{Ext}^i_R(E(k), R) \cong \mathrm{Ext}^i_R(E(k), \hat{R}) \cong \mathrm{Ext}^i_{\hat{R}}(E(k), \hat{R})$$

for all $i > 0$.

Proof.

$$0 \to \hat{R} = \hat{R} \otimes_R R \to \hat{R} \otimes_R E^0(R) \to \hat{R} \otimes_R E^1(R) \to \cdots$$

gives an injective resolution of $\hat{R}$ as an R-module. If $\mathfrak{p} \subset R$ is a prime ideal, then $E(R/\mathfrak{p}) \xrightarrow{r} E(R/\mathfrak{p})$ is locally nilpotent if $r \in \mathfrak{p}$ and is an isomorphism if $r \notin \mathfrak{p}$. Hence the same holds for $\hat{R} \otimes E(R/\mathfrak{p}) \xrightarrow{r} \hat{R} \otimes E(R/\mathfrak{p})$ and so $\hat{R} \otimes E(R/\mathfrak{p})$ is the direct sum of copies of $E(R/\mathfrak{p})$. Hence if $\mathfrak{p} \neq \mathfrak{m}$, then $\mathrm{Hom}_R(E(k), \hat{R} \otimes E(R/\mathfrak{p})) = 0$. If $\mathfrak{p} = \mathfrak{m}$, then $\hat{R} \otimes_R E(k) = E(k)$. From these remarks it follows that $\mathrm{Hom}_R(E(k), E^i(R)) \cong \mathrm{Hom}_R(E(k), \hat{R} \otimes_R E^i(R))$ and we get the first isomorphism above. For the second, we note that since $\hat{R}$ is a flat R-module, $0 \to \hat{R} \to E^0_{\hat{R}}(\hat{R}) \to E^1_{\hat{R}}(\hat{R}) \to \cdots$ is also an injective resolution of $\hat{R}$ as an R-module.

Then note that if $\mathfrak{q} \subset \hat{R}$ is a prime ideal and $\mathfrak{q} \neq \hat{\mathfrak{m}}$), then $\mathfrak{q} \cap R \neq \mathfrak{m}$. Hence again arguing with scalar multiplications, we see that $\mathrm{Hom}_R(E(k), E_{\hat{R}}(\hat{R}/\mathfrak{q})) = 0$ and $\mathrm{Hom}_{\hat{R}}(E(k), E_{\hat{R}}(\hat{R})) = 0$. Since $\mathrm{Hom}_R(E(k), E(k)) = \mathrm{Hom}_{\hat{R}}(E(k), E(k))$, we see that $\mathrm{Hom}_R(E(k), E^i_{\hat{R}}(\hat{R})) = \mathrm{Hom}_{\hat{R}}(E(k), E^i_{\hat{R}}(\hat{R}))$. This gives the second isomorphism. □

Remark 9.4.2. If r is not a zero divisor on R or on M and $\bar{R} = R/(r)$, then $\text{Tor}_1^R(\bar{R}, M) = 0$. Hence if $S \subset M$ and r is not a zero divisor on M/S, then $\text{Tor}_1^R(\bar{R}, M/S) = 0$ implies that $0 \to \bar{R} \otimes S \to \bar{R} \otimes M$ is exact.

Lemma 9.4.3. *Suppose $r \in \mathfrak{m}$ is not a zero divisor of R and let $\bar{R} = R/(r)$. If r is not a zero divisor on $\text{Ext}_R^i(E(k), R)$ for $0 \leq i \leq n$, then*

$$\text{Ext}_{\bar{R}}^{i-1}(E_{\bar{R}}(k), \bar{R}) \cong \text{Ext}_R^i(E_R(k), R) \otimes \bar{R}$$

for $1 \leq i \leq n-1$.

Proof. Apply $\text{Hom}_R(E_R(k), -)$ to $0 \to E^0(R) \to E^1(R) \to \cdots$ and get the complex

$$0 \to \hat{R}^{\mu_0} \to \hat{R}^{\mu_1} \to \cdots$$

with $\mu_i = \mu_i(\mathfrak{m}, R)$. The homology of this complex gives us the modules $\text{Ext}_R^i(E(k), R)$. We note that since r is not a zero divisor of R, depth $R \geq 1$ and so $\mu_0 = 0$.

We now let B^i, $Z^i \subset \hat{R}^{\mu_i}$ be the images and kernels of the boundary maps in the complex above.

By Lemma 9.2.2, $0 \to \bar{R} \to \text{Hom}_R(\bar{R}, E_R^1(R)) \to \text{Hom}_R(\bar{R}, E_R^2(R)) \to \cdots$ is a minimal injective resolution of $\bar{R}$ as an $\bar{R}$-module. So we compute $\text{Ext}_{\bar{R}}^i(E_{\bar{R}}(k), \bar{R})$ by applying $\text{Hom}_{\bar{R}}(E_{\bar{R}}(k), -)$ to the deleted complex $0 \to \text{Hom}_R(\bar{R}, E_R^1(R)) \to \text{Hom}_R(\bar{R}, E_R^2(R)) \to \cdots$ and compute homology. But $\text{Hom}_R(\bar{R}, E_R(k)) \cong E_{\bar{R}}(k)$. So we get the complex

$$0 \to \hat{\bar{R}}^{\mu_1} \to \hat{\bar{R}}^{\mu_2} \to \cdots$$

But then this is just the complex tensored with $\hat{\bar{R}}$ since $\hat{\bar{R}} = \hat{R} \otimes_R \bar{R}$.

Now we note that r is neither a zero divisor on $\hat{R}^{\mu_i}/Z^i \subset \hat{R}^{\mu_i+1}$ nor on $\hat{R}^{\mu_i}/B^i$ since it is neither a zero divisor on $\text{Ext}^i(E(k), R) = Z^i/B^i$ nor on $\hat{R}^{\mu_i}/Z^i$ for $0 \leq i \leq n$. So by Remark 9.4.2, $0 \to B^i \otimes \bar{R} \to \hat{R}^{\mu_i} \otimes \bar{R} = \hat{\bar{R}}^{\mu_i}$ and $0 \to Z^i \otimes \bar{R} \to \hat{\bar{R}}^{\mu_i}$ are exact.

These observations show that $Z^i \otimes \bar{R}$ and $B^i \otimes \bar{R}$ can be identified with the obvious kernels and images in $0 \to \hat{\bar{R}}^{\mu_1} \to \hat{\bar{R}}^{\mu_2} \to \cdots$ for $1 \leq i \leq n-1$. Then the exact sequence $B^i \to Z^i \to \text{Ext}^i(E(k), \bar{R}) \to 0$ tensored with $\bar{R}$ gives the isomorphism

$$\text{Ext}_{\bar{R}}^{i-1}(E_{\bar{R}}(k), \bar{R}) \cong \text{Ext}_R^i(E_R(k), R) \otimes \bar{R}$$

for $1 \leq i \leq n-1$. $\square$

Lemma 9.4.4. *If A and B are artinian modules, then $\text{Tor}_i(A, B)$ is artinian for all $i \geq 0$.*

Proof. It suffices to show that $\text{Tor}_i(A, B)^v \cong \text{Ext}^i(A, B^v)$ is finitely generated. But B^v is finitely generated and so $\mu_i(\mathfrak{m}, B^v) = \mu_i$ is finite. But then $\text{Hom}(A, E^i(B^v)) = \text{Hom}(A, E(k)^{\mu_i})$ is finitely generated and hence so is $\text{Ext}^i(A, B^v)$. $\square$

Lemma 9.4.5. *If $\mathfrak{p}$, $\mathfrak{q}$ are prime ideals and $\mathfrak{p} \not\subset \mathfrak{q}$ or $\mathfrak{q} \not\subset \mathfrak{p}$, then*

$$\mathrm{Tor}_i(E(R/\mathfrak{p}), E(R/\mathfrak{q})) = 0$$

for all $i \geq 0$.

Proof. Suppose $\mathfrak{p} \not\subset \mathfrak{q}$. If $r \in \mathfrak{p}, r \notin \mathfrak{q}$, then

$$\mathrm{Tor}_i(E(R/\mathfrak{p}), E(R/\mathfrak{q})) \xrightarrow{r} \mathrm{Tor}_i(E(R/\mathfrak{p}), E(R/\mathfrak{q}))$$

is both locally nilpotent and an isomorphism. Hence $\mathrm{Tor}_i(E(R/\mathfrak{p}), E(R/\mathfrak{q})) = 0$. □

Theorem 9.4.6. *The following are equivalent for R:*

(1) *R is Gorenstein.*

(2) *For any injective modules E and E' and any $i \geq 0$, $\mathrm{Tor}_i(E, E')$ is injective.*

(3) *For any prime ideal $\mathfrak{p} \subset R$*

$$\mathrm{Tor}_i^R(E(R/\mathfrak{p}), E(R/\mathfrak{p})) = \begin{cases} 0 & \text{if } i \neq \mathrm{ht}\,\mathfrak{p} \\ E(R/\mathfrak{p}) & \text{if } i = \mathrm{ht}\,\mathfrak{p}. \end{cases}$$

Proof. (1) $\Rightarrow$ (3). We have

$$\begin{aligned}
\mathrm{Tor}_i^R(E(R/\mathfrak{p}), E(R/\mathfrak{p}))^v &\cong \mathrm{Ext}_R^i(E(R/\mathfrak{p}), \hat{R}_\mathfrak{p}) \\
&\cong \mathrm{Ext}_{R_\mathfrak{p}}^i(E(R/\mathfrak{p}), \hat{R}_\mathfrak{p}) \\
&\cong \mathrm{Ext}_{\widehat{R_\mathfrak{p}}}^i(E(R/\mathfrak{p}), \widehat{R_\mathfrak{p}})
\end{aligned}$$

where the Matlis dual is with respect to $R_\mathfrak{p}$.

Since $\hat{R}_\mathfrak{p}$ is Gorenstein,

$$\mu_i(m(\hat{R}_\mathfrak{p}), \hat{R}_\mathfrak{p}) = \begin{cases} 1 & \text{if } i = \dim \hat{R}_\mathfrak{p} = \mathrm{ht}\,\mathfrak{p} \\ 0 & \text{otherwise} \end{cases}$$

by Theorem 9.2.27. This shows that $\mathrm{Ext}_{\widehat{R}_\mathfrak{p}}^i(E(R/\mathfrak{p}), \hat{R}_\mathfrak{p})$ is $\hat{R}_\mathfrak{p}$ for $i = \mathrm{ht}\,\mathfrak{p}$ and 0 otherwise. Then since $E(R/\mathfrak{p})^v \cong \hat{R}_\mathfrak{p}$, the claim follows by Lemma 9.4.4 since $\mathrm{Tor}_i(E(R/\mathfrak{p}), E(R/\mathfrak{p}))$ is then Matlis reflexive (over $R_\mathfrak{p}$).

(3) $\Rightarrow$ (2) is trivial.

(2) $\Rightarrow$ (1). By Lemma 9.4.5, (2) simply says that $\mathrm{Tor}_i(E(R/\mathfrak{p}), E(R/\mathfrak{p}))$ is injective for all prime ideals $\mathfrak{p}$ and all $i \geq 0$. Since

$$\mathrm{Tor}_i^R(E(R/\mathfrak{p}), E(R/\mathfrak{p})) = \mathrm{Tor}_i^R(E(R/\mathfrak{p}), E(R/\mathfrak{p}))_\mathfrak{p} = \mathrm{Tor}_i^{R_\mathfrak{p}}(E(k(\mathfrak{p})), E(k(\mathfrak{p}))),$$

we can assume $\mathfrak{p} = \mathfrak{m}$ and that $\mathrm{Tor}_i(E(k), E(k))$ is injective for all $i \geq 0$. Since by Lemma 9.4.4 this module is also artinian, it is the direct sum of finitely many copies of $E(k)$.

But then $\mathrm{Tor}_i(E(k), E(k))^v = \mathrm{Ext}^i_R(E(k), \hat{R})$ is a finitely generated $\hat{R}$-module for each i. Using Lemma 9.4.1, we see we can assume R is complete and so $\mathrm{Ext}^i(E(k), R)$ is finitely generated and free for each i. If depth $R = d$ and $r_1, \ldots, r_d$ is an R-sequence, then letting $\bar{R} = R/(r_1, \ldots, r_d)$ and making repeated use of Lemma 9.4.3, we see that we can also assume depth $R = 0$. This implies $k \subset R$ and so by Matlis duality, k is a quotient of $E(k)$. Therefore $\mathrm{Hom}(E(k), R) = \mathrm{Ext}^0(E(k), R) \neq 0$ is a finitely generated free R-module. Then $\mathrm{Hom}(E(k) \otimes E(k), E(k)) \cong \mathrm{Hom}(E(k), \hat{R})$ shows that $E(k) \otimes E(k) \neq 0$. Its dual is finitely generated and projective. Hence $E(k) \otimes E(k) \neq 0$ is an injective module. Then by Lemma 9.3.2, R is Gorenstein of Krull dimension 0. This completes the proof.   □

Exercises

1. Let R be a Gorenstein ring. Prove that every flat R-module F can be embedded in $\bigoplus_{\mathrm{ht}\, \mathfrak{p}=0} E(R/\mathfrak{p})^{(X_\mathfrak{p})}$ for some sets $X_\mathfrak{p}$.
    Hint: show that $F \otimes E(R) \cong \bigoplus_{\mathrm{ht}\, \mathfrak{p}=0} E(R/\mathfrak{p})^{(X_\mathfrak{p})}$.

2. Let R be 1-Gorenstein and M be an R-module. Prove that M is cotorsion if and only if $\mathrm{Ext}^1(E(R), M) = 0$.

3. Let R be 1-Gorenstein. Prove that every h-divisible R-module is *strongly co-torsion*, that is, $\mathrm{Ext}^i(F, M) = 0$ for all flat R-modules M and all $i \geq 1$.

9.5 Local cohomology and the dualizing module

In this section, R will again denote a commutative noetherian ring.

Definition 9.5.1. Let I be an ideal of R and M be an R-module. Then we set

$$L_I(M) = \{x \in M : I^t x = 0 \text{ for some } t \geq 0\}.$$

We note that $\mathrm{Hom}_R(R/I^t, M) \cong \{x \in M : I^t x = 0\}$. Thus

$$L_I(M) \cong \varinjlim \mathrm{Hom}_R(R/I^t, M).$$

It is clear that L_I is a left exact covariant functor and so we can compute the right derived functors $R^i L_I$ of L_I using an injective resolution of M (see Section 8.2).

Definition 9.5.2. The derived functors $R^i L_I$ are called *local cohomology functors* and are denoted by H^i_I. Thus

$$H^i_I(M) = R^i L_I(M) = R^i(\varinjlim \mathrm{Hom}(R/I^t, M))$$

$$\cong \varinjlim R^i \mathrm{Hom}(R/I^t, M)$$

$$\cong \varinjlim \mathrm{Ext}^i_R(R/I^t, M).$$

since $\varinjlim$ commutes with homology. We note that if $i > 0$ and E is an injective R-module, then $H_I^i(E) = 0$.

Now let $0 \to M \to E^0 \to E^1 \to \cdots$ be the minimal injective resolution of M. Then $E^i \cong \bigoplus_{\mathfrak{p} \in \operatorname{Spec} R} \mu^i(\mathfrak{p}, M) E(R/\mathfrak{p})$. But if $\mathfrak{m}$ is a maximal ideal, then

$$\varinjlim \operatorname{Hom}(R/\mathfrak{m}^t, E(R/\mathfrak{p})) \cong \begin{cases} 0 & \text{if } \mathfrak{p} \neq \mathfrak{m} \\ E(R/\mathfrak{m}) & \text{if } \mathfrak{p} = \mathfrak{m}. \end{cases}$$

So $L_\mathfrak{m}(E^i) \cong \oplus \mu^i(\mathfrak{m}, M) E(R/\mathfrak{m})$. Thus if M is finitely generated, then $\mu^i(\mathfrak{m}, M) < \infty$ for each i by Theorem 9.2.4 and so $L_\mathfrak{m}(E^i)$ is artinian. Hence $H_\mathfrak{m}^i(M)$ is artinian for each i. Furthermore if R is local, then depth $M = \inf\{i : \mu_i(\mathfrak{m}, M) \neq 0\}$ by Remark 9.2.9. So if $i < \operatorname{depth} M$, then $H_\mathfrak{m}^i = 0$. Now let $s = \operatorname{depth} M$. Then

$$H_\mathfrak{m}^s(M) \cong \operatorname{Ker}(L_\mathfrak{m}(E^s(M)) \to L_\mathfrak{m}(E^{s+1}(M)))$$
$$= \operatorname{Ker}(E^s(M) \to E^{s+1}(M)) \cap L_\mathfrak{m}(E^s).$$

But $\operatorname{Ker}(E^s(M) \to E^{s+1}(M))$ is an essential submodule of $E^s(M)$ and $L_\mathfrak{m}(E^s)$ is nonzero since $s = \operatorname{depth} M$. So $H_\mathfrak{m}^s(M) \neq 0$.

We summarize the above in the following.

Lemma 9.5.3. *Let R be local and M be a finitely generated R-module. Then $H_\mathfrak{m}^i(M)$ is an artinian R-module and* depth M *is the least integer s such that $H_\mathfrak{m}^s(M) \neq 0$.*

Proposition 9.5.4. *Let R be a local ring with maximal ideal $\mathfrak{m}$, and M be a finitely generated R-module. Then $H_\mathfrak{m}^i(M) \cong H_\mathfrak{m}^i(M) \otimes \hat{R} \cong H_{\hat{\mathfrak{m}}}^i(\hat{M})$.*

Proof. Since $H_\mathfrak{m}^i(M)$ is artinian by the above, $H_\mathfrak{m}^i(M) = \varinjlim L_i$ where L_i are submodules of $H_\mathfrak{m}^i(M)$ of finite length. Then

$$H_\mathfrak{m}^i(M) = \varinjlim L_i \cong \varinjlim(L_i \otimes \hat{R}) \cong (\varinjlim L_i) \otimes \hat{R}$$

and we have the first isomorphism. But

$$H_\mathfrak{m}^i(M) \otimes \hat{R} \cong \varinjlim \operatorname{Ext}^i(R/\mathfrak{m}^t, M) \otimes \hat{R}$$
$$\cong \varinjlim \operatorname{Ext}^i(\hat{R}/\hat{\mathfrak{m}}^t, \hat{M})$$
$$\cong H_{\hat{\mathfrak{m}}}^i(\hat{M})$$

since $\hat{R}$ is a flat R-module. $\square$

Definition 9.5.5. Let $\boldsymbol{x} = x_1, x_2, \ldots, x_n$ be a sequence of elements of R. Then we define a complex $\mathbf{K}_\bullet(\boldsymbol{x})$ by $K_0(\boldsymbol{x}) = R$ and $K_p(\boldsymbol{x}) = 0$ if $p > n$. If $1 \leq p \leq n$, then

$K_p(x)$ is the free R-module of rank $\binom{n}{p}$ with basis $\{e_{i_1 i_2 \ldots i_p} : 1 \le i_1 < \cdots < i_p \le n\}$. The differentiation $d_p : K_p(x) \to K_{p-1}(x)$ is defined by

$$d_p(e_{i_1 \ldots i_p}) = \sum_{j=1}^{p} (-1)^{j+1} x_{i_j} e_{i_1 \ldots \hat{i}_j \ldots i_p}$$

where $\hat{i}_j$ means that i_j is deleted and we set $d_p(e_i) = x_i$ if $p = 1$. Then $d_{p-1} \circ d_p = 0$ and we have a complex $\mathbf{K_\bullet}(x)$

$$\mathbf{K_\bullet}(x) : 0 \to K_n(x) \xrightarrow{d_n} K_{n-1}(x) \to \cdots \to K_1(x) \xrightarrow{d_1} K_0(x) \to 0$$

where each $K_p(x)$ is free and finitely generated.

The complex $\mathbf{K_\bullet}(x)$ is called the *Koszul complex*. In particular, if $n = 1$, then $\mathbf{K_\bullet}(x)$ is the complex

$$0 \to R \xrightarrow{x} R \to 0$$

with $K_1(x) = K_0(x) = R$ and $K_p(x) = 0$ if $p \ne 0, 1$. If $n = 2$, then the Koszul complex is

$$0 \to R \xrightarrow{d_2} R^2 \xrightarrow{d_1} R \to 0$$

with bases e_{12}, $\{e_1, e_2\}$, $\{1\}$ respectively and differentials $d_1(e_i) = x_i$ and $d_2(e_{12}) = x_1 e_2 - x_2 e_1$.

One may check that $\mathbf{K_\bullet}(x) = \mathbf{K_\bullet}(x_1) \otimes \mathbf{K_\bullet}(x_2) \otimes \cdots \otimes \mathbf{K_\bullet}(x_n)$ where the *tensor product of two complexes* of R-modules $\mathbf{C_\bullet}$ and $\mathbf{D_\bullet}$ is a complex $\mathbf{C_\bullet} \otimes_R \mathbf{D_\bullet}$ where

$$(\mathbf{C_\bullet} \otimes_R \mathbf{D_\bullet})_n = \oplus_{p+q=n} C_p \otimes D_q$$

with differentials

$$d : (\mathbf{C_\bullet} \otimes_R \mathbf{D_\bullet})_n \to (\mathbf{C_\bullet} \otimes_R \mathbf{D_\bullet})_{n-1}$$

defined by $d(x \otimes y) = dx \otimes y + (-1)^p x \otimes dy$.

If M is an R-module, then $\mathbf{K_\bullet}(x, M)$ is the complex $\mathbf{K_\bullet}(x) \otimes M$ whose homology groups $H_p(\mathbf{K_\bullet}(x) \otimes M)$ are denoted by $H_p(x, M)$. So $H_0(x, M) = M/(x_1 M + \cdots + x_n M)$ and $H_n(x, M) = \{y \in M : x_i y = 0 \text{ for } 1 \le i \le n\}$. In particular, if $n = 1$, then $H_0(x, M) = M/xM$ and $H_1(x, M) = \text{Ann}_M(x) = \{y \in M : xy = 0\}$. $\mathbf{K^\bullet}(x, M)$ denotes the complex $\text{Hom}(\mathbf{K_\bullet}(x), M)$ whose cohomology groups $H^p(\text{Hom}(\mathbf{K_\bullet}(x), M))$ are denoted by $H^p(x, M)$. In this case $H^0(x, M) = \text{Hom}(R/xR, M) = \{y \in M : x_i y = 0 \text{ for } 1 \le i \le n\} = H_n(x, M)$. More generally $H_p(x, M) \cong H^{n-p}(x, M)$.

Lemma 9.5.6. *If $\mathbf{C}$ is a complex of R-modules, then there is an exact sequence*

$$0 \to H_0(x, H_p(\mathbf{C})) \to H_p(x, \mathbf{C}) \to H_1(x, H_{p-1}(\mathbf{C})) \to 0.$$

Proof. $(\mathbf{K}_\bullet(x) \otimes \mathbf{C})_p = (K_0(x) \otimes C_p) \oplus (K_1(x) \otimes C_{p-1})$. So we have an exact sequence of complexes $0 \to (K_0(x) \otimes \mathbf{C})_p \to (\mathbf{K}_\bullet(x) \otimes \mathbf{C})_p \to (K_1(x) \otimes \mathbf{C})_{p-1} \to 0$ since $K_0(x) = K_1(x) = R$. Thus we have an associated exact sequence

$$K_1(x) \otimes H_p(\mathbf{C}) \xrightarrow{x \otimes \mathrm{id}} K_0(x) \otimes H_p(\mathbf{C}) \to H_p(\mathbf{K}_\bullet(x) \otimes \mathbf{C}) \to K_1(x) \otimes H_{p-1}(\mathbf{C})$$
$$\xrightarrow{x \otimes \mathrm{id}} K_0(x) \otimes H_{p-1}(\mathbf{C}).$$

But $\mathrm{Coker}\,(x \otimes \mathrm{id}_{H_p(\mathbf{C})}) \cong H_p(\mathbf{C})/x H_p(\mathbf{C}) = H_0(x, H_p(\mathbf{C}))$ and $\mathrm{Ker}(x \otimes \mathrm{id}_{H_{p-1}(\mathbf{C})})$ $\cong \mathrm{Ann}_{H_{p-1}(\mathbf{C})}(x) = H_1(x, H_{p-1}(\mathbf{C}))$ and so we are done.   □

We now show that local cohomology groups $H_{\mathfrak{m}}^i(M)$ can be expressed in terms of Koszul complexes.

Let $\mathbf{x}^t$ denote the sequence $x_1^t, x_2^t, \ldots, x_n^t$. Then for each x_i, there is a natural map $\mathbf{K}_\bullet(x_i^{t+1}) \to \mathbf{K}_\bullet(x_i^t)$ given by the diagram

$$
\begin{array}{ccccccccc}
\mathbf{K}_\bullet(x_i^{t+1}): & & 0 & \longrightarrow & R & \xrightarrow{x_i^{t+1}} & R & \longrightarrow & 0 \\
& & & & \downarrow{\scriptstyle x_i} & & \parallel & & \\
\mathbf{K}_\bullet(x_i^t): & & 0 & \longrightarrow & R & \xrightarrow{x_i^t} & R & \longrightarrow & 0.
\end{array}
$$

Since $\mathbf{K}_\bullet(\mathbf{x}) = \mathbf{K}_\bullet(x_1) \otimes \cdots \otimes \mathbf{K}_\bullet(x_n)$, we can tensor these maps to get a map of complexes $\mathbf{K}_\bullet(\mathbf{x}^{t+1}) \to \mathbf{K}_\bullet(\mathbf{x}^t)$. Hence we get a map of homology groups $H_i(\mathbf{x}^{t+1}, M) \to H_i(\mathbf{x}, M)$ and a map of cohomology groups $H^i(\mathbf{x}^t, M) \to H^i(\mathbf{x}^{t+1}, M)$. Thus we have an inverse system $\{H_i(\mathbf{x}^t, M)\}$ and a directed system $\{H^i(\mathbf{x}^t, M)\}$. We set $H_{\mathbf{x}}^i(M) = \varinjlim H^i(\mathbf{x}^t, M)$.

Lemma 9.5.7. *Let $\mathbf{x} = x_1, x_2, \ldots, x_n$ be a sequence of elements of R and M be a finitely generated R-module. Then for each t, there is an $s > t$ such that the map $H_i(\mathbf{x}^s, M) \to H_i(\mathbf{x}^t, M)$ is a zero homomorphism for each $i \geq 1$.*

Proof. By induction on n. Suppose $n = 1$. Then $i = 1$ and $H_1(x^t, M) = \mathrm{Ann}_M(x^t)$. We note that if $s \geq t$, then the map $H_1(x^s, M) \to H_1(x^t, M)$ is multiplication by x^{s-t}. But the modules $H_1(x^t, M)$ form an increasing sequence of submodules of M. Hence since M is noetherian, there is a t_0 such that $\mathrm{Ann}_M(x^{t_0})$ is maximal. But then $x^{t_0} H_1(x^t, M) = 0$ for each t. So if $s = t + t_0$, then the map $H_1(x^s, M) \to H_1(x^t, M)$ is multiplication by $x^{s-t} = x^{t_0}$ and thus is zero.

Now suppose $n \geq 1$ and let $\mathbf{y} = x_1, x_2, \ldots, x_{n-1}$. Then by Lemma 9.5.6, we have an exact sequence

$$0 \to H_0(x_n^t, H_i(\mathbf{y}^t, M)) \to H_i(\mathbf{x}^t, M) \to H_1(x_n^t, H_{i-1}(\mathbf{y}^t, M)) \to 0.$$

If $s > t$, then we have a factorization

$$H_0(x_n^s, H_i(\mathbf{y}^s, M)) \xrightarrow{f} H_0(x_n^s, H_i(\mathbf{y}^t, M)) \xrightarrow{g} H_0(x_n^t, H_i(\mathbf{y}^t, M)).$$

By induction hypothesis, there is an $s > t$ such that $H_i(y^s, M) \rightarrow H_i(y^t, M)$ is zero and thus $gf = 0$. Hence there is an $s > t$ such that $H_0(x_n^s, H_i(y^s, M)) \rightarrow H_0(x_n^t, H_i(y^t, M))$ is zero. Similarly, we have a factorization

$$H_1(x_n^s, H_{i-1}(y^s, M)) \xrightarrow{f} H_1(x_n^s, H_{i-1}(y^t, M)) \xrightarrow{g} H_1(x_n^t, H_{i-1}(y^t, M)).$$

But $H_{i-1}(y^t, M)$ is finitely generated. So by induction, there is an $s > t$ such that $g = 0$ and hence $gf = 0$.

But now it easily follows that the middle term $H_i(x^t, M)$ has the same property (see Problem 4 at the end of this section).   □

If E is an injective R-module and $i > 0$, then

$$\begin{aligned}
H^i(x^t, E) &= H^i(\text{Hom}(\mathbf{K}_\bullet(x^t), E)) \\
&\cong H^i(\text{Hom}(\mathbf{K}_\bullet(x^t), \text{Hom}(R, E))) \\
&\cong H^i(\text{Hom}(\mathbf{K}_\bullet(x^t) \otimes R, E)) \\
&\cong \text{Hom}(H_i(x^t, R), E)
\end{aligned}$$

But for each t, there is an $s > t$ such that the map $H_i(x^s, R) \rightarrow H_i(x^t, R)$ is a zero homomorphism by Lemma 9.5.7 above. Hence $H_x^i(E) = \varinjlim H^i(x^t, E) \cong \varinjlim \text{Hom}(H_i(x^t, R), E) = 0$. We are now in a position to prove the following result.

Theorem 9.5.8. *Let R be a local ring, I be an ideal of R generated by the sequence $x = x_1, x_2, \ldots, x_n$, and M be an R-module. Then*

$$H_I^i(M) \cong H_x^i(M).$$

Proof. We first note that

$$\begin{aligned}
H_I^0(M) &= \varinjlim \text{Hom}(R/I^t, M) \\
&= \varinjlim \text{Hom}(R/x^t R, M) \\
&= \varinjlim H^0(x^t, M) \\
&= H_x^0(M).
\end{aligned}$$

If $0 \rightarrow M' \rightarrow M \rightarrow M'' \rightarrow 0$ is an exact sequence of R-modules, then we have an exact sequence of complexes

$$0 \rightarrow \varinjlim \text{Hom}(\mathbf{K}_\bullet(x^t), M') \rightarrow \varinjlim \text{Hom}(\mathbf{K}_\bullet(x^t), M) \rightarrow \varinjlim \text{Hom}(\mathbf{K}_\bullet(x^t), M'') \rightarrow 0.$$

Thus we get an associated long exact sequence of the cohomology groups $H_x^i(-)$. But for each $i > 0$, $H_x^i(E) = H_I^i(E) = 0$ for every injective R-module E by the above. So $H_x^i(M) \cong H_I^i(M)$ by Theorem 8.2.11.   □

Corollary 9.5.9. *Let R be local with maximal ideal $\mathfrak{m}$ and M be an R-module. If the ideal generated by $x = x_1, x_2, \ldots, x_n$ is $\mathfrak{m}$-primary, then $H^i_{\mathfrak{m}}(M) \cong H^i_x(M)$. In particular, $H^i_{\mathfrak{m}}(M) = 0$ for $i > n$.*

Proof. Let $I = xR$. Then for some $r > 0$, $\mathfrak{m}^r \subset I \subset \mathfrak{m}$ since I is $\mathfrak{m}$-primary by Proposition 2.4.25. Thus

$$H^0_I(M) = \lim_{\rightarrow} \mathrm{Hom}(R/I^t, M) = \lim_{\rightarrow} \mathrm{Hom}(R/\mathfrak{m}^t, M) = H^0_{\mathfrak{m}}(M)$$

and so $H^i_I(M) \cong H^i_{\mathfrak{m}}(M)$ as in the theorem above. So the first part of the result follows from the theorem. But $\mathbf{K_\bullet}(x)$ has length n. So $H^i_{\mathfrak{m}}(M) = H^i_x(M) = 0$ for $i > n$.   □

Corollary 9.5.10. $H^i_{\mathfrak{m}}(M) = 0$ *for all $i > \dim R$.*

Proof. Let $d = \dim R$. Then there is a sequence $x = x_1, x_2, \ldots, x_d$ of elements of R such that $\sqrt{xR} = \mathfrak{m}$ by Proposition 2.4.17 and Theorem 2.4.33. So xR is $\mathfrak{m}$-primary. Thus the result follows from the previous corollary.   □

Theorem 9.5.11. *Let $(R, \mathfrak{m}, k)$ be a local ring and M be a finitely generated R-module. Then the following are equivalent.*

(1) *M is a maximal Cohen–Macaulay module.*

(2) *$\mu_i(\mathfrak{m}, M) = 0$ for all $i < \dim R$.*

(3) *$H^i_{\mathfrak{m}}(M) = 0$ for all $i \neq \dim R$.*

Proof. (1) $\Rightarrow$ (2) is trivial since $\mathrm{depth}\, M = \mathrm{depth}\, R$.
    (2) $\Rightarrow$ (1). $\mathrm{depth}\, M \leq \dim R$ and $\mathrm{depth}\, M = \inf\{i : \mu_i(\mathfrak{m}, M) \neq 0\}$. So $\mathrm{depth}\, M = \dim R$ by (2).
    (1) $\Rightarrow$ (3). Since $\mathrm{depth}\, M = \dim R$ by assumption, $H^i_{\mathfrak{m}}(M) = 0$ for all $i < \dim R$ by Lemma 9.5.3. So (3) follows from Corollary 9.5.10 above.
    (3) $\Rightarrow$ (1). $\mathrm{depth}\, M = \inf\{i : H^i_{\mathfrak{m}}(M) \neq 0\}$ by Lemma 9.5.3. So $\mathrm{depth}\, M \geq \dim R$ by assumption. So M is maximal Cohen Macaulay.   □

Corollary 9.5.12. *R is Cohen–Macaulay if and only if $H^i_{\mathfrak{m}}(R) = 0$ for all $i \neq \dim R$.*

Corollary 9.5.13. *The following are equivalent for a local ring of Krull dimension d with residue field k.*

(1) *R is Gorenstein.*

(2) *R is Cohen–Macaulay and $H^d_{\mathfrak{m}}(R) \cong E(k)$.*

(3) *$H^i_{\mathfrak{m}}(R) = 0$ for $i \neq d$ and $H^d_{\mathfrak{m}}(R) \cong E(k)$.*

Proof. (1) $\Leftrightarrow$ (2) by Theorem 9.2.4 since $H^d_{\mathfrak{m}}(R) \cong E(k)$ means $\mu_d(\mathfrak{m}, R) = 1$.
    (2) $\Leftrightarrow$ (3) follows from Corollary 9.5.12.   □

Definition 9.5.14. If R is a local Cohen–Macaulay ring with residue field k, then a finitely generated R-module Ω is called a *dualizing module* of R if

$$\mathrm{Ext}_R^i(k, \Omega) \cong \begin{cases} 0 & \text{if } i \neq \dim R \\ k & \text{if } i = \dim R. \end{cases}$$

Remark 9.5.15. We note that a dualizing module of R is maximal Cohen–Macaulay, and if R is a Gorenstein local ring then R is a dualizing module of R by Theorem 9.2.27.

If $\dim R = 0$ then $\Omega = E(k)$ is seen to be dualizing and in fact, if Ω is dualizing then since $\mathrm{Ext}^1(k, \Omega) = 0$ for $i > 0$, Ω is injective. Since $\mathrm{Hom}(k, \Omega) \cong k$, we then get $\Omega \cong E(k)$.

Now suppose Ω is a dualizing module of R. Then for any $\mathfrak{p} \in \mathrm{Spec}\, R$, $\Omega_\mathfrak{p}$ is a dualizing module of $R_\mathfrak{p}$. Also, it is easy to check that $\hat{\Omega}$ is a dualizing module of $\hat{R}$ and $\Omega/I\Omega$ is a dualizing module of R/I if I is generated by an R-sequence. If the sequence is maximal, then $\dim R/I = 0$ and so $\Omega/I\Omega \cong E_{R/I}(k)$. It is also well known that a dualizing module of R, if it exists, is unique up to isomorphism.

We now consider the functors $T^i(-) = \mathrm{Hom}(H_\mathfrak{m}^{d-i}(-), E(k))$. Then $T^0(-)$ is a contravariant left exact functor which converts sums to products. So if we set $\Omega = T^0(R) = \mathrm{Hom}(H_\mathfrak{m}^d(R), E(k))$, then $T^0(-) \cong \mathrm{Hom}(-, \Omega)$ by Theorem 1.3.18. We are now in a position to prove the following *local duality* theorems.

Theorem 9.5.16. *Let $(R, \mathfrak{m}, k)$ be a complete local Cohen–Macaulay ring of dimension d. Then R has a dualizing module Ω such that if M is a finitely generated R-module, then*

$$\mathrm{Ext}_R^i(M, \Omega) \cong \mathrm{Hom}(H_\mathfrak{m}^{d-i}(M), E(k))$$

and

$$H_\mathfrak{m}^i(M) \cong \mathrm{Hom}(\mathrm{Ext}_R^{d-i}(M, \Omega), E(k)).$$

Proof. We prove the existence of isomorphisms for $i \geq 0$ since if $i < 0$, then $H_\mathfrak{m}^{d-i}(M) = 0$ by Corollary 9.5.10 and $\mathrm{Ext}_R^i(M, \Omega) = 0$.

Now let T^i and Ω be defined as above. If $i = 0$, then $\mathrm{Hom}(M, \Omega) \cong T^0(R) = \mathrm{Hom}(H_\mathfrak{m}^d(M), E(k))$ from the above. But every exact sequence $0 \to M' \to M \to M'' \to 0$ has an associated long exact sequence of cohomology groups $H_\mathfrak{m}^i(-)$ (see the proof of Theorem 9.5.8) and so $\{T^i\}$ is a strongly connected sequence of functors. Moreover, $T^i(R) = \mathrm{Hom}(H_\mathfrak{m}^{d-i}(R), E(k))$. So if $i \geq 1$, then $d - i < d$ and $H_\mathfrak{m}^{d-i}(R) = 0$ by Corollary 9.5.12. Hence $T^i(R) = 0$ for all $i \geq 1$. But if F is free, then $T^i(F) = 0$ for all $i \geq 1$ for T^i takes direct sums to products since $\mathrm{Ext}^i(N, -)$ commutes with arbitrary sums if N is finitely generated (see Lemma 3.1.16). Hence $T^i(-) \cong R^n(\mathrm{Hom}(-, \Omega))$ by Theorem 8.2.12. But then $\mathrm{Hom}(\mathrm{Ext}_R^{d-i}(M, \Omega), E(k)) \cong \mathrm{Hom}(\mathrm{Hom}(H_\mathfrak{m}^i(M), E(k)), E(k)) \cong H_\mathfrak{m}^i(M)$ since $H_\mathfrak{m}^i(M)$ is artinian and R is complete.

It now remains to show that Ω is a dualizing module for R. But Ω is finitely generated since $H_{\mathfrak{m}}^d(R)$ is artinian, and $H_{\mathfrak{m}}^i(R) = 0$ if $i \neq d$ by Corollary 9.5.12 since R is Cohen–Macaulay. So $\mathrm{Ext}_R^i(k, \Omega) = 0$ if $i \neq d$. But $\mathrm{Ext}_R^d(k, \Omega) \cong \mathrm{Hom}(H_{\mathfrak{m}}^0(k), E(k)) \cong \mathrm{Hom}(k, E(k)) \cong k$. Thus Ω is a dualizing module.   □

Theorem 9.5.17. *Let* $(R, \mathfrak{m}, k)$ *be a local Cohen–Macaulay ring of dimension* d *and* M *be a finitely generated* R-*module. If* R *has a dualizing module* Ω, *then*

$$\mathrm{Ext}_R^i(M, \Omega)^\wedge \cong \mathrm{Hom}_R(H_{\mathfrak{m}}^{d-i}(M), E(k))$$

and

$$H_{\mathfrak{m}}^i(M) \cong \mathrm{Hom}_R(\mathrm{Ext}_R^{d-i}(M, \Omega), E(k)).$$

Proof. By Theorem 9.5.16, Proposition 9.5.4 and the uniqueness of Ω, we have $H_{\mathfrak{m}}^i(M) \cong H_{\hat{\mathfrak{m}}}^i(\hat{M}) \cong \mathrm{Hom}_{\hat{R}}(\mathrm{Ext}_{\hat{R}}^{d-i}(\hat{M}, \hat{\Omega}), E(k)) \cong \mathrm{Hom}_R(\mathrm{Ext}_R^{d-i}(M, \Omega), E(k))$ since $\mathrm{Ext}_R^{d-i}(M, \Omega)$ is a finitely generated R-module. So we have the second isomorphisms. But

$$\mathrm{Hom}_R(H_{\mathfrak{m}}^{d-i}(M), E(k)) \cong \mathrm{Hom}_R(\mathrm{Hom}_R(\mathrm{Ext}_R^i(M, \Omega), E(k)), E(k))$$
$$\cong \mathrm{Ext}_R^i(M, \Omega) \otimes \hat{R}$$

and so we have the first isomorphisms.   □

Remark 9.5.18. By setting $M = R$ in the theorem above, we see that if Ω is a dualizing module of R, then

$$\mathrm{Hom}_R(H_{\mathfrak{m}}^d(R), E(k)) \cong {}_R\hat{\Omega}.$$

Herzog–Kunz [110] show that the converse also holds.

Corollary 9.5.19. $\mathrm{inj\,dim}\,\Omega = \mathrm{flat\,dim}\,H_{\mathfrak{m}}^d(R) = d$ *and*

$$H_{\mathfrak{m}}^i(M) \cong \mathrm{Tor}_{d-i}(M, H_{\mathfrak{m}}^d(R)).$$

Proof. If $i > d$, then $\mathrm{Ext}_R^i(M, \Omega)^\wedge = 0$ by the above theorem and so $\mathrm{Ext}_R^i(M, \Omega) = 0$ for each finitely generated R-module M. Thus $\mathrm{inj\,dim}\,\Omega \leq d$ by Proposition 8.4.4. But $\mathrm{inj\,dim}\,\Omega = d$ by Theorem 9.2.16. So $d = \mathrm{inj\,dim}\,\Omega = \mathrm{inj\,dim}\,H_{\mathfrak{m}}^d(R)^\vee = \mathrm{flat\,dim}\,H_{\mathfrak{m}}^d(R)$.
    Finally, $H_{\mathfrak{m}}^i(M) \cong \mathrm{Tor}_{d-i}^R(M, \mathrm{Hom}_R(\Omega, E(k))) \cong \mathrm{Tor}_{d-i}^R(M, H_{\mathfrak{m}}^d(R))$.   □

Corollary 9.5.20. *If* R *is a local Gorenstein ring and* M *is a finitely generated* R-*module, then*

$$\mathrm{Ext}_R^i(M, R)^\wedge \cong \mathrm{Hom}_R(H_{\mathfrak{m}}^{d-i}(M), E(k))$$

and

$$H_{\mathfrak{m}}^i(M) \cong \mathrm{Hom}(\mathrm{Ext}_R^{d-i}(M, R), E(k)).$$

Proof. As noted before, $\Omega = R$ in this case.   □

Definition 9.5.21. If $(R, \mathfrak{m}, k)$ is local, then a finitely generated R-module K is said to be a *canonical module* of R if

$$\mathrm{Hom}_R(H_{\mathfrak{m}}^d(R), E(k)) \cong {}_R\hat{K}.$$

We note that if R is complete, then $K = \mathrm{Hom}_R(H_{\mathfrak{m}}^d(R), E(k))$ is a canonical module of R. If R is Cohen–Macaulay, then K is a canonical module of R if and only if K is a dualizing module of R by Remark 9.5.18. We have the following results.

Proposition 9.5.22. *Let $(R, \mathfrak{m}, k)$ be a local ring with a canonical module. Then the following are equivalent for an integer $d \geq 1$.*

(1) R *is Cohen–Macaulay of Krull dimension d.*

(2) flat dim $H_{\mathfrak{m}}^d(R) = d$.

(3) proj dim $H_{\mathfrak{m}}^d(R) = d$.

Proof. (1) $\Rightarrow$ (2) by Corollary 9.5.19.
(2) $\Rightarrow$ (1). If flat dim $H_{\mathfrak{m}}^d(R) = d$, then flat dim$_{\hat{R}} H_{\mathfrak{m}}^d(R) = d$ since $H_{\mathfrak{m}}^d(R)$ is artinian and so is an $\hat{R}$-module naturally. Thus $H_{\mathfrak{m}}^d(R)^v$ is a nonzero finitely generated $\hat{R}$-module of finite injective dimension. So $\hat{R}$ is Cohen–Macaulay (see Strooker [171, Theorem 13.1.7]) and hence R is Cohen–Macaulay. Furthermore, dim $R = d$ by Corollary 9.5.12.
(2) $\Rightarrow$ (3). dim $R = d$ since $2 \Rightarrow 1$. So proj dim $H_{\mathfrak{m}}^d(R) \leq d$ by Corollary 8.5.28. But then proj dim $H_{\mathfrak{m}}^d(R) = d$.
(3) $\Rightarrow$ (2) is easy.   □

Proposition 9.5.23. *Let $(R, \mathfrak{m}, k)$ be Cohen–Macaulay. Then the following are equivalent for a finitely generated R-module M.*

(1) M *is maximal Cohen–Macaulay.*

(2) $\mathrm{Ext}_R^i(M, L) = 0$ *for all $i \geq 1$ and all finitely generated R-modules L of finite injective dimension.*

Furthermore, if R admits a canonical module K, then each of the above statements is equivalent to the following.

(3) $\mathrm{Ext}_R^i(M, K) = 0$ *for all $i \geq 1$.*

(4) $\mathrm{Ext}_R^i(M, \hat{K}) = 0$ *for all $i \geq 1$.*

Proof. (1) $\Leftrightarrow$ (2). If L is a finitely generated R-module with inj dim $L < \infty$, then inj dim $L = $ depth R by Corollary 9.2.17. But then depth $M + \sup\{i : \mathrm{Ext}^i(M, L) \neq 0\} = $ depth R by Theorem 9.2.16. So the result follows.

(1) $\Leftrightarrow$ (3). By the local duality (Theorem 9.5.17),

$$\mathrm{Ext}^i(M, K) \otimes_R \hat{R} \cong \mathrm{Hom}(H_{\mathfrak{m}}^{d-i}(M), E(k)).$$

So M is maximal Cohen–Macaulay if and only if $H_{\mathfrak{m}}^{d-i}(M) = 0$ for all $i > 0$ by Theorem 9.5.11, and hence if and only if $\mathrm{Ext}^i(M, K) = 0$ for all $i > 0$.

(3) $\Leftrightarrow$ (4) is trivial since $\mathrm{Ext}^i_R(M, K) \otimes_R \hat{R} \cong \mathrm{Ext}^i_R(M, \hat{K})$. $\square$

Exercises

1. Prove that if $x = x_1, x_2, \ldots, x_n$ is a sequence of elements of R, then $\mathbf{K}_{\bullet}(x) = \mathbf{K}_{\bullet}(x_1) \otimes \cdots \otimes \mathbf{K}_{\bullet}(x_n)$.

2. Let M be an R-module and $x = x_1, \ldots, x_n$ be an M-sequence. Prove that $H_p(x, M) = 0$ for $p > 0$ and $H_0(x, M) = M/xM$.

3. Let M be an R-module and $x = x_1 \ldots, x_n$ be a sequence of elements of R. Prove that

   a) $H_0(x, M) = M/xM$.

   b) $H^0(x, M) = \mathrm{Hom}(R/xR, M) = \{y \in M : x_i y = 0 \text{ for } i = 1, \ldots, n\}$.

   c) $H_p(x, M)$ is isomorphic to $H^{n-p}(x, M)$ for all $p \geq 0$.

4. An inverse system is said to be *essentially zero* if for each $i > 0$, there exists a $j > i$ such that $M_j \to M_i$ is zero. Let $0 \to ((M'_i), (f'_{ij})) \to ((M_i), (f_{ij})) \to ((M''_i), (f''_{ij})) \to 0$ be an exact sequence of inverse systems. Prove that if $((M'_i), (f'_{ij}))$ and $((M''_i), (f''_{ij}))$ are essentially zero, then so is $((M_i), (f_{ij}))$.

5. Let Ω be a dualizing module of R. Prove that

   a) $\Omega_{\mathfrak{p}}$ is a dualizing module of $R_{\mathfrak{p}}$ for each $\mathfrak{p} \in \mathrm{Spec}\, R$.

   b) $\hat{\Omega}$ is a dualizing module of $\hat{R}$.

   c) If I is generated by an R-sequence, then $\Omega/I\Omega$ is a dualizing module of R/I.

6. (Herzog–Kunz [110] or Bruns–Herzog [32]) Prove that a dualizing module of R, if it exists, is unique up to isomorphism.

7. Let R be a commutative noetherian ring and $k(\mathfrak{p})$ denote the quotient ring of $R/\mathfrak{p}$. Prove that R is Gorenstein if and only if

   $$\mathrm{flat\,dim}_{R_{\mathfrak{p}}} E(k(\mathfrak{p})) = \mathrm{proj\,dim}_{R_{\mathfrak{p}}} E(k(\mathfrak{p})) = \mathrm{ht}\,\mathfrak{p}$$

   for all $\mathfrak{p} \in \mathrm{Spec}\, R$.

Chapter 10

Gorenstein Modules

Auslander introduced the notion of G-dimension of finitely generated modules over Cohen–Macaulay rings. It seems appropriate to call G-dimension 0 modules Gorenstein projective. Our aim in this chapter is to define and study Gorenstein injective, Gorenstein projective, and Gorenstein flat modules. The way we define Gorenstein injective modules can be dualized and allow us to define Gorenstein projective modules (that is, modules of G-dimension 0) whether the modules are finitely generated or not. These notions generalize the usual injective, projective, and flat modules.

10.1   Gorenstein injective modules

Definition 10.1.1. A module N is said to be *Gorenstein injective* if there exists a $\mathrm{Hom}(\mathcal{I}nj, -)$ exact exact sequence

$$\cdots \to E_1 \to E_0 \to E^0 \to E^1 \to \cdots$$

of injective modules such that $N = \mathrm{Ker}(E^0 \to E^1)$. We note that in the above definition, the complex $\cdots \to E_1 \to E_0 \to E^0 \to E^1 \to \cdots$ is a complete $\mathcal{I}nj$-resolution of N. Moreover, if N is a Gorenstein injective R-module, then $\mathrm{Ext}^i(E, N) = 0$ for all $i \geq 1$ and all injective R-modules E, or equivalently, every right $\mathcal{I}nj$-resolution of N is a left $\mathcal{I}nj$-resolution. As a consequence, we have the following result.

Proposition 10.1.2. *The injective dimension of a Gorenstein injective R-module is either zero or infinite.*

Proof. Suppose $0 \to N \to E^0 \to E^1 \to \cdots \to E^{n-1} \to E^n \to 0$ is a right $\mathcal{I}nj$-resolution of a Gorenstein injective R-module N. Then it is also a left $\mathcal{I}nj$-resolution of E^n by the remarks above. But then the resolution is split exact since $E^i \in \mathcal{I}nj$ for each i. Thus N is injective. $\qquad\square$

If R is noetherian, every R-module N has a (minimal) complete $\mathcal{I}nj$-resolution by Theorem 5.4.1. Furthermore, in this case we can compute left derived functors $\mathrm{Ext}_i(M, N)$ of $\mathrm{Hom}(M, N)$ by using a right $\mathcal{I}nj$-resolution of M or a left $\mathcal{I}nj$-resolution of N (see Example 8.3.5). Hence we get the following characterization of

Gorenstein injective modules where $\text{Ext}^i(M, N)$ denote the standard derived functors of Example 8.3.1.

Proposition 10.1.3. *Let R be a noetherian ring. Then the following are equivalent for an R-module N.*

(1) *N is Gorenstein injective.*

(2) *$\text{Ext}_i(Q, N) = \text{Ext}^i(Q, N) = 0$ for $i \geq 1$ and $\overline{\text{Ext}}_0(Q, N) = \overline{\text{Ext}}^0(Q, N) = 0$ for all projective or injective R-modules Q.*

(3) *$\text{Ext}_i(Q, N) = \text{Ext}^i(Q, N) = 0$ for $i \geq 1$ and $\overline{\text{Ext}}_0(Q, N) = \overline{\text{Ext}}^0(Q, N) = 0$ for all modules Q of finite projective or injective dimension.*

(4) *Every complete $\mathcal{I}nj$-resolution of N is exact and $\text{Hom}(\mathcal{I}nj, -)$ exact.*

(5) *Every left $\mathcal{I}nj$-resolution of N is exact and $\text{Ext}^i(E, N) = 0$ for all $i \geq 1$ and any injective R-module E.*

(6) *The minimal complete $\mathcal{I}nj$-resolution of N is exact and $\text{Hom}(\mathcal{I}nj, -)$ exact.*

(7) *The minimal left $\mathcal{I}nj$-resolution of N is exact and $\text{Ext}^i(E, N) = 0$ for $i \geq 1$ and any injective R-module E.*

Proof. (1) $\Rightarrow$ (2). If N is Gorenstein injective, let $\cdots \rightarrow E_1 \rightarrow E_0 \rightarrow E^0 \rightarrow E^1 \rightarrow \cdots$ be as in Definition 10.1.1 above. If Q is projective, then the homology groups vanish since the complex is exact. If Q is injective, then they vanish by definition of Gorenstein injective.

(2) $\Rightarrow$ (3). This follows by induction on the dimension using the extended long exact sequences in part (1) of Theorem 8.2.7.

(2) $\Rightarrow$ (4). Let $\cdots \rightarrow E_1 \rightarrow E_0 \rightarrow E^0 \rightarrow E^1 \rightarrow \cdots$ be a complete $\mathcal{I}nj$-resolution of N. Then the extension groups in the assumption can be computed by applying $\text{Hom}(Q, -)$ and computing homology (see Proposition 8.2.9). If we set $Q = R$, we get that the complex is exact. If Q is injective and we apply $\text{Hom}(Q, -)$, we again get an exact sequence since the extension groups vanish and so (4) follows.

(3) $\Rightarrow$ (2); (4) $\Rightarrow$ (5) and (6); (5) $\Rightarrow$ (7); and (6) $\Rightarrow$ (7) are trivial.

(7) $\Rightarrow$ (1) follows from the definitions of left $\mathcal{I}nj$-resolution and Gorenstein injective.   □

Theorem 10.1.4. *Let R be noetherian and $0 \rightarrow N' \rightarrow N \rightarrow N'' \rightarrow 0$ be an exact sequence of R-modules. If N', N'' are Gorenstein injective, then so is N. If N', N are Gorenstein injective, then so is N''. If N and N'' are Gorenstein injective, then N' is Gorenstein injective if and only if $\text{Ext}^1(E, N') = 0$ for all injective R-modules E.*

Proof. If N' is Gorenstein injective, then $\text{Ext}^1(E, N') = 0$ for all injectives E. But $\text{Ext}^1(E, N') = 0$ implies that $0 \rightarrow \text{Hom}(E, N') \rightarrow \text{Hom}(E, N) \rightarrow \text{Hom}(E, N'') \rightarrow 0$ is exact. So we get the extended long exact sequence of part (1) of Theorem 8.2.7. Hence if any two of N', N or N'' are Gorenstein injective, then so is the third.   □

Remark 10.1.5. We note that if R is noetherian and N is Gorenstein injective, then any complete $\mathcal{I}nj$-resolution $\cdots \to E_1 \to E_0 \to E^0 \to E^1 \to \cdots$ of N is exact and each of $K^i = \mathrm{Ker}(E^i \to E^{i+1})$, $K_{i+2} = \mathrm{Ker}(E_{i+1} \to E_i)$ for $i \geq 0$, $K_1 = \mathrm{Ker}(E_0 \to E^0)$ is Gorenstein injective by Proposition 10.1.3. In particular, E^0/N and $\mathrm{Ker}(E_0 \to N)$ are Gorenstein injective and $E_0 \to N$ is surjective. So we have the following result recalling that a module is said to be *reduced* if it has no nonzero injective submodules.

Proposition 10.1.6. *Let R be noetherian. If N is a reduced Gorenstein injective R-module and $E \to N$ is its injective cover, then $E \to N$ is surjective, $K = \mathrm{Ker}(E \to N)$ is reduced and Gorenstein injective and $K \subset E$ is an injective envelope.*

Proof. It only remains to argue that $K \subset E$ is essential. But if $E' \subset E$ is injective and $K \cap E' = 0$, then E' is isomorphic to a submodule of $E/K \cong N$. So $E' = 0$ since N is reduced. $\square$

Corollary 10.1.7. *If N is a nonzero reduced Gorenstein injective R-module, then N has infinite injective and projective dimensions.*

Proof. The first part is a special case of Proposition 10.1.2 or simply note that if $K^i = 0$, then $N = 0$ and so N has infinite injective dimension.

Now let $0 \to K_{i+1} \to E_i(N) \to E_{i-1}(N) \to \cdots \to E_0(N) \to N \to 0$ be the minimal left $\mathcal{I}nj$-resolution of N. Then this is also a minimal right $\mathcal{I}nj$-resolution of K_{i+1} for $i \geq 0$. So $\mathrm{Ext}^1(N, K_1) \cong \mathrm{Ext}^{i+1}(N, K_{i+1})$. But $0 \to K_1 \to E_0 \to N \to 0$ does not split and so $\mathrm{Ext}^1(N, K_1) \neq 0$. Thus $\mathrm{Ext}^{i+1}(N, K_{i+1}) \neq 0$ for any $i \geq 1$. Hence N has infinite projective dimension. $\square$

Proposition 10.1.8. *Let R be n-Gorenstein and M be an R-module. If $0 \to M \to E^0 \to E^1 \to \cdots \to E^{i-1} \to C^i \to 0$ is a right $\mathcal{I}nj$-resolution of M, then $E^{i-1} \to C^i$ is an injective precover for each $i \geq n+1$. If furthermore the resolution is minimal, then C^i is reduced for each $i \geq n+1$, and $E^{i-1} \to C^i$ is an injective cover for all $i > n+1$.*

Proof. $E^n \to E^{n+1} \to \cdots$ is a left $\mathcal{I}nj$-resolution by Theorem 9.1.11 and so $E^{i-1} \to C^i$ is an injective precover for each $i \geq n+1$.

Now suppose the resolution is minimal. If E is an injective submodule of C^{n+1}, then we have a factorization $E \to E^n \to C^{n+1}$ of the inclusion $E \subset C^{n+1}$ since $E^n \to C^{n+1}$ is an injective precover from the above. If $E \neq 0$, then this would contradict the minimality of the resolution. So C^{n+1} is reduced.

We now note that $C^{n+1} = \mathrm{Ker}(E^{n+1} \to E^{n+2})$ and $E^{n+1} \to C^{n+2}$ is an injective precover. So E^{n+1} has a summand E (say $E^{n+1} = E \oplus E'$) with $E \to C^{n+2}$ an injective cover and E' in the kernel of $E^{n+1} \to C^{n+2}$. But then $E' \subset C^{n+1}$. So $E' = 0$ since C^{n+1} is reduced. Thus $E^{n+1} \to C^{n+2}$ is a cover. The same argument works for $E^i \to C^{i+1}$ when $i \geq n+2$. $\square$

Corollary 10.1.9. *Let M, N be R-modules and*

$$0 \to M \to E^0 \to E^1 \to \cdots \to E^i \to C^{i+1} \to 0$$

and

$$0 \to N \to H^0 \to H^1 \to \cdots \to H^i \to D^{i+1} \to 0$$

be right $\mathcal{I}nj$-resolutions of M and N respectively where $i \geq n$. If $f : C^{i+1} \to D^{i+1}$ is any homomorphism, then there exist maps $E^n \to H^n, \ldots, E^i \to H^i$ such that the diagram

$$
\begin{array}{ccccccccc}
E^n & \longrightarrow & E^{n+1} & \longrightarrow & \cdots & \longrightarrow & E^i & \longrightarrow & C^{i+1} & \longrightarrow & 0 \\
\downarrow & & \downarrow & & & & \downarrow & & \downarrow {\scriptstyle f} & & \\
H^n & \longrightarrow & H^{n+1} & \longrightarrow & \cdots & \longrightarrow & H^i & \longrightarrow & D^{i+1} & \longrightarrow & 0
\end{array}
$$

is a commutative diagram.

Furthermore, if the resolutions are minimal and f is an isomorphism, then each of the maps $E^{n+1} \to H^{n+1}, \ldots, H^i \to E^i$ are also isomorphisms.

Proof. The result follows from the proposition above and definitions of precovers and covers.   □

Corollary 10.1.10. *If the minimal right $\mathcal{I}nj$-resolution $0 \to M \to E^0(M) \to E^1(M) \to \cdots$ is eventually periodic, then the complex $E^{n+1}(M) \to E^{n+2}(M) \to \cdots$ is also periodic.*

Proof. Any isomorphism $C^i \to C^{i+m}$ for $i > n + 1$ and $m > 0$ induces an isomorphism $C^{n+1} \to C^{n+1+m}$ by the corollary above.   □

We note that Corollary 10.1.10 above is similar to a result of Eisenbud [46] concerning minimal projective resolutions over hypersurface rings.

Proposition 10.1.11. *Let R be noetherian and* inj dim $_R R = n$. *If*

$$\cdots \to E_2 \to E_1 \to E_0 \to M \to 0$$

is a left $\mathcal{I}nj$-resolution of an R-module M and $C_i = \mathrm{Coker}(E_{i+1} \to E_i)$, then $C_i \to E_{i-1}$ is an injection for $i \geq n - 1$. If furthermore, the resolution is minimal, then C_i is reduced for each $i \geq n$ and $C_i \to E_{i-1}$ is an injective envelope for $i \geq n+1$.

Proof. $\cdots \to E_{n+1} \to E_n \to E_{n-1} \to E_{n-2}$ is exact by Theorem 8.4.36. So $C_i \to E_{i-1}$ is an injection for each $i \geq n - 1$. Now suppose $i \geq n$ and the resolution is minimal. Then $E_{i-1} \to C_{i-1}$ is an injective cover with kernel C_i. But the kernel of an injective cover is reduced. So C_i is reduced for $i \geq n$.

Now if $i > n$ and E is an injective submodule of E_{i-1} such that $E \cap C_i = 0$, then $E_{i-1} \to C_{i-1}$ maps E isomorphically into C_{i-1}. But C_{i-1} is reduced by the above. So $E = 0$ and hence $C_i \hookrightarrow E_{i-1}$ is an injective envelope.   □

Corollary 10.1.12. *Let M, N be R-modules and*

$$0 \to C_{i+1} \to E_i(M) \to \cdots \to E_0(M) \to M \to 0$$

and

$$0 \to D_{i+1} \to E_i(N) \to \cdots \to E_0(N) \to N \to 0$$

be left $\mathcal{I}nj$-resolutions of M and N respectively with $i \geq n - 2$ and $f : C_{i+1} \to D_{i+1}$ be any homomorphism, then there exists a commutative diagram

$$
\begin{array}{ccccccc}
0 & \longrightarrow & C_{i+1} & \longrightarrow & E_i(M) & \longrightarrow \cdots \longrightarrow & E_{n-2}(M) \\
& & \downarrow{\scriptstyle f} & & \downarrow & & \downarrow \\
0 & \longrightarrow & D_{i+1} & \longrightarrow & E_i(N) & \longrightarrow \cdots \longrightarrow & E_{n-2}(N).
\end{array}
$$

Furthermore, if the resolutions are minimal and f is an isomorphism, then so are each of the maps $E_i(M) \to E_i(N), \ldots, E_n(M) \to E_n(N)$.

Proof. This follows from the proposition above and the definition of an injective envelope.   □

We are now in a position to show that there is an abundant supply of Gorenstein injective modules.

Theorem 10.1.13. *Suppose R is n-Gorenstein and M is an R-module. Then*

(1) *If $0 \to M \to E^0 \to E^1 \to \cdots$ is a right $\mathcal{I}nj$-resolution of M and $C^i = \mathrm{Ker}(E^i \to E^{i+1})$ for $i \geq 0$, then C^i is Gorenstein injective for $i \geq n$, and is reduced for $i \geq n + 1$ if the resolution is minimal.*

(2) *If $\cdots \to E_1 \to E_0 \to M \to 0$ is a left $\mathcal{I}nj$-resolution of M and $C_i = \mathrm{Coker}(E_{i+1} \to E_i)$ for $i \geq 0$, then C_i is Gorenstein injective for $i \geq n - 1$, and is reduced for $i \geq n$ if the resolution is minimal.*

Proof. By Theorem 9.1.11, if $i \geq n$, then any right $\mathcal{I}nj$-resolution $0 \to C^i \to E^i \to E^{i+1} \to \cdots$ of C^i is also a left $\mathcal{I}nj$-resolution. But then any left $\mathcal{I}nj$-resolution of C^i $\cdots \to E_1 \to E_0 \to C^i \to 0$ is exact by Theorem 8.4.36. Pasting these sequences together, we get a $\mathrm{Hom}(\mathcal{I}nj, -)$ exact exact sequence

$$\cdots \to E_1 \to E_0 \to E^i \to E^{i+1} \to \cdots$$

of injective modules. So C^i is Gorenstein injective. If $i \geq n + 1$ and the right $\mathcal{I}nj$-resolution of M is minimal, then C^n is reduced by Proposition 10.1.8.

The proof of (2) follows similarly from Theorem 9.1.11 and Proposition 10.1.11.□

Remark 10.1.14. If G is a finite group, then it is easy to see that the group ring $\mathbb{Z}G$ is 1-Gorenstein for

$$0 \to \mathbb{Z}G \to \mathbb{Q}G \to \mathbb{Q}G/\mathbb{Z}G \to 0$$

is an exact sequence of left and right $\mathbb{Z}G$-modules. But $\mathbb{Q}G \cong \mathbb{Q} \otimes_\mathbb{Z} \mathbb{Z}G \cong \text{Hom}_\mathbb{Z}(\mathbb{Z}G, \mathbb{Q})$ is an injective $\mathbb{Z}G$-module (see Brown [31]). Similarly $\mathbb{Q}G/\mathbb{Z}G$ is injective.

From this we can argue that a $\mathbb{Z}G$-module is Gorenstein injective if and only if it is a divisible $\mathbb{Z}$-module. The condition is necessary since every injective $\mathbb{Z}G$-module is divisible and M is a quotient of an injective module. Conversely, if M is divisible, then $\text{Hom}_\mathbb{Z}(\mathbb{Z}G, M)$ is injective. But $\text{Hom}_\mathbb{Z}(\mathbb{Z}G, M) \cong \mathbb{Z}G \otimes_\mathbb{Z} M$ and there is a surjection $\mathbb{Z}G \otimes_\mathbb{Z} M \to M$. So M is Gorenstein injective by Theorem 10.1.13 above.

Proposition 10.1.15. *The following are equivalent for an R-module M.*

(1) $\text{Ext}^1(M, N) = 0$ *for all Gorenstein injective modules N.*

(2) $\text{Ext}^i(M, N) = 0$ *for all $i \geq 1$ and all Gorenstein injective modules N.*

(3) $\overline{\text{Ext}}^0(M, N) = 0$ *for all Gorenstein injective modules N.*

(4) $\overline{\text{Ext}}_0(M, N) = 0$ *for all Gorenstein injective modules N.*

(5) $\text{Ext}_1(M, N) = 0$ *for all Gorenstein injective modules N.*

(6) $\text{Ext}_i(M, N) = 0$ *for all $i \geq 1$ and all Gorenstein injective modules N.*

If furthermore R is n-Gorenstein, then each of the above statements is equivalent to

(7) *M has finite injective dimension.*

(8) *M has finite projective dimension.*

(9) *M has finite injective dimension at most n.*

(10) *M has finite projective dimension at most n.*

Proof. (1) $\Rightarrow$ (2). We consider the exact sequence $0 \to N \to E^0(N) \to N'' \to 0$. Then $\text{Ext}^1(M, N'') \cong \text{Ext}^2(M, N)$. But $\text{Ext}^1(M, N'') = 0$ by assumption since N'' is Gorenstein injective. So $\text{Ext}^2(M, N) = 0$. So (2) follows by induction.

(2) $\Rightarrow$ (1) and (6) $\Rightarrow$ (5) are trivial.

(1) $\Leftrightarrow$ (3). We now consider the exact sequences $0 \to N \to E^0(N) \to N'' \to 0$ and $0 \to N' \to E_0(N) \to N \to 0$. Then $\overline{\text{Ext}}^0(M, N'') \cong \text{Ext}^1(M, N)$ and $\overline{\text{Ext}}^0(M, N) \cong \text{Ext}^1(M, N')$ by part (1) of Theorem 8.2.7. So the result follows since N' and N'' are Gorenstein injective.

(3) $\Leftrightarrow$ (4) and (3) $\Leftrightarrow$ (5) follow similarly from the extended long exact sequence in Theorem 8.2.7.

(5) $\Rightarrow$ (6). Using the exact sequence $0 \to N' \to E_0(N) \to N \to 0$ and Theorem 8.2.7, we get that $\text{Ext}_2(M, N) \cong \text{Ext}_1(M, N')$ and so $\text{Ext}_2(M, N) = 0$. The result then follows by induction.

$(6) \Rightarrow (9)$. Let $0 \to M \to E^0(M) \to \cdots \to E^{n-1}(M) \to C^n \to 0$ be the minimal right $\mathcal{I}nj$-resolution of M. Then C^n is Gorenstein injective by Theorem 10.1.13. So $\operatorname{Ext}_n(M, C^n) = 0$. But then $\operatorname{Hom}(E^{n+1}(M), C^n) \to \operatorname{Hom}(E^n(M), C^n) \to \operatorname{Hom}(E^{n-1}(M), C^n)$ is exact and so C^n is a retract of $E^n(M)$. Thus C^n is injective.

(7), (8), (9), and (10) are equivalent by Theorem 9.1.10 and $(8) \Rightarrow (1)$ by Proposition 10.1.3. $\qquad\square$

Exercises

1. Prove that (2) implies (3) in Proposition 10.1.3.

2. Let N be a reduced Gorenstein injective left R-module. Prove that

   a) the kernels K_i, K^i in the complete minimal $\mathcal{I}nj$-resolution of N are reduced and Gorenstein injective.

   b) for any i, $K_i = 0$ if and only if $N = 0$ if and only if $K^i = 0$.

3. Prove Corollary 10.1.9.

4. Prove Corollary 10.1.12.

5. Prove part 2 of Theorem 10.1.13.

6. Let R be an n-Gorenstein commutative ring. Prove that if G is a Gorenstein injective R-module, then $S^{-1}G$ is a Gorenstein injective $S^{-1}R$-module for each multiplicative subset S of R.

7. Let R be a commutative noetherian ring of finite Krull dimension. Prove that an R-module M is Gorenstein injective if and only if $\operatorname{Hom}(F, M)$ is Gorenstein injective for all flat R-modules F.
   Hint: proj dim $F < \infty$ in this case.

8. An R-module N is said to be *mock finitely generated* if for any finitely generated R-module M, each of $\operatorname{Ext}^i(M, N), \operatorname{Ext}_i(M, N), \overline{\operatorname{Ext}}_0^R(M, N)$, and $\overline{\operatorname{Ext}}_R^0(M, N)$ are finitely generated R-modules where $\operatorname{Ext}_i(M, N)$ are the left derived functors obtained by using a left $\mathcal{I}nj$-resolution of N or a right $\mathcal{I}nj$-resolution of M. Prove the following

   a) If $0 \to N' \to N \to N'' \to 0$ is a $\operatorname{Hom}(\mathcal{I}nj, -)$ exact exact sequence of R-modules, then if any two of N', N, N'' are mock finitely generated then so is the third.

   b) If N is mock finitely generated, and C is a cosyzygy of a finitely generated R-module, then $\operatorname{Ext}_i^R(C, N)$ is finitely generated for all $i \geq 1$.

   c) Let R be a commutative noetherian ring. Then every finitely generated R-module is mock finitely generated (see Enochs–Jenda [68, Proposition 3.2]).

9. Let R be a Gorenstein commutative ring, M a Gorenstein injective R-module, and C a cosyzygy of a finitely generated R-module. Prove that

   a) $\operatorname{Ext}_i^R(C, M)_{\mathfrak{p}} \cong \operatorname{Ext}_i^{R_{\mathfrak{p}}}(C_{\mathfrak{p}}, M_{\mathfrak{p}})$ for all $i \geq 1$ for all $\mathfrak{p} \in \operatorname{Spec} R$.

   b) $\operatorname{Ext}_R^i(C, M)_{\mathfrak{p}} \cong \operatorname{Ext}_{R_{\mathfrak{p}}}^i(C_{\mathfrak{p}}, M_{\mathfrak{p}})$ for all $i \geq 1$ for all $\mathfrak{p} \in \operatorname{Spec} R$.

10. Let R and M be as in the previous problem, and N be a finitely generated R-module. Prove that $\operatorname{Ext}_i^R(N, M)_{\mathfrak{p}} \cong \operatorname{Ext}_i^{R_{\mathfrak{p}}}(N_{\mathfrak{p}}, M_{\mathfrak{p}})$ for all $i \geq 0$ and all $\mathfrak{p} \in \operatorname{Spec} R$.

11. Let R be a local ring with maximal ideal $\mathfrak{m}$ and residue field k. Let $\cdots \to E_1 \to E_0 \to M \to 0$ be the minimal left $\mathcal{I}nj$-resolution of M and $v_i(\mathfrak{p}, M)$ denote the number of components of E_i that are isomorphic to $E(R/\mathfrak{p})$ where $\mathfrak{p} \in \operatorname{Spec} R$. Prove that

    a) $v_i(\mathfrak{m}, M) = \dim_k \operatorname{Ext}_i^R(k, M)$

    b) If R is Gorenstein and M is a reduced Gorenstein injective R-module, then for each $\mathfrak{p} \in \operatorname{Spec} R$,

$$v_i(\mathfrak{p}, M) = \dim_{k(\mathfrak{p})} \operatorname{Ext}_i^{R_{\mathfrak{p}}}(k(\mathfrak{p}), M_{\mathfrak{p}}) = \dim_{k(\mathfrak{p})} \operatorname{Ext}_i^R(R/\mathfrak{p}, M)_{\mathfrak{p}}.$$

In particular, if M is mock finitely generated, then $v_i(\mathfrak{p}, M) < \infty$ for all i.

10.2 Gorenstein projective modules

We now dualize the notion of Gorenstein injective modules introduced in the previous section.

Definition 10.2.1. A module M is said to be *Gorenstein projective* if there is a $\operatorname{Hom}(-, \mathcal{P}roj)$ exact exact sequence

$$\cdots \to P_1 \to P_0 \to P^0 \to P^1 \to \cdots$$

of projective modules such that $M = \operatorname{Ker}(P^0 \to P^1)$.

Remark 10.2.2. The complex above is a complete $\mathcal{P}roj$-resolution of M. We note that if M is Gorenstein projective, then $\operatorname{Ext}^i(M, P) = 0$ for all $i \geq 1$ and all projective R-modules P and so by induction, $\operatorname{Ext}^i(M, L) = 0$ for all $i \geq 1$ and all R-modules L of finite projective dimension. In particular, every left $\mathcal{P}roj$-resolution of M is $\operatorname{Hom}(-, \mathcal{P}roj)$ exact. So we have a dual result to Proposition 10.1.2.

Proposition 10.2.3. *The projective dimension of a Gorenstein projective module is either zero or infinite.*

We will be needing the following two results the first of which is a generalization of Lemma 3.1.16.

Lemma 10.2.4. *Let R be left coherent, M be a finitely presented R-module, and $\lim\limits_{\rightarrow} N_j$ be a directed limit of R-modules. Then*

$$\mathrm{Ext}^i_R(M, \lim\limits_{\rightarrow} N_j) \cong \lim\limits_{\rightarrow} \mathrm{Ext}^i_R(M, N_j) \text{ for all } i \geq 0.$$

Proof. This follows as in the proof of Lemma 3.1.16 by Remark 2.3.12.   □

Lemma 10.2.5. *If R is left coherent and M is a finitely presented R-module, then a complex $0 \rightarrow M \rightarrow P^0 \rightarrow P^1 \rightarrow \cdots$ with P^i finitely generated and projective is a right $\mathcal{F}lat$-resolution if and only if the dual complex $\cdots \rightarrow P^{1*} \rightarrow P^{0*} \rightarrow M^* \rightarrow 0$ is exact.*

Proof. If F is a flat R-module, then we can write $F = \lim\limits_{\rightarrow} F_j$ with each F_j finitely generated and free. So if the dual sequence is exact, then $\cdots \rightarrow \mathrm{Hom}(P^1, F_j) \rightarrow \mathrm{Hom}(P^0, F_j) \rightarrow \mathrm{Hom}(M, F_j) \rightarrow 0$ is exact. But $\mathrm{Hom}(N, -)$ commutes with direct limits when N is finitely presented by the previous lemma and so the claim follows since the direct limit functor preserves exactness. The converse is trivial.   □

Although modules M always have projective precovers, they many not have projective preenvelopes. However, if R is left coherent and M is a finitely presented right R-module, then M has a finitely generated projective preenvelope. Thus we can construct a right $\mathcal{P}roj_{fg}$-resolution of M. So we can compute left derived functors $\mathrm{Ext}_n(M, N)$ by using a right $\mathcal{P}roj_{fg}$-resolution of M or a flat resolution of N (see Example 8.3.11).

If furthermore R is right coherent, then M also has a left $\mathcal{P}roj_{fg}$-resolution (see Example 8.3.3) and thus in this case the module M has a complete $\mathcal{P}roj_{fg}$-resolution.

We are now in a position to state the following result which is dual to Proposition 10.1.3.

Proposition 10.2.6. *Let R be left coherent and M be a finitely presented right R-module. Then the following are equivalent.*

(1) *M is Gorenstein projective.*

(2) *$\mathrm{Ext}_i(M, Q) = \mathrm{Ext}^i(M, Q) = 0$ for $i \geq 1$ and $\overline{\mathrm{Ext}}_0(M, Q) = \overline{\mathrm{Ext}}^0(M, Q) = 0$ for all projective or injective R-modules Q.*

(3) *$\mathrm{Ext}_i(M, Q) = \mathrm{Ext}^i(M, Q) = 0$ for $i \geq 1$ and $\overline{\mathrm{Ext}}^0(M, Q) = \overline{\mathrm{Ext}}_0(M, Q) = 0$ for all modules Q of finite projective or injective dimension.*

(4) *Every right $\mathcal{P}roj_{fg}$-resolution of M is exact and $\mathrm{Ext}^i(M, P) = 0$ for all $i \geq 1$ and all projective R-modules P.*

If furthermore R is right coherent, then each of the above statements are equivalent to the following

(5) *Every complete* $\mathcal{P}roj_{fg}$*-resolution of M is exact and* $\mathrm{Hom}(-, \mathcal{P}roj)$ *exact.*

(6) *There exists an exact sequence* $\cdots \to P_1 \to P_0 \to P^0 \to P^1 \to \cdots$ *of projective modules such that* $M = \mathrm{Ker}(P_0 \to P^0)$ *and* $\mathrm{Hom}(-, R)$ *leaves the sequence exact.*

(7) $\mathrm{Ext}_i(M, R) = \mathrm{Ext}^i(M, R) = 0$ *for* $i \geq 1$ *and* $\overline{\mathrm{Ext}_0}(M, R) = \overline{\mathrm{Ext}}^0(M, R) = 0$.

(8) *Every complete* $\mathcal{P}roj_{fg}$*-resolution of M is exact and remains exact when* $\mathrm{Hom}(-, R)$ *is applied to it.*

(9) *Every right* $\mathcal{P}roj_{fg}$*-resolution of M is exact and* $\mathrm{Ext}^i(M, R) = 0$ *for all* $i \geq 1$.

(10) *M is reflexive and* $\mathrm{Ext}^i(M, R) = \mathrm{Ext}^i(M^*, R) = 0$ *for all* $i > 0$.

(11) *Every complete* $\mathcal{P}roj_{fg}$*-resolution of M is exact and* $\mathrm{Hom}(-, \mathcal{F}lat)$ *exact.*

Proof. The equivalence of (1) through (4) and (5) through (9) follow as in Proposition 10.1.3.

(4) $\Rightarrow$ (9) and (11) $\Rightarrow$ (8) are trivial.

(9) $\Rightarrow$ (4). Let P be a projective R-module. Then P is a direct summand of a free R-module, say $F = P \oplus P'$. So $\mathrm{Ext}^i_R(M, F) \cong \mathrm{Ext}^i_R(M, P) \oplus \mathrm{Ext}^i_R(M, P')$. But $\mathrm{Ext}^i_R(M, F) \cong \varinjlim \mathrm{Ext}^i_R(M, R)$ by Lemma 10.2.4 above. So $\mathrm{Ext}^i_R(M, P) = 0$ and thus we are done.

(9) $\Leftrightarrow$ (10). If $0 \to M \to P^0 \to P^1 \to \cdots$ is a right $\mathcal{P}roj_{fg}$-resolution of M, then it is exact and so $0 \to M^{**} \to P^{0**} \to P^{1**} \to \cdots$ is also exact since finitely generated projective modules are reflexive. In particular, M is reflexive and $\mathrm{Ext}^i(M^*, R) = 0$ for all $i > 0$. Conversely, if $\mathrm{Ext}^i(M^*, R) = 0$ and M is reflexive, then $0 \to M \to P^0 \to P^1 \to \cdots$ is exact.

(8) $\Rightarrow$ (11). Let $\cdots \to P_1 \to P_0 \to P^0 \to P^1 \to \cdots$ be a complete $\mathcal{P}roj_{fg}$-resolution of M. Then $\cdots \to P^{1*} \to P^{0*} \to P^*_0 \to P^*_1 \to \cdots$ is exact and so sequences $\cdots \to P_1 \to P_0$ and $P^0 \to P^1 \to \cdots$ are right $\mathcal{F}lat$-resolutions by Lemma 10.2.5 above. $\qquad\square$

Corollary 10.2.7. *Let R be a local Cohen–Macaulay ring. Then every finitely generated Gorenstein projective R-module is maximal Cohen–Macaulay.*

Proof. This follows from the proposition above and Proposition 9.5.23. $\qquad\square$

Theorem 10.2.8. *Let R be left coherent and* $0 \to M' \to M \to M'' \to 0$ *be an exact sequence of finitely presented right R-modules. If* M', M'' *are Gorenstein projective, then so is M. If* M, M'' *are Gorenstein projective, then so is* M'. *If* M, M' *are Gorenstein projective, then* M'' *is Gorenstein projective if and only if* $\mathrm{Ext}^1(M'', P) = 0$ *for all finitely generated projective R-modules P.*

Proof. $0 \to \operatorname{Hom}(M'', P) \to \operatorname{Hom}(M, P) \to \operatorname{Hom}(M', P) \to 0$ is exact if $\operatorname{Ext}^1(M'', P) = 0$. So we have the extended long exact sequence of part (2) of Theorem 8.2.7 since $\mathcal{P}roj_{fg}$ is a preenveloping class (see Example 8.3.10). Thus if any two of M', M, M'' are Gorenstein projective, then so is the third. □

Remark 10.2.9. There are results concerning Gorenstein projective modules analogous to those of Gorenstein injective modules.

Even though finitely presented right R-modules over left coherent rings R have projective precovers and preenvelopes, they may fail to have covers and envelopes. However, if they do have covers and envelopes, for example when R is a local ring (see Theorem 5.3.3 and Proposition 6.6.8), then all the results in Section 7.1 have counterparts for Gorenstein projective modules where in this setting *reduced* means no nonzero projective summands. We now summarize some of these results below omitting obvious corollaries.

We first note that if R is left and right coherent and M is a finitely presented Gorenstein projective right R-module, then any complete $\mathcal{P}roj_{fg}$-resolution $\cdots \to P_1 \to P_0 \to P^0 \to P^1 \to \cdots$ of M is exact and each $C^i = \operatorname{Ker}(P^i \to P^{i+1})$, $C_{i+2} = \operatorname{Ker}(P_{i+1} \to P_i)$ for $i \geq 0$, $C_1 = \operatorname{Ker}(P_0 \to P^0)$ are Gorenstein projective by Proposition 10.2.6. In particular, P^0/M and $\operatorname{Ker}(P_0 \to M)$ are Gorenstein projective and $M \to P^0$ is a monomorphism. So we have the following result.

Proposition 10.2.10. *Let R be a local left and right coherent ring, M be a reduced finitely presented Gorenstein projective R-module, and $M \to P$ be its flat envelope. Then $M \to P$ is a monomorphism, $C = \operatorname{Coker}(M \to P)$ is reduced and Gorenstein projective, and $P \to C$ is a projective cover.*

Proof. $M \to P$ is a monomorphism and C is Gorenstein projective from the above. Now let P' be a projective summand of C. Then $P = P' \oplus P''$ for some projective P''. But then $M \to P''$ is a flat preenvelope of M since M is reduced. So $P' = 0$. Thus C is reduced.

Now let $\psi : P_0 \to C$ be a projective cover. Then we consider the following commutative diagram.

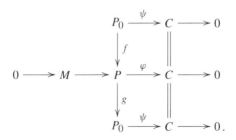

But $g \circ f$ is an automorphism. So $P = \operatorname{Im} f \oplus \operatorname{Ker} g$ and $\psi \circ g = \psi$. Therefore $\operatorname{Ker} g \subseteq \operatorname{Ker} \psi = M$. But M is reduced. So $\operatorname{Ker} g = 0$ and thus $P \cong P_0$ and we are done. □

Corollary 10.2.11. *If M is a nonzero reduced finitely presented Gorenstein projective R-module, then M has infinite projective and injective dimensions.*

Proposition 10.2.12. *Let R be n-Gorenstein and M be an R-module. If $0 \to C_i \to P_{i-1} \to \cdots \to P_1 \to P_0 \to M \to 0$ is a projective resolution of M, then $C_i \hookrightarrow P_{i-1}$ is a flat preenvelope for $i \geq n+1$. If furthermore, flat envelopes exist and the resolution is minimal, then C_i is reduced for each $i \geq n+1$, and $C_i \hookrightarrow P_{i-1}$ is a flat envelope for all $i > n+1$.*

Proof. The first part follows from Theorem 9.1.11. The rest follows using an argument dual to that of Proposition 10.1.8.   □

Lemma 10.2.13. *Let R be n-Gorenstein. Then the class $\mathcal{L}$ of R-modules of finite injective dimension is preenveloping.*

Proof. Let M be an R-module with Card $M \leq \aleph_\alpha$ and let $L \in \mathcal{L}$. Let $f : M \to L$ be any map and let $x_0 \in M$. By Proposition 7.4.5, there is a submodule L' of L with Card $\leq$ Card R such that $f(x_0) \in L'$ and $L', L/L' \in \mathcal{L}$. Now consider the map $M \to L \to L/L'$ and choose $x_1 \in M$. Then using Proposition 7.4.5 again, we can find a submodule L'' of L with Card $L'' \leq$ Card R such that $f(x_1) \in L'', L' \subset L''$, and $L'', L''/L' \in \mathcal{L}$. So $f(x_0), f(x_1) \in L''$ and $L/L'' \in \mathcal{L}$. But $\mathcal{L}$ is closed under direct limits. So if we well order M and proceed in this manner, we see that we can find a submodule $\bar{L}$ of L with Card $\bar{L} \leq \aleph_\alpha$.Card R such that $f(M) \subset \bar{L}$ and $\bar{L}, L/\bar{L} \in \mathcal{L}$. Hence the result follows by Proposition 6.2.1.   □

We note that the preenvelopes guaranteed by the lemma above are monomorphisms since $\mathcal{L}$ contains injective modules.

Theorem 10.2.14. *Let R be n-Gorenstein, $\cdots \to P_1 \to P_0 \to M \to 0$ be a projective resolution of an R-module M, and $C_i = \mathrm{Coker}(P_{i+1} \to P_i)$ for $i \geq 0$, then C_i is Gorenstein projective for $i \geq n$, and is reduced for $i \geq n+1$ if flat envelopes exist and the resolution is minimal.*

Proof. Let $i \geq n$. Then $\mathrm{Ext}^j(C_i, P) = 0$ for all $j \geq 1$ and for all projective R-modules P by Proposition 10.2.12 above. So it suffices to show that C_i has an exact right $\mathcal{P}roj$-resolution. But by the remark above, C_i has an injective $\mathcal{L}$-preenvelope $C_i \to L$. Now let $0 \to K \to Q \to L \to 0$ be exact with Q projective. Then $K \in \mathcal{L}$. But $\mathrm{Ext}^j(C_i, L) = 0$ for all $j \geq 1$ and for all $L \in \mathcal{L}$ by induction. So in particular $\mathrm{Ext}^1(C_i, K) = 0$. Hence $\mathrm{Hom}(C_i, Q) \to \mathrm{Hom}(C_i, L) \to 0$ is exact and so the map $C_i \to L$ above can be lifted to an injective projective preenvelope $C_i \to Q$. Then $\mathrm{Ext}^j(Q/C_i, P) = 0$ for all $j \geq 1$ and all projectives P. But Q/C_i has an injective $\mathcal{L}$-preenvelope and so by the above it also has an injective projective preenvelope. We proceed in this manner to get an exact right $\mathcal{P}roj$-resolution of C_i. Hence C_i is Gorenstein projective.

The second part follows from Proposition 10.2.12 above.   □

Proposition 10.2.15. *Let R be a left noetherian ring with* $\operatorname{inj\,dim}_R R = n$, *and M be a finitely presented right R-module. If $0 \to M \to P^0 \to P^1 \to \cdots$ is a right $\mathcal{P}roj_{fg}$-resolution of M and $C^i = \operatorname{Ker}(P^i \to P^{i+1})$, then $P^{i-1} \to C^i$ is surjective for $i \geq n - 1$. If furthermore, projective covers exist and the resolution is minimal, then C^i is reduced for each $i \geq n$ and $P^{i-1} \to C^i$ is a projective cover for $i \geq n+1$.*

Proof. The first part follows from Theorem 8.4.36. If $i \geq n$ and the resolution is minimal, then $C^{i-1} \to P^{i-1}$ is a projective envelope with cokernel C^i. But then C^i is reduced. It remains to show $P^{i-1} \to C^i$ is a cover. But this follows as in Proposition 10.2.10 above since C^{i-1} is reduced. $\qquad\square$

Theorem 10.2.16. *Let R be n-Gorenstein and M be a finitely presented left R-module. If $0 \to M \to P^0 \to P^1 \to \cdots$ is a right $\mathcal{P}roj_{fg}$-resolution of M and $C^i = \operatorname{Ker}(P^i \to P^{i+1})$ for $i \geq 0$, then C^i is Gorenstein projective for $i \geq n - 1$, and is reduced for $i \geq n$ if the resolution is minimal.*

Proof. This follows from Theorem 9.1.11 and Proposition 10.2.15 above. $\qquad\square$

Proposition 10.2.17. *Let R be left coherent. Then the following are equivalent for a right R-module N.*

(1) $\operatorname{Ext}^1(M, N) = 0$ *for all finitely presented Gorenstein projective right R-modules M.*

(2) $\operatorname{Ext}^i(M, N) = 0$ *for all $i \geq 1$ and all finitely presented Gorenstein projective right R-modules M.*

(3) $\overline{\operatorname{Ext}}^0(M, N) = 0$ *for all finitely presented Gorenstein projective right R-module M.*

(4) $\overline{\operatorname{Ext}}_0(M, N) = 0$ *for all finitely presented Gorenstein projective right R-modules M.*

(5) $\operatorname{Ext}_1(M, N) = 0$ *for all finitely presented Gorenstein projective right R-modules M.*

(6) $\operatorname{Ext}_i(M, N) = 0$ *for all $i \geq 1$ and all finitely presented Gorenstein projective right R-modules M.*

If furthermore R is n-Gorenstein and N is a finitely presented right R-module, then each of the above statements is equivalent to

(7) *N has finite injective dimension at most n.*

(8) *N has finite projective dimension at most n.*

Proof. This follows as in Proposition 10.1.15 using corresponding dual results.

We now provide a proof for (6) $\Rightarrow$ (8) for completeness. Let N be a finitely generated right R-module and $0 \to C_n \to P_{n-1} \to P_{n-2} \to \cdots \to P_1 \to P_0 \to$

$N \to 0$ be a projective resolution of N with P_i finitely generated. Then C_n is finitely generated. But C_n is Gorenstein projective by Theorem 10.2.14. So $\mathrm{Ext}_n(C_n, N) = 0$. Thus $\mathrm{Hom}(C_n, P_{n+1}) \to \mathrm{Hom}(C_n, P_n) \to \mathrm{Hom}(C_n, P_{n-1})$ is exact. So C_n is a summand of P_n and thus N has projective dimension at most n. □

Remark 10.2.18. A similar argument to Remark 10.1.14 shows that a finitely generated $\mathbb{Z}G$-module is Gorenstein projective if and only if it is a free $\mathbb{Z}$-module.

Exercises

1. Prove Proposition 10.2.3.

2. Prove Corollary 10.2.7.

3. Prove Corollary 10.2.11.

4. Prove Proposition 10.2.12.

5. Prove Remark 10.2.18.

6. (Enochs–Jenda [63]). Let R be n-Gorenstein and N be a Gorenstein injective R-module. Then prove that the following are equivalent.

   a) N is mock finitely generated.
   b) For every finitely generated Gorenstein projective R-module M, each of $\mathrm{Ext}_i^R(M, N)$, $\overline{\mathrm{Ext}}_0^R(M, N)$, $\overline{\overline{\mathrm{Ext}}}_0^R(M, N)$, and $\mathrm{Ext}_R^i(M, N)$ are finitely generated for all $i \geq 1$.
   c) For every finitely generated Gorenstein projective R-module M, $\mathrm{Ext}_R^1(M, N)$ is finitely generated.

7. Let R be n-Gorenstein and M be a finitely generated R-module. Prove that if $0 \to M \to E^0 \to E^1 \to \cdots$ is a right $\mathcal{I}nj$-resolution of M and $C^i = \mathrm{Ker}(E^i \to E^{i+1})$, then C^i is a mock finitely generated Gorenstein injective R-module for all $i \geq n$.

8. Let R be a complete local ring. Prove that the following are equivalent for a nonzero artinian R-module M.

   1) M is Gorenstein injective.
   2) M^v is Gorenstein projective.
   3) $\mathrm{Hom}(E(k), M)$ is a nonzero Gorenstein projective R-module.

9. (Enochs–Jenda [69]). Let R be a complete local ring. Prove that the following are equivalent for nonzero R-module M.

   1) M is a finitely generated Gorenstein injective R-module.
   2) M is of finite length and M^v is Gorenstein projective.
   3) M is of finite length and $\mathrm{Hom}(E(k), M)$ is a nonzero Gorenstein projective R-module.

   In this case, R is either 0-Gorenstein, that is *quasi-Frobenius*, or inj dim $R = \infty$.

10.3 Gorenstein flat modules

Definition 10.3.1. A module M is said to be *Gorenstein flat* if there exists an $\mathcal{I}nj \otimes -$ exact exact sequence

$$\cdots \to F_1 \to F_0 \to F^0 \to F^1 \to \cdots$$

of flat modules such that $M = \operatorname{Ker}(F^0 \to F^1)$.

It follows from this definition that $\operatorname{Tor}_i(E, M) = 0$ for all $i \geq 1$ and any injective module E. We will show in Theorem 10.3.8 below that over n-Gorenstein rings, this condition in fact characterizes Gorenstein flat modules.

If M is a finitely presented Gorenstein projective R-module over a left and right coherent ring, then M has a complete $\mathcal{P}roj_{fg}$-resolution $\cdots \to P_1 \to P_0 \to P^0 \to P^1 \to \cdots$. But then subcomplexes $\cdots \to P_1 \to P_0 \to M \to 0$ and $0 \to M \to P^0 \to P^1 \to \cdots$ are $\operatorname{Hom}(-, \mathcal{F}lat)$ exact by Proposition 10.2.6. Thus if E is an injective right R-module, then

$$\cdots \to E \otimes P_1 \to E \otimes P_0 \to E \otimes P^0 \to E \otimes P^1 \to \cdots$$

is exact as in Example 8.3.9. Hence we have the following result.

Proposition 10.3.2. *Let R be left and right coherent. Then every finitely presented Gorenstein projective R-module is Gorenstein flat.*

Proposition 10.3.3. *If R is noetherian and M is a Gorenstein flat R-module, then the character module M^+ is a Gorenstein injective right R-module.*

Proof. If M is Gorenstein flat, then there exists an exact sequence $\cdots \to F_1 \to F_0 \to F^0 \to F^1 \to \cdots$ of flat R-modules such that $\cdots \to E \otimes F_1 \to E \otimes F_0 \to E \otimes F^0 \to E \otimes F^1 \to \cdots$ is exact for all injective right R-modules E where $M = \operatorname{Ker}(F^0 \to F^1)$. But then $\cdots \to (E \otimes F^1)^+ \to (E \otimes F^0)^+ \to (E \otimes F_0)^+ \to (E \otimes F_1)^+ \to \cdots$ is exact. So $\cdots \to \operatorname{Hom}(E, F^{1+}) \to \operatorname{Hom}(E, F^{0+}) \to \operatorname{Hom}(E, F_0^+) \to \operatorname{Hom}(E, F_1^+) \to \cdots$ is exact for all injectives E. But $\cdots \to F^{1+} \to F^{0+} \to F_0^+ \to F_1^+ \to \cdots$ is an exact sequence of injective right R-modules with $M^+ = \operatorname{Ker}(F_0^+ \to F_1^+)$. Hence M^+ Gorenstein injective. $\square$

Corollary 10.3.4. *The flat dimension of a Gorenstein flat module is either zero or infinite.*

Lemma 10.3.5. *Let R be right coherent, M be an R-module, and $0 \to M \to F^0 \to F^1 \to \cdots$ be a right $\mathcal{F}lat$-resolution. Then $\operatorname{Tor}_i(L, M) = 0$ for all $i \geq 1$ and all right R-modules L of finite injective dimension if and only if the sequence $\cdots \to L \otimes F_1 \to L \otimes F_0 \to L \otimes F^0 \to L \otimes F^1 \to \cdots$ is exact for all such L where $\cdots \to F_1 \to F_0 \to M \to 0$ is a flat resolution of M.*

Proof. By Example 8.3.9, the sequence $0 \to E \otimes M \to E \otimes F^0 \to E \otimes F^1 \to \cdots$ is exact since E is injective. So if **F** denotes the complex $\cdots \to F_1 \to F_0 \to F^0 \to F^1 \to \cdots$, then $E \otimes \mathbf{F}$ is an exact complex since $\mathrm{Tor}_i(E, M) = 0$ for all $i \geq 1$.

We now proceed to argue that the complex $L \otimes \mathbf{F}$ is exact by induction on $m = \mathrm{inj\,dim}\, L$. The case $m = 0$ is the above. If $m = 1$, let $0 \to L \to E^0 \to E^1 \to 0$ be exact with E^0, E^1 injective. Then we have an exact sequence $0 \to L \otimes \mathbf{F} \to E^0 \otimes \mathbf{F} \to E^1 \otimes \mathbf{F} \to 0$ of complexes with the last two exact. Hence $L \otimes \mathbf{F}$ is exact. We then argue by induction in the obvious manner. The converse is trivial. □

Lemma 10.3.6. *Let R be n-Gorenstein and M be an R-module. Suppose there exists an exact sequence $0 \to M \to F^0 \to F^1 \to \cdots \to F^n$ with F^i flat. Then for each finitely generated R-module N, any map $N \to M$ has a factorization $N \to C \to M$ where C is a finitely generated Gorenstein projective R-module.*

Proof. Consider the exact sequence $0 \to M \to F^0 \to F^1 \to \cdots \to F^{n-1} \to L \to 0$. Let $0 \to N \to P^0 \to P^1 \to \cdots \to P^{n-1} \to D \to 0$ be a right $\mathcal{P}roj_{fg}$-resolution of N and $0 \to C \to P_{n-1} \to P_{n-2} \to \cdots \to P_1 \to P_0 \to D \to 0$ be exact with P_i finitely generated projective. Then C is Gorenstein projective by Theorem 10.2.16.

Now let $f : N \to M$ be a map. Then we can form the following commutative diagram with obvious commutativity.

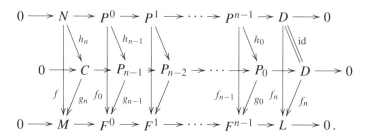

By homotopy, $f - g_n \circ h_n$ can be factored through P^0. So f can be factored through $C \oplus P^0$. Thus the result follows since $C \oplus P^0$ is Gorenstein projective. □

The argument in the proof of the following result is a modification of an analogous result in Lazard's thesis [135].

Lemma 10.3.7. *Let R be any ring and M be an R-module. If for each finitely generated R-module N, any map $N \to M$ has a factorization $N \to C \to M$ where C is a finitely generated Gorenstein projective R-module, then M is a direct limit of some inductive system $((C_i), (f_{ji}))$ where each C_i is a finitely generated Gorenstein projective R-module.*

Proof. We construct a set of pairs (C, f) where C is a finitely generated Gorenstein projective R-module and $f : C \to M$ is a map which includes every such $C \to M$ up to isomorphism.

Let $D = \oplus C$, over all such pairs (C, f), and let $g : D \to M$ be given by the maps in the pairs (C, f). Now set $\overline{D} = \oplus_{i=1}^{\infty} D_i$ where $D_i = D$ for each i and let $\overline{D} \to M$ agree with g on each D_i. Consider the directed set (S, U) where U is the sum of a finite number of summands C (for various pairs (C, f) and various i) and S is a finitely generated submodule of U with $S \subset \text{Ker}(\overline{D} \to M)$. Order the pairs by $(S, U) \leq (S', U')$ if and only if $S \subset S'$, $U \subset U'$. Then easily $\varinjlim U/S \cong M$.

But U/S is finitely generated. So by the hypothesis, the map $U/S \to M$ has a factorization $U/S \xrightarrow{\bar{h}} \overline{C} \xrightarrow{\bar{f}} M$ where we can suppose $(\overline{C}, \overline{f})$ is in our original set of pairs. We now note that each of the summands C in U is a summand of some D_i. So let n_0 be a nonnegative integer such that $n_0 \neq i$ for all such i, and $\overline{U}$ be the sum of U and $\overline{C}$ as a summand of D_{n_0}. We write $\overline{U} = U \oplus \overline{C}$.

Now let h be the map $U \to U/S \xrightarrow{\bar{h}} \overline{C}$ and define $\overline{S} = \{(u, -h(u)) : u \in U\}$. Then $\overline{U}/\overline{S} \cong \overline{C}$ and $\overline{S} \subset \text{Ker}(\overline{D} \to M)$. If $u \in S$, then $h(u) = 0$ and so $S \subset \overline{S}$. Hence $(U, S) \leq (\overline{U}, \overline{S})$. Note that $\overline{U}/\overline{S}$ is Gorenstein projective since $\overline{C}$ is. Thus we have constructed a cofinal subset of the pairs (S, U) such that U/S is finitely generated and Gorenstein projective. This completes the proof.   □

We are now in a position to prove the main result of this section.

Theorem 10.3.8. *Let R be n-Gorenstein and M be an R-module. Then the following are equivalent.*

(1) *M is Gorenstein flat.*

(2) *There exists an exact sequence $0 \to M \to F^0 \to F^1 \to \cdots \to F^n$ with each F^i flat.*

(3) *If N is any finitely generated R-module, then any map $N \to M$ can be factored through a finitely generated Gorenstein projective R-module.*

(4) *$M \cong \varinjlim C_i$ for some inductive system $((C_i), (f_{ji}))$ where each C_i is a finitely generated Gorenstein projective R-module.*

(5) *$\text{Tor}_i(E, M) = 0$ for all $i \geq 1$ and all injective right R-modules E.*

(6) *$\text{Tor}_i(E, M) = 0$ for $1 \leq i \leq n$ and all injective right R-modules E.*

(7) *M^+ is a Gorenstein injective right R-module.*

(8) *$\text{Tor}_1(L, M) = 0$ for all right R-modules L with $\text{inj dim } L < \infty$.*

(9) *$\text{Tor}_i(L, M) = 0$ for all $i \geq 1$ and all right R-modules L with $\text{inj dim } L < \infty$.*

Proof. (1) $\Rightarrow$ (2), (5) $\Rightarrow$ (6), and (9) $\Rightarrow$ (8) are trivial, (2) $\Rightarrow$ (3) is Lemma 10.3.6, (3) $\Rightarrow$ (4) is Lemma 10.3.7, and (1) $\Rightarrow$ (7) follows from Proposition 10.3.3.

(4) $\Rightarrow$ (5). Every finitely generated Gorenstein projective is Gorenstein flat by Proposition 10.3.2 and so (5) follows since Tor commutes with direct limits.

(6) $\Rightarrow$ (5). proj dim $E \le n$ for each right injective R-module E by Theorem 9.1.10 since R is n-Gorenstein. So $\mathrm{Tor}_i(E, M) = 0$ for $i \ge n + 1$ and so (5) follows.

(5) $\Rightarrow$ (9). Let $0 \to L \to E^0 \to E^1 \to \cdots \to E^m \to 0$ be exact with each E^i injective. Then $\mathrm{Tor}_i(L, M) \cong \mathrm{Tor}_{m+i}(E^m, M)$ and so (9) follows.

(9) $\Rightarrow$ (1). Let $\mathbf{F}$ be as in Lemma 10.3.5. Then $E \otimes \mathbf{F}$ is exact for each injective right R-module E. Furthermore, inj dim $R \le n$ (as a right R-module). So the complex $\mathbf{F} \cong R \otimes \mathbf{F}$ is exact by Lemma 10.3.5 and so M is Gorenstein flat.

(8) $\Rightarrow$ (9). Suppose inj dim $L < \infty$ and $0 \to L' \to P \to L \to 0$ is exact with P projective. Then inj dim $L' < \infty$ by Theorem 9.1.10 and so (9) follows by dimension shifting.

(7) $\Rightarrow$ (8). Suppose M^+ is Gorenstein injective. Then there is an exact sequence $E_n \to E_{n-1} \to \cdots \to E_1 \to E_0 \to M^+ \to 0$ where each E_i is injective. So we have an exact sequence

$$0 \to M^{++} \to E_0^+ \to E_1^+ \to \cdots \to E_n^+$$

with each E_i^+ flat. But M^{++} satisfies (2) and hence (9) by the above. So $\mathrm{Tor}_i(L, M^{++}) = 0$ for all $i \ge 1$ and for all right R-modules L such that inj dim $L < \infty$. But $0 \to M \to M^{++} \to M^{++}/M \to 0$ is pure exact. So $\mathrm{Tor}_1(L, M^{++}/M) = 0$ for all such L. But then $\mathrm{Tor}_i(L, M^{++}/M) = 0$ for all $i \ge 1$ and all such L since (8) is equivalent to (9). Hence (8) follows. $\square$

Corollary 10.3.9. *If R is n-Gorenstein and G is a Gorenstein injective left (right) R-module, then G^+ is a Gorenstein flat right (left) R-module.*

Proof. Let G be a Gorenstein injective left R-module. Then there is an exact sequence $E_n \to \cdots \to E_1 \to E_0 \to G \to 0$ with the E_i's injective. Then $0 \to G^+ \to E_0^+ \to E_1^+ \to \cdots \to E_n^+$ is exact with each E_i^+ flat. So by Theorem 10.3.8, G^+ is Gorenstein flat. $\square$

Corollary 10.3.10. *If R is Gorenstein, then every Gorenstein projective R-module is Gorenstein flat.*

Proof. This is trivial since Gorenstein projective modules satisfy part (2) of the theorem. $\square$

Corollary 10.3.11. *If R is Gorenstein, then a finitely generated R-module is Gorenstein flat if and only if it is Gorenstein projective.*

Proof. Let M be a finitely generated Gorenstein flat R-module. Then the identity map $M \to M$ has a factorization $M \to C \to M$ where C is a finitely generated Gorenstein projective R-module by the theorem above. So M is isomorphic to a direct summand of C and so is Gorenstein projective. The converse follows from Proposition 10.3.2 or Corollary 10.3.10 above. $\square$

Corollary 10.3.12. *If R is Gorenstein, then arbitrary products and sums and any inductive limit of Gorenstein flat modules are Gorenstein flat.*

Proof. (5) of the theorem shows that any inductive limit and arbitrary sums are Gorenstein flat. The products are Gorenstein flat by (2). □

Corollary 10.3.13. *Let R be n-Gorenstein and M be an R-module. Then*

(1) *If $\cdots \to F_1 \to F_0 \to M \to 0$ is a flat resolution of M and $C_i = \mathrm{Coker}(F_{i+1} \to F_i)$ for $i \geq 0$, then C_i is Gorenstein flat for $i \geq n$.*

(2) *If $0 \to M \to F^0 \to F^1 \to \cdots$ is a right $\mathcal{F}$lat-resolution of M and $C^i = \mathrm{Ker}(F^i \to F^{i+1})$ for $i \geq 0$, then C^i is Gorenstein flat for $i \geq n - 1$.*

Proof. (1) $\mathrm{Tor}_j(L, C_i) \cong \mathrm{Tor}_{i+j}(L, M)$. But if $\mathrm{inj\ dim}\,L < \infty$, then $\mathrm{flat\ dim}\,L \leq n$ since R is n-Gorenstein and so $\mathrm{Tor}_{i+j}(L, M) = 0$ for all $j \geq 1$ and all $i \geq n$. Thus if $i \geq n$, then $\mathrm{Tor}_j(L, C_i) = 0$ for all $i \geq 1$. Hence C_i is Gorenstein flat for each $i \geq n$ by Theorem 10.3.8. (2) follows from (1) and Theorem 8.4.36. □

Theorem 10.3.14. *Suppose R is Gorenstein and $0 \to M' \to M \to M'' \to 0$ is an exact sequence of R-modules. Then if M' and M'' are Gorenstein flat, so is M. If M and M'' are Gorenstein flat, so is M'. If M' and M are Gorenstein flat, then M'' is Gorenstein flat if and only if $0 \to E \otimes M' \to E \otimes M$ is exact for any injective module E.*

Proof. This follows from (5) and (8) of the theorem. □

Exercises

1. Prove Corollary 10.3.4.

2. Prove part (2) of Corollary 10.3.13.

3. Prove Theorem 10.3.14.

4. Prove that if G is a finite group, then a $\mathbb{Z}G$-module is Gorenstein flat if and only if it is torsion free as a $\mathbb{Z}$-module.

5. Prove that the following are equivalent for an n-Gorenstein ring R.

   a) R is quasi-Frobenius.
   b) Every R-module is Gorenstein injective.
   c) Every R-module is Gorenstein projective.
   d) Every R-module is Gorenstein flat.

6. Prove that the following are equivalent for an n-Gorenstein ring R.

   a) $\mathrm{gl\ right}\,\mathcal{I}nj\text{-}\dim_R \mathcal{M} < \infty$.
   b) Every Gorenstein injective R-module is injective.
   c) Every Gorenstein projective R-module is projective.
   d) Every Gorenstein flat R-module is flat.

10.4 Foxby classes

Throughout this section, R will denote a local Cohen–Macaulay ring of Krull dimension d admitting a dualizing module Ω and with residue field k.

Definition 10.4.1. $\mathcal{G}_0(R)$ will denote the class of R-modules M such that $\mathrm{Tor}_i(\Omega, M) = \mathrm{Ext}^i(\Omega, \Omega \otimes M) = 0$ for all $i \geq 1$ and such that the natural map $M \to \mathrm{Hom}(\Omega, \Omega \otimes M)$ is an isomorphism. $\mathcal{J}_0(R)$ will denote the class of R-modules N such that $\mathrm{Ext}^i(\Omega, N) = \mathrm{Tor}_i(\Omega, \mathrm{Hom}(\Omega, N)) = 0$ for all $i \geq 1$ and such that the natural map $\Omega \otimes \mathrm{Hom}(\Omega, N) \to N$ is an isomorphism. $\mathcal{G}_0(R)$ and $\mathcal{J}_0(R)$ are called *Foxby classes*. This notation will also be used to denote the corresponding full subcategories.

Remark 10.4.2. The functor $\Omega \otimes - : \mathcal{G}_0(R) \to \mathcal{J}_0(R)$ gives an equivalence between the two categories. Similarly, $\mathrm{Hom}(\Omega, -) : \mathcal{J}_0(R) \to \mathcal{G}_0(R)$ gives an equivalence. It follows, for example, that if $M_1, M_2 \in \mathcal{G}_0(R)$, then

$$\mathrm{Hom}(M_1, M_2) \cong \mathrm{Hom}(M_1, \mathrm{Hom}(\Omega, \Omega \otimes M_2)) \cong \mathrm{Hom}(\Omega \otimes M_1, \Omega \otimes M_2),$$

and if $M_1, M_2 \in \mathcal{J}_0(R)$, then

$$\mathrm{Hom}(M_1, M_2) \cong \mathrm{Hom}(\mathrm{Hom}(\Omega, M_1), \mathrm{Hom}(\Omega, M_2)).$$

We state the following result for completeness.

Proposition 10.4.3. *The following are equivalent for a ring R.*

(1) *R is Gorenstein.*

(2) *$M \in \mathcal{G}_0(R)$ for all R-modules M.*

(3) *$M \in \mathcal{J}_0(R)$ for all R-modules M.*

Proof. (1) $\Rightarrow$ (2), (3) trivially since $\Omega = R$ if R is Gorenstein.

(2), (3) $\Rightarrow$ (1). If $M \in \mathcal{G}_0(R)$ (or $M \in \mathcal{J}_0(R)$) for all M, then $\mathrm{Tor}_i(\Omega, M) = 0$ (or $\mathrm{Ext}^i(\Omega, M) = 0$) for all R-modules M and all $i \geq 1$. Thus Ω is projective and so finitely generated and free. But inj dim $\Omega = d$. So R is Gorenstein. □

Lemma 10.4.4. $\mathrm{Ext}^i_R(\Omega, \Omega)^{\wedge} \cong \begin{cases} 0 & \text{if } i \neq 0 \\ \hat{R} & \text{if } i = 0. \end{cases}$

Proof. Since Ω is maximal Cohen–Macaulay, $H^i_{m(R)}(\Omega) = 0$ for all $i \neq d$ by Theorem 9.5.11 and so $\mathrm{Ext}^i_R(\Omega, \Omega)^{\wedge} \cong \mathrm{Hom}(H^{d-i}_{m(R)}(\Omega), E(k)) = 0$ for all $i \geq 1$ by Theorem 9.5.17. Thus the case $i \neq 0$ follows.

Now let $0 \to \Omega \to E^0 \to E^1 \to \cdots$ be the minimal injective resolution of Ω. Then $H^d_{m(R)}(\Omega) \cong \mathrm{Ker}(L_{m(R)}(E^d(\Omega)) \to L_{m(R)}(E^{d+1}(\Omega)))$ (see Definition 9.5.2). But $E^d(\Omega) \cong E(k)$ and $E^{d+1}(\Omega) = 0$. So $H^d_{m(R)}(\Omega) \cong E(k)$. Thus $\mathrm{Hom}_R(\Omega, \Omega)^{\wedge} \cong \mathrm{Hom}_R(E(k), E(k))^{\wedge} \cong \hat{R}$. □

Proposition 10.4.5. *If P is projective, then $P \in \mathcal{G}_0(R)$.*

Proof. Since $\text{Ext}^i(\Omega, \Omega) = 0$ for $i \geq 1$ by the lemma above, $\text{Ext}^i(\Omega, \Omega \otimes P) = 0$ for all projective R-modules P and all $i \geq 1$. Moreover, $P \to \text{Hom}(\Omega, \Omega \otimes P)$ is an isomorphism for any projective P since it is an isomorphism when $P = R$. Hence the result follows. $\square$

Proposition 10.4.6. *If E is injective, then $E \in \mathcal{G}_0(R)$.*

Proof. Let I be an ideal generated by a maximal R-sequence $r_1, r_2, \ldots, r_d$. Then $\text{Hom}(R/I^t, E_R(k)) = E_{R/I^t}(k)$. But $\Omega/I^t\Omega \cong E_{R/I^t}(k)$ (see Remark 9.5.15). So $\text{Hom}(\Omega/I^t\Omega, E_{R/I^t}(k)) \cong R/I^t$ by Matlis duality. Hence $\Omega \otimes_R \text{Hom}_R(\Omega, E_{R/I^t}(k))$ $\cong \text{Hom}_R(\hat{R}, E_{R/I^t}(k))$ by Lemma 10.4.4. But $\text{Hom}_R(\hat{R}, E_{R/I^t}(k)) \cong E_{R/I^t}(k)$. So $\Omega \otimes_R \text{Hom}_R(\Omega, E_{R/I^t}(k)) \cong E_{R/I^t}(k)$. Now taking inductive limits, we get that $\Omega \otimes \text{Hom}(\Omega, E(k)) \cong E(k)$.

Now if $E = E(R/\mathfrak{p})$ for an arbitrary prime ideal $\mathfrak{p}$, then $\Omega \otimes \text{Hom}(\Omega, E) \to E$ is an isomorphism by localizing at $\mathfrak{p}$ and appealing to the case $E = E(k)$ above. Thus $\Omega \otimes \text{Hom}(\Omega, E) \to E$ is an isomorphism for any injective R-module E. But if E is injective, $\text{Ext}^i(\Omega, E) = 0$ and $\text{Tor}_i(\Omega, \text{Hom}(\Omega, E)) \cong \text{Hom}(\text{Ext}^i(\Omega, \Omega), E) = 0$ for all $i \geq 1$ by Lemma 10.4.4, and so we are done. $\square$

Notation. We will let $\mathcal{W}$ denote the class of all modules W such that $W \cong \Omega \otimes P$ for some projective P, and let $\mathcal{V}$ be the class of all modules V such that $V \cong \text{Hom}(\Omega, E)$ for some injective module E.

Proposition 10.4.7. *$\mathcal{V}$ is a preenveloping class and $\mathcal{W}$ is a precovering class.*

Proof. Let M be an R-module and embed $\Omega \otimes M$ in an injective R-module E. Then the composition of the maps $M \to \text{Hom}(\Omega, \Omega \otimes M)$ and $\text{Hom}(\Omega, \Omega \otimes M) \subset \text{Hom}(\Omega, E)$ is a $\mathcal{V}$-preenvelope. For if $V \in \mathcal{V}$, and $M \to V$ is a map, then we have a map $\Omega \otimes M \to \Omega \otimes V \cong \Omega \otimes \text{Hom}(\Omega, E')$ for some injective E'. But $\Omega \otimes \text{Hom}(\Omega, E') \cong E'$ since $E' \in \mathcal{G}_0(R)$ by Proposition 10.4.6 above. So the map $\Omega \otimes M \to E'$ can be extended to a map $E \to E'$. This gives a map $\text{Hom}(\Omega, E) \to \text{Hom}(\Omega, E')$ such that the composition $M \to \text{Hom}(\Omega, E) \to \text{Hom}(\Omega, E')$ is the map $M \to \text{Hom}(\Omega, E') \cong V$.

Now note that for any M, if $W \in \mathcal{W}$, then $W \to M$ is a $\mathcal{W}$-precover if and only if $\text{Hom}(\Omega, W) \to \text{Hom}(\Omega, M) \to 0$ is exact. So $\Omega^{(\text{Hom}(\Omega, M))} \to M$ is a $\mathcal{W}$-precover. $\square$

Proposition 10.4.8. *The following are equivalent for an R-module M.*

(1) $M \in \mathcal{G}_0(R)$

(2) *There exists an exact sequence*

$$\cdots \to P_1 \to P_0 \to V^0 \to V^1 \to \cdots$$

of R-modules with each $P_i \in \mathcal{P}roj$, $V^i \in \mathcal{V}$, such that $M = \text{Ker}(V^0 \to V^1)$ and $\Omega \otimes -$ leaves the sequence exact.

(3) M has an exact right $\mathcal{V}$-resolution and the functor $\Omega \otimes -$ leaves any projective resolution of M exact.

Proof. (1) $\Rightarrow$ (2). Let $\cdots \rightarrow P_1 \rightarrow P_0 \rightarrow M \rightarrow 0$ be a projective resolution of M. Then $\cdots \rightarrow \Omega \otimes P_1 \rightarrow \Omega \otimes P_0 \rightarrow \Omega \otimes M \rightarrow 0$ is exact since $\mathrm{Tor}_i(\Omega, M) = 0$ for all $i \geq 1$ and $\Omega \otimes -$ is right exact. Now let $0 \rightarrow \Omega \otimes M \rightarrow E^0 \rightarrow E^1 \rightarrow \cdots$ be an injective resolution of $\Omega \otimes M$ and set $V^i = \mathrm{Hom}(\Omega, E^i)$. Then we get a complex

$$0 \rightarrow \mathrm{Hom}(\Omega, \Omega \otimes M) \rightarrow V^0 \rightarrow V^1 \rightarrow \cdots$$

which is exact since $\mathrm{Ext}^i(\Omega, \Omega \otimes M) = 0$ for all $i \geq 1$ and $\mathrm{Hom}(\Omega, -)$ is left exact. So the result follows since $M \cong \mathrm{Hom}(\Omega, \Omega \otimes M)$.

(2) $\Leftrightarrow$ (3). We note that $0 \rightarrow M \rightarrow V^0 \rightarrow V^1 \rightarrow \cdots$ is a right $\mathcal{V}$-resolution if and only if $\cdots \rightarrow \mathrm{Hom}(V^0, \mathrm{Hom}(\Omega, E)) \rightarrow \mathrm{Hom}(M, \mathrm{Hom}(\Omega, E)) \rightarrow 0$ is exact for all injective R-modules E if and only if $\cdots \rightarrow \mathrm{Hom}(\Omega \otimes V^1, E) \rightarrow \mathrm{Hom}(\Omega \otimes V^0, E) \rightarrow \mathrm{Hom}(\Omega \otimes E, E) \rightarrow 0$ is exact for all injective modules E and if and only if $\Omega \otimes -$ makes the complex $0 \rightarrow M \rightarrow V^0 \rightarrow V^1 \rightarrow \cdots$ exact. Finally, it is easy to see that $\Omega \otimes -$ leaves $\cdots \rightarrow P_1 \rightarrow P_0 \rightarrow M \rightarrow 0$ exact if and only if it leaves every projective resolution exact.

(2) $\Rightarrow$ (1). $\Omega \otimes -$ leaves $\cdots \rightarrow P_1 \rightarrow P_0 \rightarrow M \rightarrow 0$ exact means $\mathrm{Tor}_i(\Omega, M) = 0$ for all $i \geq 1$. But $V^i \cong \mathrm{Hom}(\Omega, E^i)$ for some injective E^i. So $\Omega \otimes V^i \cong \Omega \otimes \mathrm{Hom}(\Omega, E^i) \cong E^i$ by Proposition 10.4.6. Thus the natural map $V^i \rightarrow \mathrm{Hom}(\Omega, \Omega \otimes V^i)$ is an isomorphism, and $0 \rightarrow \Omega \otimes M \rightarrow \Omega \otimes V^0 \rightarrow \Omega \otimes V^1 \rightarrow \cdots$ is an injective resolution of $\Omega \otimes M$ since it is exact by assumption. But then the complex $0 \rightarrow \mathrm{Hom}(\Omega, \Omega \otimes M) \rightarrow \mathrm{Hom}(\Omega, \Omega \otimes V^0) \rightarrow \mathrm{Hom}(\Omega, \Omega \otimes V^1) \rightarrow \cdots$ is equivalent to the exact sequence $0 \rightarrow M \rightarrow V^0 \rightarrow V^1 \rightarrow \cdots$. So the natural map $M \rightarrow \mathrm{Hom}(\Omega, \Omega \otimes M)$ is an isomorphism and $\mathrm{Ext}^i(\Omega, \Omega \otimes M) = 0$ for all $i \geq 1$. $\square$

Similarly, we have the following result noting that a complex $\cdots \rightarrow W_1 \rightarrow W_0 \rightarrow M \rightarrow 0$ with each $W_i \in \mathcal{W}$ is a left $\mathcal{W}$-resolution of M if and only if $\mathrm{Hom}(\Omega, -)$ makes the complex exact.

Proposition 10.4.9. *The following are equivalent for an R-module M.*

(1) $M \in \mathcal{J}_0(R)$.

(2) *There exists an exact sequence*

$$\cdots \rightarrow W_1 \rightarrow W_0 \rightarrow E^0 \rightarrow E^1 \rightarrow \cdots$$

of R-modules with each E^i injective, $W_i \in \mathcal{W}$, such that $M = \mathrm{Ker}(E^0 \rightarrow E^1)$ and $\mathrm{Hom}(\Omega, -)$ leaves the sequence exact.

(3) *M has an exact left $\mathcal{W}$-resolution and the functor $\mathrm{Hom}(\Omega, -)$ leaves every injective resolution of M exact.*

Theorem 10.4.10. *Let* $0 \to M' \to M \to M'' \to 0$ *be an exact sequence of R-modules. Then if any two of* M', M, M'' *are in* $\mathcal{G}_0(R)$ *(or* $\mathcal{J}_0(R)$*), then so is the third.*

Proof. If $M'' \in \mathcal{G}_0(R)$, then $\mathrm{Tor}_1(\Omega, M'') = 0$ and so $0 \to \Omega \otimes M' \to \Omega \otimes M \to \Omega \otimes M'' \to 0$ is exact. If $M \in \mathcal{G}_0(R)$, then $\mathrm{Tor}_1(\Omega, M) = 0$ and we have an exact sequence $0 \to \mathrm{Tor}_1(\Omega, M'') \to \Omega \otimes M' \to \Omega \otimes M$. So $0 \to \mathrm{Hom}(\Omega, \mathrm{Tor}_1(\Omega, M'')) \to \mathrm{Hom}(\Omega, \Omega \otimes M') \to \mathrm{Hom}(\Omega, \Omega \otimes M)$ is exact. But then if $M, M' \in \mathcal{G}_0(R)$, then $0 \to \mathrm{Hom}(\Omega, \mathrm{Tor}_1(\Omega, M'')) \to M' \to M$ is exact and so $\mathrm{Tor}_1(\Omega, M'') = 0$. Hence if any two of M', M, M'' are in $\mathcal{G}_0(R)$, then $0 \to \Omega \otimes M' \to \Omega \otimes M \to \Omega \otimes M'' \to 0$ is exact. But this is equivalent to $0 \to \mathrm{Hom}(M'', \mathrm{Hom}(\Omega, E)) \to \mathrm{Hom}(M, \mathrm{Hom}(\Omega, E)) \to \mathrm{Hom}(M', \mathrm{Hom}(\Omega, E)) \to 0$ is exact for all injective R-modules E. Thus $0 \to \mathrm{Hom}(M'', V) \to \mathrm{Hom}(M, V) \to \mathrm{Hom}(M', V) \to 0$ is exact for all $V \in \mathcal{V}$. So by the Horseshoe Lemma (8.2.1), right $\mathcal{V}$-resolutions of M' and M'' can be combined to form a right $\mathcal{V}$-resolution of M. Similarly for projective resolutions of M' and M''. Then we can paste these resolutions together along $0 \to M' \to M \to M'' \to 0$ to get a short exact sequence of complexes which remains exact when we apply $\Omega \otimes -$ to it. So if any two of the complexes are exact, then so is the third. But then the result follows by Proposition 10.4.8. An analogous proof gives the result for $\mathcal{J}_0(R)$. $\square$

Corollary 10.4.11. *If* $\mathrm{proj\,dim}\, M < \infty$, *then* $M \in \mathcal{G}_0(R)$, *and if* $\mathrm{inj\,dim}\, M < \infty$, *then* $M \in \mathcal{J}_0(R)$.

Proof. This follows from the theorem and Propositions 10.4.5 and 10.4.6. $\square$

Lemma 10.4.12. *If* $M \in \mathcal{G}_0(R)$, *then* $\mathrm{Ext}^i(M, V) = 0$ *for all* $i \geq 1$ *and all* $V \in \mathcal{V}$.

Proof. We first note that $M \cong \mathrm{Hom}(\Omega, \Omega \otimes M)$. Now let $V \cong \mathrm{Hom}(\Omega, E)$ with E injective. Then

$$\mathrm{Ext}^i(M, V) \cong \mathrm{Ext}^i(M, \mathrm{Hom}(\Omega, E)) \cong \mathrm{Hom}(\mathrm{Tor}_i(\Omega, \mathrm{Hom}(\Omega \otimes M)), E) = 0$$

for all $i \geq 1$ since $\Omega \otimes M \in \mathcal{J}_0(R)$. $\square$

Theorem 10.4.13. *Every R-module* $M \in \mathcal{G}_0(R)$ *has a* $\mathcal{V}$-*envelope.*

Proof. If $M \in \mathcal{G}_0(R)$, then $M \cong \mathrm{Hom}(\Omega, \Omega \otimes M)$ and so M has a one to one $\mathcal{V}$-preenvelope by the proof of Proposition 10.4.7. Thus $\mathcal{V}$-preenvelopes of M are injections. Now let $((V_\alpha), (\varphi_{\beta\alpha}))$ be an inductive system of $\mathcal{V}$-preenvelopes of M. Then we have an exact sequence $0 \to M \to \varinjlim V_\alpha$. But $\varinjlim V_\alpha \in \mathcal{V} \subset \mathcal{G}_0(R)$. So $(\varinjlim V_\alpha)/M \in \mathcal{G}_0(R)$ by Theorem 10.4.10 and hence for any map $M \to V$ with $V \in \mathcal{V}$, we have a factorization $M \to \varinjlim V_\alpha \to V$ by the lemma above. Thus M has a $\mathcal{V}$-envelope by Lemma 6.6.1. $\square$

We now need the following result which holds for any local ring.

Proposition 10.4.14. *If R is any local ring, then the class $\mathcal{L}$ of modules of finite projective dimension is preenveloping.*

Proof. If $\text{proj dim } L < \infty$, then $\text{proj dim } L \leq \dim R$ (see Raynaud–Gruson [157, Theorem 3.2.6]). Moreover, every flat module has projective dimension at most $\dim R$ by Corollary 8.5.28. So the class of modules of finite projective dimension is closed under products.

Now for any R-module M, there is an infinite cardinal $\aleph_\alpha$ (depending on M) such that if $L \in \mathcal{L}$ and $S \subset L$ is a submodule with $\text{Card } S \leq \text{Card } M$, then there is a pure submodule L' of L (hence $L' \in \mathcal{L}$) containing S with $\text{Card } L' \leq \aleph_\alpha$ by Lemma 5.3.12. So M has an $\mathcal{L}$-preenvelope by Corollary 6.2.2.   □

We note that an $\mathcal{L}$-preenvelope need not be a monomorphism.

We are now in a position to prove the following result.

Lemma 10.4.15. *If $M \in \mathcal{G}_0(R)$ and $0 \to C \to P_{d-1} \to \cdots \to P_1 \to P_0 \to M \to 0$ is exact with each P_i projective, then C is Gorenstein projective.*

Proof. We first note that any projective resolution of M remains exact when we apply $\Omega \otimes -$ by Proposition 10.4.8 since $M \in \mathcal{G}_0(R)$. But if P and Q are projective, then $\text{Ext}^i(\Omega \otimes Q, \Omega \otimes P) = 0$ for all $i \geq 1$ since $\text{Ext}^i(\Omega, \Omega) = 0$ for all $i \geq 1$ by Lemma 10.4.4. Furthermore, $\text{inj dim } \Omega = d$ and so $\text{inj dim } \Omega \otimes P \leq d$. Hence $\text{Ext}^{d+i}(\Omega \otimes M, \Omega \otimes P) = 0$ for all $i \geq 1$ and all projective P. Thus the exact sequence

$$\cdots \to \Omega \otimes P_{d+1} \to \Omega \otimes P_d \to \Omega \otimes P_{d-1} \to \cdots \to \Omega \otimes P_0 \to \Omega \otimes M \to 0$$

remains exact beginning with the term $\text{Hom}(\Omega \otimes P_d, \Omega \otimes P)$ when $\text{Hom}(-, \Omega \otimes P)$ is applied to it with P projective. But $\text{Hom}(\Omega \otimes P_i, \Omega \otimes P) \cong \text{Hom}(P_i, P)$ for each i. So $0 \to \text{Hom}(C, P) \to \text{Hom}(P_d, P) \to \text{Hom}(P_{d+1}, P) \to \cdots$ is exact for all projective R-modules P.

Now it remains to show that C has an exact right $\mathcal{P}roj$-resolution. But by Proposition 10.4.14, C has an $\mathcal{L}$-preenvelope $C \to L$ which is a monomorphism since $C \subset P_{d-1}$. Now let $Q \to L$ be a projective precover, and $K = \text{Ker}(Q \to L)$. Then $K \in \mathcal{L}$. But $\text{Ext}^i(C, P) = 0$ for all $i \geq 1$ and all projectives P from the above. So $\text{Ext}^i(C, L) = 0$ for all $i \geq 1$ and for all $L \in \mathcal{L}$ by induction. In particular, $\text{Ext}^1(C, K) = 0$. Hence $C \to L$ can be lifted to a monomorphism $C \to Q$ which is still an $\mathcal{L}$-preenvelope. We now need to argue that Q/C has a projective preenvelope that is a monomorphism. But $\text{Ext}^i(Q/C, P) = 0$ for all projective P for all $i \geq 1$. Thus $\text{Ext}^i(Q/C, L) = 0$ for all $L \in \mathcal{L}$ and all $i \geq 1$. But by Theorem 10.4.10, $Q/C \in \mathcal{G}_0(R)$. Now embed $\Omega \otimes Q/C$ into an injective E. Then $Q/C \cong \text{Hom}(\Omega, \Omega \otimes Q/C) \subset \text{Hom}(\Omega, E)$. But flat dim $\text{Hom}(\Omega, E) < \infty$ and so $\text{Hom}(\Omega, E) \in \mathcal{L}$. Thus any $\mathcal{L}$-preenvelope of Q/C is a monomorphism. So we have an $\mathcal{L}$-preenvelope $Q/C \to Q^1$ which is a monomorphism with Q^1 projective by the argument above. Now let $Q^0 = Q$. Then we have an exact right $\mathcal{P}roj$-resolution $0 \to C \to Q^0 \to Q^1$. We now proceed in this manner to construct an exact right $\mathcal{P}roj$-resolution $0 \to C \to Q^0 \to Q^1 \to Q^2 \to \cdots$. $\qquad \square$

Lemma 10.4.16. *If M is a Gorenstein projective R-module, then $M \in \mathcal{G}_0(R)$.*

Proof. Let $\cdots \to P_1 \to P_0 \to P^0 \to P^1 \to \cdots$ be a complete $\mathcal{P}roj$-resolution of M. This exact sequence is left exact by $\mathrm{Hom}(-, L)$ whenever L is a module of finite projective dimension. In particular, $\mathrm{Hom}(-, \mathrm{Hom}(\Omega, E))$ leaves the sequence exact for any injective R-module E. Hence so does $\mathrm{Hom}(\Omega \otimes -, E)$. Thus $\Omega \otimes -$ leaves the sequence exact since E is arbitrary. So $\mathrm{Tor}_i(\Omega, M) = 0$ for all $i \geq 1$.

We note that $0 \to \Omega \otimes M \to \Omega \otimes P^0 \to \Omega \otimes P^1$ is exact and so $0 \to \mathrm{Hom}(\Omega, \Omega \otimes M) \to \mathrm{Hom}(\Omega, \Omega \otimes P^0) \to \mathrm{Hom}(\Omega, \Omega \otimes P^1)$ is exact. But $\mathrm{Hom}(\Omega, \Omega \otimes P^i) \cong P^i$ since $P^i \in \mathcal{G}_0(R)$ by Proposition 10.4.5. So the natural map $M \to \mathrm{Hom}(\Omega, \Omega \otimes M)$ is an isomorphism.

Now consider the short exact sequence $0 \to M \to P^0 \to N \to 0$. Then N is also Gorenstein projective and so $N \to \mathrm{Hom}(\Omega, \Omega \otimes N)$ is an isomorphism from the above. But $0 \to \Omega \otimes M \to \Omega \otimes P^0 \to \Omega \otimes N \to 0$ is exact. So applying $\mathrm{Hom}(\Omega, -)$ we get that $0 \to M \to P^0 \to N \to \mathrm{Ext}^1(\Omega, \Omega \otimes M) \to \mathrm{Ext}^1(\Omega, \Omega \otimes P^0)$ is exact. But $\mathrm{Ext}^1(\Omega, \Omega \otimes P^0) = 0$ since $\mathrm{Ext}^1(\Omega, \Omega) = 0$. Thus $\mathrm{Ext}^1(\Omega, \Omega \otimes M) = 0$ and hence likewise $\mathrm{Ext}^1(\Omega, \Omega \otimes N) = 0$. But $\mathrm{Ext}^2(\Omega, \Omega \otimes P^0) = 0$. So we get $\mathrm{Ext}^2(\Omega, \Omega \otimes M) = 0$ and by induction we get that $\mathrm{Ext}^i(\Omega, \Omega \otimes M) = 0$ for all $i \geq 1$. $\qquad \square$

Proposition 10.4.17. *$M \in \mathcal{G}_0(R)$ if and only if for some $n \geq 0$, there exists an exact sequence $0 \to C_n \to C_{n-1} \to \cdots \to C_1 \to C_0 \to M \to 0$ with each C_i Gorenstein projective. If there is such a sequence, then there is one with $n \leq d$.*

Proof. This follows from Theorem 10.4.10, Lemmas 10.4.15 and 10.4.16. $\qquad \square$

Theorem 10.4.18. *The following are equivalent for an R-module C.*

(1) *C is Gorenstein projective.*

(2) *$C \in \mathcal{G}_0(R)$ and $\mathrm{Ext}^i(C, L) = 0$ for all $i \geq 1$ and all L such that $\mathrm{proj\,dim}\, L < \infty$.*

(3) *C has an exact right $\mathcal{P}roj$-resolution and $\mathrm{Ext}^i(C, L) = 0$ for all $i \geq 1$ and all L such that $\mathrm{proj\,dim}\, L < \infty$.*

(4) *There exists an exact sequence $0 \to C \to P^0 \to P^1 \to \cdots \to P^{d-1} \to B \to 0$ with each P^i projective and $B \in \mathcal{G}_0(R)$.*

(5) *There exists an exact sequence $0 \to C_n \to C_{n-1} \to \cdots \to C_1 \to C_0 \to C \to 0$ with each C_i Gorenstein projective for some $n \geq 0$ and $\mathrm{Ext}^i(C, L) = 0$ for all $i \geq 1$ and all L such that $\mathrm{proj\,dim}\, L < \infty$.*

Proof. (1) $\Rightarrow$ (2). $C \in \mathcal{G}_0(R)$ by Lemma 10.4.16 and the second part is standard.

(2) $\Rightarrow$ (3). We apply the arguments we used concerning the C in the proof of Lemma 10.4.15 to get that C has an exact right $\mathcal{P}roj$-resolution.

(3) $\Rightarrow$ (1) by definition and (1) $\Rightarrow$ (4) is trivial since if $\cdots \rightarrow P_1 \rightarrow P_0 \rightarrow P^0 \rightarrow P^1 \rightarrow \cdots \rightarrow P^d \rightarrow P^{d+1} \rightarrow \cdots$ is a complete $\mathcal{P}roj$-resolution of C, then $B = \mathrm{Ker}(P^d \rightarrow P^{d+1})$ is Gorenstein projective and so is in $\mathcal{G}_0(R)$.

(4) $\Rightarrow$ (1) by Lemma 10.4.15 and (2) $\Leftrightarrow$ (5) follows from Proposition 10.4.17.$\square$

Corollary 10.4.19. *If $C = C_1 \oplus C_2$, then C is Gorenstein injective if and only if C_1 and C_2 are.*

Similar arguments give the following results.

Lemma 10.4.20. *If $M \in \mathcal{G}_0(R)$ and $0 \rightarrow M \rightarrow E^0 \rightarrow \cdots \rightarrow E^{d-1} \rightarrow G \rightarrow 0$ is exact with E^i injective, then G is Gorenstein injective.*

Proposition 10.4.21. *If $\mathrm{inj\,dim}\, L < \infty$ for an R-module L, then $\mathrm{inj\,dim}\, L \leq d$.*

Proof. Since $\mathrm{inj\,dim}\, L < \infty$, we have that $L \in \mathcal{G}_0(R)$ by Corollary 10.4.11. Then if $0 \rightarrow L \rightarrow E^0 \rightarrow E^1 \rightarrow \cdots \rightarrow E^{d-1} \rightarrow G \rightarrow 0$ is exact with $E^0, E^1, \ldots, E^{d-1}$ injective, G is Gorenstein injective by Lemma 10.4.20. But $\mathrm{inj\,dim}\, G < \infty$. So G is injective by Proposition 10.1.2. Hence $\mathrm{inj\,dim}\, L \leq d$. $\square$

Lemma 10.4.22. *If M is Gorenstein injective, then $M \in \mathcal{G}_0(R)$.*

Proposition 10.4.23. *$M \in \mathcal{G}_0(R)$ if and only if for some $n \geq 0$, there exists an exact sequence $0 \rightarrow M \rightarrow G^0 \rightarrow G^1 \rightarrow \cdots \rightarrow G^n \rightarrow 0$ with each G^i Gorenstein injective. If there is such a sequence, then there is one with $n \leq d$.*

Theorem 10.4.24. *The following are equivalent for an R-module G.*

(1) *G is Gorenstein injective.*

(2) *$G \in \mathcal{G}_0(R)$ and $\mathrm{Ext}^i(L, G) = 0$ for all $i \geq 1$ and all R-modules L such that $\mathrm{inj\,dim}\, L < \infty$.*

(3) *G has an exact left $\mathcal{I}nj$-resolution and $\mathrm{Ext}^i(L, G) = 0$ for all $i \geq 1$ and all L such that $\mathrm{inj\,dim}\, L < \infty$.*

(4) *There exists an exact sequence $0 \rightarrow K \rightarrow E_{d-1} \rightarrow \cdots \rightarrow E_0 \rightarrow G \rightarrow 0$ with each E_i injective and $K \in \mathcal{G}_0(R)$.*

(5) *There exists an exact sequence $0 \rightarrow G \rightarrow G^0 \rightarrow G^1 \rightarrow \cdots \rightarrow G^n \rightarrow 0$ with each G^i Gorenstein injective for some $n \geq 0$ and $\mathrm{Ext}^i(L, G) = 0$ for all $i \geq 1$ and all L such that $\mathrm{inj\,dim}\, L < \infty$.*

Corollary 10.4.25. *If $G = G_1 \oplus G_2$, then G is Gorenstein injective if and only if G_1 and G_2 are.*

We now need the following result which holds for any commutative noetherian ring.

Lemma 10.4.26. *Let R be a commutative noetherian ring. If $\cdots \to F^{-1} \to F^0 \to F^1 \to \cdots$ is an exact sequence of flat R-modules such that $E \otimes -$ leaves the sequence exact when E is an injective R-module, then $\mathrm{Hom}(-, K)$ leaves the sequence exact when K is cotorsion and of finite flat dimension.*

Proof. If K is flat and cotorsion, then K is a summand of an R-module $\mathrm{Hom}(E, E')$ where E, E' are injective R-modules by Lemma 5.3.27. But $\mathrm{Hom}(-, \mathrm{Hom}(E, E^1)) \cong \mathrm{Hom}(E \otimes -, E^1)$. So $\mathrm{Hom}(-, \mathrm{Hom}(E, E^1))$ leaves the sequence exact. Hence $\mathrm{Hom}(-, K)$ for such a K leaves the sequence exact.

Now suppose $0 \to K' \to K \to K'' \to 0$ is an exact sequence such that $\mathrm{Hom}(F, -)$ leaves the sequence exact whenever F is flat. Then applying each of $\mathrm{Hom}(-, K'), \mathrm{Hom}(-, K)$ and $\mathrm{Hom}(-, K'')$ to the exact sequence $\mathbf{F} : \cdots \to F^{-1} \to F^0 \to F^1 \to \cdots$, we get the short exact sequence $0 \to \mathrm{Hom}(\mathbf{F}, K') \to \mathrm{Hom}(\mathbf{F}, K) \to \mathrm{Hom}(\mathbf{F}, K'') \to 0$ of complexes. Hence if any two of $\mathrm{Hom}(-, K'), \mathrm{Hom}(-, K)$ and $\mathrm{Hom}(-, K'')$ leave $\mathbf{F}$ exact, so does the third. Now if K has finite flat dimension, then K has a minimal left $\mathcal{F}lat$-resolution $0 \to F_n \to F_{n-1} \to \cdots \to F_0 \to K \to 0$ by Theorem 7.4.4. If furthermore K is cotorsion, then each F_i is cotorsion by Corollary 5.3.26. So each $\mathrm{Hom}(-, F_i)$ leaves the sequence $\mathbf{F}$ exact and hence we get by induction that $\mathrm{Hom}(-, K)$ leaves the sequence exact. $\qquad\square$

Corollary 10.4.27. *If M is a Gorenstein flat R-module and K is cotorsion and of finite flat dimension, then $\mathrm{Ext}^i(M, K) = 0$ for all $i \geq 1$.*

Theorem 10.4.28. *Let $\mathcal{L}$ be the class of R-modules of finite injective dimension. Then an R-module M is Gorenstein flat if and only if $M \in \mathcal{G}_0(R)$ and $\mathrm{Tor}_i(L, M) = 0$ for all $i \geq 1$ and all $L \in \mathcal{L}$.*

Proof. Suppose M is Gorenstein flat. Then there is an $\mathcal{I}nj \otimes -$ exact exact sequence $\cdots \to F_1 \to F_0 \to F^0 \to F^1 \to \cdots$ with each F_i, F^i flat such that $M = \mathrm{Ker}(F^0 \to F^1)$. Hence $L \otimes -$ leaves the sequence exact for each $L \in \mathcal{L}$. So $\mathrm{Tor}_i(L, M) = 0$ for all $i \geq 1$ and all $L \in \mathcal{L}$. In particular, $\mathrm{Tor}_i(\Omega, M) = 0$ for all $i \geq 1$. Thus we have an exact sequence

$$0 \to \mathrm{Hom}(\Omega, \Omega \otimes M) \to \mathrm{Hom}(\Omega, \Omega \otimes F^0) \to \mathrm{Hom}(\Omega, \Omega \otimes F^1) \to \mathrm{Hom}(\Omega, \Omega \otimes F^2)$$

where $F^i \cong \mathrm{Hom}(\Omega, \Omega \otimes F^i)$ since F^i is in $\mathcal{G}_0(R)$. Hence $M \cong \mathrm{Hom}(\Omega, \Omega \otimes M)$. It now remains to show that $\mathrm{Ext}^i(\Omega, \Omega \otimes M) = 0$ for each $i \geq 1$. We consider the exact sequence $0 \to M \to F^0 \to Y \to 0$. Then clearly Y is Gorenstein flat. So $\mathrm{Tor}_i(\Omega, Y) = 0$ for all $i \geq 1$ and $Y \cong \mathrm{Hom}(\Omega, \Omega \otimes Y)$ by the above. Thus we have an exact sequence

$$0 \to \mathrm{Hom}(\Omega, \Omega \otimes M) \to \mathrm{Hom}(\Omega, \Omega \otimes F^0) \to \mathrm{Hom}(\Omega, \Omega \otimes Y) \to \mathrm{Ext}^1(\Omega, \Omega \otimes M) \to 0.$$

So $\text{Ext}^1(\Omega, \Omega \otimes M) = 0$. Similarly $\text{Ext}^1(\Omega, \Omega \otimes Y) = 0$. But then $\text{Ext}^2(\Omega, \Omega \otimes M) = 0$. So repeating the argument gives $\text{Ext}^i(\Omega, \Omega \otimes M) = 0$ for all $i \geq 1$. Hence $M \in \mathcal{G}_0(R)$.

Conversely, suppose $\cdots \to F_1 \to F_0 \to M \to 0$ is a flat resolution of M. Then $E \otimes -$ leaves the resolution exact for all injectives E since $\text{Tor}_i(E, M) = 0$ for all $i \geq 1$ and all such E by assumption. To show that M is Gorenstein flat, we only need to construct the other half of the complete flat resolution of M.

Let $Y = \Omega \otimes M$. Then $\text{Ext}^i(\Omega, Y) = \text{Ext}^i(\Omega, \Omega \otimes M) = 0$ for all $i \geq 1$ since $M \in \mathcal{G}_0(R)$ by assumption. Now let $0 \to Y \to E \to X \to 0$ be exact with E the injective envelope of Y. Then we have an exact sequence $0 \to \text{Hom}(\Omega, Y) \to \text{Hom}(\Omega, E) \to \text{Hom}(\Omega, X) \to 0$. But $M \cong \text{Hom}(\Omega, \Omega \otimes M) \cong \text{Hom}(\Omega, Y)$ and $L = \text{Hom}(\Omega, E)$ has finite flat dimension. So we consider the exact sequence $0 \to M \to L \to W \to 0$. Since $\mathcal{F}lat$ is covering by Theorem 7.4.4, we have an exact sequence $0 \to K \to F \to L \to 0$ where $F \to L$ is a flat cover. Thus we can form the following pullback diagram

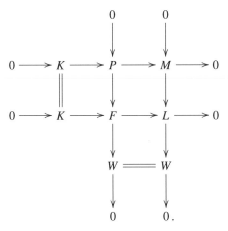

But K is cotorsion and flat dim $K < \infty$. So $\text{Hom}(-, K)$ leaves $\cdots \to F_2 \to F_1 \to F_0$ exact by Lemma 10.4.26. Hence $\text{Ext}^1(M, K) = 0$ and thus M is a direct summand of P. So M can be embedded into a flat R-module F. But then M can be embedded into a flat preenvelope $M \to F$ giving an exact sequence $0 \to M \to F \to C \to 0$. So we get that $0 \to \text{Hom}(C, F^1) \to \text{Hom}(F, F^1) \to \text{Hom}(M, F^1) \to 0$ is exact for all flat R-modules F. In particular, $0 \to \text{Hom}(C, \text{Hom}(E, E(k))) \to \text{Hom}(F, \text{Hom}(E, E(k))) \to \text{Hom}(M, \text{Hom}(E, E(k))) \to 0$ is exact. Thus $E \otimes -$ leaves $0 \to M \to F \to C \to 0$ exact. But $M, F \in \mathcal{G}_0(R)$. So $C \in \mathcal{G}_0(R)$ by Theorem 10.4.10. Furthermore, it is now easy to check that $\text{Tor}_i(L, C) = 0$ for all $L \in \mathcal{L}$ and all $i \geq 1$. We then repeat the argument above to get the desired complete flat resolution of M. Hence M is Gorenstein flat. $\qquad \square$

Corollary 10.4.29. *Any Gorenstein projective R-module is Gorenstein flat.*

Proof. Let M be Gorenstein projective. Then $M \in \mathcal{G}_0(R)$ by Lemma 10.4.16. If $L \in \mathcal{L}$, then $\text{Hom}(L, E(k))$ has finite projective dimension since it has finite flat dimension and so $\mathrm{Hom}(\mathrm{Tor}_i(L, M), E(k)) \cong \mathrm{Ext}^i(M, \mathrm{Hom}(L, E(k))) = 0$ for $i \geq 1$. Thus $\mathrm{Tor}_i(L, M) = 0$ for all $i \geq 1$ and all $L \in \mathcal{L}$. Hence M is Gorenstein flat by the theorem. □

Corollary 10.4.30. *Any direct limit of a family of Gorenstein flat modules is Gorenstein flat.*

Proof. This also follows from the theorem since $\mathcal{G}_0(R)$ is closed under direct limits.□

We are now in a position to state the Cohen–Macaulay version of Theorem 10.3.8.

Theorem 10.4.31. *Let $\mathcal{L}$ be the class of R-modules of finite injective dimension. Then the following are equivalent for an R-module M.*

(1) *M is Gorenstein flat.*

(2) *$M \in \mathcal{G}_0(R)$ and there exists an exact sequence $0 \to M \to F^0 \to \cdots \to F^{d-1} \to F^d$ with each F^i flat.*

(3) *$M \in \mathcal{G}_0(R)$ and if N is any finitely generated R-module, then any map $N \to M$ can be factored through a finitely generated Gorenstein projective R-module.*

(4) *$M \in \mathcal{G}_0(R)$ and $M \cong \underrightarrow{\lim}\, C_i$ for some inductive system $((C_i), (f_{ji}))$ where each C_i is a finitely generated Gorenstein projective R-module.*

(5) *$M \in \mathcal{G}_0(R)$ and $\mathrm{Tor}_i(E, M) = 0$ for all $i \geq 1$ and all injective R-modules E.*

(6) *$M \in \mathcal{G}_0(R)$ and $\mathrm{Tor}_i(L, M) = 0$ for all $i \geq 1$ and all $L \in \mathcal{L}$.*

(7) *$M \in \mathcal{G}_0(R)$ and $\mathrm{Tor}_1(L, M) = 0$ for all $L \in \mathcal{L}$.*

(8) *M^+ is Gorenstein injective.*

Proof. (1) $\Rightarrow$ (2) and (6) $\Rightarrow$ (7) are trivial.

(2) $\Rightarrow$ (3) follows as in Lemma 10.3.6 using Lemma 10.4.15.

(3) $\Rightarrow$ (4) by Lemma 10.3.7.

(4) $\Rightarrow$ (5), (5) $\Rightarrow$ (6) follow as in (4) $\Rightarrow$ (5), (5) $\Rightarrow$ (9) of Theorem 10.3.8.

(6) $\Leftrightarrow$ (1) is Theorem 10.4.28.

(7) $\Rightarrow$ (6). $L \in \mathcal{J}_0(R)$ since $L \in \mathcal{L}$. So there is an exact sequence $0 \to L' \to W \to L \to 0$ by Proposition 10.4.9 where $W \cong \Omega \otimes P$ with P projective. Furthermore, $\mathrm{Tor}_i(\Omega, M) = 0$ for all $i \geq 1$ since $M \in \mathcal{G}_0(R)$. Hence $\mathrm{Tor}_i(W, M) = 0$ for all $i \geq 1$. But $L' \in \mathcal{L}$ since $W \in \mathcal{L}$. So (6) follows by dimension shifting.

(8) $\Leftrightarrow$ (1). If M^+ is Gorenstein injective, then there exists an exact sequence $E_d \to E_{d-1} \to \cdots \to E_1 \to E_0 \to M^+ \to 0$ with each E_i injective. So M^{++} is Gorenstein flat and thus the result follows as in Theorem 10.3.8. The converse follows by Proposition 10.3.3. □

Corollary 10.4.32. *If M is Gorenstein injective, then M^+ is Gorenstein flat.*

Corollary 10.4.33. *A finitely generated R-module is Gorenstein flat if and only if it is Gorenstein projective.*

Exercises

1. Prove that the functor $\Omega \otimes - : \mathcal{G}_0(R) \to \mathcal{G}_0(R)$ gives an equivalence between the two categories and so does $\mathrm{Hom}(\Omega, -) : \mathcal{G}_0(R) \to \mathcal{G}_0(R)$.

2. Prove Proposition 10.4.9.

3. Prove the second part of Theorem 10.4.10.

4. Let $\mathcal{X}$ denote the class of R-modules X such that $X \cong \Omega \otimes F$ for some flat R-module F. Prove that $\mathcal{X} \subset \mathcal{G}_0(R)$.

5. Prove that $\mathcal{G}_0(R)$ contains flat R-modules. Conclude that if flat dim $M < \infty$, then $M \in \mathcal{G}_0(R)$.

6. Prove that the class $\mathcal{V}$ is precovering.

7. Prove Lemma 10.4.20.

8. Prove Lemma 10.4.22.

9. Prove Proposition 10.4.23.

10. Prove Theorem 10.4.24.

11. Prove Corollaries 10.4.32 and 10.4.33.

Chapter 11

Gorenstein Covers and Envelopes

In this chapter we consider the existence of precovers, covers, preenvelopes and envelopes for the various Gorenstein related classes of modules.

11.1 Gorenstein injective precovers and covers

We now show that over n-Gorenstein rings, Gorenstein injective modules are precovering and covering.

We start with the following result.

Theorem 11.1.1. *If R is n-Gorenstein, then every R-module has a Gorenstein injective precover.*

Proof. Let M be an R-module and $\cdots \to E_{n-1} \to \cdots \to E_1 \to E_0 \to M \to 0$ be a left $\mathcal{I}nj$-resolution of M. Now let $0 \to K \to E^0 \to E^1 \to \cdots \to E^{n-1} \to \cdots$ be a right $\mathcal{I}nj$-resolution of K. Then $C = \operatorname{Coker}(E^{n-2} \to E^{n-1})$ is Gorenstein injective by Theorem 10.1.13. So if G is Gorenstein injective and $\cdots \to H_{n-1} \to \cdots \to H_1 \to H_0 \to G \to 0$ is a left $\mathcal{I}nj$-resolution of G with $L = \operatorname{Ker}(H_{n-1} \to H_{n-2})$, then given a map $f : G \to M$ we can construct the following commutative diagram

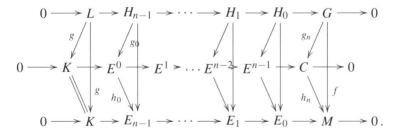

By homotopy, $f - h_n \circ g_n$ can be factored through E_0. So $f : G \to M$ can be factored through the Gorenstein injective module $C \oplus E_0$. Thus $C \oplus E_0 \to M$ is a Gorenstein injective precover. $\qquad\square$

Lemma 11.1.2. *Let R be n-Gorenstein. Then every inductive limit of Gorenstein injective modules is Gorenstein injective.*

Proof. If M is Gorenstein injective, then M has an exact left $\mathcal{I}nj$-resolution $\cdots \to$ $E_1(M) \to E_0(M) \to M \to 0$. So if $((G_i), (\varphi_{ji}))$ is an inductive system with G_i Gorenstein injective, then we get an inductive system $(E_\ell(G_i), E_\ell(\varphi_{ji}))$ for any $\ell \geq 1$. Hence we have an exact sequence

$$\cdots \to \varinjlim E_1(G_i) \to \varinjlim E_0(G_i) \to \varinjlim G_i \to 0.$$

But $\varinjlim E_\ell(G_i)$ is injective for each ℓ since R is noetherian. So $\varinjlim G_i$ is Gorenstein injective by Theorem 10.1.13. $\square$

Theorem 11.1.3. *Let R be n-Gorenstein. Then every R-module has a Gorenstein injective cover.*

Proof. This follows from Theorem 11.1.1, Lemma 11.1.2 and Corollary 5.2.7. $\square$

Exercises

1. Prove that if R is n-Gorenstein, then R is Gorenstein injective if and only if every Gorenstein injective cover is surjective.

2. Let R be a local Cohen–Macaulay ring of Krull dimension d admitting a dualizing module, M be an R-module, and $0 \to K \to E_{d-1} \to \cdots \to E_1 \to E_0 \to M \to 0$ be a left $\mathcal{I}nj$-resolution of M. Prove that if $K \in \mathcal{J}_0(R)$, then M has a Gorenstein injective cover.

11.2   Gorenstein injective preenvelopes

We first consider the existence of a Gorenstein injective preenvelope $M \to G$ of a module M. Since injective modules are Gorenstein injective, such a preenvelope is necessarily an injection. We can find such a preenvelope if we can exhibit an exact sequence $0 \to M \to G \to L \to 0$ with G Gorenstein injective and with inj dim $L < \infty$. For then by Proposition 10.1.3, $\text{Ext}^1(L, H) = 0$ when H is Gorenstein injective. So $\text{Hom}(G, H) \to \text{Hom}(M, H) \to \text{Ext}^1(L, H) = 0$ is exact, showing that $M \to G$ is a preenvelope.

We now recall from Proposition 1.4.14 that if

$$
\begin{array}{ccccccccccc}
0 & \longrightarrow & C^0 & \overset{\delta^0}{\longrightarrow} & C^1 & \overset{\delta^1}{\longrightarrow} & C^2 & \longrightarrow & \cdots & \longrightarrow & C^n & \longrightarrow & 0 \\
& & \downarrow{\scriptstyle f^0} & & \downarrow{\scriptstyle f^1} & & \downarrow{\scriptstyle f^2} & & & & \downarrow{\scriptstyle f^n} & & \\
0 & \longrightarrow & B^0 & \overset{\delta^0}{\longrightarrow} & B^1 & \overset{\delta^1}{\longrightarrow} & B^2 & \longrightarrow & \cdots & \longrightarrow & B^n & \longrightarrow & 0
\end{array}
$$

is a commutative diagram of R-modules with exact rows, then $0 \to C^0 \to B^0 \oplus C^1 \to B^1 \oplus C^2 \to \cdots \to B^{n-1} \oplus C^n \to B^n \to 0$ is an exact sequence where the map $B^{i-1} \oplus C^i \to B^i \oplus C^{i+1}$ is $(x, y) \mapsto (\delta(x) + (-1)^i f^i(y), \delta(y))$ (where $B^i = C^i = 0$ for $i < 0$ and $i > n$).

Theorem 11.2.1. *If R is an n-Gorenstein ring, then every R-module M has a Gorenstein injective preenvelope $M \to G$ such that if $0 \to M \to G \to L \to 0$ is exact, then* inj dim $L \leq n - 1$ *whenever $n \geq 1$.*

Proof. Let $0 \to M \to E^0 \to \cdots \to E^{n-1} \to H \to 0$ be a partial injective resolution of M. By Theorem 10.1.13, H is Gorenstein injective. Hence by the definition of a Gorenstein injective module, there exists a $\mathrm{Hom}(\mathcal{I}nj, -)$ exact exact sequence $0 \to G \to E_{n-1} \to \cdots \to E_0 \to H \to 0$ where G is also Gorenstein injective by Theorem 10.1.4. But then the diagram

$$
\begin{array}{ccccccccccc}
0 & \longrightarrow & M & \longrightarrow & E^0 & \longrightarrow & \cdots & \longrightarrow & E^{n-1} & \longrightarrow & H & \longrightarrow & 0 \\
 & & \big\downarrow & & \big\downarrow & & & & \big\downarrow & & \big\| & & \\
0 & \longrightarrow & G & \longrightarrow & E_{n-1} & \longrightarrow & \cdots & \longrightarrow & E_0 & \longrightarrow & H & \longrightarrow & 0
\end{array}
$$

can be completed to a commutative diagram. So by the remark above, we get an exact sequence $0 \to M \to G \oplus E^0 \to E_{n-1} \oplus E^1 \to \cdots \to E_0 \oplus H \to H \to 0$. This exact sequence has the exact subcomplex $0 \to 0 \to 0 \to \cdots \to H \stackrel{\pm \mathrm{id}}{\to} H \to 0$. Forming the quotient complex, we get an exact sequence $0 \to M \to G \oplus E^0 \to E^{n-1} \oplus E_1 \to \cdots \to E_0 \to 0$ by Remark 1.4.15. But then if $0 \to M \to G \oplus E^0 \to L \to 0$ is exact, we see that inj dim $L \leq n - 1$. Since $G \oplus E^0$ is Gorenstein injective by Theorem 10.1.4, we see that $M \to G \oplus E^0$ is the desired Gorenstein injective preenvelope.   □

Corollary 11.2.2. *If R is n-Gorenstein, then the following are equivalent for any R-module M.*

(1) *M is Gorenstein injective.*

(2) *$\mathrm{Ext}^i(L, M) = 0$ for all R-modules L with* proj dim $L < \infty$ *and all $i \geq 1$.*

(3) *$\mathrm{Ext}^1(L, M) = 0$ for all R-modules L with* proj dim $L < \infty$.

(4) *$\mathrm{Ext}^i(E, M) = 0$ for all injective R-modules E and all $i \geq 1$.*

Proof. (1) $\Rightarrow$ (2) by Proposition 10.1.3 and (2) $\Rightarrow$ (3), (4) are trivial.

(3) $\Rightarrow$ (1). By the theorem above, there is an exact sequence $0 \to M \to G \to L \to 0$ with G Gorenstein injective and inj dim $L < \infty$. So proj dim $L < \infty$ by Theorem 9.1.10. But then the exact sequence splits and so M is Gorenstein injective.

(4) $\Rightarrow$ (1). (4) means that any right $\mathcal{I}nj$-resolution of M is a left $\mathcal{I}nj$-resolution. So M is Gorenstein injective by Theorem 10.1.13.   □

Remark 11.2.3. It follows from Proposition 10.1.15 and Corollary 11.2.2 above that $(\mathcal{L}, \mathcal{G}or\mathcal{I}nj)$ is a cotorsion theory over any Iwanaga–Gorenstein ring R where $\mathcal{L}$ consists of all L with inj dim $L < \infty$ (or equivalently, proj dim $L < \infty$) and $\mathcal{G}or\mathcal{I}nj$ denotes the class of all Gorenstein injective R-modules.

Corollary 11.2.4. *The following are equivalent for a Gorenstein injective preenvelope* $M \rightarrow G$ *of Theorem* 11.2.1.

(1) $\mathrm{inj\,dim}\,M < \infty$.

(2) *G is injective.*

(3) $M \rightarrow G$ *is an injective preenvelope.*

Proof. (1) $\Leftrightarrow$ (2). Let $0 \rightarrow M \rightarrow G \rightarrow L \rightarrow 0$ be an exact sequence with G Gorenstein injective and $\mathrm{inj\,dim}\,L < \infty$ (see Theorem 11.2.1). If $\mathrm{inj\,dim}\,M < \infty$, then $\mathrm{inj\,dim}\,G < \infty$ and thus G is injective by Proposition 10.1.2. Conversely, if G is injective, then $\mathrm{inj\,dim}\,M < \infty$ since $\mathrm{inj\,dim}\,L < \infty$.

(2) $\Rightarrow$ (3) is trivial since injectives are Gorenstein injective. (3) $\Rightarrow$ (2) is also trivial. $\square$

If R is Gorenstein, then the class $\mathcal{G}or\mathcal{I}nj$ is preenveloping by the above and hence by Proposition 8.1.3 every R-module has a right $\mathcal{G}or\mathcal{I}nj$-resolution. This resolution is exact and is usually called a *Gorenstein injective resolution*.

Proposition 11.2.5. *Let R be an n-Gorenstein ring and $\mathcal{L}$ be the class of R-modules of finite projective dimension. Then the following are equivalent for an R-module M and integer $r \geq 0$.*

(1) *right* $\mathcal{G}or\mathcal{I}nj$- $\dim M \leq r$.

(2) *There exists an exact sequence* $0 \rightarrow M \rightarrow G^0 \rightarrow \cdots \rightarrow G^r \rightarrow 0$ *with each G^i Gorenstein injective.*

(3) $\mathrm{Ext}^i(L, M) = 0$ *for all $i \geq r + 1$ and all $L \in \mathcal{L}$.*

(4) $\mathrm{Ext}^{r+1}(L, M) = 0$ *for all $L \in \mathcal{L}$.*

(5) $\mathrm{Ext}^i(E, M) = 0$ *for all $i \geq r + 1$ and all injective R-modules E.*

(6) *Every rth $\mathcal{I}nj$-cosyzygy of M is Gorenstein injective.*

(7) *Every rth $\mathcal{G}or\mathcal{I}nj$-cosyzygy of M is Gorenstein injective.*

Proof. (1) $\Rightarrow$ (2); (3) $\Rightarrow$ (4) and (5); and (7) $\Rightarrow$ (6) are trivial.

(2) $\Rightarrow$ (3). $\mathrm{Ext}^i(L, M) \cong \mathrm{Ext}^{i-r}(L, G^r) = 0$ for all $i > r$ by Corollary 11.2.2. So (3) follows.

(4) $\Rightarrow$ (1). Let $0 \rightarrow M \rightarrow G^0 \rightarrow G^1 \rightarrow \cdots$ be a right $\mathcal{G}or\mathcal{I}nj$-resolution and $G = \mathrm{Ker}(G^r \rightarrow G^{r+1})$. Then $\mathrm{Ext}^1(L, G) \cong \mathrm{Ext}^{r+1}(L, M)$ for all $L \in \mathcal{L}$. So G is Gorenstein injective by Corollary 11.2.2. (5) $\Rightarrow$ (1) is similar.

(5) $\Rightarrow$ (7). If G is an rth $\mathcal{G}or\mathcal{I}nj$-cosyzygy, then $\mathrm{Ext}^{i-r}(E, G) \cong \mathrm{Ext}^i(E, M)$ for all $i > 1$. So G is Gorenstein injective again by Corollary 11.2.2.

(6) $\Rightarrow$ (5). If G is an rth $\mathcal{I}nj$-cosyzygy, then $\mathrm{Ext}^i(E, M) \cong \mathrm{Ext}^{i-r}(E, G) = 0$ for all injectives E and all $i > r$ since G is Gorenstein injective. $\square$

Over Cohen–Macaulay rings, we have the following result.

Theorem 11.2.6. *Let R be a local Cohen–Macaulay ring of Krull dimension d admitting a dualizing module. Then every R-module $M \in \mathcal{J}_0(R)$ has a Gorenstein injective preenvelope $M \to G$ such that if $0 \to M \to G \to L \to 0$ is exact, then* inj dim $L \leq d - 1$.

Proof. The proof follows as in Theorem 11.2.1 using Lemma 10.4.20 instead of Theorem 10.1.13. □

Corollary 11.2.7. *An R-module M is Gorenstein injective if and only if $M \in \mathcal{J}_0(R)$ and* $\mathrm{Ext}^1(L, M) = 0$ *for all R-modules L with* inj dim $L < \infty$.

Proof. If $M \in \mathcal{J}_0(R)$, then M has a Gorenstein injective preenvelope $\psi : M \to G$ with inj dim Coker $\psi \leq d - 1$. So $\mathrm{Ext}^1(L, M) = 0$ for all L such that inj dim $L < \infty$ means M is a summand of G and hence M is Gorenstein injective. The converse follows from Theorem 10.4.24. □

Proposition 11.2.8. *Let R be as in Theorem* 11.2.6. *Then the following are equivalent for an R-module $M \in \mathcal{J}_0(R)$.*

(1) $\mathrm{Ext}^i(M, N) = 0$ *for all $i \geq 1$ and all Gorenstein injective R-modules N.*

(2) $\mathrm{Ext}^1(M, N) = 0$ *for all Gorenstein injective R-modules N.*

(3) *M has finite injective dimension.*

(4) *M has injective dimension at most d.*

Proof. (1) $\Leftrightarrow$ (2) is part of Proposition 10.1.15.

(1) $\Rightarrow$ (4). Let $0 \to M \to E^0 \to E^1 \to \cdots$ be an injective resolution of M and K be a dth $\mathcal{I}nj$-cosyzygy of M. Then K is Gorenstein injective by Lemma 10.4.20. So $\mathrm{Ext}^d(M, K) = 0$ by assumption. Thus $\mathrm{Hom}(E^{d+1}, K) \to \mathrm{Hom}(E^d, K) \to \mathrm{Hom}(E^{d-1}, K)$ is exact and so K is a summand of E^d and hence inj dim $M \leq d$.

(4) $\Rightarrow$ (3) is trivial, and (3) $\Rightarrow$ (1) by Proposition 10.1.3. □

Exercises

1. Prove that if R is n-Gorenstein, then right $\mathcal{G}or\mathcal{I}nj$- dim $M \leq$ inj dim M for each R-module M.

2. Prove that if R is n-Gorenstein and M is an R-module, then right $\mathcal{G}or\mathcal{I}nj$- dim $M \leq n$ and right $\mathcal{G}or\mathcal{I}nj$- dim $M =$ inj dim M if and only if inj dim $M < \infty$.

3. Let R be a local Cohen–Macaulay ring of finite Krull dimension and $\mathcal{L}$ be the class of R-modules of finite injective dimension. State and prove a result corresponding to Proposition 11.2.5.

4. Let R and $\mathcal{L}$ be as in Problem 3 above. State and prove a result corresponding to Problem 2 above.

5. Prove Theorem 11.2.6.

11.3   Gorenstein injective envelopes

We now want to show the existence of Gorenstein injective envelopes.

Lemma 11.3.1. *If R is n-Gorenstein and $\mathcal{L}$ is the class of R-modules L such that* $\operatorname{inj} \dim L < \infty$, *then $\mathcal{L}$ is closed under inductive limits.*

Proof. By Theorem 9.1.10 we have $\operatorname{inj} \dim L \leq n$ for all $L \in \mathcal{L}$. Since R is left noetherian we have $\operatorname{inj} \dim \varinjlim L_i \leq \sup_i \operatorname{inj} \dim L_i$ for any inductive system (L_i) of left R-modules. Hence $\operatorname{inj} \dim \varinjlim L_i \leq n$ if all $L_i \in \mathcal{L}$. $\qquad\square$

Theorem 11.3.2. *If R is n-Gorenstein, then every R-module M has a Gorenstein injective envelope $M \to G$ such that if $0 \to M \to G \to L \to 0$ is exact, then* $\operatorname{inj} \dim L \leq n - 1$ *whenever $n \geq 1$.*

Proof. By Remark 11.2.3, $(\mathcal{L}, \mathcal{G}or\mathcal{I}nj)$ is a cotorsion theory which has enough injectives by Theorem 11.2.1. So the result follows from Theorem 7.2.6 and the lemma above. $\qquad\square$

Using this theorem we get the following result.

Corollary 11.3.3. *If $M \to G$ is a Gorenstein injective envelope, then* $\operatorname{inj} \dim M < \infty$ *if and only if $M \to G$ is an injective envelope.*

Proof. This follows from Corollary 11.2.4. $\qquad\square$

Theorem 11.3.4. *Let R be n-Gorenstein. If M is an R-module and $0 \to M \to G^0 \to G^1 \to \cdots \to G^n \to \cdots$ is a minimal right $\mathcal{G}or\mathcal{I}nj$-resolution, then G^i is injective for $i \geq 1$ and $G^i = 0$ for $i > n$.*

Proof. This follows from Corollary 11.3.3 and Theorem 11.3.2. $\qquad\square$

Definition 11.3.5. Let A be an R-submodule of B. Then $A \subset B$ is called a *Gorenstein extension* if $\operatorname{proj} \dim B/A < \infty$, and a *Gorenstein injective extension* if B is furthermore Gorenstein injective. We note that every Gorenstein injective envelope over an n-Gorenstein ring is a Gorenstein injective extension. The Gorenstein injective envelope of an R-module M is denoted by $G(M)$.

Proposition 11.3.6. *Let R be n-Gorenstein. Then the following are equivalent for an R-module M.*

(1) *$M \subset G$ is a Gorenstein injective extension.*

(2) *$M \subset G$ is a Gorenstein injective preenvelope with* $\operatorname{proj} \dim G/M < \infty$.

(3) *$G \cong G(M) \oplus E$ for some injective R-module E where the isomorphism leaves M fixed.*

Proof. (1) ⇒ (2). If $M \subset G$ is a Gorenstein injective extension, then proj dim $G/M < \infty$ by definition and so $\text{Ext}^1(G/M, G') = 0$ for all Gorenstein injective R-modules G' by Corollary 11.2.2. Thus $M \subset G$ is a Gorenstein injective preenvelope.

(2) ⇒ (3). (2) implies that we have the following commutative diagram with exact rows

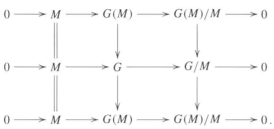

But $G(M) \to G \to G(M)$ is an automorphism. So $G \cong G(M) \oplus G'$ for some Gorenstein injective R-module G'. But then G' is also a summand of G/M. Hence G' is injective by Proposition 10.1.2.

(3) ⇒ (2) and (2) ⇒ (1) are trivial.   □

Proposition 11.3.7. *Let R be n-Gorenstein and M be a submodule of a Gorenstein injective R-module G. If $\text{proj dim } M < \infty$, then $G \cong E(M) \oplus G'$ for some Gorenstein injective R-module G'.*

Proof. By Corollary 11.3.3, $E(M) \cong G(M)$ and so we have the following commutative diagram

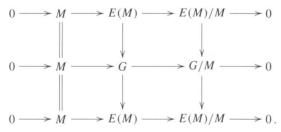

But then $E(M) \to G \to E(M)$ is an automorphism and so the result follows.   □

Corollary 11.3.8. *The following are equivalent for a Gorenstein injective R-module G.*

(1) *G is reduced.*

(2) *$G \cong G(M)$ for every submodule M of G such that $\text{proj dim } G/M < \infty$.*

(3) *G has no nontrivial submodules of finite projective dimension.*

Proof. (1) ⇒ (2) follows from Proposition 11.3.6.

(2) ⇒ (3). Let M be a submodule of finite projective dimension. Then $G \cong E(M) \oplus G'$ for some Gorenstein injective G' by Proposition 11.3.7. But

proj dim $E(M) < \infty$. So $G' \subset G$ is a Gorenstein injective extension. Thus $G \cong G(G') \cong G'$ by assumption. Hence $E(M) = 0$.

(3) $\Rightarrow$ (1) follows easily since every injective module has finite projective dimension. $\qquad\square$

Proposition 11.3.9. *If R is n-Gorenstein and $M_i \to G_i$ is a Gorenstein injective envelope of M_i for each $i \in I$, then $\oplus M_i \to \oplus G_i$ is a Gorenstein injective envelope.*

Proof. We first argue that $\oplus G_i$ is Gorenstein injective. Since each G_i is Gorenstein injective, there is an exact sequence $E_i^0 \to \cdots \to E_i^n \to G_i \to 0$ with each E_i injective. But then we have the exact sequence $\oplus E_i^0 \to \cdots \to \oplus E_i^n \to \oplus G_i \to 0$. So by Theorem 10.1.13, we see that $\oplus G_i$ is Gorenstein injective. Then by Corollary 6.4.4, we get that $\oplus M_i \to \oplus G_i$ is a Gorenstein injective envelope. $\qquad\square$

Proposition 11.3.10. *Let R be n-Gorenstein and $M \subset G$ be a Gorenstein injective envelope. If $f : G \to G$ is linear and $M \subset \operatorname{Ker} f$, then f is locally nilpotent on G (that is, for each $x \in G$ there is an $m \geq 1$ such that $f^m(x) = 0$).*

Proof. By the preceding result we know that $M \oplus M \oplus M \oplus \cdots \subset G \oplus G \oplus G \oplus \cdots$ is a Gorenstein injective envelope. Let $\psi : G \oplus G \oplus \cdots \to G \oplus \cdots$ be the map such that $\psi(x_0, x_1, x_2, \dots) = (x_0, x_1 - f(x_0), x_2 - f(x_1), \dots)$. By our hypothesis on f, ψ is the identity on $M \oplus M \oplus \cdots$. Since we have an envelope, ψ must be an automorphism. Let $x \in G$. Then $(x, 0, 0, \dots)$ must be in the image of ψ. Let $\psi(x_0, x_1, x_2, \dots) = (x, 0, 0, \dots)$. Then we see that $x_0 = x$, $x_1 = f(x)$, $x_2 = f^2(x)$, $\dots$. But for large m, $x_m = 0$, that is $f^m(x) = 0$. $\qquad\square$

Corollary 11.3.11. *If R is a commutative n-Gorenstein ring and $M \subset G$ is a Gorenstein injective envelope where $IM = 0$ for some ideal $I \subset R$, then for each $x \in G$, $I^m x = 0$ for some $m \geq 1$.*

Proof. If $r \in I$, we consider the function $f : G \to G$ with $f(x) = rx$. By applying Proposition 11.3.10 above to this f, we see that for each $x \in G$, $r^m x = 0$ for some $m \geq 1$. Then since I is finitely generated we see that for each $x \in G$, $I^m x = 0$ for some $m \geq 1$. $\qquad\square$

Exercises

1. Prove Theorem 11.3.4.

2. Let $A \subset B \subset C$ be modules and consider the three inclusions $A \subset B$, $B \subset C$, and $A \subset C$. Prove that if any two of these are Gorenstein extensions, then so is the third.

3. Let R be n-Gorenstein and N be a submodule of M. Prove that $N \subset M$ is a Gorenstein extension if and only if $G(M) \cong G(N) \oplus E$ for some injective R-module E where the isomorphism leaves N fixed.

4. Let R be a local Cohen–Macaulay ring of Krull dimension d admitting a dualizing module. Prove that every $N \in \mathcal{J}_0(R)$ has a Gorenstein injective envelope $\psi : N \to G$ such that inj dim Coker $\psi \le d - 1$.

5. Let R be as in Problem 4 above. Prove that every $N \in \mathcal{J}_0(R)$ has a minimal right $\mathcal{G}or\mathcal{I}nj$-resolution of the form $0 \to N \to G^0 \to E^1 \to E^2 \to \cdots \to E^{d-1} \to E^d \to 0$ where G^0 is Gorenstein injective and each E^i is injective.

6. Let R be Iwanaga–Gorenstein. Prove that the class $\mathcal{L}$ of R-modules of finite projective dimension is covering. And moreover, if $0 \to K \to P \to M \to 0$ is exact where $P \to M$ is an $\mathcal{L}$-precover, then K is Gorenstein injective.

7. Prove that if R is Iwanaga–Gorenstein then the minimal left $\mathcal{L}$-resolution of an R-module M is of the form

$$\cdots \to E_2 \to E_1 \to L_0 \to M \to 0$$

where $L_0 \in \mathcal{L}$ and E_i is injective for each i.

8. Prove that if R is a local Cohen–Macaulay ring of Krull dimension d admitting a dualizing module and if $\mathcal{L}$ is the class of R-modules of finite injective dimension, then every $N \in \mathcal{J}_0(R)$ has an $\mathcal{L}$-cover. And moreover, if $\psi : P \to N$ is such a cover, then Ker ψ is Gorenstein injective.

11.4 Gorenstein essential extensions

Definition 11.4.1. Let A be a submodule of an R-module B. Then A is said to be a *Gorenstein essential submodule* of B if for each submodule N of B such that proj dim $N < \infty$, $N \cap A = 0$ implies $N = 0$. If A is a Gorenstein essential submodule of B and $A \subset B$ is a Gorenstein extension, then we say that $A \subset B$ is a *Gorenstein essential extension*.

It is trivial to see that every essential submodule is Gorenstein essential, and if $A \subset B \subset C$ are modules such that $A \subset C$ is Gorenstein essential, then A, B are Gorenstein essential submodules of B, C respectively.

Lemma 11.4.2. *Let R be n-Gorenstein. Then the following are equivalent for an R-module G.*

(1) *G is Gorenstein injective.*

(2) *G has no proper Gorenstein essential extensions.*

(3) *G is a direct summand of all Gorenstein extensions of itself.*

Proof. (1) $\Rightarrow$ (2). Let $G \subset H$ be a Gorenstein essential extension of a Gorenstein injective R-module G. Then $H \cong G \oplus E$ for some injective R-module E by Proposition 11.3.6. But proj dim $E < \infty$, $G \cap E = 0$, and G is a Gorenstein essential submodule of H. So $E = 0$.

(2) $\Rightarrow$ (3). Suppose G has a proper Gorenstein extension H, and let $\mathcal{C}$ be the collection of all nonzero submodules N of H such that proj dim $N < \infty$ and $N \cap G = 0$. $\mathcal{C} \neq \emptyset$ for otherwise H would be a proper Gorenstein essential extension of G. Now order $\mathcal{C}$ by inclusion and we note that $\mathcal{C}$ is an inductive system since R is n-Gorenstein. Hence $\mathcal{C}$ has a maximal element N_0 by Zorn's lemma. We now argue that in fact $H = G \oplus N_0$.

We first note that $H/(G + N_0) \cong (H/G)/(N_0/G \cap N_0) \cong (H/G)/N_0$. So $H/(G + N_0)$ has finite projective dimension since H/G and N_0 do. Hence $(G + N_0)/N_0 \subset H/N_0$ is a Gorenstein extension. But $G \cong (G + N_0)/N_0$ has no proper Gorenstein essential extensions. Hence if $H \neq G + N_0$, then $(G + N_0)/N_0 \subsetneq H/N_0$ is not a Gorenstein essential extension. So there is a submodule M of H such that $N_0 \subsetneq M$, proj dim $M/N_0 < \infty$, $(M/N_0) \cap ((G + N_0)/N_0) = 0$. But then proj dim $M < \infty$ and $M \cap G = 0$. So M is in $\mathcal{C}$ contradicting the maximality of N_0. Hence $H = G \oplus N_0$.

(3) $\Rightarrow$ (1). By Theorem 11.2.1, H has an injective Gorenstein injective preenvelope $H \to G$ with proj dim $G/H < \infty$. So $G \subset H$ is a Gorenstein extension. Thus G is a direct summand of H and so we are done.   □

Lemma 11.4.3. *If E is an injective submodule of an R-module M and $S \subset M$ is maximal with respect to $E \cap S = 0$, then $M = E \oplus S$.*

Proof. $(E \oplus S)/S$ is a summand of M/S since E is injective. But then $M/S \cong (T/S) \oplus E$ for some R-module T. Hence $E \cong M/S$ since S is maximal with respect to $E \cap S = 0$. Thus $M \cong E \oplus S$.   □

Definition 11.4.4. $A \subset C$ is said to be a *minimal Gorenstein injective extension* if it is a Gorenstein injective extension and whenever $A \subset B \subsetneq C$ is such that $A \subset B$ is a Gorenstein extension, then B is not Gorenstein injective. $A \subset B$ is said to be a *maximal Gorenstein essential extension* if $A \subset B$ is a Gorenstein essential extension and whenever $A \subset B \subsetneq C$ is such that $A \subsetneq C$ is a Gorenstein extension, then A is not a Gorenstein essential submodule of C.

Theorem 11.4.5. *Let R be n-Gorenstein. Then the following are equivalent for a submodule M of an R-module G.*

(1) *$M \subset G$ is a Gorenstein injective envelope.*

(2) *G is a Gorenstein essential Gorenstein injective extension of M.*

(3) *G is a minimal Gorenstein injective extension of M.*

(4) *G is a maximal Gorenstein essential extension of M.*

Proof. (1) $\Rightarrow$ (2). $M \subset G$ is a Gorenstein injective extension by Theorem 11.2.1. We now show that M is a Gorenstein essential submodule.

Let $N \subset G$ be a submodule of finite projective dimension such that $N \cap M = 0$. Then $E(N)$ is a submodule of G by Proposition 11.3.7. Now let $G' \supset M$ be maximal in G with respect to $G' \cap E(N) = 0$. Thus $G \cong E(N) \oplus G'$ by Lemma 11.4.3 above and $G/M \cong E(N) \oplus G'/M$. But then $M \subset G'$ is a Gorenstein injective extension. So $G' \cong G(M) \oplus E = G \oplus E$ for some injective R-module E by Proposition 11.3.6. Hence $E(N) = 0$ and so M is a Gorenstein essential submodule.

$(2) \Rightarrow (3)$. Let H be a Gorenstein injective extension of M contained in G. Then $\text{proj dim } G/H < \infty$ and so $G \cong H \oplus E$ for some injective R-module E. But $E \cap H = 0$ and $\text{proj dim } E < \infty$. So $E = 0$ since M is a Gorenstein essential submodule of G. Thus $H = G$.

$(3) \Rightarrow (1)$. By Proposition 11.3.6, $G(M)$ is a direct summand of G and so $G(M) \cong G$ by minimality.

$(2) \Leftrightarrow (4)$ follows from the equivalence of (1) and (2) in Lemma 11.4.2. $\quad\square$

Exercises

1. Let $A \subset B \subset C$ be modules and suppose $A \subset C$ is Gorenstein essential. Prove that $A \subset B$ and $B \subset C$ are also Gorenstein essential.

2. Prove the equivalence of parts (2) and (4) of Theorem 11.4.5.

3. Argue that if A is an essential submodule of B and B is a Gorenstein essential submodule of C, then A is a Gorenstein essential submodule of C.

4. Let R be n-Gorenstein. Prove that if $A \subset B$ is an essential Gorenstein extension, then $G(A) \cong G(B)$.

11.5 Gorenstein projective precovers and covers

We now consider the existence of Gorenstein projective precovers. We first note that since projective modules are Gorenstein projective, such precovers are necessarily surjective.

Theorem 11.5.1. *If R is n-Gorenstein, then every module M has a Gorenstein projective precover $C \to M$ such that if $0 \to L \to C \to M \to 0$ is exact, then $\text{proj dim } L \leq n - 1$ whenever $n \geq 1$.*

Proof. The proof is dual to the proof of Theorem 11.2.1. It begins with an appeal to Theorem 10.2.14 and then follows by a dual argument. $\quad\square$

Remark 11.5.2. We note that if M is finitely generated, then a Gorenstein projective precover constructed in the above may be chosen to be also finitely generated.

Corollary 11.5.3. *If R is n-Gorenstein, then the following are equivalent for any R-module M.*

(1) *M is Gorenstein projective.*

(2) *$\text{Ext}^i(M, L) = 0$ for all R-modules L with $\text{proj dim } L < \infty$ and all $i \geq 1$.*

(3) $\text{Ext}^1(M, L) = 0$ *for all R-modules L with* $\text{proj dim } L < \infty$.

Furthermore, if M is finitely generated, then each of the above statements is equivalent to

(4) $\text{Ext}^i(M, L) = 0$ *for all finitely generated R-modules L with* $\text{proj dim } L < \infty$ *and all* $i \geq 1$.

(5) $\text{Ext}^1(M, L) = 0$ *for all finitely generated R-modules L with* $\text{proj dim } L < \infty$.

(6) $\text{Ext}^i(M, P) = 0$ *for all projective R-modules P and all* $i \geq 1$.

Proof. (1) $\Rightarrow$ (2) by Remark 10.2.2 and (3) $\Rightarrow$ (1) follows from the theorem above.
(2) $\Rightarrow$ (3); (2) $\Rightarrow$ (4) $\Rightarrow$ (5); and (2) $\Rightarrow$ (6) are trivial.
(5) $\Rightarrow$ (1) follows from the theorem choosing the precover to be finitely generated and (6) $\Rightarrow$ (1) follows from Theorem 10.2.16.   □

Corollary 11.5.4. *If R is a commutative local Gorenstein ring, then a finitely generated R-module is Gorenstein projective if and only if it is maximal Cohen–Macaulay.*

Proof. This is left to the reader.   □

Remark 11.5.5. Dual to Corollary 11.2.4, we have that for a Gorenstein projective precover $C \to M$ of Theorem 11.5.1, $\text{proj dim } M < \infty$ if and only if C is projective and if and only if $C \to M$ is a projective precover.

We are now in a position to prove a Gorenstein version of Theorem 10.2.8.

Theorem 11.5.6. *Let R be n-Gorenstein and* $0 \to M' \to M \to M'' \to 0$ *be an exact sequence of R-modules. If* M', M'' *are Gorenstein projective, then so is M, and if* M, M'' *are Gorenstein projective, then so is* M'.

Proof. We consider the following exact sequence

$$\cdots \to \text{Ext}^1(M'', L) \to \text{Ext}^1(M, L) \to \text{Ext}^1(M', L) \to \text{Ext}^2(M'', L) \to \cdots$$

with $\text{proj dim } L < \infty$. So if M', M'' are Gorenstein projective, then $\text{Ext}^1(M'', L) = \text{Ext}^1(M', L) = 0$ and hence $\text{Ext}^1(M, L) = 0$. Thus M is Gorenstein projective by Corollary 11.5.3. Similarly for M'.   □

The class of Gorenstein projective R-modules, denoted $\mathcal{G}or\mathcal{P}roj$, is precovering over n-Gorenstein rings by Theorem 11.5.1 and hence every R-module has a left $\mathcal{G}or\mathcal{P}roj$-resolution. This resolution is is exact and is usually called a *Gorenstein projective resolution.* We now have the following result which is dual to Proposition 11.2.5.

Proposition 11.5.7. *Let R be n-Gorenstein and* $\mathcal{L}$ *be the class of R-modules of finite projective dimension. Then the following are equivalent for an R-module M and integer* $r \geq 0$.

(1) left $\mathcal{G}or\mathcal{P}roj$- dim $M \leq r$.

(2) *There exists an exact sequence* $0 \to C_r \to C_{r-1} \to \cdots \to C_1 \to C_0 \to M \to 0$ *with each* C_i *Gorenstein projective.*

(3) $\text{Ext}^i(M, L) = 0$ *for all* $i \geq r + 1$ *and all* $L \in \mathcal{L}$.

(4) $\text{Ext}^{r+1}(M, L) = 0$ *for all* $L \in \mathcal{L}$.

(5) *Every* r*th* $\mathcal{G}or\mathcal{P}roj$*-syzygy of* M *is Gorenstein projective.*

(6) *Every* r*th* $\mathcal{P}roj$*-syzygy of* M *is Gorenstein projective.*

Furthermore, if M *is finitely generated, then each of the above statements is equivalent to*

(7) $\text{Ext}^i(M, P) = 0$ *for all projective R-modules* P *and all* $i \geq r + 1$.

Proof. (1) $\Rightarrow$ (2); (3) $\Rightarrow$ (4) and (7); (5) $\Rightarrow$ (1) and (6) are trivial.

(2) $\Rightarrow$ (3). $\text{Ext}^i(M, L) \cong \text{Ext}^{i-r}(C_r, L) = 0$ for all $i > r$ and all $L \in \mathcal{L}$. So (3) follows.

(4) $\Rightarrow$ (5). Let K be an rth $\mathcal{G}or\mathcal{P}roj$-syzygy. Then $\text{Ext}^{r+1}(M, L) \cong \text{Ext}^1(K, L)$ for all $L \in \mathcal{L}$. But $\text{Ext}^1(K, L) = 0$ for all $L \in \mathcal{L}$ implies K is Gorenstein projective by Corollary 11.5.3.

(6) $\Rightarrow$ (3). Let $\cdots \to P_1 \to P_0 \to M \to 0$ be a projective resolution of M and $K = \text{Ker}(P_{r-1} \to P_{r-2})$. Then $\text{Ext}^i(M, L) \cong \text{Ext}^{i-r}(K, L)$ for all $L \in \mathcal{L}$ and so (3) now easily follows.

(7) $\Rightarrow$ (6). If K is an rth $\mathcal{P}roj_{fg}$-syzygy of M, then $\text{Ext}^{i-r}(K, P) \cong \text{Ext}^i(M, P)$ for all $i > r$ and K is finitely generated. But $\text{Ext}^{i-r}(K, P) = 0$ for all $i > r$ by assumption. So K is Gorenstein projective by Corollary 11.5.3. $\square$

Corollary 11.5.8. *The following properties hold for any R-module* M.

(1) left $\mathcal{G}or\mathcal{P}roj$- dim $M \leq n$.

(2) left $\mathcal{G}or\mathcal{P}roj$- dim $M = \text{proj dim } M$ *if and only if* $\text{proj dim } M < \infty$.

We can now prove a Gorenstein version of Proposition 10.2.17.

Proposition 11.5.9. *Let R be n-Gorenstein. Then the following are equivalent for an R-module N*

(1) $\text{Ext}^1(M, N) = 0$ *for all Gorenstein projective R-modules* M.

(2) $\text{Ext}^i(M, N) = 0$ *for all Gorenstein projective R-modules* M *and all* $i \geq 1$.

(3) N *has finite projective dimension.*

(4) N *has finite flat dimension.*

(5) $\text{Tor}_i(M, N) = 0$ *for all* $i \geq 1$ *and all Gorenstein flat right R-modules* M.

(6) $\text{Tor}_1(M, N) = 0$ *for all Gorenstein flat right R-modules* M.

Proof. (1) $\Leftrightarrow$ (2). Assume (1). We consider the exact sequence $0 \to M' \to P \to M \to 0$ with P projective. Then M' is Gorenstein projective by Theorem 11.5.6. So $\mathrm{Ext}^2(M, N) \cong \mathrm{Ext}^1(M', N) = 0$ and thus (2) follows by induction. The converse is trivial.

(2) $\Leftrightarrow$ (3). Assume (2). Let $\cdots \to P_1 \to P_0 \to N \to 0$ be a projective resolution of N and $K = \mathrm{Ker}(P_{n-1} \to P_{n-2})$. Then K is Gorenstein projective by Theorem 10.2.14. So $\mathrm{Ext}^n(K, N) = 0$ by assumption. Thus $\mathrm{Hom}(K, P_{n+1}) \to \mathrm{Hom}(K, P_n) \to \mathrm{Hom}(K, P_{n-1})$ is exact. So K is a summand of P_n and thus proj dim $N \leq n$. The converse follows from Remark 10.2.2.

(1) $\Leftrightarrow$ (4) by Theorem 9.1.10.

(5) $\Leftrightarrow$ (4). Assume (5). $\mathrm{Tor}_i(M, N) = 0$ means $\mathrm{Ext}^i(M, N^+) = 0$. Thus $\mathrm{Ext}^i(M, N^+) = 0$ for all $i \geq 1$ and all Gorenstein projective R-modules M since Gorenstein projectives are Gorenstein flat. So flat dim $N^+ < \infty$ by the above. Thus inj dim $N^+ < \infty$ since R is n-Gorenstein. Therefore flat dim $N < \infty$. The converse follows from Theorem 10.3.8 and Theorem 9.1.10.

(6) $\Leftrightarrow$ (5). Assume (6). We consider the sequence $0 \to M' \to P \to M \to 0$ with P projective. Then M' is Gorenstein flat by Theorem 10.3.14. So $\mathrm{Tor}_2(M, N) \cong \mathrm{Tor}_1(M', N) = 0$ and so (5) follows by induction. The converse is trivial. $\square$

Remark 11.5.10. By Proposition 11.5.9 above and Corollary 11.5.3, we see that $(\mathcal{G}or\mathcal{P}roj, \mathcal{L})$ is a cotorsion theory over any Iwanaga–Gorenstein ring R and has enough injectives and projectives by Proposition 7.1.7 and Theorem 11.5.1.

In general, modules over n-Gorenstein rings do not have Gorenstein projective covers. This can be seen by considering the 1-Gorenstein ring $\mathbb{Z}$ where the Gorenstein projective modules are the free modules. When the ring R is local, all finitely generated left R-modules have projective covers by Theorem 5.3.3. So it is natural to raise the analogous question in the Gorenstein situation.

If R is local, commutative and Gorenstein, Auslander announced that all finitely generated R-modules have finitely generated Gorenstein projective covers. We will defer a proof of this fact to the next section where we will prove a more general result (concerning modules over a local Cohen–Macaulay ring admitting a dualizing module). In order to do this, we will need the following result.

Theorem 11.5.11. *Let R be a local Cohen–Macaulay ring of Krull dimension d admitting a dualizing module. Then every R-module $M \in \mathcal{G}_0(R)$ has a Gorenstein projective precover $\psi : C \to M$ with proj dim $\mathrm{Ker}\,\psi \leq d - 1$.*

Proof. The proof is like the proof of Theorem 11.5.1 above. We start by appealing to Lemma 10.4.15 and follow the same argument. $\square$

Corollary 11.5.12. *An R-module M is Gorenstein projective if and only if $M \in \mathcal{G}_0(R)$ and $\mathrm{Ext}^1(M, L) = 0$ for all R-modules L of finite projective dimension.*

Remark 11.5.13. Using Theorem 11.5.11 and Corollary 11.5.12 above, we get results corresponding to Remark 11.5.5 and Theorem 11.5.6 for R-modules $M, M', M'' \in \mathcal{G}_0(R)$ with identical proofs.

Proposition 11.5.14. *The following are equivalent for an R-module $N \in \mathcal{G}_0(R)$.*

(1) $\text{Tor}_i(N, M) = 0$ *for all $i \geq 1$ and all Gorenstein flat R-modules M.*

(2) $\text{Tor}_1(N, M) = 0$ *for all Gorenstein flat R-modules M.*

(3) *N has finite injective dimension.*

Proof. (1) $\Leftrightarrow$ (2) follows as in Proposition 11.5.9.

(1) $\Leftrightarrow$ (3). (1) means $\text{Ext}^i(M, N^+) = 0$ for all $i \geq 1$ and all Gorenstein projective R-modules M by Corollary 10.4.29. Thus flat dim $N^+ < \infty$ by a result dual to Proposition 11.2.8 above and so inj dim $N < \infty$. The converse by Theorem 10.4.28.□

Proposition 11.5.15. *An R-module K is cotorsion and has finite projective dimension if and only if $K \in \mathcal{G}_0(R)$ and $\text{Ext}^i(M, K) = 0$ for all Gorenstein flat R-modules M and all $i \geq 1$.*

Proof. If proj dim $K < \infty$, then $K \in \mathcal{G}_0(R)$. The second part follows from Corollary 10.4.27. Conversely, $\text{Ext}^1(F, K) = 0$ for all flat R-modules and so K is cotorsion, and proj dim $K < \infty$ by Corollary 10.4.29 and a result dual to Proposition 11.2.8. □

Exercises

1. Prove Theorem 11.5.1.

2. Prove that if R is Iwanaga–Gorenstein, then every finitely generated R-module M has an $\mathcal{L}$-preenvelope $M \to K$ with K finitely generated. And moreover, if $0 \to M \to K \to C \to 0$ is exact, then C is a finitely generated Gorenstein projective R-module.

3. Prove Corollary 11.5.4.
   Hint: Use Proposition 9.5.23.

4. Prove the second part of Theorem 11.5.6.

5. Prove Corollary 11.5.8.

6. Prove Theorem 11.5.11.

7. Prove Corollary 11.5.12.

8. Let R be a local Cohen–Macaulay ring of finite Krull dimension and $\mathcal{L}$ be the class of R-modules of finite projective dimension. State and prove a result corresponding to Proposition 11.5.7.

9. Let R be as in Problem 7 above. State and prove a result corresponding to Proposition 11.5.9.

11.6   Auslander's last theorem (Gorenstein projective covers)

In this section, we let R be a local Cohen–Macaulay ring of Krull dimension d admitting a dualizing module Ω. We will prove that any finitely generated R-module M with $M \in \mathcal{G}_0(R)$ (see Chapter 10, Section 4) has a Gorenstein projective cover. We recall from Proposition 10.4.3 that if R is Gorenstein, then $\mathcal{G}_0(R)$ is the class of all R-modules. So in this case, all finitely generated R-modules have Gorenstein projective covers. This result was first announced by Auslander.

We will now be concerned with the Matlis duals M^v of modules M in the Foxby classes $\mathcal{G}_0(R)$ and $\mathcal{J}_0(R)$. So we start with the following result.

Proposition 11.6.1. *If R is complete and M is finitely generated, then $M \in \mathcal{G}_0(R)$ if and only if $M^v \in \mathcal{J}_0(R)$.*

Proof. Suppose $M \in \mathcal{G}_0(R)$. Then since $\mathrm{Tor}_i(\Omega, M) = 0$, we get $\mathrm{Ext}^i(\Omega, M^v) = (\mathrm{Tor}_i(\Omega, M))^v = 0$ for $i \geq 1$. To get $\mathrm{Tor}_i(\Omega, \mathrm{Hom}(\Omega, M^v)) = 0$ we only need to establish that

$$(\mathrm{Tor}_i(\Omega, \mathrm{Hom}(\Omega, M^v)))^v = \mathrm{Ext}^i(\Omega, \mathrm{Hom}(\Omega, M^v)^v) = 0$$

for $i \geq 1$. But $\mathrm{Hom}(\Omega, M^v)^v \cong (\Omega \otimes M)^{vv} \cong \Omega \otimes M$ and $\mathrm{Ext}^i(\Omega, \Omega \otimes M) = 0$ for $i \geq 1$ by hypothesis.

Now we need to establish that $\Omega \otimes \mathrm{Hom}(\Omega, M^v) \to M^v$ is an isomorphism. We have that $M \to \mathrm{Hom}(\Omega, \Omega \otimes M)$ is an isomorphism and so $\mathrm{Hom}(\Omega, \Omega \otimes M)^v \to M^v$ is an isomorphism. Also $\Omega \otimes \mathrm{Hom}(\Omega, M^v) \cong \Omega \otimes (\Omega \otimes M)^v$ and so we only need establish that the natural map $\Omega \otimes (\Omega \otimes M)^v \to \mathrm{Hom}(\Omega, \Omega \otimes M)^v$ is an isomorphism. But the two functors $- \otimes (\Omega \times M)^v$ and $\mathrm{Hom}(-, \Omega \otimes M)^v$ are right exact and the natural transformation $- \otimes (\Omega \otimes M)^v \to \mathrm{Hom}(-, \Omega \otimes M)^v$ is an isomorphism when evaluated at R^n for any $n \geq 1$. So a standard argument gives that $\Omega \otimes (\Omega \otimes M)^v \to \mathrm{Hom}(\Omega, \Omega \otimes M)^v$ is an isomorphism. Thus we have $M^v \in \mathcal{J}_0(R)$.

Now let $N = M^v$ and assume $N \in \mathcal{J}_0(R)$. We want to show $M = N^v \in \mathcal{J}_0(R)$. We have $\mathrm{Tor}_i(\Omega, M)^v = \mathrm{Ext}^i(\Omega, M^v) = \mathrm{Ext}^i(\Omega, N)$. Since $N \in \mathcal{J}_0(R)$, $\mathrm{Ext}^i(\Omega, N) = 0$ and so $\mathrm{Tor}_i(\Omega, M) = 0$. But also, $\mathrm{Tor}_i(\Omega, \mathrm{Hom}(\Omega, N)) = 0$ for $i \geq 1$. So $0 = \mathrm{Tor}_i(\Omega, \mathrm{Hom}(\Omega, N))^v = \mathrm{Ext}^i(\Omega, \mathrm{Hom}(\Omega, N)^v)$. But $\mathrm{Hom}(\Omega, N)^v = \mathrm{Hom}(\Omega, M^v)^v \cong (\Omega \otimes M)^{vv} \cong \Omega \otimes M$. So $\mathrm{Ext}^i(\Omega, \Omega \otimes M) = 0$ for $i \geq 1$.

To establish that $M \to \mathrm{Hom}(\Omega, \Omega \otimes M)$ is an isomorphism, we note that $\Omega \otimes \mathrm{Hom}(\Omega, N) \to N$ is an isomorphism, and so $N^v \to (\Omega \otimes \mathrm{Hom}(\Omega, N))^v \cong \mathrm{Hom}(\Omega, \mathrm{Hom}(\Omega, N)^v)$ is an isomorphism. So we need that the natural map $\Omega \otimes M = \Omega \otimes \mathrm{Hom}(N, E(k)) \to \mathrm{Hom}(\mathrm{Hom}(\Omega, N), E(k))$ is an isomorphism. This is the case since the natural transformation

$$- \otimes \mathrm{Hom}(N, E(k)) \to \mathrm{Hom}(\mathrm{Hom}(-, N), E(k))$$

is an isomorphism on all R^n, $n \geq 1$, and since both functors are right exact.   □

Corollary 11.6.2. *If C is a finitely generated module, then C is Gorenstein projective if and only if C^v is Gorenstein injective.*

Proof. If C is Gorenstein projective, then there is an exact sequence $0 \to C \to P_{d-1} \to \cdots \to P_0 \to M \to 0$ with $P_0, \ldots, P_{d-1}$ finitely generated projective modules. Hence by Proposition 10.4.17 $M \in \mathcal{G}_0(R)$. So $M^v \in \mathcal{J}_0(R)$ by the proposition above. But $0 \to M^v \to P_0^v \to \cdots \to P_{d-1}^v \to C^v \to 0$ is exact and $P_0^v, \ldots, P_{d-1}^v$ are injective. So C^v is Gorenstein injective by Proposition 10.4.23.

Conversely, suppose C^v is Gorenstein injective. Since C is finitely generated, C^v is artinian. We claim that if $E \to C^v$ is an injective cover of C, then E is also artinian. That is, E is of the form $E(k)^n$ for some $n \geq 0$. To see this, recall that C has a flat preenvelope $C \to F$. Since $C \to F$ can be factored $C \to R^n \to F$ for some $n \geq 0$, we can assume $F = R^n$. But then $E(k)^n = (R^n)^v \to C^v$ is an injective precover and E is isomorphic to a summand of $E(k)^n$ and thus is artinian. Hence we see that if $E_{d-1} \to \cdots \to E_1 \to E_0 \to C^v \to 0$ is a partial minimal left $\mathcal{I}nj$-resolution of C^v, then it is exact with $E_{d-1}, \ldots, E_0$ artinian. So if $0 \to N \to E_{d-1} \to \cdots \to E_1 \to E_0 \to C^v \to 0$ is exact, we get $N \in \mathcal{G}_0(R)$ by Proposition 10.4.23 and so by Proposition 11.6.1 above $N^v \in \mathcal{G}_0(R)$. But $0 \to C(= C^{vv}) \to E_0^v \to E_1^v \to \cdots \to E_{d-1}^v \to N^v \to 0$ is exact and $E_0^v, \ldots, E_{d-1}^v$ are free. So by Proposition 10.4.17 C is Gorenstein projective.   □

Theorem 11.6.3. *Let R be complete and M be a finitely generated R-module. If $M \in \mathcal{G}_0(R)$, then M has a Gorenstein projective cover $C \to M$. If $C \to M$ is such a cover, then C is finitely generated and $\operatorname{proj} \dim \operatorname{Ker}(C \to M) < \infty$.*

Proof. By Theorem 11.5.11 there is an exact sequence $0 \to L \to C \to M \to 0$ with C Gorenstein projective and with $\operatorname{proj} \dim L < \infty$. By the proof of that result, we see that we can assume C is finitely generated. But then $0 \to M^v \to C^v \to L^v \to 0$ is exact. By Corollary 11.6.2, C^v is Gorenstein injective. Since $\operatorname{proj} \dim L < \infty$, we have $\operatorname{inj} \dim L^v < \infty$. Since $\operatorname{Ext}^1(L^v, G) = 0$ whenever G is Gorenstein injective (Proposition 10.1.3), we have that $\operatorname{Hom}(C^v, G) \to \operatorname{Hom}(M^v, G) \to 0$ is exact. That is, $M^v \to C^v$ is a Gorenstein injective preenvelope. By Proposition 11.6.1 and Exercise 4 of Section 11.3, M^v has a Gorenstein injective envelope $\varphi : M^v \to G$ and G is a retract of C^v and so is artinian. But then $\operatorname{Coker} \varphi$ is a retract of L^v. So $\operatorname{inj} \dim \operatorname{Coker} \varphi < \infty$ since $\operatorname{inj} \dim L^v < \infty$. So with $\bar{L} = \operatorname{Coker} \varphi^v$, we have $\operatorname{proj} \dim \bar{L} < \infty$. Thus with $\bar{L} = (G/M^v)^v$, we have that $0 \to \bar{L} \to G^v \to M \to 0$ is exact with $\operatorname{proj} \dim \bar{L} < \infty$. But G^v is Gorenstein projective if and only if $G \cong G^{vv}$ is Gorenstein injective by Corollary 11.6.2. So by Proposition 10.2.6, $G^v \to M$ is a Gorenstein projective precover. But M and G are Matlis reflexive (and M^v and G^v are also) and $M^v \to G$ is an envelope. So $G^v \to M$ is a cover. This completes the proof.   □

We now show that we can drop the hypothesis of completeness from the preceding theorem.

Theorem 11.6.4. *Let M be finitely generated such that $M \in \mathcal{G}_0(R)$, and let $\varphi : C \to M$ be a surjective linear map where C is Gorenstein projective. If M has a Gorenstein projective cover, then $\varphi : C \to M$ is a Gorenstein projective cover if and only if* proj dim Ker $\varphi < \infty$ *and* Ker φ *contains no nonzero projective summands of C.*

Proof. M has a Gorenstein projective precover $\psi : C' \to M$ by Theorem 11.5.11 where proj dim Ker $\psi < \infty$. If $\varphi : C \to M$ is a cover, then $C \to M$ is a retract of $C' \to M$ (over M) and so Ker φ is a retract of Ker ψ. Hence proj dim Ker $\varphi < \infty$. If Ker φ contains a projective (so free) summand F of C, then there is easily an $f : C \to C$ with $\varphi \circ f = \varphi$ which is not an automorphism of C. Hence the two conditions are necessary.

Now assume the conditions. Then if $\psi : C' \to M$ is a cover we have a commutative diagram

$$
\begin{array}{ccc}
C' & & \\
\downarrow & \searrow & \\
C & \to & M \\
\downarrow & \nearrow & \\
C'. & &
\end{array}
$$

Since $C \to M$ is a cover, $C' \to C \to C'$ is an automorphism of C. So Ker$(C \to C')$ is a summand of C containing Ker$(C \to M)$. So Ker$(C \to C')$ is Gorenstein projective. But Ker$(C \to C')$ is also isomorphic to a summand of Ker φ and so proj dim Ker$(C \to C') < \infty$. Hence by Proposition 10.2.3, Ker$(C \to C')$ is projective and so is zero by hypothesis. Hence $C \to C'$ is an isomorphism and thus $C \to M$ is also a cover.   □

Proposition 11.6.5. *For each finitely generated M with $M \in \mathcal{G}_0(R)$, there exists a Gorenstein projective precover $\varphi : C \to M$ with C finitely generated and* proj dim Ker $\varphi < d - 1$ *such that C has a direct sum decomposition $C = U \oplus F$ where U has no nonzero free summands and $F \to M/\varphi(U)$ is a projective cover.*

Proof. By Theorem 11.5.11, there exists a Gorenstein projective precover $\psi : C' \to M$ with proj dim Ker $\psi < d - 1$. From the proof of that theorem, it is easy to see that if M is finitely generated, then C' can be chosen to be finitely generated. Let $C' = U \oplus \bar{F}$ where U has no nonzero free summands. Since ψ is necessarily a surjection, $\bar{F} \to M/\psi(U)$ is a surjection. But then $\bar{F}$ has a direct sum decomposition $\bar{F} = F \oplus F'$ such that $F \to M/\psi(U)$ is a projective cover and such that the map $F' \to M/\psi(U)$ is zero. This implies the map $F' \to M$ has a lifting $g : F' \to U$. Now let $C = U \oplus F$ and $\varphi = \psi_{|C}$. Then we see that there is a retraction of $C' = U \oplus F \oplus F' \to M$ to $C \to M$ over M. In matrix notation, the retraction is given by

$$
\begin{pmatrix}
\mathrm{id}_U & 0 & 0 \\
0 & \mathrm{id}_F & 0 \\
g & 0 & 0
\end{pmatrix}.
$$

This gives that Ker φ is a retract of Ker ψ and so proj dim Ker $\varphi < d - 1$. Hence $\varphi : C = U \oplus F \rightarrow M$ is the required precover.   □

Lemma 11.6.6. *If $\varphi : C \rightarrow M$ is linear and $C = U \oplus F$ where U has no nonzero free summands and $F \rightarrow M/\varphi(U)$ is a projective cover, then* Ker φ *contains no nonzero free summands of C.*

Proof. By contradiction. Suppose $x = (x_1, x_2) \in U \oplus F$ generates a free summand of C contained in Ker φ. Then $\varphi(x_1) = \varphi(-x_2) \in \varphi(U)$. Since $F \rightarrow M/\varphi(U)$ is a projective cover and maps $-x_2$ to 0, we have $-x_2 \in m(R)F$. Since $x \in C$ generates a free summand of C, we have $\sigma \in C^* = \text{Hom}(C, R)$ with $\sigma(x) = 1$. Then $\sigma(x_1) + \sigma(x_2) = 1$. But $x_2 \in m(R)F$ and so $\sigma(x_2) \in m(R)$. Hence $\sigma(x_1)$ is a unit of R. This implies U has a rank one free summand, contradicting our hypothesis.   □

Lemma 11.6.7. *If $\varphi : C \rightarrow M$ is a Gorenstein projective precover of M with* proj dim Ker $\varphi < \infty$ *and if M has a cover, then $C \rightarrow M$ is a cover if and only if* Ker φ *contains no nonzero free summand of C.*

Proof. If $C' \rightarrow M$ is a cover, then we have a decomposition $C = C' \oplus F$ with $F \subset$ Ker φ. Then proj dim $F < \infty$ and F is Gorenstein projective. So by Proposition 10.2.3, F is projective and thus free. So if $\varphi : C \rightarrow M$ satisfies the hypothesis, $F = 0$ and hence $\varphi : C \rightarrow M$ is a cover.

If $\varphi : C \rightarrow M$ does not satisfy the hypothesis, there is clearly an $f : C \rightarrow C$ with $\varphi \circ f = \varphi$ and f not an isomorphism (we can choose f with $F \subset$ Ker f).   □

We now let $\hat{R}$ denote the completion of R. For a finitely generated module M, we let $\hat{M}$ denote the completion of M. Then $\hat{M} \cong \hat{R} \otimes_R M$. If M and N are finitely generated and $f : M \rightarrow N$ is linear, we have $\hat{f} : \hat{M} \rightarrow \hat{N}$. Since $\hat{R}$ is a faithfully flat R-module, $\hat{f}$ is an isomorphism if and only if f is an isomorphism. Furthermore, $\text{Hom}_{\hat{R}}(\hat{M}, \hat{N})$ can be identified with the completion of $\text{Hom}_R(M, N)$ and the completion of $M \otimes_R N$ is $\hat{M} \otimes_{\hat{R}} \hat{N}$ (see Theorems 3.2.5 and 2.1.11). Hence we have the following result.

Lemma 11.6.8. *Given finitely generated R-modules M and N, N is isomorphic to a summand of M if and only if $\hat{N}$ is isomorphic (as an $\hat{R}$-module) to a summand of $\hat{M}$.*

Proof. N is isomorphic to a summand of M if and only if id_N is in the image of $\text{Hom}_R(N, M) \times \text{Hom}_R(M, N) \rightarrow \text{Hom}_R(N, N)$ (with the map $(f, g) \mapsto g \circ f$), or equivalently, if and only if $\text{Hom}_R(N, M) \otimes_R \text{Hom}_R(M, N) \rightarrow \text{Hom}_R(N, N)$ is surjective. Now apply the remarks above.   □

We note that this lemma says M has a nonzero free summand if and only if $\hat{M}$ does.

Theorem 11.6.9. *If M is a finitely generated R-module and $M \in \mathcal{G}_0(R)$, then M has a Gorenstein projective cover $\varphi : C \to M$. Furthermore C is finitely generated and* proj dim Ker $\varphi < d - 1$.

Proof. By Proposition 11.6.5, there is a precover $\varphi : C \to M$ with C finitely generated and proj dim Ker $\varphi < d-1$ and such that C has a direct sum decomposition $C = U \oplus F$ where U has no nonzero free summands and $F \to M/\varphi(U)$ is a projective cover. We want to show that in fact $\varphi : C \to M$ is a cover.

By our remarks above about completions and by Lemma 11.6.8, we see that $\hat{\varphi} : \hat{C} \to \hat{M}$ inherits all the properties of φ. So $\hat{F} \to \hat{M}/\hat{\varphi}(\hat{U})$ is a projective cover since its kernel is contained in $\widehat{m(R)F} = m(\hat{R})\hat{F}$. But then since we know by Theorem 11.6.3 that $\hat{M}$ has a cover, we see that in fact $\hat{\varphi} : \hat{C} \to \hat{M}$ is a cover by Lemmas 11.6.6 and 11.6.7.

Now let $f : C \to C$ be such that $\varphi \circ f = \varphi$. Then $\hat{\varphi} \circ \hat{f} = \hat{\varphi}$ and $\hat{f}$ is an automorphism of $\hat{C}$ since $\hat{\varphi} : \hat{C} \to \hat{M}$ is a cover. This implies that f is an automorphism of C and so $\varphi : C \to M$ is the desired Gorenstein projective cover. □

Corollary 11.6.10. *If $C \to M$ is a Gorenstein projective cover, then* proj dim $M < \infty$ *if and only if $C \to M$ is a projective cover.*

Theorem 11.6.11. *Let M be a finitely generated R-module and $M \in \mathcal{G}_0(R)$. Then the minimal left Gorenstein projective resolution of M is of the form*

$$\cdots \to P_2 \to P_1 \to C_0 \to M \to 0$$

where C_0 is Gorenstein projective and P_i is projective for each $i \geq 1$ and $P_i = 0$ for $i > d$.

Exercises

1. Prove that $M \in \mathcal{G}_0(R)$ if and only if $M^v \in \mathcal{J}_0(R)$.

2. Prove Corollary 11.6.10.

3. Prove Theorem 11.6.11.

11.7   Gorenstein flat covers

Our aim in this section is to prove that Gorenstein flat covers exist for all modules over n-Gorenstein rings. The proof will follow from the following results. But first we note that the class of Gorenstein flat modules is closed under direct limits and so to find a Gorenstein flat cover it suffices to find a Gorenstein flat precover by Corollary 5.2.7.

The following lemma uses $\bar{\lambda}$-dimension defined in 8.6.11.

Lemma 11.7.1. *Let R be n-Gorenstein and $\mathcal{F}$ be the class of Gorenstein flat left R-modules, then $\bar{\lambda}_{\mathcal{F}}(P) = \infty$ for every pure injective left R-module P.*

Proof. Let N be any right R-module and let $N \subset G$ be a Gorenstein injective envelope. Then we have the exact sequence $0 \to (G/N)^+ \to G^+ \to N^+ \to 0$ where G^+ is a Gorenstein flat left R-module by Corollary 10.3.9.

$\operatorname{Ext}^1(F, (G/N)^+) \cong \operatorname{Tor}_1(F, G/N)^+ = 0$ if F is a Gorenstein flat left R-module since $\operatorname{Tor}_1(F, G/N) = 0$ by Theorems 11.3.2 and 10.3.8. Hence in the language of Chapter 7, Section 1, $G^+ \to N^+$ is a special Gorenstein flat precover (see Definition 7.1.6).

Now let P be a pure injective left R-module and set $N = P^+$. Then we have a special Gorenstein flat precover $G^+ \to N^+ = P^{++}$. Since P is pure injective, it is a direct summand of P^{++} and so P has a Gorenstein flat precover. But then by Corollary 10.3.12 and Corollary 5.2.7, P has a Gorenstein flat cover $F \to P$. So there exists a commutative diagram

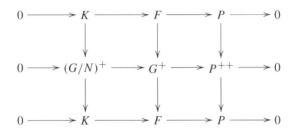

with exact rows and $P \to P^{++} \to P$ the identity on P. Since $F \to P$ is a Gorenstein flat cover, we see that F is isomorphic to a direct summand of G^+ and K is isomorphic to a direct summand of $(G/N)^+$. Since $(G/N)^+$ is pure injective, so is K. Since $\operatorname{Ext}^1(F', (G/N)^+) = 0$ for F' Gorenstein flat, we have $\operatorname{Ext}^1(F', K) = 0$ also. Hence $0 \to K \to F \to P \to 0$ is exact with $F \to P$ a special Gorenstein flat cover and K pure injective. But then we can repeat the argument with K replacing P. Proceeding in this manner we see that $\bar{\lambda}_{\mathcal{F}}(P) = \infty$. $\square$

Corollary 11.7.2. *For every R-module L of finite injective dimension, $\bar{\lambda}_{\mathcal{F}}(L) = \infty$ where $\mathcal{F}$ is the class of Gorenstein flat R-modules.*

Proof. If L is injective then L is pure injective and so the result holds by the lemma. If $\operatorname{inj} \dim L < \infty$, then we see that a repeated application of Theorem 8.6.16 gives the result. $\square$

Theorem 11.7.3. *If R is n-Gorenstein, then every R-module M has a Gorenstein flat cover $F \to M$ such that if $0 \to L \to F \to M \to 0$ is exact, then $\operatorname{proj} \dim L < \infty$.*

Proof. We will argue that for every R-module M, $\bar{\lambda}_{\mathcal{F}}(M) = \infty$ with $\mathcal{F}$ the class of Gorenstein flat R-modules. By Theorem 11.5.1, there is an exact sequence $0 \to L \to C \to M \to 0$ with C Gorenstein projective and with $\operatorname{proj} \dim L < \infty$.

By Theorem 9.1.10, we also have inj dim $L < \infty$. By Corollary 11.7.2, $\bar{\lambda}_{\mathcal{F}}(L) = \infty$. By Corollary 10.3.10, C is Gorenstein flat and so easily $\bar{\lambda}_{\mathcal{F}}(C) = \infty$. Then Theorem 8.6.13 says $\bar{\lambda}_{\mathcal{F}}(M) = \infty$. So M has a special Gorenstein flat precover $\bar{F} \xrightarrow{\varphi} M$ with $\mathrm{Ext}^1(F', \mathrm{Ker}\,\varphi) = 0$ for all Gorenstein flat R-modules F'. But then $\mathrm{Ext}^1(C, \mathrm{Ker}\,\varphi) = 0$ for all Gorenstein projective R-modules C. So proj dim $\mathrm{Ker}\,\varphi < \infty$ by Proposition 11.5.9. $\square$

Corollary 11.7.4. *If $F \to M$ is a Gorenstein flat cover, then* flat dim $M < \infty$ *if and only if $F \to M$ is a flat cover.*

We note that it follows from the above that the class of Gorenstein flat R-modules, denoted $\mathcal{Gor}\mathcal{F}lat$, over an n-Gorenstein ring is covering and hence every R-module has a minimal left $\mathcal{Gor}\mathcal{F}lat$-resolution. This resolution is exact and is usually called a *minimal Gorenstein flat resolution*.

Proposition 11.7.5. *Let R be n-Gorenstein and $\mathcal{L}$ be the class of R-modules of finite projective dimension. Then the following are equivalent for an R-module M and integer $r \geq 0$.*

(1) left $\mathcal{Gor}\mathcal{F}lat$- dim $M \leq r$.

(2) $\mathrm{Tor}_i(L, M) = 0$ *for all $i \geq r + 1$ for all $L \in \mathcal{L}$.*

(3) $\mathrm{Tor}_{r+1}(L, M) = 0$ *for all $L \in \mathcal{L}$.*

(4) $\mathrm{Tor}_i(E, M) = 0$ *for all injective R-modules E and all $i \geq r + 1$.*

(5) $\mathrm{Tor}_{r+1}(E, M) = 0$ *for all injective R-modules E.*

(6) *Every rth $\mathcal{Gor}\mathcal{F}lat$-syzygy is Gorenstein flat.*

(7) *Every rth $\mathcal{F}lat$-syzygy is Gorenstein flat.*

(8) right $\mathcal{Gor}\mathcal{I}nj$- dim $M^+ \leq r$.

Proof. (1) $\Leftrightarrow$ (2). Let $\cdots \to F_1 \to F_0 \to M \to 0$ be a left $\mathcal{Gor}\mathcal{F}lat$-resolution of M and $K = \mathrm{Ker}(F_{r-1} \to F_{r-2})$. Then $\mathrm{Tor}_{i+r}(L, M) \cong \mathrm{Tor}_i(L, K)$ for all $i \geq 1$ and all $L \in \mathcal{L}$. So K is Gorenstein flat if and only if $\mathrm{Tor}_{i+r}(L, M) = 0$ for all $i \geq 1$ and all $L \in \mathcal{L}$ by Theorem 10.3.8.

The equivalence of (1), (3), (4), (5), (6) and (7) follows similarly from Theorem 10.3.8.

1 $\Leftrightarrow$ 8. We consider the exact sequence $0 \to K \to F_{r-1} \to \cdots \to F_1 \to F_0 \to M \to 0$ above. Then $0 \to M^+ \to F_0^+ \to \cdots \to F_{r-1}^+ \to K^+ \to 0$ is exact. So left $\mathcal{Gor}\mathcal{F}lat$- dim $M \leq r$ if and only if K^+ is Gorenstein injective by Theorem 10.3.8, and if and only if right $\mathcal{Gor}\mathcal{I}nj$- dim $M^+ \leq r$ by Proposition 11.2.5 since each F_i^+ is Gorenstein injective by Theorem 10.3.8. $\square$

Theorem 11.7.6. *Let R be n-Gorenstein and M be an R-module. Then the minimal left $\mathcal{G}or\,\mathcal{F}lat$-resolution of M is of the form*

$$0 \to F_n \to F_{n-1} \to \cdots \to F_1 \to G_0 \to M \to 0$$

where G_0 is Gorenstein flat and F_i is flat for each $i = 1, \ldots, n$. In particular, left $\mathcal{G}or\,\mathcal{F}lat$-dim $M \leq n$.

We note that it follows from the above that if R is n-Gorenstein and $\psi : G \to M$ is a Gorenstein flat cover, then flat dim Ker $\psi \leq n - 1$.

Lemma 11.7.7. *Let R be left coherent. Then every R-module M has an embedding $M \subset G$ such that G is cotorsion and G/M is flat.*

Proof. By Theorem 7.4.4, let $0 \to K \to F \to M \to 0$ be exact where $F \to M$ is a flat cover of M. Then we have the following pushout diagram

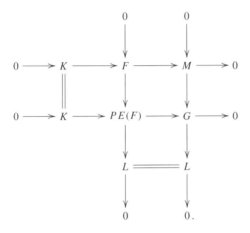

But $PE(F)$ and K are cotorsion by Lemmas 5.3.23 and 5.3.25. So G is cotorsion (see Definition 5.3.22). Moreover, $PE(F)$ is flat by Proposition 6.7.1 and so L is flat. □

Theorem 11.7.8. *Let R be a local Cohen–Macaulay ring of Krull dimension d admitting a dualizing module. Then every R-module $M \in \mathcal{G}_0(R)$ has a Gorenstein flat cover. Moreover, if $F \to M$ is such a cover, then flat dim Ker $\psi \leq d - 1$.*

Proof. By Theorem 11.5.11, there exists an exact sequence $0 \to L \to P \to M \to 0$ where $P \to M$ is a Gorenstein projective precover and proj dim $L \leq d - 1$. But by the lemma above there is an exact sequence $0 \to L \to C \to K \to 0$ such that K is flat and C is cotorsion. But proj dim $C < \infty$. Hence $\mathrm{Ext}^i(G, C) = 0$ for all Gorenstein flat G and all $i \geq 1$ by Corollary 10.4.27.

We now consider the following pushout diagram

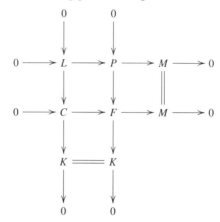

where F is Gorenstein flat since P and K are. But then $F \to M$ is a Gorenstein flat precover since $\mathrm{Ext}^1(G, C) = 0$ for all Gorenstein flat G. Hence M has a Gorenstein flat cover.

Now we note from the above that flat dim $L \leq d - 1$ and K is flat. So flat dim $C \leq d - 1$. If $\psi : F' \to M$ is a Gorenstein flat cover, then Ker ψ is a summand of C and so flat dim Ker $\psi \leq d - 1$. $\qquad\square$

Exercises

1. Complete the proof of Proposition 11.7.5.

2. Prove that if R is n-Gorenstein, then left $\mathcal{G}or\,\mathcal{F}lat$- dim M = flat dim M if and only if flat dim $M < \infty$.

3. Prove Theorem 11.7.6.

4. Prove that if $M \in \mathcal{G}_0(R)$ and $F \to M$ is a Gorenstein flat cover, then flat dim $M < \infty$ if and only if $F \to M$ is a flat cover.

5. State and prove a result corresponding to Proposition 11.7.5 for Gorenstein flat dimensions over Cohen–Macaulay rings admitting a dualizing module.

11.8   Gorenstein flat and projective preenvelopes

We start with the following result.

Lemma 11.8.1. *Let R be n-Gorenstein, $\{X_k\}$ be a representative set of indecomposable injective right R-modules, and $N = \oplus X_K$. If $\aleph_\alpha$ is an infinite cardinal, then there is an infinite cardinal $\aleph_\beta$ such that if M is a Gorenstein flat left R-module and $S \subset M$ is a submodule with Card $S \leq \aleph_\alpha$, then there is an R-module T with $S \subset T \subset M$ and Card $T \leq \aleph_\beta$ such that $S \subset T$ induces the zero map*

$$\mathrm{Tor}_i(N, S) \to \mathrm{Tor}_i(N, T) \quad \text{for all } i \geq 1.$$

Proof. We first note that M is Gorenstein flat if and only if $\mathrm{Tor}_i(N, M) = 0$ for all $i \geq 1$ by Theorem 10.3.8.

Let $\cdots \to P_1 \to P_0 \to N \to 0$ be a projective resolution of N. Then $\mathrm{Card}\,\mathrm{Tor}_i(N, S) \leq \mathrm{Card}\,P_i \otimes S$. Hence given $\aleph_\alpha$, there exists an infinite cardinal $\aleph_\delta \geq \aleph_\alpha$ such that if $\mathrm{Card}\,S \leq \aleph_\alpha$, then $\mathrm{Card}\,\mathrm{Tor}_i(N, S) \leq \aleph_\delta$ for all $i \geq 1$.

Now note that $0 = \mathrm{Tor}_i(N, M) \cong \varinjlim \mathrm{Tor}_i(N, S')$ where the limit is over all S' with $S \subset S' \subset M$ and S'/S finitely generated. Hence for $z \in \mathrm{Tor}_i(N, S)$, there is an S' such that $\mathrm{Tor}_i(N, S) \to \mathrm{Tor}_i(N, S')$ maps z to zero. Choosing one such S' for each z and letting T be the sum of all the chosen S''s, we see that $\mathrm{Tor}_i(N, S) \to \mathrm{Tor}_i(N, T)$ is the zero map for all $i \geq 1$. It is now easy to see that we can choose $\aleph_\beta \geq \aleph_\alpha$ so that $\mathrm{Card}\,T \leq \aleph_\beta$ whenever $\mathrm{Card}\,S \leq \aleph_\alpha$, no matter what choice of S''s we make. $\square$

Theorem 11.8.2. *Let R be n-Gorenstein. Then every R-module has a Gorenstein flat preenvelope.*

Proof. Let M be a Gorenstein flat left R-module. We use the notation in Lemma 11.8.1 above. Let $\alpha_0 = \alpha$ and $\alpha_1 = \beta$. Let α_1 play the role of α and α_2 be the new β guaranteed by the lemma. Repeating the procedure, we get $\aleph_{\alpha_0}, \aleph_{\alpha_1}, \ldots$ with obvious properties. Then given a submodule $S \subset M$ with $\mathrm{Card}\,S \leq \aleph_\alpha = \aleph_{\alpha_0}$, let $S_0 = S$ and find $S_1, S_2, \ldots$ with $S_1 \subset S_2 \subset M$ such that $\mathrm{Card}\,S_j \leq \aleph_{\alpha_j}$ and such that $S_j \subset S_{j+1}$ induces the zero map $\mathrm{Tor}_i(N, S_j) \to \mathrm{Tor}_i(N, S_{j+1})$ for all $i \geq 1$. So let $T = \bigcup_{j=1}^\infty S_j$. Then $\mathrm{Tor}_i(N, T) = \varinjlim_j \mathrm{Tor}_i(N, S_j) = 0$ for all $i \geq 1$. Thus T is Gorenstein flat. Furthermore, if we set $\aleph_\beta = \sup_j \aleph_{\alpha_j}$, then $\mathrm{Card}\,T \leq \aleph_\beta$.

So for each infinite cardinal $\aleph_\alpha$, there exists an infinite cardinal $\aleph_\beta$ such that if M is Gorenstein flat and $S \subset M$ is a submodule, then there is a Gorenstein flat R-submodule T of M containing S such that $\mathrm{Card}\,T \leq \aleph_\beta$. Hence every R-module has a Gorenstein flat preenvelope by Corollary 6.2.2. $\square$

Corollary 11.8.3. *Every finitely generated R-module has a Gorenstein projective preenvelope which is also finitely generated.*

Proof. This follows from Theorem 11.8.2 and Corollary 10.3.10 since if M is finitely generated, then every Gorenstein flat preenvelope $M \to G$ can be factored $M \to C \to G$ where C is a finitely generated Gorenstein projective R-module by Theorem 10.3.8. $\square$

Exercises

1. Let R be a local Cohen–Macaulay ring admitting a dualizing module. Prove that any finitely generated R-module M has a Gorenstein projective preenvelope $M \to C$ with C finitely generated such that $\mathrm{Hom}(C, F) \to \mathrm{Hom}(M, F) \to 0$ is exact for all Gorenstein flat R-modules F.

2. Prove that every R-module in $\mathcal{G}_0(R)$ has a Gorenstein flat preenvelope.

Chapter 12

Balance over Gorenstein and Cohen–Macaulay Rings

In Chapter 11, we studied the existence of Gorenstein precovers and preenvelopes. We can therefore apply methods of relative homological algebra of Chapter 8 and compute derived functors. In particular, we can study the question of balance of $\mathrm{Hom}(-, -)$ and tensor.

We will again let $\mathcal{G}or\mathcal{I}nj$, $\mathcal{G}or\mathcal{P}roj$, $\mathcal{G}or\mathcal{F}lat$ denote the classes of Gorenstein injective, Gorenstein projective, and Gorenstein flat modules, respectively. As usual, we will let $_R\mathcal{M}$ and $\mathcal{M}_R$ denote the classes of left and right R-modules, respectively. For a class $\mathcal{F}$, we will again let $\mathcal{F}_{fg}$ denote the class of finitely generated modules in $\mathcal{F}$. These terms will also be used to denote the corresponding full subcategories.

12.1   Balance of $\mathrm{Hom}(-, -)$

It was shown in Chapter 11 that if R is Gorenstein, then $\mathcal{G}or\mathcal{I}nj$ and $\mathcal{G}or\mathcal{F}lat$ are precovering and preenveloping classes, and $\mathcal{G}or\mathcal{P}roj$ is precovering by Theorems 11.1.1, 11.2.1, 11.5.1, 11.7.3, and 11.8.2, while $\mathcal{G}or\mathcal{P}roj_{fg}$ is preenveloping for $\mathcal{M}_{fg}$ by Corollary 11.8.3. Thus we have the following result.

Theorem 12.1.1. *Let R be Gorenstein. Then $\mathrm{Hom}(-, -)$ is left balanced on $_R\mathcal{M} \times {}_R\mathcal{M}$ by $\mathcal{G}or\mathcal{I}nj \times \mathcal{G}or\mathcal{I}nj$ and $\mathcal{G}or\mathcal{F}lat \times \mathcal{G}or\mathcal{F}lat$, on $_R\mathcal{M}_{fg} \times {}_R\mathcal{M}$ by $\mathcal{G}or\mathcal{P}roj_{fg} \times \mathcal{G}or\mathcal{F}lat$ and $\mathcal{G}or\mathcal{P}roj_{fg} \times \mathcal{G}or\mathcal{P}roj$, and on $\mathcal{M}_{fg} \times \mathcal{M}_{fg}$ by $\mathcal{G}or\mathcal{P}roj_{fg} \times \mathcal{G}or\mathcal{P}roj_{fg}$.*

Proof. This follows from the remarks above, Proposition 8.1.3, Corollary 10.3.10, and Remark 11.5.2. $\qquad\qquad\square$

Lemma 12.1.2. *If R is n-Gorenstein, then every right $\mathcal{G}or\mathcal{I}nj$-resolution is $\mathrm{Hom}(\mathcal{G}or\mathcal{P}roj, -)$ exact.*

Proof. By Theorem 11.2.1 there exists a Gorenstein injective preenvelope $N \to G$ of N with $L = \mathrm{Coker}\,(N \to G)$ of finite injective dimension. So let $0 \to L \to E^0 \to E^1 \to \cdots \to E^n \to 0$ be an injective resolution of L.

Since C is Gorenstein projective, there exists an exact sequence $0 \to C \to P^0 \to P^1 \to \cdots \to P^{n-1} \to P^n \to \cdots$ with each P^i projective. Now set $D = \mathrm{Ker}(P^{n-1} \to P^n)$. Then given a linear map $C \to L$, there is a commutative diagram

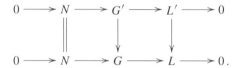

But D is Gorenstein projective and so $\mathrm{Ext}^i(D, L) = 0$ for all $i \geq 1$ by Remark 10.2.2 since $\mathrm{proj\,dim}\, L < \infty$. Hence the map $D \to E^{n-1}$ has a factorization $D \to E^{n-2} \to E^{n-1}$. Then as usual, this gives an extension $P^0 \to L$ which can be lifted to a map $P^0 \to G$ whose restriction gives a required lifting $C \to G$. So $0 \to \mathrm{Hom}(C, N) \to \mathrm{Hom}(C, G) \to \mathrm{Hom}(C, L) \to 0$ is exact for all Gorenstein projective R-modules C.

Now if $0 \to N \to G' \to L' \to 0$ is any exact sequence with $N \to G'$ a Gorenstein injective preenvelope of N, then there exists a commutative diagram

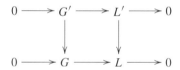

Since $N \to G$ and $N \to G'$ are both preenvelopes, we see that

$$
\begin{array}{ccccc}
0 & \longrightarrow & G' & \longrightarrow & L' & \longrightarrow & 0 \\
 & & \downarrow & & \downarrow & & \\
0 & \longrightarrow & G & \longrightarrow & L & \longrightarrow & 0
\end{array}
$$

gives an equivalence of complexes (the two rows). This in turn gives the equivalence of complexes

$$
\begin{array}{ccccc}
0 & \longrightarrow & \mathrm{Hom}(C, G') & \longrightarrow & \mathrm{Hom}(C, G') & \longrightarrow & 0 \\
 & & \downarrow & & \downarrow & & \\
0 & \longrightarrow & \mathrm{Hom}(C, G) & \longrightarrow & \mathrm{Hom}(C, L) & \longrightarrow & 0
\end{array}
$$

So if $\mathrm{Hom}(C, G) \to \mathrm{Hom}(C, L) \to 0$ is exact, then so is $\mathrm{Hom}(C, G') \to \mathrm{Hom}(C, L') \to 0$. Thus $0 \to \mathrm{Hom}(C, N) \to \mathrm{Hom}(C, G') \to \mathrm{Hom}(C, L') \to 0$ is exact for all Gorenstein projective R-modules C. Hence the result follows. $\square$

Dually, we have the following result.

Lemma 12.1.3. *If R is n-Gorenstein, then every left $\mathcal{G}or\mathcal{P}roj$-resolution and left $\mathcal{G}or\mathcal{F}lat$-resolution is* $\mathrm{Hom}(-, \mathcal{G}or\mathcal{I}nj)$ *exact.*

The two lemmas above give the following result.

Theorem 12.1.4. *Let R be Gorenstein, then* $\mathrm{Hom}(-, -)$ *is right balanced on* $_R\mathcal{M} \times {}_R\mathcal{M}$ *by* $\mathcal{G}or\mathcal{P}roj \times \mathcal{G}or\mathcal{I}nj$.

By Theorem 12.1.4, we can compute right derived functors of $\mathrm{Hom}(M, N)$ using a left $\mathcal{G}or\mathcal{P}roj$-resolution of M or a right $\mathcal{G}or\mathcal{I}nj$-resolution of N. We will denote these derived functors by $\mathrm{Gext}^i(M, N)$. It is easy to check that:

1) $\mathrm{Gext}^0(-, -) \cong \mathrm{Hom}(-, -)$.

2) $\mathrm{Gext}^i(C, -) = 0$ for all $i \geq 1$ and all $C \in \mathcal{G}or\mathcal{P}roj$.

3) $\mathrm{Gext}^i(-, G) = 0$ for all $i \geq 1$ and all $G \in \mathcal{G}or\mathcal{I}nj$.

4) If the exact sequence $0 \to M' \to M \to M'' \to 0$ of R-modules is $\mathrm{Hom}(\mathcal{G}or\,\mathcal{P}roj, -)$ exact, then by part (2) of Theorem 8.2.3 there is a long exact sequence

$$\cdots \to \mathrm{Gext}^i(M, -) \to \mathrm{Gext}^i(M', -) \to \mathrm{Gext}^{i+1}(M'', -) \to \cdots.$$

5) If the exact sequence $0 \to N' \to N \to N'' \to 0$ of R-modules is $\mathrm{Hom}(-, \mathcal{G}or\,\mathcal{I}nj)$ exact, then by part (1) of Theorem 8.2.5 there is a long exact sequence

$$\cdots \to \mathrm{Gext}^i(-, N) \to \mathrm{Gext}^i(-, N'') \to \mathrm{Gext}^{i+1}(-, N') \to \cdots.$$

6) There are natural transformations

$$\mathrm{Gext}^i(-, -) \to \mathrm{Ext}^i(-, -)$$

which are also natural in the long exact sequences as in (4) and (5) above.

Now by Theorem 12.1.1, let $\mathrm{Gext}_i(M, N)$ denote the left derived functors of $\mathrm{Hom}(M, N)$ computed using a right $\mathcal{G}or\mathcal{I}nj$-resolution of M or a left $\mathcal{G}or\mathcal{I}nj$-resolution of N. Then again it is easy to check the following properties:

1) There is a natural map $\mathrm{Gext}_0(M, N) \overset{\sigma}{\to} \mathrm{Hom}(M, N)$. If we let $\mathrm{Ker}\,\sigma = \overline{\overline{\mathrm{Gext}}}_0(M, N)$ and $\mathrm{Coker}\,\sigma = \overline{\mathrm{Gext}}^0(M, N)$, then we have an exact sequence

$$0 \to \overline{\overline{\mathrm{Gext}}}_0(M, N) \to \mathrm{Gext}_0(M, N) \to \mathrm{Hom}(M, N) \to \overline{\mathrm{Gext}}^0(M, N) \to 0.$$

2) If M or N are Gorenstein injective, then

$$\mathrm{Gext}_0(M, N) = \mathrm{Hom}(M, N) \text{ and } \overline{\mathrm{Gext}}_0(M, N) = \overline{\mathrm{Gext}}^0(M, N) = 0.$$

3) $\mathrm{Gext}_i(M, N) = 0$ for all $i \geq 1$ and for all $N \in \mathcal{G}or\mathcal{I}nj$.

4) $\mathrm{Gext}_i(M, N) = 0$ for all $i \geq 1$ and for all $M \in \mathcal{G}or\mathcal{I}nj$.

5) There are natural transformations

$$\text{Ext}_i(-, -) \to \text{Gext}_i(-, -)$$

where $\text{Ext}_i(M, N)$ denote left derived functors obtained using a right $\mathcal{I}nj$-resolution of M or a left $\mathcal{I}nj$-resolution of N (see Example 8.3.5). These transformations are also natural in the corresponding long exact sequences below.

6) If the exact sequence $0 \to M' \to M \to M'' \to 0$ is Hom$(-, \mathcal{G}or\mathcal{I}nj)$ exact, then by part (2) of Theorem 8.2.5 there is a long exact sequence

$$\cdots \to \text{Gext}_i(M, -) \to \text{Gext}_i(M', -) \to \text{Gext}_{i-1}(M'', -) \to \cdots .$$

7) If the exact sequence $0 \to N' \to N \to N'' \to 0$ is Hom$(\mathcal{G}or\mathcal{I}nj, -)$ exact, then by part (1) of Theorem 8.2.3 there is a long exact sequence

$$\cdots \to \text{Gext}_i(-, N) \to \text{Gext}_i(-, N'') \to \text{Gext}_{i-1}(-, N') \to \cdots .$$

Likewise, we can compute left derived functors of Hom(M, N) using a right $\mathcal{G}or\mathcal{F}lat$-resolution of M and a left $\mathcal{G}or\mathcal{F}lat$-resolution of N by Theorem 12.1.1. These functors are again denoted Gfext$_i(M, N)$ and have the same properties as above noting that we now get natural transformations

$$\text{Ext}_i(-, -) \to \text{Gfext}_i(-, -)$$

where $\text{Ext}_i(-, -)$ denote the left derived functors obtained by using a right $\mathcal{F}lat$-resolution of M or a left $\mathcal{F}lat$-resolution of N of Example 8.3.6. The other cases in Theorem 12.1.1 are similar and correspond to Example 8.3.11.

If R is a local Cohen–Macaulay ring admitting a dualizing module, then every $M \in \mathcal{G}_0(R)$ has a Gorenstein projective and Gorenstein flat precover and Gorenstein flat preenvelope by Theorems 11.5.11, 11.7.8, and Exercise 2 of Section 11.8, respectively, and each $M \in \mathcal{J}_0(R)$ has a Gorenstein injective preenvelope by Theorem 11.2.6. Thus we have the following result.

Theorem 12.1.5. Hom$(-, -)$ *is left balanced on* $\mathcal{G}_0(R) \times \mathcal{G}_0(R)$ *by* $\mathcal{G}or\mathcal{F}lat \times \mathcal{G}or\mathcal{F}lat$, *on* $\mathcal{G}_0(R)_{fg} \times \mathcal{G}_0(R)$ *by* $\mathcal{G}or\mathcal{P}roj_{fg} \times \mathcal{G}or\mathcal{F}lat$ *and* $\mathcal{G}or\mathcal{P}roj_{fg} \times \mathcal{G}or\mathcal{P}roj$, *and on* $\mathcal{G}_0(R)_{fg} \times \mathcal{G}_0(R)_{fg}$ *by* $\mathcal{G}or\mathcal{P}roj_{fg} \times \mathcal{G}or\mathcal{P}roj_{fg}$.

Lemma 12.1.6. *Let* $M \in \mathcal{J}_0(R)$. *Then every right* $\mathcal{G}or\mathcal{I}nj$-*resolution of* M *is* Hom$(\mathcal{G}or\mathcal{P}roj_{fg}, -)$ *exact.*

Proof. Let C be a finitely generated Gorenstein projective R-module. Then there exists an exact sequence $0 \to C \to P^0 \to P^1 \to \cdots$ with each P^i finitely generated and projective. So if $C' = \text{Ker}(P^{d-1} \to P^d)$, then C' is finitely generated Gorenstein projective and so $\text{Ext}^i(C', L) = 0$ for all $i \geq 1$ and all R-modules L of finite injective dimension by Proposition 10.2.6. Thus the result now follows as in Lemma 12.1.2. □

Remark 12.1.7. This lemma together with Exercise 3 and Remark 11.5.2 give the following result which corresponds to Theorem 12.1.4 above.

Theorem 12.1.8. $\mathrm{Hom}(-,-)$ *is right balanced on* $\mathcal{G}_0(R)_{fg} \times \mathcal{J}_0(R)$ *by* $\mathcal{G}or\mathcal{P}roj_{fg} \times \mathcal{G}or\mathcal{I}nj$.

Remark 12.1.9. Using Theorems 12.1.5 and 12.1.8, we can compute derived functors $\mathrm{Gfext}_i(M, N)$ (which have the same properties as those of Example 8.3.6) and $\mathrm{Gext}^i(M, N)$, respectively.

Exercises

1. Prove Theorem 12.1.1.

2. Prove Lemma 12.1.3 and Theorem 12.1.4.

3. Prove that if $M \in \mathcal{G}_0(R)$ then every left $\mathcal{G}or\mathcal{F}lat$-resolution or left $\mathcal{G}or\mathcal{P}roj$-resolution of M is $\mathrm{Hom}(-, \mathcal{G}or\mathcal{I}nj)$ exact.

4. Prove Theorem 12.1.8.

5. Prove Remark 12.1.9.

6. Let R be a Cohen–Macaulay ring of Krull dimension d with a dualizing module, and C be the full subcategory of R-modules whose dth $\mathcal{I}nj$-syzygies are in $\mathcal{G}_0(R)$. Prove that $\mathrm{Hom}(-, -)$ is left balanced on $\mathcal{J}_0(R) \times \mathsf{C}$ by $\mathcal{G}or\mathcal{I}nj \times \mathcal{G}or\mathcal{I}nj$. Conclude that in this case derived functors $\mathrm{Gext}_i(M, N)$ can be computed.
Hint: Use Exercise 2 of Section 11.1.

12.2   Balance of $- \otimes -$

We start with the following result

Lemma 12.2.1. *If R is n-Gorenstein, then every left $\mathcal{G}or\mathcal{P}roj$-resolution or left $\mathcal{G}or \mathcal{F}lat$-resolution is $\mathcal{G}or\mathcal{F}lat \otimes -$ exact.*

Proof. If $\cdots \to C_1 \to C_0 \to M \to 0$ is a left $\mathcal{G}or\mathcal{P}roj$ (or $\mathcal{G}or\mathcal{F}lat$)-resolution of an R-module M, then $\cdots \to F \otimes C_1 \to F \otimes C_0 \to F \otimes M \to 0$ is exact if and only if $0 \to (F \otimes M)^+ \to (F \otimes C_0)^+ \to (F \otimes C_1)^+ \to \cdots$ is exact. But the latter is equivalent to the sequence $0 \to \mathrm{Hom}(M, F^+) \to \mathrm{Hom}(C_0, F^+) \to \mathrm{Hom}(C_1, F^+) \to \cdots$ which is exact by Proposition 10.3.3 and Lemma 12.1.3.   □

This gives the following result.

Theorem 12.2.2. *Let R be n-Gorenstein. Then $- \otimes -$ is left balanced on $\mathcal{M}_R \times {}_R\mathcal{M}$ by $\mathcal{G}or\mathcal{P}roj \times \mathcal{G}or\mathcal{P}roj$ and $\mathcal{G}or\mathcal{F}lat \times \mathcal{G}or\mathcal{F}lat$.*

Lemma 12.2.3. *If R is n-Gorenstein, then every right $\mathcal{G}or\mathcal{F}lat$-resolution is $\mathcal{G}or\mathcal{I}nj \otimes -$ exact and every right $\mathcal{G}or\mathcal{I}nj$-resolution is $\mathcal{G}or\mathcal{F}lat \otimes -$ exact.*

Theorem 12.2.4. *Let R be n-Gorenstein. Then $-\otimes-$ is right balanced on $\mathcal{M}_R \times {}_R\mathcal{M}$ by $\mathcal{G}or\mathcal{F}lat \times \mathcal{G}or\mathcal{I}nj$ and on $\mathcal{M}_{R_{fg}} \times {}_R\mathcal{M}$ by $\mathcal{G}or\mathcal{P}roj_{fg} \times \mathcal{G}or\mathcal{I}nj$.*

Proof. Follows easily from the lemmas above.   □

Using Theorem 12.2.2, we can compute the left derived functors of $M \otimes N$ using Gorenstein projective modules. These functors will be denoted by $\mathrm{Gtor}_i(M, N)$. We then have the following easy properties:

1) $\mathrm{Gtor}_0(-, -) \cong -\otimes-$.

2) $\mathrm{Gtor}_i(C, -) = 0$ for all $i \geq 1$ and all Gorenstein projective right R-modules C.

3) $\mathrm{Gtor}_i(-, D) = 0$ for all $i \geq 1$ and all Gorenstein projective left R-modules D.

4) If the exact sequence $0 \to M' \to M \to M'' \to 0$ of right R-modules is $\mathrm{Hom}(\mathcal{G}or\mathcal{P}roj, -)$ exact, then by part (1) of Theorem 8.2.3 there is a long exact sequence

$$\cdots \to \mathrm{Gtor}_{i+1}(M'', -) \to \mathrm{Gtor}_i(M', -) \to \mathrm{Gtor}_i(M, -) \to \cdots$$

5) Same as (4) for an exact sequence $0 \to N' \to N \to N'' \to 0$ of left R-modules.

6) There are natural transformations

$$\mathrm{Tor}_i(-, -) \to \mathrm{Gtor}_i(-, -)$$

which are natural in the long exact sequences as in (4) and (5) above.

By Theorem 12.2.2, we also have left derived functors $\mathrm{Gtor}_i(M, N)$ that can be obtained using Gorenstein flat modules. These functors have the same properties as above. Again by Theorem 12.2.2, we can compute left derived functors $\mathrm{gtor}_i(M, N)$ using left $\mathcal{G}or\mathcal{F}lat$-resolutions with similar properties as above. But $\mathrm{gtor}_i(M, F) = \mathrm{Gtor}_i(M, F) = 0$ for all $i \geq 1$ and for all Gorenstein flat R-modules F by Lemma 12.2.1. Hence $\mathrm{gtor}_i(-, -) \cong \mathrm{Gtor}_i(-, -)$ since $\{\mathrm{gtor}_i\}$ and $\{\mathrm{Gtor}_i\}$ are covariantly left strongly connected sequences (see Definition 8.2.10).

We can also compute right derived functors $\mathrm{Gtor}^i(M, N)$ using a right $\mathcal{G}or\mathcal{F}lat$-resolution of M or a right $\mathcal{G}or\mathcal{I}nj$-resolution of N by Theorem 12.2.4. We have the following properties

1) There is a natural transformation

$$M \otimes N \overset{\sigma}{\to} \mathrm{Gtor}^0(M, N).$$

Again, we can let $\mathrm{Ker}\,\sigma = \overline{\mathrm{Gtor}_0}(M, N)$ and $\mathrm{Coker}\,\sigma = \overline{\mathrm{Gtor}^0}(M, N)$, then we have an exact sequence

$$0 \to \overline{\mathrm{Gtor}_0}(M, N) \to M \otimes N \to \mathrm{Gtor}^0(M, N) \to \overline{\mathrm{Gtor}^0}(M, N) \to 0.$$

2) If M or N are Gorenstein flat, then

$$M \otimes N = \mathrm{Gtor}^0(M, N) \text{ and } \overline{\mathrm{Gtor}_0}(M, N) = \overline{\mathrm{Gtor}^0}(M, N) = 0$$

3) $\mathrm{Gtor}^i(M, N) = 0$ for all $i \geq 1$ and for all $M \in \mathcal{G}or\mathcal{F}lat$

4) $\mathrm{Gtor}^i(M, N) = 0$ for all $i \geq 1$ and for all $N \in \mathcal{G}or\mathcal{F}lat$.

5) There are natural transformations

$$\mathrm{Gtor}^i(M, N) \to \mathrm{Tor}^i(M, N)$$

where $\mathrm{Tor}^i(M, N)$ denote right derived functors obtained using a right $\mathcal{F}lat$-resolution of M or a right $\mathcal{I}nj$-resolution of N (see Example 8.3.9). These transformations are again natural in the long exact sequences that follow below.

6) If the exact sequence $0 \to M' \to M \to M'' \to 0$ is $\mathrm{Hom}(-, \mathcal{G}or\mathcal{F}lat)$ exact, then by part (1) of Theorem 8.2.5 there is a long exact sequence

$$\cdots \to \mathrm{Gtor}^i(M, -) \to \mathrm{Gtor}^i(M'', -) \to \mathrm{Gtor}^{i+1}(M', -) \to \cdots.$$

7) Same as (6) for an exact sequence $0 \to N' \to N \to N'' \to 0$ of left R-modules.

Exercises

1. Prove Lemma 12.2.3 and Theorem 12.2.4.

2. State and prove a result corresponding to Lemma 12.2.1 for Cohen–Macaulay rings admitting a dualizing module.

3. Prove that $- \otimes -$ is left balanced on $\mathcal{G}_0(R) \times \mathcal{G}_0(R)$ by $\mathcal{G}or\mathcal{P}roj \times \mathcal{G}or\mathcal{P}roj$ and $\mathcal{G}or\mathcal{F}lat \times \mathcal{G}or\mathcal{F}lat$.

4. Prove that $- \otimes -$ is right balanced on $\mathcal{G}_0(R) \times \mathcal{I}_0(R)$ by $\mathcal{G}or\mathcal{F}lat \times \mathcal{G}or\mathcal{I}nj$ and on $\mathcal{G}_0(R)_{fg} \times \mathcal{I}_0(R)$ by $\mathcal{G}or\mathcal{P}roj_{fg} \times \mathcal{G}or\mathcal{I}nj$.

12.3 Dimensions over n-Gorenstein rings

Theorem 12.3.1. *Let R be left and right noetherian. Then the following are equivalent.*

(1) *R is n-Gorenstein.*

(2) *Every nth $\mathcal{I}nj$-cosyzygy of an R-module (left or right) is Gorenstein injective.*

(3) *Every nth $\mathcal{P}roj$-syzygy of an R-module (left or right) is Gorenstein projective.*

(4) *Every nth $\mathcal{F}lat$-syzygy of an R-module (left or right) is Gorenstein flat.*

(5) *Every nth $\mathcal{P}roj_{fg}$-syzygy of a finitely generated R-module (left or right) is Gorenstein projective.*

Proof. (1) $\Rightarrow$ (2) follows from Exercise 2 of Section 11.2 and Proposition 11.2.5.

(2) $\Rightarrow$ (1). Let E be an injective module, N be an R-module and G be an *n*th $\mathcal{I}nj$-cosyzygy of N. Then $\mathrm{Ext}^{n+1}(E, N) \cong \mathrm{Ext}^1(E, G)$. But $\mathrm{Ext}^1(E, G) = 0$ since G is Gorenstein injective by assumption. Hence proj dim $E \leq n$ for all injective R-modules E. Similarly for right R-modules. So R is *n*-Gorenstein by Theorem 9.1.11.

The proofs of the equivalence of (1), (3), (5) are by arguments dual to the above.

(1) $\Rightarrow$ (4) by and Proposition 11.7.5 and Theorem 11.7.6.

(4) $\Rightarrow$ (1). If an *n*th $\mathcal{F}lat$-syzygy of any right R-module M is Gorenstein flat, then $\mathrm{Tor}_{n+1}(M, E) = 0$ for all right R-modules M and all injective R-modules E. Hence proj dim $_R E \leq n$ for all such E. Similarly for R-modules M. Thus R is *n*-Gorenstein. $\square$

Corollary 12.3.2. *If R is Iwanaga–Gorenstein and* inj dim $_R R = n$*, then*

$$\mathrm{gl\,right\,}\mathcal{G}or\mathcal{I}nj\text{-}\dim{}_R\mathcal{M} = \mathrm{gl\,left\,}\mathcal{G}or\mathcal{P}roj\text{-}\dim{}_R\mathcal{M} = \mathrm{gl\,left\,}\mathcal{G}or\mathcal{F}lat\text{-}\dim{}_R\mathcal{M}$$
$$= \mathrm{gl\,left\,}\mathcal{G}or\mathcal{P}roj_{fg}\text{-}\dim{}_R\mathcal{M}_{fg} = n.$$

Proposition 12.3.3. *If R is n-Gorenstein, then the following are equivalent for a right R-module L.*

(1) proj dim $L < \infty$ *(and so $\leq n$).*

(2) *The natural transformation* $\mathrm{Gext}^i(L, -) \to \mathrm{Ext}^i(L, -)$ *is an isomorphism for $i \geq 0$.*

(3) $\mathrm{Tor}_i(L, -) \to \mathrm{Gtor}_i(L, -)$ *is an isomorphism for $i \geq 0$.*

Proof. (1) $\Rightarrow$ (2). (1) means that $\mathrm{Ext}^i(L, G) = 0$ for $i \geq 1$ and for all Gorenstein injective right R-modules G by Proposition 10.1.15. Now let $N \to G$ be a Gorenstein injective preenvelope of N and $H = G/N$. Then we have a commutative diagram

$$
\begin{array}{ccccccc}
\mathrm{Gext}^i(L, G) & \longrightarrow & \mathrm{Gext}^i(L, H) & \longrightarrow & \mathrm{Gext}^{i+1}(L, N) & \longrightarrow & 0 \\
\downarrow & & \downarrow & & \downarrow & & \\
\mathrm{Ext}^i(L, G) & \longrightarrow & \mathrm{Ext}^i(L, H) & \longrightarrow & \mathrm{Ext}^{i+1}(L, N) & \longrightarrow & 0.
\end{array}
$$

If $i = 0$, then $\mathrm{Gext}^0(-, -) = \mathrm{Ext}^0(-, -) = \mathrm{Hom}(-, -)$ and so the first two maps are isomorphisms. Hence $\mathrm{Gext}^1(L, N) \to \mathrm{Ext}^1(L, N)$ is an isomorphism. The result now follows by induction on i.

(2) $\Rightarrow$ (1). right $\mathcal{G}or\mathcal{I}nj$-dim $N \leq n$ and so $\mathrm{Gext}^{n+1}(L, N) = 0$ for all right R-modules N. But then $\mathrm{Ext}^{n+1}(L, N) = 0$ for all N and so proj dim $L < \infty$.

(1) $\Rightarrow$ (3). We consider the exact sequence $0 \to H \to C \to N \to 0$ where $C \to N$ is a Gorenstein projective precover. If proj dim $L < \infty$, then we have a

commutative diagram

$$0 \longrightarrow \mathrm{Tor}_{i+1}(L, N) \longrightarrow \mathrm{Tor}_i(L, H) \longrightarrow \mathrm{Tor}_i(L, C) \longrightarrow \mathrm{Tor}_i(L, N)$$
$$\downarrow \qquad\qquad \downarrow \qquad\qquad \downarrow \qquad\qquad \downarrow$$
$$0 \longrightarrow \mathrm{Gtor}_{i+1}(L, N) \longrightarrow \mathrm{Gtor}_i(L, H) \longrightarrow \mathrm{Gtor}_i(L, C) \longrightarrow \mathrm{Gtor}_i(L, N)$$

by Theorem 10.3.8 and Corollary 10.3.10. But $\mathrm{Tor}_0(-, -) = \mathrm{Gtor}_0(-, -) = -\otimes -$ and the result follows as in the above.

(3) $\Rightarrow$ (1). $\mathrm{Gtor}_{n+1}(L, N) = 0$ for all R-modules N since left $\mathcal{G}or\mathcal{P}roj$- dim $N \leq n$. So the result follows. $\qquad\square$

We also have the following result.

Proposition 12.3.4. *Let R be n-Gorenstein and N be an R-module. Then*

(1) *If M is an R-module, then $\mathrm{Gext}^1(M, N) \to \mathrm{Ext}^1(M, N)$ is an injection.*

(2) *If M is a right R-module, then $\mathrm{Tor}_1(M, N) \to \mathrm{Gtor}_1(M, N)$ is a surjection.*

Proof. (1) Let $N \to G$ be a Gorenstein injective preenvelope and $H = \mathrm{Coker}(N \to G)$. Then we have the following commutative diagram

$$\mathrm{Hom}(M, G) \longrightarrow \mathrm{Hom}(M, H) \longrightarrow \mathrm{Gext}^1(M, N) \longrightarrow \mathrm{Gext}^1(M, G) = 0$$
$$\| \qquad\qquad \| \qquad\qquad \downarrow$$
$$\mathrm{Hom}(M, G) \longrightarrow \mathrm{Hom}(M, H) \longrightarrow \mathrm{Ext}^1(M, N)$$

with exact rows. So (1) follows.

(2) Follows similarly. $\qquad\square$

Corollary 12.3.5. *Let R be n-Gorenstein. Then the following are equivalent for an exact sequence $0 \to N \to L \to M \to 0$ of R-modules.*

(1) *The sequence $0 \to N \to L \to M \to 0$ corresponds to an element of $\mathrm{Gext}^1(M, N) \subset \mathrm{Ext}^1(M, N)$.*

(2) *$\mathrm{Hom}(L, G) \to \mathrm{Hom}(N, G) \to 0$ is exact for all Gorenstein injective R-modules G.*

(3) *$\mathrm{Hom}(C, L) \to \mathrm{Hom}(C, M) \to 0$ is exact for all Gorenstein projective R-modules C.*

Proof. (1) $\Rightarrow$ (2). If $0 \to N \to L \to M \to 0$ corresponds to an element of $\mathrm{Gext}^1(M, N)$, then there is a commutative diagram

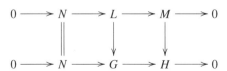

with exact rows such that $N \to G$ is a Gorenstein injective preenvelope. But if $N \to G'$ is a map with G' Gorenstein injective, then $N \to G'$ can be lifted to G. But then $N \to G'$ can be lifted to L and the result follows.

(2) $\Rightarrow$ (1). $\operatorname{Hom}(L, G) \to \operatorname{Hom}(N, G) \to 0$ exact for all Gorenstein injective R-modules G implies that we have a commutative diagram

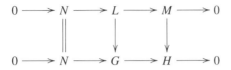

with exact rows where $N \to G$ is a Gorenstein injective preenvelope. This shows that $0 \to N \to L \to M \to 0$ corresponds to an element of $\operatorname{Gext}^1(M, N)$.

(1) $\Leftrightarrow$ (3) follows by a dual argument. $\square$

Theorem 12.3.6. *The following are equivalent for a noetherian ring R and integer $n \geq 2$.*

(1) *R is n-Gorenstein.*

(2) *Every left and right R-module has a $\mathcal{G}or\mathcal{I}nj$-precover and preenvelope and left $\mathcal{G}or\mathcal{I}nj$-dimension at most $n - 2$.*

(3) *Every left and right R-module M has a $\mathcal{G}or\mathcal{F}lat$-precover and preenvelope and right $\mathcal{G}or\mathcal{F}lat$-dimension at most $n - 2$.*

(4) *Every left and right finitely generated R-module M has a $\mathcal{G}or\mathcal{P}roj_{fg}$-precover and preenvelope and right $\mathcal{G}or\mathcal{P}roj_{fg}$-dimension at most $n - 2$.*

Proof. (1) $\Leftrightarrow$ (2). If R is n-Gorenstein, then a right $\mathcal{G}or\mathcal{I}nj$-resolution of M is of the form $0 \to M \to G^0 \to G^1 \to \cdots \to G^n \to 0$. So $\operatorname{Gext}_i(M, N) = 0$ for $i \geq n - 1$ for all R-modules N. Let $\cdots \to G_1 \to G_0 \to N \to 0$ be a minimal left $\mathcal{G}or\mathcal{I}nj$-resolution of N and $K_n = \operatorname{Ker}(G_{n-1} \to G_{n-2})$. Then $\operatorname{Gext}_{n-1}(M, N) = 0$ implies that $K_n \subset G_{n-1}$ has a factorization $K_n \to G_n \to G_{n-1}$ and thus K_n is a direct summand of G_n and hence is Gorenstein injective. So $K_n = G_n$. But if we set $M = R$ in the above, we get that $0 \to G_n \to G_{n-1} \to K_{n-1} \to 0$ is an exact sequence. So K_{n-1} is Gorenstein injective by Theorem 10.1.4 and thus $G_n = 0$ by minimality. But then again G_{n-2}/G_{n-1} is Gorenstein injective. It is now easy to see that $G_{n-2}/G_{n-1} \hookrightarrow K_{n-2}$ is a Gorenstein injective precover and so $G_{n-1} = 0$. Thus (2) follows. Conversely, left $\mathcal{G}or\mathcal{I}nj$- dim $N \leq n - 2$ means $\operatorname{Gext}_{n-1}(M, N) = 0$ for all modules M. But then right $\mathcal{G}or\mathcal{I}nj$- dim $M \leq n$ for all M and so the result follows from Theorem 12.3.1.

(1) $\Rightarrow$ (3) If R is n-Gorenstein, then $\operatorname{Gtor}^n(M, N) = \operatorname{Gtor}^{n-1}(M, N) = 0$ for all modules M and N. So if $0 \to M \to F^0 \to F^1 \to \cdots$ is a right $\mathcal{G}or\mathcal{F}lat$-resolution of M, then $F^{n-2} \otimes N \to F^{n-1} \otimes N \to F^n \otimes N \to F^{n+1} \otimes N$ is exact for all finitely generated R-modules N. So $K = \operatorname{Ker}(F^n \to F^{n+1})$ is pure in F^n by Lemma 8.4.23. But then $\operatorname{Tor}_1(L, F^n/K) = 0$ for all modules L of finite injective

dimension. So F^n/K is Gorenstein flat by Theorem 10.3.8 since R is Gorenstein. Hence K is Gorenstein flat by Theorem 10.3.14. But $F^{n-2} \to F^{n-1} \to K \to 0$ is exact by taking $N = R$ in the above. So $F = \text{Ker}(F^{n-2} \to F^{n-1})$ is also Gorenstein flat. Hence $0 \to M \to F^0 \to \cdots \to F^{n-3} \to F \to 0$ is a right $\mathcal{G}or\mathcal{F}lat$-resolution of M.

  (3) $\Rightarrow$ (4) is trivial.

  (4) $\Rightarrow$ (1). right $\mathcal{G}or\mathcal{P}roj_{fg}$-dim $M \leq n - 2$ means $\text{Gfext}_{n-1}(M, N) = 0$ for all modules $N \in \mathcal{M}_{fg}$. So if $\cdots \to F_1 \to F_0 \to N \to 0$ is a left $\mathcal{G}or\mathcal{P}roj_{fg}$-resolution and $K = \text{Ker}(F_{n-1} \to F_{n-2})$. Then K is a summand of F_n. So left $\mathcal{G}or\mathcal{P}roj_{fg}$-dim $N \leq n$. Thus the result follows from Theorem 12.3.1 above. □

Corollary 12.3.7. *If R is Iwanaga–Gorenstein and* inj dim $_R R = n$, *then*

$$\text{gl left } \mathcal{G}or\mathcal{I}nj\text{-dim } _R\mathcal{M} = \text{gl right } \mathcal{G}or\mathcal{F}lat\text{-dim } _R\mathcal{M}$$
$$= \text{gl right } \mathcal{G}or\mathcal{P}roj_{fg}\text{-dim } _R\mathcal{M}_{fg} = n - 2,$$

or is zero if $n \leq 1$.

Proof. The case $n \geq 2$ follows from the theorem above. The case $n = 1$ follows using the same arguments. If $n = 0$, then every R-module is Gorenstein injective, Gorenstein flat, and Gorenstein projective. □

Exercises

1. Prove the equivalence of parts 1, 3, 5 of Theorem 12.3.1.

2. Prove Corollary 12.3.2.

3. Prove part 2 of Proposition 12.3.4.

4. Complete the proof of Corollary 12.3.5.

5. Let R be a left and right noetherian and suppose $\mathcal{G}or\mathcal{I}nj$ is preenveloping, and $\mathcal{G}or\mathcal{F}lat$, $\mathcal{G}or\mathcal{P}roj$ are precovering over R. Prove that the following are equivalent for an integer $n \geq 0$.

   a) R is n-Gorenstein.
   b) $\text{Gext}^{n+k}(M, N) = 0$ for all R-modules (left and right) M, N and $k \geq 1$.
   c) $\text{Gext}^{n+1}(M, N) = 0$ for all R-modules (left and right) M, N.
   d) $\text{Gtor}_{n+k}(M, N) = 0$ for all R-modules (left and right) M, N and all $k \geq 1$.
   e) $\text{Gtor}_{n+1}(M, N) = 0$ for all R-modules (left and right) M, N.

6. Let R be left and right noetherian and suppose $\mathcal{G}or\mathcal{I}nj$, $\mathcal{G}or\mathcal{F}lat$ are precovering and preenveloping. Then the following are equivalent for an integer $n \geq 2$.

   a) R is n-Gorenstein.

b) $\mathrm{Gext}_{n-1}(M, N) = 0$ for all R-modules (left and right) M, N.

c) $\mathrm{Gtor}^n(M, N) = \mathrm{Gtor}^{n-1}(M, N) = 0$ for all finitely generated modules M and all modules N.

d) $\mathrm{Gfext}_{n-1}(M, N) = 0$ for all R-modules (left and right) M, N.

7. Let R be Iwanaga–Gorenstein. Prove that the following are equivalent

   a) R is 1-Gorenstein.

   b) gl left $\mathcal{G}or\mathcal{I}nj$- dim $_R\mathcal{M} = 0$.

   c) gl right $\mathcal{F}lat$- dim $_R\mathcal{M} = 0$.

   d) gl right $\mathcal{G}or\mathcal{P}roj$- dim $_R\mathcal{M}_{fg} = 0$.

8. (Enochs–Jenda [62]). Let R be a commutative Gorenstein ring. Then the following are equivalent for an integer $n \geq 2$.

   a) R is n-Gorenstein.

   b) left $\mathcal{G}or\mathcal{F}lat$- dim $\mathrm{Hom}(E, M) \leq n - 2$ for all R-modules M and all injectives E.

   c) right $\mathcal{G}or\mathcal{I}nj$- dim $M \otimes E \leq n - 2$

   d) left $\mathcal{G}or\mathcal{P}roj$- dim $\mathrm{Hom}(M, R) \leq n - 2$ for all finitely generated R-modules M.

12.4 Dimensions over Cohen–Macaulay rings

In this section, we consider Gorenstein dimensions over Cohen–Macaulay rings.

Theorem 12.4.1. *Let R be a Cohen–Macaulay local ring of Krull dimension d admitting a dualizing module Ω, and let CM denote the class of maximal Cohen–Macaulay R-modules. Then the following are equivalent for an integer $n \geq 0$.*

(1) inj dim $\Omega \leq n$.

(2) gl right $\mathcal{G}or\mathcal{I}nj$- dim $\mathcal{J}_0(R) \leq n$.

(3) *Every nth $\mathcal{I}nj$-cosyzygy of an R-module in $\mathcal{J}_0(R)$ is Gorenstein injective.*

(4) gl left $\mathcal{G}or\mathcal{P}roj$- dim $\mathcal{G}_0(R) \leq n$.

(5) *Every nth $\mathcal{P}roj$-syzygy of an R-module in $\mathcal{G}_0(R)$ is Gorenstein projective.*

(6) gl left $\mathcal{G}or\mathcal{F}lat$- dim $\mathcal{G}_0(R) \leq n$.

(7) *Every nth $\mathcal{F}lat$-syzygy of an R-module in $\mathcal{G}_0(R)$ is Gorenstein flat.*

(8) gl left CM- dim $\mathcal{M}_{fg} \leq n$.

(9) *Every nth $\mathcal{P}roj_{fg}$-syzygy of a finitely generated R-module is maximal Cohen–Macaulay.*

Proof. (1) $\Rightarrow$ (2). If inj dim $\Omega = m$, then dim $R = m$ and so by Theorem 11.2.6 and Exercise 4 of Section 11.2 gl right $\mathcal{G}or\mathcal{I}nj$- dim $\mathcal{J}_0(R) \leq m$.

(2) $\Rightarrow$ (3). Let $M \in \mathcal{J}_0(R)$ and $0 \to M \to G^0 \to \cdots \to G^n \to 0$ be a right $\mathcal{G}or\mathcal{I}nj$-resolution of M. If L is an R-module of finite injective dimension, then $\mathrm{Ext}^{n+i}(L, M) \cong \mathrm{Ext}^i(L, G^n) = 0$ for all $i \geq 1$ by Theorem 10.4.24. Now let $0 \to M \to E^0 \to E^1 \to \cdots$ be a right $\mathcal{I}nj$-resolution of M and $G = \mathrm{Coker}(E^{n-2} \to E^{n-1})$. Then $\mathrm{Ext}^i(L, G) = 0$ for all $i \geq 1$ and all R-modules L of finite injective dimension. So G is Gorenstein injective again by Theorem 10.4.24.

(3) $\Rightarrow$ (1). Since R is Cohen–Macaulay, Ω is the dualizing module and so inj dim $\Omega = \dim R$ and thus $\Omega \in \mathcal{J}_0(R)$. But then the nth $\mathcal{I}nj$-cosyzygy of Ω is Gorenstein injective by assumption and so inj dim $\Omega \leq n$ since inj dim $\Omega < \infty$.

The proof of (1) $\Rightarrow$ (4) $\Rightarrow$ (5) is dual to the above using Theorems 11.5.11 and 10.4.18.

(5) $\Rightarrow$ (1). inj dim $\Omega = \mathrm{proj\,dim}\, H_{\mathfrak{m}}^d(R) = d$ by Proposition 9.5.22. So $H_{\mathfrak{m}}^d(R) \in \mathcal{G}_0(R)$ and so the nth $\mathcal{P}roj$-syzygy of $H_{\mathfrak{m}}^d(R)$ is Gorenstein projective and hence projective since proj dim $H_{\mathfrak{m}}^d(R) < \infty$. So proj dim $H_{\mathfrak{m}}^d(R) \leq n$ and thus inj dim $\Omega \leq n$. $7 \Rightarrow 1$ is similar.

(1) $\Rightarrow$ (6) follows from Theorem 11.7.8.

(6) $\Rightarrow$ (7). Let $M \in \mathcal{G}_0(R)$ and $0 \to F_n \to \cdots \to F_1 \to F_0 \to M \to 0$ be a left $\mathcal{G}or\mathcal{F}lat$-resolution of M. If inj dim $L < \infty$, then $\mathrm{Tor}_{n+i}(L, M) \cong \mathrm{Tor}_i(L, F_n) = 0$ for all $i \geq 1$ by Theorem 10.4.28. So if C is an nth $\mathcal{F}lat$-syzygy of an R-module $M \in \mathcal{G}_0(R)$, then $\mathrm{Tor}_i(L, C) = 0$ for all $i \geq 1$ and for all R-modules L of finite injective dimension. So C is Gorenstein flat again by Theorem 10.4.28 since $C \in \mathcal{G}_0(R)$.

(1) $\Rightarrow$ (8). Let $M \in \mathcal{M}_{fg}$ and let C be the nth CM-syzygy of M. Then $\mathrm{Ext}^{n+i}(M, \Omega) \cong \mathrm{Ext}^i(C, \Omega)$ for each $i \geq 1$. But then $\mathrm{Ext}^i(C, \Omega) = 0$ for all $i \geq 1$ since inj dim $\Omega \leq n$. So C is maximal Cohen–Macaulay by Proposition 9.5.23.

(8) $\Rightarrow$ (9). This follows as in the above since if C is the nth $\mathcal{P}roj_{fg}$-syzygy, then $\mathrm{Ext}^i(C, \Omega) \cong \mathrm{Ext}^{n+i}(M, \Omega)$ for all $i \geq 1$.

(9) $\Rightarrow$ (1). (9) means that $\mathrm{Ext}^{n+i}(M, \Omega) = 0$ for all $i \geq 1$ and all $M \in \mathcal{M}_{fg}$. But then inj dim $\Omega \leq n$. $\square$

Corollary 12.4.2.   gl right $\mathcal{G}or\mathcal{I}nj$- dim $\mathcal{J}_0(R) =$ gl left $\mathcal{G}or\mathcal{F}lat$- dim $\mathcal{G}_0(R) =$ gl left $\mathcal{G}or\mathcal{P}roj$- dim $\mathcal{G}_0(R) =$ gl left CM- dim $\mathcal{M}_{fg} = \dim R$.

Exercises

1. Prove (1) $\Rightarrow$ (4) $\Rightarrow$ (5) in Theorem 12.4.1.

2. Let dim $R = d$ and let $\mathcal{J}_0(R)$ be closed under taking dth $\mathcal{G}or\mathcal{I}nj$-syzygies. Then prove that the following are equivalent for an integer $n \geq 2$.

    a) inj dim $\Omega \leq n$.
    b) $\mathrm{Gext}_{n+k}(M, N) = 0$ for all $M, N \in \mathcal{J}_0(R)$ and $k \geq -1$.

c) Every nth $\mathcal{G}or\mathcal{I}nj$-syzygy of an R- module $N \in \mathcal{J}_0(R)$ is Gorenstein injective.

d) gl left $\mathcal{G}or\mathcal{I}nj$- dim $\mathcal{J}_0(R) \leq n - 2$.

Hint: Use Exercise 6 of Section 12.1.

3. Let dim $R = d$ and let $\mathcal{G}_0(R)$ be closed under taking dth $\mathcal{G}or\mathcal{F}lat$-cosyzygies. Then prove that the following are equivalent for an integer $n \geq 2$.

a) inj dim $\Omega \leq n$.

b) right $\mathcal{G}or\mathcal{F}lat$- dim $\mathcal{G}_0(R) \leq n - 2$.

c) right $\mathcal{G}or\mathcal{P}roj$- dim $\mathcal{G}_0(R)_{fg} \leq n - 2$.

4. Under the assumptions of Exercises 2 and 3 above, prove that

$$\text{gl left } \mathcal{G}or\mathcal{I}nj\text{- dim } \mathcal{J}_0(R) = \text{gl right } \mathcal{G}or\mathcal{F}lat\text{- dim } \mathcal{G}_0(R)$$
$$= \text{gl right } \mathcal{G}or\mathcal{P}roj\text{- dim } \mathcal{G}_0(R)_{fg} = d - 2,$$

or is zero if $d \leq 1$.

12.5 Ω-Gorenstein modules

We conclude this chapter by introducing new classes of modules over Cohen–Macaulay rings admitting a dualizing module which generalize Gorenstein modules. We will again let R denote a Cohen–Macaulay ring of Krull dimension d admitting a dualizing module Ω.

We recall that $\mathcal{W}$ denotes the class of all modules W such that $W \cong \Omega \otimes P$ for some projective R-module P and $\mathcal{V}$ denotes the class of all modules V such that $V \cong \text{Hom}(\Omega, E)$ for some injective R-module E. In Proposition 10.4.8, we showed that $\mathcal{W}$ is precovering and $\mathcal{V}$ is preenveloping. If $M \in \mathcal{G}_0(R)$ and $\cdots \to E_1 \to E_0 \to \Omega \otimes M \to 0$ is a left $\mathcal{I}nj$-resolution of $\Omega \otimes M$, then we have a complex $\cdots \to V_1 \to V_0 \to M \to 0$ where $V_i = \text{Hom}(\Omega, E_i)$ for each i since $M \cong \text{Hom}(\Omega, \Omega \otimes M)$. But if E is an injective R-module, then the complex $\cdots \to \text{Hom}(\text{Hom}(\Omega, E), V_1) \to \text{Hom}(\text{Hom}(\Omega, E), V_0) \to \text{Hom}(\text{Hom}(\Omega, E), M) \to 0$ is equivalent to the exact sequence $\cdots \to \text{Hom}(E, E_1) \to \text{Hom}(E, E_0) \to \text{Hom}(E, \Omega \otimes M) \to 0$ for $\text{Hom}(\text{Hom}(\Omega, E), V_i) \cong \text{Hom}(E, E_i)$ and $\text{Hom}(\text{Hom}(\Omega, E), M) \cong \text{Hom}(\text{Hom}(\Omega, E), \text{Hom}(\Omega, \Omega \otimes M)) \cong \text{Hom}(E, \Omega \otimes M)$ by Remark 10.4.2 since $\Omega \otimes M \in \mathcal{J}_0(R)$. Thus the complex $\cdots \to V_1 \to V_0 \to M \to 0$ is a left $\mathcal{V}$-resolution of M. In fact, we have the following result.

Lemma 12.5.1. $\mathcal{V}$ *is closed under inductive limits.*

Proof. This follows from Lemma 10.2.4. □

Theorem 12.5.2. *Every R-module has a $\mathcal{V}$-cover.*

Proof. Let $\{E_k\}$ be a representative set of indecomposable injective R-modules and set $V_k = \text{Hom}(\Omega, E_k)$. Then $\{V_k\}$ is a representative set for $\mathcal{V}$. For if $V \in \mathcal{V}$, then $V \cong \text{Hom}(\Omega, E) \cong \text{Hom}(\Omega, \oplus E_k) \cong \oplus V_k$. So for an R-module M, set $S(M) = \oplus_{k \in I} V_k^{(\text{Hom}(V_k, M))}$. Then $S(M) \in \mathcal{V}$ by the lemma above and so the evaluation map $S(M) \to M$ is a $\mathcal{V}$-precover . But then M has a $\mathcal{V}$-cover by Lemma 12.5.1 and Corollary 5.2.7. □

Let $\mathcal{X}$ be the class of all R-modules X such that $X \cong \Omega \otimes F$ for some flat R-module F. Suppose $N \in \mathcal{J}_0(R)$. If $0 \to \text{Hom}(\Omega, N) \to F^0 \to F^1 \to \cdots$ is a right $\mathcal{F}lat$-resolution of $\text{Hom}(\Omega, N)$, then we have a complex $0 \to N \to X^0 \to X^1 \to \cdots$ where $X^i \cong \Omega \otimes F^i$ for each i since $\Omega \otimes \text{Hom}(\Omega, N) \cong N$. Now if $X \in \mathcal{X}$, then the complex $\cdots \to \text{Hom}(X^1, X) \to \text{Hom}(X^0, X) \to \text{Hom}(N, X) \to 0$ is equivalent to the exact sequence $\cdots \to \text{Hom}(F^1, F) \to \text{Hom}(F^0, F) \to \text{Hom}(\text{Hom}(\Omega, N), F) \to 0$ where $X \cong \Omega \otimes F$ with F flat since $F, F^i \in \mathcal{J}_0(R)$ and so $\text{Hom}(\Omega \otimes F^i, \Omega \otimes F) \cong \text{Hom}(F^i, F)$ and $\text{Hom}(N, \Omega \otimes F) \cong \text{Hom}(\Omega \otimes \text{Hom}(\Omega, N), \Omega \otimes F) \cong \text{Hom}(\text{Hom}(\Omega, N), F)$ since $\text{Hom}(\Omega, N) \in \mathcal{J}_0(R)$. Thus the complex $0 \to N \to X^0 \to X^1 \to \cdots$ is a right $\mathcal{X}$-resolution of N. $\mathcal{F}lat$ is covering by Theorem 7.4.4. Thus we have a left $\mathcal{F}lat$-resolution $\cdots \to F_1 \to F_0 \to \text{Hom}(\Omega, N) \to 0$ of $\text{Hom}(\Omega, N)$ and consequently a complex $\cdots \to X_1 \to X_0 \to N \to 0$ where $X_i = \Omega \otimes F_i$ for each i. But then this complex is an exact left $\mathcal{X}$-resolution since $\text{Hom}(\Omega \otimes F, N) \cong \text{Hom}(\Omega \otimes F, \Omega \otimes \text{Hom}(\Omega, N)) \cong \text{Hom}(F, \text{Hom}(\Omega, N))$ and $\text{Tor}_i(\Omega, \text{Hom}(\Omega, N)) = 0$ for all $i \geq 1$. So we have the following result.

Proposition 12.5.3. *Every $N \in \mathcal{J}_0(R)$ has a right $\mathcal{X}$-resolution and an exact left $\mathcal{X}$-resolution. In particular, $\mathcal{X}$ is precovering and preenveloping on $\mathcal{J}_0(R)$.*

Remark 12.5.4. We note that if $N \in \mathcal{J}_0(R)_{fg}$, then a similar argument shows that N has a right $\mathcal{W}_{fg}$-resolution and so in particular $\mathcal{W}_{fg}$ is preenveloping on $\mathcal{J}_0(R)_{fg}$ where $\mathcal{W}_{fg}$ denotes the class of modules $\Omega \otimes P$ with P finitely generated and projective.

Theorem 12.5.5. *$\text{Hom}(-, -)$ is left balanced on $\mathcal{M} \times \mathcal{M}$ by $\mathcal{V} \times \mathcal{V}$, on $\mathcal{J}_0(R) \times \mathcal{J}_0(R)$ by $\mathcal{X} \times \mathcal{X}$, and on $\mathcal{J}_0(R)_{fg} \times \mathcal{J}_0(R)_{fg}$ by $\mathcal{W}_{fg} \times \mathcal{W}_{fg}$.*

Proof. The result follows from Proposition 10.4.7, Theorem 12.5.2, Proposition 12.5.3 and Remark 12.5.4 above. □

Theorem 12.5.6. *$\text{Hom}(-, -)$ is right balanced on $\mathcal{J}_0(R) \times \mathcal{J}_0(R)$ by $\mathcal{P}roj \times \mathcal{V}$ and on $\mathcal{J}_0(R) \times \mathcal{J}_0(R)$ by $\mathcal{W} \times \mathcal{I}nj$.*

Proof. Let $M \in \mathcal{J}_0(R)$ and $\cdots \to P_1 \to P_0 \to M \to 0$ be a projective resolution of M. Then $\cdots \to \Omega \otimes P_1 \to \Omega \otimes P_0 \to \Omega \otimes M \to 0$ is exact and so easily $0 \to \text{Hom}(M, V) \to \text{Hom}(P_0, V) \to \text{Hom}(P_1, V) \to \cdots$ is exact for all $V \in \mathcal{V}$. Now the first part of the theorem easily follows since each module in $\mathcal{J}_0(R)$ has an exact right $\mathcal{V}$-resolution by Proposition 10.4.8. The second part follows dually. □

Theorem 12.5.7. $-\otimes-$ *is left balanced on* $\mathcal{G}_0(R)\times\mathcal{J}_0(R)$ *by* $\mathcal{P}roj\times\mathcal{W}$ *and* $\mathcal{F}lat\times\mathcal{X}$.

Proof. We simply note that if $M\in\mathcal{G}_0(R)$, then $\operatorname{Tor}_i(\Omega,M)=0$ for all $i\geq 1$ and each $N\in\mathcal{J}_0(R)$ has an exact left $\mathcal{W}$-resolution and an exact left $\mathcal{X}$-resolution by Proposition 10.4.8 and 12.5.3. □

Lemma 12.5.8. *Let* $M\in\mathcal{J}_0(R)$ *and* $0\to M\to X^0\to X^1\to\cdots$ *be a right* $\mathcal{X}$-resolution of M. Then the complex $0\to V\otimes M\to V\otimes X^0\to V\otimes X^1\to\cdots$ is exact for all $V\in\mathcal{V}$.

Proof. We first note that the sequence $0\to V\otimes M\to V\otimes X^0\to V\otimes X^1\to\cdots$ is equivalent to the sequence $0\to E\otimes\operatorname{Hom}(\Omega,M)\to E\otimes F^0\to E\otimes F^1\to\cdots$ where $V\cong\operatorname{Hom}(\Omega,E)$ and $X^i\cong\Omega\otimes F^i$ with E injective and F^i flat. But $0\to\operatorname{Hom}(\Omega,M)\to F^0\to F^1\to\cdots$ is a right $\mathcal{F}lat$-resolution of $\operatorname{Hom}(\Omega,M)$ and so we are done. □

Theorem 12.5.9. $-\otimes-$ *is right balanced on* $\mathcal{G}_0(R)\times\mathcal{J}_0(R)$ *by* $\mathcal{V}\times\mathcal{X}$.

Proof. Let $M\in\mathcal{G}_0(R)$ and $0\to M\to V^0\to V^1\to\cdots$ be a right $\mathcal{V}$-resolution of M. Then $0\to\Omega\otimes M\to\Omega\otimes V^0\to\Omega\otimes V^1\to\cdots$ is exact and so $0\to X\otimes M\to X\otimes V^0\to X\otimes V^1\to\cdots$ is exact for all $X\in\mathcal{X}$. The result now follows from the lemma above. □

Lemma 12.5.10. $M\in\mathcal{V}$ *if and only if* $M^+\in X$.

Proof. If $M\in\mathcal{V}$, then $M\cong\operatorname{Hom}(\Omega,E)$ for some injective E and so $M\in\mathcal{G}_0(R)$ by Remark 10.4.2. Furthermore, $M^+\cong\Omega\otimes E^+\in X$. Conversely, if $M^+\cong\Omega\otimes F$ with F flat, then $M^{++}\cong\operatorname{Hom}(\Omega,F^+)\in\mathcal{V}$. But $\Omega\otimes M$ is a pure submodule of $\Omega\otimes M^{++}\cong F^+$. So $\Omega\otimes M$ is injective by Proposition 5.3.8. But $M^+\in\mathcal{J}_0(R)$. It is now easy to check that $M\in\mathcal{G}_0(R)$ and so $M\cong\operatorname{Hom}(\Omega,\Omega\otimes M)\in\mathcal{V}$. □

A dual argument gives the following result.

Lemma 12.5.11. $M\in\mathcal{X}$ *if and only if* $M^+\in\mathcal{V}$.

Lemma 12.5.12. *Let* $G\in\mathcal{G}_0(R)$. *Then* right $\mathcal{V}$-dim $G\leq n$ *if and only if for any left* $\mathcal{V}$-resolution $\cdots\to V_1\to V_0\to M\to 0$ *of each* $M\in{}_R\mathcal{M}$, $\operatorname{Hom}(G,V_n)\to\operatorname{Hom}(G,\operatorname{Ker}(V_{n-1}\to V_{n-2}))$ *is exact where* $V_{-1}=M$.

Proof. The proof is as in Lemma 8.4.34 noting that $\mathcal{V}$-preenvelopes of modules in $\mathcal{G}_0(R)$ are injections. □

Theorem 12.5.13. *The following are equivalent for an integer* $n\geq 0$.

(1) inj dim $\Omega\leq n$.

(2) right $\mathcal{V}$-dim $R\leq n$.

(3) *If* $\cdots \to E_1 \to E_0 \to M \to 0$ *is a left* $\mathcal{I}nj$*-resolution of an R-module M,
then the complex* $\cdots \to \operatorname{Hom}(\Omega, E_1) \to \operatorname{Hom}(\Omega, E_0) \to \operatorname{Hom}(\Omega, M) \to 0$
is exact at $\operatorname{Hom}(\Omega, E_i)$ *for all* $i \geq n - 1$ *where* $E_{-1} = M$.

(4) *If* $0 \to M \to F^0 \to F^1 \to \cdots$ *is a right* $\mathcal{F}lat$*-resolution of an R-module M,
then the complex* $0 \to \Omega \otimes M \to \Omega \otimes F^0 \to \Omega \otimes F^1 \to \cdots$ *is exact at* $\Omega \otimes F^i$
for all $i \geq n - 1$ *where* $F^{-1} = M$.

(5) *If* $0 \to M \to F^0 \to F^1 \to \cdots$ *is a right* $\mathcal{F}lat$*-resolution of an R-module
$M \in \mathcal{G}_0(R)$, then the complex* $0 \to \Omega \otimes M \to \Omega \otimes F^0 \to \Omega \otimes F^1 \to \cdots$ *is
exact at* $\Omega \otimes F^i$ *for all* $i \geq n - 1$ *where* $F^{-1} = M$.

(6) *If* $0 \to M \to P^0 \to P^1 \to \cdots$ *is a right* $\mathcal{P}roj_{fg}$*-resolution of a finitely
generated R-module M, then the complex* $0 \to \Omega \otimes M \to \Omega \otimes P^0 \to \Omega \otimes P^1 \to$
$\cdots$ *is exact at* $\Omega \otimes P^i$ *for all* $i \geq n - 1$ *where* $P^{-1} = M$.

(7) *If* $0 \to M \to E^0 \to E^1 \to \cdots$ *is a right* $\mathcal{I}nj$*-resolution of an R-module
M, then for each* $V \in \mathcal{V}$ *the complex* $0 \to \operatorname{Hom}(V, M) \to \operatorname{Hom}(V, E^0) \to$
$\operatorname{Hom}(V, E^1) \to \cdots$ *is exact at* $\operatorname{Hom}(V, E^i)$ *for all* $i \geq n$.

(8) *If* $0 \to M \to E^0 \to E^1 \to \cdots$ *is a right* $\mathcal{I}nj$*-resolution of an R-module* $M \in$
$\mathcal{G}_0(R)$, then for each $V \in \mathcal{V}$ the complex* $0 \to \operatorname{Hom}(V, M) \to \operatorname{Hom}(V, E^0) \to$
$\operatorname{Hom}(V, E^1) \to \cdots$ *is exact at* $\operatorname{Hom}(V, E^i)$ *for all* $i \geq n$.

(9) *Every left* $\mathcal{V}$*-resolution* $\cdots \to V_1 \to V_0 \to M \to 0$ *of M is exact at* V_i *for all*
$i \geq n - 1$ *where* $V_{-1} = M$.

(10) *Each* $M \in \mathcal{G}_0(R)$ *has an exact right* $\mathcal{V}$*-resolution* $0 \to M \to V^0 \to V^1 \to \cdots$
such that for each $V \in \mathcal{V}$ *the complex* $0 \to \operatorname{Hom}(V, M) \to \operatorname{Hom}(V, V^0) \to$
$\operatorname{Hom}(V, V^1) \to \cdots$ *is exact at* $\operatorname{Hom}(V, V^i)$ *for all* $i \geq n$.

(11) *If* $\cdots \to P_1 \to P_0 \to M \to 0$ *is a left* $\mathcal{P}roj$*-resolution of an R-module M,
then for each* $X \in \mathcal{X}$ *the complex* $0 \to \operatorname{Hom}(M, X) \to \operatorname{Hom}(P_0, X) \to$
$\operatorname{Hom}(P_1, X) \to \cdots$ *is exact at* $\operatorname{Hom}(P_i, X)$ *for all* $i \geq n$.

(12) *Each* $M \in \mathcal{J}_0(R)$ *has an exact left* $\mathcal{W}$*-resolution* $\cdots \to W_1 \to W_0 \to M \to 0$
such that for each $W \in \mathcal{W}$ *the complex* $0 \to \operatorname{Hom}(M, W) \to \operatorname{Hom}(W_0, W) \to$
$\operatorname{Hom}(W_1, W) \to \cdots$ *is exact at* $\operatorname{Hom}(W_i, W)$ *for all* $i \geq n$.

(13) *Every right* $\mathcal{X}$*-resolution* $0 \to M \to X^0 \to X^1 \to \cdots$ *of* $M \in \mathcal{J}_0(R)$ *is exact
at* X^i *for all* $i \geq n - 1$ *where* $X^{-1} = M$.

(14) *If* $\cdots \to F_1 \to F_0 \to M \to 0$ *is a left* $\mathcal{F}lat$*-resolution of an R-module M,
then for each* $V \in \mathcal{V}$ *the complex* $\cdots \to V \otimes F_1 \to V \otimes F_0 \to V \otimes M \to 0$
is exact at $V \otimes F_i$ *for all* $i \geq n$.

(15) *Each* $M \in \mathcal{J}_0(R)$ *has an exact left* $\mathcal{X}$*-resolution* $\cdots \to X_1 \to X_0 \to M \to 0$
such that for each $V \in \mathcal{V}$ *the complex* $\cdots \to V \otimes X_1 \to V \otimes X_0 \to V \otimes M \to 0$
is exact at $V \otimes X_i$ *for all* $i \geq n$.

(16) *Every right* $\mathcal{W}_{fg}$*-resolution* $0 \to M \to W^0 \to W^1 \to \cdots$ *of* $M \in \mathcal{J}_0(R)_{fg}$ *is
exact at* W^i *for all* $i \geq n - 1$ *where* $W^{-1} = M$.

Proof. (1) $\Leftrightarrow$ (2). Let $0 \to \Omega \to E^0 \to E^1 \to \cdots \to E^n \to 0$ be an injective resolution of Ω. Then $0 \to \text{Hom}(\Omega, \Omega) \to \text{Hom}(\Omega, E^0) \to \cdots \to \text{Hom}(\Omega, E^n) \to 0$ is exact since $\text{Ext}^i(\Omega, \Omega) = 0$ for all $i \geq 1$. But $\text{Hom}(\Omega, \Omega) \cong R$ and it is easy to check that this sequence is a right $\mathcal{V}$-resolution. So (2) follows. Conversely, let $0 \to R \to V^0 \to V^1 \to V^n \to 0$ be a right $\mathcal{V}$-resolution of R (see Proposition 10.4.8). Then $0 \to \Omega \to \Omega \otimes V^0 \to \cdots \to \Omega \otimes V^{n-1} \to \Omega \otimes V^n \to 0$ is exact since $\mathcal{V} \subset \mathcal{G}_0(R)$ and so $\text{Tor}_i(\Omega, V) = 0$ for all $i \geq 1$ and all $V \in \mathcal{V}$. But $\Omega \otimes V^i$ is injective for each i. Thus (1) follows.

(1) $\Leftrightarrow$ (3) follows as in Theorem 8.4.36 and (1) $\Leftrightarrow$ (4) $\Leftrightarrow$ (6) follows as in Theorem 8.4.31.

(1) $\Leftrightarrow$ (7). It follows from Proposition 8.4.3 that proj dim $V \leq n$ if and only if (7) holds. But then the result follows since proj dim $\text{Hom}(\Omega, E) \leq n$ for all injective E if and only if inj dim $\Omega \leq n$.

(4) $\Rightarrow$ (5), (7) $\Rightarrow$ (8) are trivial.

(2) $\Leftrightarrow$ (9). This follows as in Theorem 8.4.36 using Lemma 12.5.12 and Theorem 12.5.5.

(5) $\Rightarrow$ (1). Let K be the nth $\mathcal{F}lat$-syzygy of $\text{Hom}(\Omega, E)$ with E an injective R-module. Then $K \in \mathcal{G}_0(R)$. So if $0 \to K \to F^0 \to F^1 \to \cdots$ is a right $\mathcal{F}lat$-resolution, then $\cdots \to \text{Hom}(\Omega \otimes F^1, E) \to \text{Hom}(\Omega \otimes F^0, E) \to \text{Hom}(\Omega \otimes K, E) \to 0$ is exact at $\text{Hom}(\Omega \otimes F^i, E)$ for $i \geq n - 1$ by assumption. Therefore, $\text{Ext}_i(K, \text{Hom}(\Omega, E)) = 0$ for all $i \geq n - 1$. Now computing $\text{Ext}_i(-, -)$ using a left $\mathcal{F}lat$-resolution $\cdots \to F_n \to F_{n-1} \to \cdots \to F_1 \to F_0 \to \text{Hom}(\Omega, E) \to 0$, we see that $K = \text{Ker}(F_{n-1} \to F_{n-2})$ is a direct summand of F_n and thus flat dim $\text{Hom}(\Omega, E) \leq n$. But then (1) follows.

(1) $\Rightarrow$ (10). The existence of such a resolution follows from Proposition 10.4.8. We now simply note that if $V' \in \mathcal{V}$, then

$$\text{Ext}^i(V, V') \cong \text{Ext}^i(\text{Hom}(\Omega, E), \text{Hom}(\Omega, E'))$$
$$\cong \text{Hom}(\text{Tor}_i(\text{Hom}(\Omega, E), \Omega), E')$$
$$= 0$$

for all $i \geq 1$ since $E \in \mathcal{J}_0(R)$. So inj dim $\Omega \leq n$ if and only if proj dim $V \leq n$ for all $V \in \mathcal{V}$ and thus $\text{Ext}^i(V, M) = 0$ for all $i > n$ and all $V \in \mathcal{V}$. Hence (10) follows.

(8), (10) $\Rightarrow$ (1). $V \in \mathcal{G}_0(R)$ and so its nth $\mathcal{P}roj$-syzygy is in $\mathcal{G}_0(R)$. Hence $\text{Ext}^i(V, M) = 0$ for all $i > n$ and all $M \in \mathcal{G}_0(R)$ means proj dim $V < \infty$.

(1) $\Leftrightarrow$ (11) follows similarly since inj dim $\Omega \leq n$ if and only if inj dim $X \leq n$ for all $X \in \mathcal{X}$.

(1) $\Leftrightarrow$ (12). We note that inj dim $\Omega \leq n$ if and only if inj dim $W \leq n$ for all $W \in \mathcal{W}$, and if $W, W' \in \mathcal{W}$, then $\text{Ext}^i(W', W) = 0$ for all $i \geq 1$ since $\text{Ext}^i(\Omega, \Omega) = 0$ for all $i \geq 1$. Moreover, the nth $\mathcal{I}nj$-cosyzygy of W is in $\mathcal{G}_0(R)$. Thus the result follows from Proposition 10.4.9 and an argument dual to the one in (1) $\Leftrightarrow$ (10).

(4) $\Rightarrow$ (13). Let $M \in \mathcal{J}_0(R)$ and $0 \to M \to X^0 \to X^1 \to \cdots$ be a right $\mathcal{X}$-resolution of M. Then $0 \to \text{Hom}(\Omega, M) \to F^0 \to F^1 \to \cdots$ is

a right $\mathcal{F}lat$-resolution of $\mathrm{Hom}(\Omega, M)$ where $X^i \cong \Omega \otimes F^i$. But the complex $0 \to \Omega \otimes \mathrm{Hom}(\Omega, M) \to \Omega \otimes F^0 \to \Omega \otimes F^1 \to \cdots$ is equivalent to the complex $0 \to M \to X^0 \to X^1 \to \cdots$. So (13) easily follows.

(13) $\Rightarrow$ (5). Let $M \in \mathcal{J}_0(R)$ and $0 \to M \to F^0 \to F^1 \to \cdots$ be a right $\mathcal{F}lat$-resolution. Then $0 \to \Omega \otimes M \to \Omega \to F^0 \to \Omega \otimes F^1 \to \cdots$ is a right $\mathcal{X}$-resolution of $\Omega \otimes M$. But $\Omega \otimes M \in \mathcal{J}_0(R)$. So (5) follows from (13).

(1) $\Leftrightarrow$ (14) is trivial since flat dim $V \leq n$ if and only if $\mathrm{Tor}_i(V, M) = 0$ for all $i \geq n + 1$ and all R-modules M.

(1) $\Leftrightarrow$ (15). We simply note that $\mathrm{Hom}(\mathrm{Tor}_i(V, X), \mathbb{Q}/\mathbb{Z}) \cong \mathrm{Ext}^i(V, X^+) = 0$ for all $i \geq 1$ since $X^+ \in \mathcal{V}$ by Lemma 12.5.11. Thus $\mathrm{Tor}_i(V, X) = 0$ for all $V \in \mathcal{V}$ and $X \in \mathcal{X}$.

(1) $\Leftrightarrow$ (16) is left to the reader.   □

Definition 12.5.14. An R-module M is said to be Ω-*Gorenstein injective* if there exists a $\mathrm{Hom}(\mathcal{V}, -)$ exact exact sequence

$$\cdots \to V_1 \to V_0 \to V^0 \to V^1 \to \cdots$$

of modules in $\mathcal{V}$ such that $M = \mathrm{Ker}(V^0 \to V^1)$.

Dually, an R-module M is said to be Ω-*Gorenstein projective* if there exists a $\mathrm{Hom}(-, \mathcal{W})$ exact exact sequence

$$\cdots \to W_1 \to W_0 \to W^0 \to W^1 \to \cdots$$

of modules in $\mathcal{W}$ such that $M = \mathrm{Ker}(W^0 \to W^1)$.

A module M is said to be Ω-*Gorenstein flat* if there exists a $\mathcal{V} \otimes -$ exact exact sequence

$$\cdots \to X_1 \to X_0 \to X^0 \to X^1 \to \cdots$$

of modules in $\mathcal{X}$ such that $M = \mathrm{Ker}(X^0 \to X^1)$.

We note that every $V \in \mathcal{V}$ is Ω-Gorenstein injective, $W \in \mathcal{W}$ is Ω-Gorenstein projective, and $X \in \mathcal{X}$ is Ω-Gorenstein flat.

Proposition 12.5.15. *Classes $\mathcal{V}$, $\mathcal{W}$, and $\mathcal{X}$ are closed under direct summands.*

Proof. Let $V \in \mathcal{V}$ and G be a direct summand of V. Set $V \cong \mathrm{Hom}(\Omega, E) \cong G \oplus G'$ for some R-module G' and injective R-module E. But then $\Omega \otimes G$ is an injective R-module. Furthermore, $G \in \mathcal{J}_0(R)$ since $V \in \mathcal{J}_0(R)$. So $G \cong \mathrm{Hom}(\Omega, \Omega \otimes G)$. That is, $G \in \mathcal{V}$. The proofs for $\mathcal{W}$ and $\mathcal{X}$ are similar.   □

Lemma 12.5.16. *If M is Ω-Gorenstein injective, then $\mathrm{Ext}^i(V, M) = 0$ for all $V \in \mathcal{V}$ and all $i \geq 1$.*

Proof. This is trivial since $\mathrm{Ext}^i(V, V') = 0$ for all $V, V' \in \mathcal{V}$ and all $i \geq 1$.   □

Remark 12.5.17. It follows from the above that the right $\mathcal{V}$-dimension of an Ω-Gorenstein injective module and the left $\mathcal{W}$-dimension of an Ω-Gorenstein projective module are either zero or infinite.

We now show that there is an abundant supply of Ω-Gorenstein modules.

Theorem 12.5.18. *Let M be an R-module. Then*

(1) *If $0 \to M \to V^0 \to V^1 \to \cdots$ is an exact right $\mathcal{V}$-resolution of M and $C^i = \mathrm{Ker}(V^i \to V^{i+1})$, then C^i is Ω-Gorenstein injective for $i \geq d$.*

(2) *If $\cdots \to V_1 \to V_0 \to M \to 0$ is a left $\mathcal{V}$-resolution of M and $C_i = \mathrm{Coker}\,(V_{i+1} \to V_i)$, then C_i is Ω-Gorenstein injective for $i \geq d - 1$.*

Proof. This follows as in the proof of Theorem 10.1.13 using Theorem 12.5.13 above. $\square$

Theorem 12.5.19. *Let M be a finitely generated R-module and suppose $M \in \mathcal{J}_0(R)$. Then*

(1) *If $\cdots \to W_1 \to W_0 \to M \to 0$ is an exact left $\mathcal{W}$-resolution of M and $C_i = \mathrm{Coker}\,(W_{i+1} \to W_i)$, then C_i is Ω-Gorenstein projective for $i \geq d$.*

(2) *If $0 \to M \to W^0 \to W^1 \to \cdots$ is a right $\mathcal{W}_{fg}$-resolution and $C^i = \mathrm{Ker}(W^i \to W^{i+1})$, then C^i is Ω-Gorenstein projective for $i \geq d - 1$.*

Proof. This follows as in the proof of Theorem 10.2.16 using Theorem 12.5.13 and Remark 12.5.4. $\square$

Remark 12.5.20. If M is Ω-Gorenstein flat, then M^+ is Ω-Gorenstein injective. For if $X \in \mathcal{X}$, then $X^+ \in \mathcal{V}$ by Lemma 12.5.11. Thus if $\cdots \to X_1 \to X_0 \to X^0 \to X^1 \to \cdots$ is the exact sequence in the definition, then $\cdots \to X^{1+} \to X^{0+} \to X_0^+ \to X_1^+ \to \cdots$ is an exact sequence of modules in $\mathcal{V}$. Moreover, applying $\mathrm{Hom}(V, -)$ gives the sequence $\cdots \to \mathrm{Hom}(V, \mathrm{Hom}(\Omega, F^{1+}) \to \mathrm{Hom}(V, \mathrm{Hom}(\Omega, F^{0+})) \to \mathrm{Hom}(V, \mathrm{Hom}(\Omega, F_0^+)) \to \cdots$ where $X \cong \Omega \otimes F$ with F flat. This is equivalent to the complex $\cdots \to (E \otimes F^1)^+ \to (E \otimes F^0)^+ \to (E \otimes F_0)^+ \to (E \otimes F_1)^+ \to \cdots$ where $V \cong \mathrm{Hom}(\Omega, E)$ with E injective. But the exact sequence $\cdots \to V \otimes X_1 \to V \otimes X_0 \to V \otimes X^0 \to V \otimes X^1 \to \cdots$ is equivalent to $\cdots \to E \otimes F_1 \to E \otimes F_0 \to E \otimes F^0 \to E \otimes F^1 \to \cdots$ and thus $M^+ = \mathrm{Ker}(X_0^+ \to X_1^+)$ is Ω-Gorenstein injective. We will show in Theorem 12.5.25 below that the converse holds if $M \in \mathcal{J}_0(R)$.

Proposition 12.5.21. *The left $\mathcal{X}$-dimension of an Ω-Gorenstein flat module is either zero or infinite.*

Proof. Let M be Ω-Gorenstein flat. If the left $\mathcal{X}$-dimension of M is finite, then the right $\mathcal{V}$-dimension of M^+ is finite. But M^+ is Ω-Gorenstein injective by Remark 12.5.20 above. So $M^+ \in \mathcal{V}$ by Remark 12.5.17. Thus $M \in \mathcal{X}$ by Lemma 12.5.11. $\square$

Proposition 12.5.22. *Let* $M \in \mathcal{G}_0(R)$. *Then* M *is* Ω-*Gorenstein injective if and only if* $\Omega \otimes M$ *is Gorenstein injective.*

Proof. Suppose $\cdots \rightarrow V_1 \rightarrow V_0 \rightarrow V^0 \rightarrow V^1 \rightarrow \cdots$ is an exact sequence of modules in $\mathcal{V}$ with $M = \mathrm{Ker}(V^0 \rightarrow V^1)$ such that the sequence remains exact when $\mathrm{Hom}(V, -)$ is applied to it whenever $V \in \mathcal{V}$. Now applying $\Omega \otimes -$ to the complex gives the complex $\cdots \rightarrow E_1 \rightarrow E_0 \rightarrow E^0 \rightarrow E^1 \rightarrow \cdots$ of injective modules where $V_i \cong \mathrm{Hom}(\Omega, E_i)$, $V^i \cong \mathrm{Hom}(\Omega, E^i)$, and $\Omega \otimes M = \mathrm{Ker}(E^0 \rightarrow E^1)$. But $\mathrm{Tor}_i(\Omega, V) = 0$ for all $i \geq 1$ and all $V \in \mathcal{V}$ since $V \in \mathcal{G}_0(R)$. Furthermore, $M \in \mathcal{G}_0(R)$ by assumption. So the complex $\cdots \rightarrow E_1 \rightarrow E_0 \rightarrow E^0 \rightarrow E^1 \rightarrow \cdots$ is exact by Theorem 10.4.10. But if $V, V' \in \mathcal{V}$, then $\mathrm{Hom}(V, V') \cong \mathrm{Hom}(E, E')$ where $V \cong \mathrm{Hom}(\Omega, E)$, $V' \cong \mathrm{Hom}(\Omega, E')$ with E, E' injective. Thus $\mathrm{Hom}(E, -)$ leaves the complex $\cdots \rightarrow E_1 \rightarrow E_0 \rightarrow E^0 \rightarrow E^1 \rightarrow \cdots$ exact. That is, $\Omega \otimes M$ is Gorenstein injective.

Conversely, if $\Omega \otimes M$ is Gorenstein injective and $\cdots \rightarrow E_1 \rightarrow E_0 \rightarrow E^0 \rightarrow E^1 \rightarrow \cdots$ is the exact sequence of injective modules for $\Omega \otimes M$, then $\mathrm{Hom}(\Omega, -)$ leaves the sequence exact since $\mathrm{inj\,dim}\, \Omega < \infty$. Thus we get an exact sequence $\cdots \rightarrow V_1 \rightarrow V_0 \rightarrow V^0 \rightarrow V^1 \rightarrow \cdots$ of modules in $\mathcal{V}$. It is now easy to check that this sequence remains exact if $\mathrm{Hom}(V, -)$ is applied to it whenever $V \in \mathcal{V}$. Thus $M \cong \mathrm{Hom}(\Omega, \Omega \otimes M)$ is Ω-Gorenstein injective. $\square$

Proposition 12.5.23. *Let* $M \in \mathcal{G}_0(R)$. *Then* M *is* Ω-*Gorenstein projective if and only if* $\mathrm{Hom}(\Omega, M)$ *is Gorenstein projective.*

Proof. The proof is dual to the proof of the proposition above noting that $\mathrm{Hom}(W', W) \cong \mathrm{Hom}(P', P)$ where $W' \cong \Omega \otimes P'$, $W \cong \Omega \otimes P$ with P, P' projective and $\mathrm{Ext}^i(\Omega, M) = \mathrm{Ext}^i(\Omega, W) = 0$ for all $i \geq 1$ and all $W \in \mathcal{W}$ since $W \in \mathcal{G}_0(R)$. $\square$

Lemma 12.5.24. *Let* $M \in \mathcal{G}_0(R)$ *and* $\cdots \rightarrow X_1 \rightarrow X_0 \rightarrow X^0 \rightarrow X^1 \rightarrow \cdots$ *be a complete* $\mathcal{X}$-*resolution of* M. *Then* $\mathrm{Tor}_i(L, M) = 0$ *for all* $i \geq 1$ *and all* R-*modules* L *of finite* $\mathcal{V}$-*dimension if and only if the sequence* $\cdots \rightarrow L \otimes X_1 \rightarrow L \otimes X_0 \rightarrow L \otimes X^0 \rightarrow L \otimes X^1 \rightarrow \cdots$ *is exact for all such* L.

Proof. We recall from the proof of Theorem 12.5.13 that $\mathrm{Tor}_i(V, X) = 0$ for all $i \geq 1$, $V \in \mathcal{V}$, and $X \in \mathcal{X}$. So the complex $\cdots \rightarrow V \otimes X_1 \rightarrow V \otimes X_0 \rightarrow V \otimes X^0 \rightarrow V \otimes X^1 \rightarrow \cdots$ is exact for all $V \in \mathcal{V}$ by assumption and Lemma 12.5.8 above. One then proceeds by induction on right $\mathcal{V}$-dimension of L as in the proof of Lemma 10.3.7. The converse is trivial. $\square$

Theorem 12.5.25. *The following are equivalent for an* R-*module* $M \in \mathcal{G}_0(R)$.

(1) M *is* Ω-*Gorenstein flat.*

(2) $\mathrm{Hom}(\Omega, M)$ *is Gorenstein flat.*

(3) M^+ *is* Ω-*Gorenstein injective.*

(4) $\mathrm{Tor}_i(V, M) = 0$ *for all* $i \geq 1$ *and all* $V \in \mathcal{V}$.

(5) $\mathrm{Tor}_i(L, M) = 0$ *for all* $i \geq 1$ *and all* L *of finite right* $\mathcal{V}$-*dimension*.

(6) $\mathrm{Tor}_1(L, M) = 0$ *for all* L *of finite right* $\mathcal{V}$-*dimension*.

Proof. (1) $\Leftrightarrow$ (2). Let $\cdots \to X_1 \to X_0 \to X^0 \to X^1 \to \cdots$ be as in Definition 12.5.14. Then we have an exact sequence $\cdots \to F_1 \to F_0 \to F^0 \to F^1 \to \cdots$ of flat modules where $X_i \cong \Omega \otimes F_i$, $X^i \cong \Omega \otimes F^i$, and $\mathrm{Hom}(\Omega, M) = \mathrm{Ker}(F^0 \to F^1)$ since $\mathrm{Ext}^i(\Omega, M) = \mathrm{Ext}^i(\Omega, X) = 0$ for all $i \geq 1$, all $X \in \mathcal{X}$ and $\mathrm{Hom}(\Omega, \Omega \otimes F) \cong F$ for all flat F. But $\cdots \to V \otimes X_1 \to V \otimes X_0 \to V \otimes X^0 \to V \otimes X^1 \to \cdots$ is exact for all $V \in \mathcal{V}$ by assumption and $V \otimes X \cong \mathrm{Hom}(\Omega, E) \otimes \Omega \otimes F \cong E \otimes F$ where $V \cong \mathrm{Hom}(\Omega, E)$. Thus the complex $\cdots \to E \otimes F_1 \to E \otimes F_0 \to E \otimes F^0 \to E \otimes F^1 \to \cdots$ is exact for all injective R-modules E. So $\mathrm{Hom}(\Omega, M)$ is Gorenstein flat. The converse is now standard.

(2) $\Leftrightarrow$ (3). $\mathrm{Hom}(\Omega, M)$ is Gorenstein flat if and only if $\mathrm{Hom}(\Omega, M)^+$ is Gorenstein injective if and only if $\Omega \otimes M^+$ is Gorenstein injective. But $M^+ \in \mathcal{G}_0(R)$. So the result follows by Proposition 12.5.22 above.

(1) $\Rightarrow$ (4) is trivial since $\mathrm{Tor}_i(V, X) = 0$ for all $V \in \mathcal{V}$ and $X \in \mathcal{X}$, (4) $\Rightarrow$ (5) follows by dimension shifting, and (5) $\Rightarrow$ (6) is trivial.

(6) $\Rightarrow$ (5) Suppose L has finite right $\mathcal{V}$-dimension and consider the exact sequence $0 \to L' \to P \to L \to 0$ with P projective. Then P has a finite $\mathcal{V}$-dimension. Thus L' also has finite $\mathcal{V}$-dimension and so the result follows by dimension shifting.

(5) $\Rightarrow$ (1). Setting $R = L$ in Lemma 12.5.24, we get that the complete $\mathcal{X}$-resolution $\cdots \to X_1 \to X_0 \to X^0 \to X^1 \to \cdots$ is exact and so M is Ω-Gorenstein flat. $\qquad\square$

Theorem 12.5.26. *Let $M \in \mathcal{J}_0(R)$. Then*

(1) *If $\cdots \to X_1 \to X_0 \to M \to 0$ is an exact left $\mathcal{X}$-resolution of M and $C_i = Coker(X_{i+1} \to X_i)$, then C_i is Ω-Gorenstein flat for all $i \geq d$.*

(2) *If $0 \to M \to X^0 \to X^1 \to \cdots$ is a right $\mathcal{X}$-resolution and $C^i = Ker(X^i \to X^{i+1})$, then C^i is Ω-Gorenstein flat for $i \geq d - 1$.*

Proof. (1) If $V \in \mathcal{V}$, then $\mathrm{Tor}_j(V, M) = 0$ for all $j \geq d + 1$. But if C_i is the ith $\mathcal{X}$-syzygy of M, then $\mathrm{Tor}_j(V, C_i) \cong \mathrm{Tor}_{i+j}(V, M) = 0$ for all $i \geq d$, $j \geq 1$, $V \in \mathcal{V}$ since $\mathrm{Tor}_i(V, X) = 0$ for all $i \geq 1$, $V \in \mathcal{V}$, $X \in \mathcal{X}$. So the result follows from Theorem 12.5.25 above.

(2) This follows from Theorem 12.5.13 as in Corollary 10.3.13. $\qquad\square$

Theorem 12.5.27. *Every R-module has an Ω-Gorenstein injective precover.*

Proof. If $\cdots \to V_1 \to V_0 \to M \to 0$ is a left $\mathcal{V}$-resolution of M, then $K = \mathrm{Ker}(V_{d-1} \to V_{d-2})$ is Ω-Gorenstein injective by Theorem 12.5.18 above. So K has an exact sequence $\cdots \to V_1 \to V_0 \to V^0 \to V^1 \to \cdots$ such that each $V^i \in \mathcal{V}$ and $\mathrm{Hom}(V, -)$ leaves the sequence exact whenever $V \in \mathcal{V}$. But then it is easy to see that each cosyzygy in the sequence is also Ω-Gorenstein injective. Thus the result follows as in Theorem 11.1.1.   □

Theorem 12.5.28. *Every R-module has an Ω-Gorenstein injective cover.*

Proof. This is now standard.   □

Theorem 12.5.29. *Every R-module $M \in \mathcal{G}_0(R)$ has an Ω-Gorenstein injective preenvelope $M \to G$ such that if $0 \to M \to G \to L \to 0$ is exact, then* right $\mathcal{V}$-dim $L \leq d - 1$ *whenever $d \geq 1$.*

Proof. The proof is similar to the proof of Theorem 11.2.1 using Theorem 12.5.18 and Lemma 12.5.16.   □

Corollary 12.5.30. *The following are equivalent for an R-module $M \in \mathcal{G}_0(R)$.*

(1) *M is Ω-Gorenstein injective.*

(2) $\mathrm{Ext}^i(V, M) = 0$ *for all $V \in \mathcal{V}$ and for all $i \geq 1$.*

(3) $\mathrm{Ext}^i(L, M) = 0$ *for all R-modules L such that* right $\mathcal{V}$-dim $L < \infty$ *and for all $i \geq 1$.*

(4) $\mathrm{Ext}^1(L, M) = 0$ *for all R-modules L such that* right $\mathcal{V}$-dim $L < \infty$.

Proof. (1) $\Rightarrow$ (2) is Lemma 12.5.16 above, (2) $\Rightarrow$ (3) follows by dimension shifting, and (3) $\Rightarrow$ (4) is trivial.

(4) $\Rightarrow$ (1) follows from the theorem above as in Corollary 11.2.2 since by Theorem 12.5.18 a direct summand of an Ω-Gorenstein injective in $\mathcal{G}_0(R)$ is also such. □

Corollary 12.5.31. *Let $M \in \mathcal{G}_0(R)$. Then the following are equivalent for a Ω-Gorenstein injective preenvelope $M \to G$ with* right $\mathcal{V}$-dim $G/M < \infty$.

(1) right $\mathcal{V}$-dim $M < \infty$.

(2) $G \in \mathcal{V}$.

(3) $M \to G$ *is a right $\mathcal{V}$-preenvelope.*

Proof. (1) $\Leftrightarrow$ (2) Let $0 \to M \to G \to L \to 0$ be exact. Then right $\mathcal{V}$-dim $L < \infty$ by Theorem 12.5.29. So if right $\mathcal{V}$-dim $M < \infty$, then right $\mathcal{V}$-dim $G < \infty$ and thus $G \in \mathcal{V}$ by Remark 12.5.17. The converse is now trivial.

(2) $\Leftrightarrow$ (3) is trivial.   □

Proposition 12.5.32. *Let $\mathcal{L}$ be the class of R-modules of finite right $\mathcal{V}$-dimension and $\Omega\mathcal{G}or\mathcal{I}nj$ denote the class of Ω-Gorenstein injective R-modules. Then the following are equivalent for an R-module $M \in \mathcal{G}_0(R)$ and integer $r \geq 0$.*

(1) right $\Omega\mathcal{G}or\mathcal{I}nj$-dim $M \leq r$.

(2) *There exists an exact sequence* $0 \to M \to G^0 \to G^1 \to \cdots \to G^{r-1} \to G^r \to 0$ *with each* G^i Ω-*Gorenstein injective.*

(3) $\mathrm{Ext}^i(L, M) = 0$ *for all* $L \in \mathcal{L}$ *and all* $i \geq r + 1$.

(4) $\mathrm{Ext}^{r+1}(L, M) = 0$ *for all* $L \in \mathcal{L}$.

(5) *Every* rth $\Omega\mathcal{G}or\mathcal{I}nj$-*cosyzygy of* M *is* Ω-*Gorenstein injective.*

(6) *Every* rth $\mathcal{V}$-*cosyzygy of* M *is* Ω-*Gorenstein injective.*

Proof. The result follows as in Proposition 11.2.5 noting that $\mathrm{Ext}^1(L, V) = 0$ for all $L \in \mathcal{L}$, $V \in \mathcal{V}$ since V is Ω-Gorenstein injective. $\square$

Corollary 12.5.33. *Let* $M \in \mathcal{G}_0(R)$. *Then*

(1) right $\Omega\mathcal{G}or\mathcal{I}nj$-dim $M \leq d$.

(2) right $\Omega\mathcal{G}or\mathcal{I}nj$-dim $M = $ right $\mathcal{V}$-dim M *if and only if* right $\mathcal{V}$-dim $M < \infty$.

Proof. (1) follows from Theorem 12.5.18 and Proposition 12.5.32 above.
  (2) follows from (1) above and Corollary 12.5.31. $\square$

Proposition 12.5.34. *The following are equivalent for an* R-*module* $M \in \mathcal{G}_0(R)$.

(1) $\mathrm{Ext}^i(M, N) = 0$ *for all* $i \geq 1$ *and all* Ω-*Gorenstein injective* R-*modules* N.

(2) $\mathrm{Ext}^1(M, N) = 0$ *for all* Ω-*Gorenstein injective* R-*modules* N.

(3) M *has right* $\mathcal{V}$-*dimension at most* d.

(4) M *has finite right* $\mathcal{V}$-*dimension.*

Proof. (1) $\Rightarrow$ (2) and (3) $\Rightarrow$ (4) are trivial, and (2) $\Rightarrow$ (1) follows by induction.
  (1) $\Rightarrow$ (3). Let $0 \to M \to V^0 \to V^1 \to \cdots$ be exact with each $V^i \in \mathcal{V}$ and $K = \mathrm{Ker}(V^d \to V^{d+1})$. Then K is Ω-Gorenstein injective by Theorem 12.5.18. So $\mathrm{Hom}(V^{d+1}, K) \to \mathrm{Hom}(V^d, K) \to \mathrm{Hom}(V^{d-1}, K)$ is exact and thus K is a summand of V^d.
  (4) $\Rightarrow$ (1) by Corollary 12.5.30. $\square$

Lemma 12.5.35. *Let* $\mathcal{L}$ *be the class of* R-*modules in* $\mathcal{G}_0(R)$ *such that* right $\mathcal{V}$-dim $L < \infty$. *Then* $\mathcal{L}$ *is closed under inductive limits.*

Proof. right $\mathcal{V}$-dim $L \leq d$ by the proposition above and so the result follows since $\mathcal{V}$ is closed under inductive limits by Lemma 12.5.1. $\square$

Theorem 12.5.36. *Every* R-*module* $M \in \mathcal{G}_0(R)$ *has a* Ω-*Gorenstein injective envelope* $M \to G$ *such that if* $0 \to M \to G \to L \to 0$ *is exact, then* right $\mathcal{V}$-dim $L \leq d - 1$ *whenever* $d \geq 1$.

Corollary 12.5.37. *Let* $M \in \mathcal{G}_0(R)$ *and* $M \to G$ *be a* Ω-*Gorenstein injective envelope, then* right $\mathcal{V}$-dim $M < \infty$ *if and only if* $M \to G$ *is a* $\mathcal{V}$-*envelope.*

Theorem 12.5.38. *If $M \in \mathcal{G}_0(R)$ and $0 \to M \to G^0 \to G^1 \to \cdots$ is a minimal right Ω-Gorenstein injective resolution, then $G^i \in \mathcal{V}$ for each $i \geq 1$ and $G^i = 0$ for $i > d$.*

Proof. This follows from Theorem 12.5.36 and Corollary 12.5.37 above.   □

Remark 12.5.39. There are analogous results for Ω-Gorenstein projective and Ω-Gorenstein flat precovers and preenvelopes and balance.

Exercises

1. Prove Lemma 12.5.1.

2. Prove Remark 12.5.4.

3. Prove Theorem 12.5.5.

4. Prove the second part of Theorem 12.5.6.

5. Prove (1) $\Leftrightarrow$ (3), (1) $\Leftrightarrow$ (4) $\Leftrightarrow$ (6), (2) $\Leftrightarrow$ (9), (1) $\Leftrightarrow$ (12), (1) $\Leftrightarrow$ (15), and (1) $\Leftrightarrow$ (16) in Theorem 12.5.13.

6. Complete the proof of Theorem 12.5.15.

7. Prove Remark 12.5.17.

8. Prove Theorems 12.5.18 and 12.5.19.

9. Prove that an R-module M is Gorenstein injective if and only if $M \in \mathcal{G}_0(R)$ and $\mathrm{Hom}(\Omega, M)$ is Ω-Gorenstein injective.

10. Prove that an R-module M is Gorenstein projective (flat) if and only if $M \in \mathcal{G}_0(R)$ and $\Omega \otimes M$ is Ω-Gorenstein projective (flat).

11. Prove part (2) of Theorem 12.5.26.

12. Complete the proof of Theorem 12.5.27.

13. Prove that every inductive limit of Ω-Gorenstein injective modules is Ω-Gorenstein injective.

14. Prove Theorem 12.5.28.

15. Prove Proposition 12.5.32.

16. Prove Theorem 12.5.36.

Bibliographical Notes

Chapters 1–3. This material, except possibly for Section 1.1, can be found scattered over many excellent books. Among the more notable such books are Anderson–Fuller [6], Atiyah–McDonald [10], Balcerzyk–Jozefiak [19] and [20], Bourbaki [29] and [30], Bruns–Herzog [32], Faith [86], Kaplansky [127], Jacobson [118], Kasch [128], Matsumura [146] and [147], Nagata [150], Rotman [160], Sharpe–Vamos [166], Strenström [170], Strooker [171], Weibel [178], Wisbauer [180], and Zariski–Samuel [187]. A good reference for Section 1.1 is Halmos [106].

Chapter 4. This chapter is based on Banaschewski [21], Enochs [50] and [51]. Additional properties and examples for interested readers can be found in Cheatam [36], Jenda [123] and Matlis [145]. Torsion free covers relative to torsion theories are studied in Banaschewski [21], Golan–Teply [102], and Teply [174].

Chapter 5. The material on the existence of covers is based on Enochs [49] and [52]. Basic properties of flat and injective covers in this chapter originate from Cheatham–Enochs–Jenda [38], Enochs [53], and Enochs–Jenda [68]. Interested readers may find additional properties of injective covers in Ahsan–Enochs [3] and [2]. Covers are known as "approximations" in Auslander's school where the main concern is on finitely generated modules. This terminology first appears in Auslander–Smalø [16].

Chapter 6. The material on the existence of preenvelopes and envelopes is based on Enochs [52], Enochs–Jenda [80], and Enochs–Jenda [70]. Properties of flat and pure injective envelopes in this chapter come from Enochs [52], Enochs–Jenda–Xu [80], Enochs–Jenda [76], and Warfield [177]. Background material on pure injective modules can be found in Fuchs [95], Warfield [177] or in a standard text such as Wisbauer [180]. Strongly copure injective modules are studied in Enochs–Jenda [72] and [120] and results on absolutely pure modules can be found in Adams [1], Maddox [138], and Megibben [149]. Further studies of flat envelopes are carried out in Akatsa [4], Asensio Mayor–Martinez Hernandez [7], [8], [9], Martinez Hernandez–Saorin–de Valle [140] and Martinez Hernandez [141].

Chapter 7. The notion of cotorsion theory was introduced by Salce [163] and the results guaranteeing that under certain conditions cotorsion theories have enough injectives and projectives are due to the work of Eklof–Trlifaj [48]. The proof for the existence of flat covers is due to Bican–El Bashir–Enochs [28]. Additional material on cotorsion theory and Wakamatsu lemma type of results can be found in Göbel–

Shelah [101] and Wakamatsu [175], respectively. The source for some of the material on set–theoretic homological algebra is Eklof [47].

Chapter 8. Additional background material for this chapter, with possibly different notation and terminology, can be found in Cartan–Eilenberg [34], Eilenberg–Moore [44], Hilton–Stammbach [111], Lazard [134], Maclane [137], Matsumura [147], Rotman [160], Warfield [177], and Wisbauer [180]. The chapter is based on Enochs–Jenda [77]. Some results on injective precovers and flat preenvelopes originate from Enochs–Jenda [74], [120], and [121] and additional material on absolutely pure modules may be found in Strenström [168] and Würfel [181]. The material on minimal pure injective resolutions of flat modules is based on Enochs [54], [55], and [56], Enochs–Jenda [76], Gruson–Jensen [105], and Jensen [124] while the section on λ and μ-dimensions is from Enochs–Jenda–Oyonarte [78]. Earlier studies of relative homological algebra can be found in Hochschild [112], Heller [109], and Butler–Horrocks [33].

Chapter 9. Some of the material on Iwanaga–Gorenstein rings can be found in Iwanaga [116] and Jenda [121]. The material on minimal injective resolutions originates from Bass' work in [25] which can be found in many standard books. The material on flat and injective modules and torsion products is based on Enochs–Jenda [73] and Enochs [57]. Section 5 contains standard material on local cohomology and the dualizing module which may be found in Bruns–Herzog [32], Foxby [91], Hartshorne [108], Herzog–Kunz [110], Koszul [130], Sharp [165], Strooker [171], and Weibel [178].

Chapter 10. Gorenstein injective and projective modules were introduced and studied in Enochs–Jenda [63] while Gorenstein flat modules were first studied in Enochs–Jenda–Torrecillas [79]. Foxby classes were introduced in Foxby [94] and the material on Foxby classes included here is based on Enochs–Jenda–Xu [81] and Enochs–Jenda [60].

Chapter 11. The material on Gorenstein injective covers and envelopes and Gorenstein projective precovers originates from Enochs–Jenda–Xu [82] while Gorenstein essential extensions can be found in Enochs–Jenda [61]. The material on Gorenstein projective covers is based on Enochs–Jenda–Xu [81] and Gorenstein flat covers is based on Enochs–Jenda–Oyonarte [78], Enochs–Jenda–Xu [81] and Xu–Enochs [185]. Gorenstein flat and projective preenvelopes can be found in Enochs–Jenda [65]. Additional material on $\mathcal{L}$–covers and envelopes can be found in Enochs–Jenda [66], Aldrich–Enochs–Jenda–Oyonarte [5], and Igusa–Smalø–Todorov [113].

Chapter 12. The Gorenstein balance of Hom and Tensor is the subject of Enochs–Jenda [64] and some of the material in Section 3 originate from this paper and Enochs–Jenda [62]. The material on Gorenstein dimensions over Cohen–Macaulay rings is based on Foxby [94] and Enochs–Jenda [60]. The last section originates from Enochs–Jenda [58] and [59].

Bibliography

[1] Adams, D. D., Absolutely pure modules, Ph.D. Thesis, University of Kentucky, 1978.

[2] Ahsan, J., Enochs, E. E., Torsion-free injective covers, Comm. Algebra 12 (9) (1984), 1139–1146.

[3] Ahsan, J., Enochs, E. E., Rings admitting torsion injective covers, Portugal. Math. 40 (1981), 257–261.

[4] Akatsa, V. K. A., Flat envelopes and negative torsion functors, Ph.D. Thesis, University of Kentucky, 1991.

[5] Aldrich, S., Enochs, E. E., Jenda, O. M. G., Oyonarte, L., Envelopes and covers by modules of finite injective and projective dimension and their orthogonal classes, preprint.

[6] Anderson, F. W., Fuller, K. R., Rings and categories of modules, Grad. Texts in Math. 13, Springer-Verlag, New York 1973.

[7] Asensio Mayor, J., Martinez Hernandez, J., On flat and projective envelopes, J. Algebra 160 (1993), 434–440.

[8] Asensio Mayor, J., Martinez Hernandez, J., Monomorphic flat envelopes in commutative rings, Arch. Math. (Basel) 54 (1990), 430–435.

[9] Asensio Mayor, J., Martinez Hernandez, J., Flat envelopes in commutative rings, Israel J. Math. 62 (1988), 123–128.

[10] Atiyah, M. F., MacDonald, I. G., Introduction to Commutative Algebra, Addison-Wesley, Reading, Mass., 1969.

[11] Auslander, M., Buchsbaum, D., Homological dimension in local rings, Trans. Amer. Math. Soc. 85 (1957), 390–405.

[12] Auslander, M., Buchsbaum, D., Homological dimension in Noetherian rings II, Trans. Amer. Math. Soc. 88 (1958), 194–206.

[13] Auslander, M., Anneaux de Gorenstein et torsion en algèbre commutative, séminaire d'algèbre commutative, Ecole Normale Supérieure de Jeunes Filles, Paris 1966/67.

[14] Auslander, M., Bridger, M., Stable module theory, Mem. Amer. Math. Soc. 94, Providence, R.I., 1969.

[15] Auslander, M., Buchweitz, R., The homological theory of maximal Cohen–Macaulay approximations, Mém. Soc. Math. France 38 (1989), 5–37.

[16] Auslander, M., Smalø, S. O., Preprojective modules over artin rings, J. Algebra 66 (1980), 61–122.

[17] Auslander, M., Reiten, I., Applications of contravariantly finite subcategories, Adv. Math. 86 (1991), 111–152.

[18] Baer, R., Abelian groups which are direct summands of every containing group, Bull. Amer. Math. Soc. 46 (1940), 800– 806.

[19] Balcerzyk, S., Jozefiak, T., Commutative noetherian and Krull rings, Horwood-PWN, Warszawa 1989.

[20] Balcerzyk, S., Jozefiak, T., Commutative rings, Horwood-PWN, Warszawa 1989.

[21] Banaschewski, B., On coverings of modules, Math. Nachr. 31 (1966), 57–71.

[22] Bartijn, J., Strooker, J. R., Modifications monomiales, Sém. d'Algèbre P. Dubreil et M.-P. Malliavin, Proceedings, Paris 1982 (35ème Année), Lecture Notes in Math. 1029, Springer-Verlag, Berlin 1983, 192–217.

[23] Bass, H., Finitistic dimension and a homological generalization of semi-primary rings, Trans. Amer. Math. Soc. 95 (1960), 446–448.

[24] Bass, H., Injective dimension in noetherian rings, Trans. Amer. Math. Soc. 102 (1962), 189–209.

[25] Bass, H., On the ubiquity of Gorenstein rings, Math. Z. 82 (1963), 8–28.

[26] Belshoff, R., Enochs, E. E., Xu J., The existence of flat covers, Proc. Amer. Math. Soc. 122 (1994), 985–991.

[27] Bernecker, H., Flatness and absolute purity, J. Algebra 44 (1977), 411–419.

[28] Bican, L., El Bashir, R., Enochs, E. E., Modules have flat covers, preprint.

[29] Bourbaki, N., Elements of Mathematics. Commutative Algebra, Hermann, Paris, Addison-Wesley, Reading, Mass., 1972.

[30] Bourbaki, N., Algebra I. Chapters 1–3, Springer-Verlag, Berlin 1989.

[31] Brown, K., Cohomology of groups, Grad. Texts in Math. 87, Springer-Verlag, New York 1982.

[32] Bruns, W., Herzog, J., Cohen–Macaulay rings, Cambridge Stud. Adv. Math. 39, Cambridge University Press, Cambridge 1993.

[33] Butler, M. C. R., Horrocks, G., Classes of extensions and resolutions, Philos. Trans. Roy. Soc. London Ser. A 254 (1961/1962), 155–222.

[34] Cartan, H., Eilenberg S., Homological Algebra, Princeton University Press, Princeton 1956.

[35] Chase, S., Direct products of modules, Trans. Amer. Math. Soc. 97 (1960), 457–473.

[36] Cheatham, T., The quotient field as a torsion-free covering module, Israel J. Math. 33 (1979), 173-176.

[37] Cheatham, T., Enochs, E., Injective hulls of flat modules, Comm. Algebra 8 (1980), 1989–1995.

[38] Cheatham, T., Enochs, E. E., Jenda O. M. G., The structure of injective covers of special modules, Israel J. Math. 63 (1988), 237–242.

[39] Cohen, I. S., On the structure and ideal theory of complete local rings, Trans. Amer. Math. Soc. 59 (1946), 54–106.

[40] Colby, R., Rings which have flat injective modules, J. Algebra 35 (1975), 239–252.

[41] Dieudonné, J., Remarks on quasi-Frobenius rings, Illinois J. Math. 12 (1958), 346–354.

[42] Eakin, P., The converse to a well-known theorem on Noetherian rings, Math. Ann. 177 (1968), 278–282.

[43] Eckmann, B., Schopf, A., Über injective Moduln, Arch. Math. (Basel) 4 (1953), 75–78.

[44] Eilenberg, S., Moore J. C., Foundations of Relative Homological Algebra, Amer. Math. Soc. Mem. 55, Providence, R.I., 1965.

[45] Eilenberg, S., Nakayama, T., On the dimension of modules and algebras II, Nagoya Math. J. 9 (1955), 1–16.

[46] Eisenbud, D., Homological algebra on a complete intersection with an application to group representations, Trans. Amer. Math. Soc. 260 (1980), 35–64.

[47] Eklof, P. C., Homological algebra and set theory, Trans. Amer. Math. Soc. 227 (1977), 207–225.

[48] Eklof, P. C., Trlifaj, J., How to make Ext vanish, preprint.

[49] Enochs, E. E., Covers by flat modules and submodules of flat modules, J. Pure Appl. Algebra 57 (1989), 33–38.

[50] Enochs, E. E., Torsion free covering modules, Proc. Amer. Math. Soc. 14 (1963) 884–889.

[51] Enochs, E. E., Torsion free coverings modules II, Arch. Math. (Basel) 22 (1971), 37–52.

[52] Enochs, E. E., Injective and flat covers, envelopes and resolvents, Israel J. Math. 39 (1981), 189–209.

[53] Enochs, E. E., Flat covers and flat cotorsion modules, Proc. Amer. Math. Soc. 92 (1984), 179–184.

[54] Enochs, E. E., Minimal pure injective resolutions of flat modules, J. Algebra 105 (1987), 351–364.

[55] Enochs, E. E., Minimal pure injective resolutions of complete rings, Math. Z. 200 (1989), 239–243.

[56] Enochs, E., E., The first term in a minimal pure injective resolution, Math. Scand. 65 (1989), 41–49.

[57] Enochs, E. E., Remarks on commutative noetherian rings whose flat modules have flat injective envelopes, Portugal. Math. 45 (1988), 151–156.

[58] Enochs, E. E., Jenda, O. M. G., On D-Gorenstein modules, Proceedings of the Euroconference Interactions between Ring Theory and Representation of Algebras, preprint.

[59] Enochs, E. E., Jenda, O. M. G., Ω-Gorenstein covers and envelopes, preprint.

[60] Enochs, E. E., Jenda, O. M. G., Gorenstein injective, projective, and flat dimensions over Cohen–Macaulay rings, International Conference on Algebra and its Applications, Ohio University, 1999.

[61] Enochs, E. E., Jenda, O. M. G., Gorenstein injective modules over Gorenstein rings, Comm. Algebra 26 (1998), 3489–3496.

[62] Enochs, E. E., Jenda, O. M. G., Gorenstein injective and flat dimensions, Math. Japon. 44 (1996), 261–268.

[63] Enochs, E. E., Jenda, O. M. G., Gorenstein injective and projective modules, Math. Z. 220 (1995), 611–633.

[64] Enochs, E. E., Jenda, O. M. G., Gorenstein balance of Hom and Tensor, Tsukuba J. Math. 19 (1995), 1–13.

[65] Enochs, E. E., Jenda, O. M. G., Gorenstein flat preenvelopes and resolvents, Nanjing Math. J. 12 (1995), 1–9.

[66] Enochs, E. E., Jenda, O. M. G., Resolutions by Gorenstein injective and projective modules and modules of finite injective dimension over Gorenstein rings, Comm. Algebra 23 (1995), 869–877.

[67] Enochs, E. E., Jenda, O. M. G., On Cohen–Macaulay rings, Comment. Math. Univ. Carolin. 35 (1994), 223–230.

[68] Enochs, E. E., Jenda, O. M. G., Mock finitely generated Gorenstein injective modules and isolated singularities, J. Pure Appl. Algebra 96 (1994), 259–269.

[69] Enochs, E. E., Jenda, O. M. G., On Gorenstein injective modules, Comm. Algebra 21 (1993), 3489–3501.

[70] Enochs, E. E., Jenda, O. M. G., Copure injective resolutions, flat resolvents and dimensions, Comment. Math. Univ. Carolin. 34 (1993), 203–211.

[71] Enochs, E. E., Jenda, O. M. G., h-divisible and cotorsion modules over one-dimensional Gorenstein rings, J. Algebra 161 (1993), 444–454.

[72] Enochs, E. E., Jenda, O. M. G., Copure injective modules, Quaestiones Math. 14 (1991), 401–409.

[73] Enochs, E. E., Jenda, O. M. G., Tensor and torsion products of injective modules, J. Pure Appl. Algebra 76 (1991), 143–149.

[74] Enochs, E. E., Jenda, O. M. G., Resolvents and dimensions of modules and rings, Arch. Math. (Basel) 56 (1991), 528–532.

[75] Enochs, E. E., Jenda, O. M. G., Syzygies of resolvents over Gorenstein rings, Comm. Algebra 18 (1990), 3187–3194.

[76] Enochs, E. E., Jenda, O. M. G., Homological properties of pure injective resolutions, Comm. Algebra 16 (1988), 2069–2082.

[77] Enochs, E. E., Jenda, O. M. G., Balanced functors applied to modules, J. Algebra 92 (1985), 303–310.

[78] Enochs, E. E., Jenda, O. M. G., and Oyonarte, L., λ and μ-dimensions, preprint.

[79] Enochs, E. E., Jenda, O. M. G., Torrecillas, B., Gorenstein flat modules, Nanjing Univ. J. Math. Biquarterly 10 (1) (1993), 1–9.

[80] Enochs, E. E., Jenda, O. M. G., Xu, J., The existence of envelopes, Rend. Sem. Mat. Univ. Padova 90 (1993), 45–51.

[81] Enochs, E. E., Jenda, O. M. G., Xu, J., Foxby duality and Gorenstein injective and projective modules, Trans. Amer. Math. Soc. 348 (1996), 3223–3234.

[82] Enochs, E. E., Jenda, O. M. G., Xu, J., Covers and envelopes over Gorenstein rings, Tsukuba J. Math. 20 (1996), 487–503.

[83] Enochs, E. E., Jenda, O. M. G., Xu, J., Lifting group representations to maximal Cohen–Macaulay representations, J. Algebra 188 (1997), 58–68.

[84] Enochs, E. E., Jenda, O. M. G., Xu, J., A generalization of Auslander's last theorem, J. Algebras Representation Theory 2 (1999), 259–268.

[85] Enochs, E. E., Xu, J., Gorenstein flat covers of modules over Gorenstein rings, J. Algebra (1996), 288–313.

[86] Faith, C., Algebra II: Ring Theory, Springer-Verlag, Berlin–New York 1976.

[87] Faith, C., Rings with ascending condition on annihilators, Nagoya Math. J. 27 (1966), 179–191.

[88] Faith, C., Walker, E. A., Direct-sum representations of injective modules, J. Algebra 5 (1967), 203–221.

[89] Fieldhouse, D. J., Character modules, Comment. Math. Helv. 46 (1971), 274–276.

[90] Fossum, R., Foxby, H.-B., Griffith P., Reiten, I., Minimal injective resolutions with applications to dualizing modules and Gorenstein modules, Inst. Hautes Études Sci. Publ. Math. 45 (1975), 193–215.

[91] Foxby, H.-B., Gorenstein modules and related modules, Math. Scand. 31 (1972) 267–284.

[92] Foxby, H.-B., Duality homomorphisms for modules over certain Cohen–Macaulay rings, Math. Z. 132 (1973), 215–226.

[93] Foxby, H.-B., Quasi-perfect modules over Cohen–Macaulay rings, Math. Nachr. 66 (1975), 103–110.

[94] Foxby, H.-B., Gorenstein dimensions over Cohen–Macaulay rings, Commutative Algebra: Extended abstracts of an international conference, July 27–August 1, 1994, Vechta, Germany (Bruns, W. et al., eds.), Vechtaer Universitätsschriften 13, Runge, Cloppenburg 1994, 59–63.

[95] Fuchs, L., Algebraically compact modules over noetherian rings, Indian J. Math. 9 (9167), 357–374.

[96] Fuchs, L., Infinite Abelian groups, Vol I, Academic Press, New York 1970.

[97] Fuchs, L., Cotorsion modules over Noetherian hereditary rings, Houston J. Math. 3 (1977), 33–46.

[98] Fuchs, L., Notes on abelian groups II, Acta Math. Acad. Sci. Hung. 11 (1960), 117–125.

[99] Gabriel, P., Objets injectifs dans les catégories abéliennes, Sém. P. Dubreil, M.-L. Dubreil-Jacotin et C. Pisot (Algèbre et Théorie des Nombres), Année 1958/59, Fasc. 2, Exposé 17.

[100] Gilmer, R., Multiplicative Ideal Theory, Queen's University, 1968.

[101] Göbel, R., Shelah, S., Cotorsion theories and splitters, Trans. Amer. Math. Soc., to appear.

[102] Golan, J. S., Teply, M. L., Torsion-free covers, Israel J. Math. 15 (1973), 273–256.

[103] Griffith, P., A representation theorem for complete local rings, J. Pure Appl. Algebra 7 (1976), 303–315.

[104] Grothendieck, A., Local cohomology, Lecture Notes in Math. 41, Springer-Verlag, Berlin 1967.

[105] Gruson, L., Jensen, C. U., Dimensions cohomologiques reliés aux foncteurs, $\varprojlim^{(i)}$, Sém. d'Algèbre P. Dubreil et M.-P. Malliavin, Proceedings, Paris 1980 (33ème Année), Lecture Notes in Math. 867, Springer-Verlag, Berlin 1981, 234–294.

[106] Halmos, P. R., Naive set theory, D. Van Nostrand Company, Princeton 1960.

[107] Harrison, D. K., Infinite abelian groups and homological methods, Ann. of Math. 69 (1959), 366–391.

[108] Hartshorne, R., Residues and duality, Lecture Notes in Math. 20, Springer-Verlag, Berlin 1966.

[109] Heller, A. Homological algebra in abelian categories, Ann. of Math. 68 (1958), 484–525.

[110] Herzog, J., Kunz, E., Der kanonische Modul eines Cohen–Macaulay Rings, Lecture Notes in Math. 238, Springer-Verlag, Berlin 1971.

[111] Hilton, P. J., Stammbach, U., A course in homological algebra, Grad. Texts in Math. 4, Springer-Verlag, New York 1971.

[112] Hochschild, G., Relative homological algebra, Trans. Amer. Math. Soc. 82 (1956), 246–269.

[113] Igusa, K., Smalø, S. O., Todorov, G., Finite projectivity and contravariant finiteness, Proc. Amer. Math. Soc. 109 (1990), 937–941.

[114] Ischebeck, F., Eine Dualität Zwischen den Funktoren Ext und Tor, J. Algebra 11 (1969), 510–531.

[115] Ishikawa, T., On injective modules and flat modules, J. Math. Soc. Japan 17 (1965), 291–296.

[116] Iwanaga, Y., On rings with finite self-injective dimension II, Tsukuba J. Math. 4 (1980), 107–113.

[117] Jacobson, N., Basic Algebra I, W. H. Freeman and Company, Second Edition, San Francisco 1985.

[118] Jacobson, N., Basic Algebra II, W. H. Freeman and Company, San Francisco 1980.

[119] Jenda, O. M. G., Injective and flat dimensions, Math. Japon. 31 (1986), 205–209.

[120] Jenda, O. M. G., Injective resolvents and preenvelopes, Quaestiones Math. 9 (1986), 301–309.

[121] Jenda, O. M. G., On Gorenstein rings, Math. Z. 197 (1988), 119–122.

[122] Jenda, O. M. G., The dual of the grade of a module, Arch. Math. (Basel) 51 (1988), 297–302.

[123] Jenda, O. M. G., On the torsion free cover of the $k[x_1, \ldots, x_n]$-module k, Portugal. Math., 49 (1992), 397–402.

[124] Jensen, C. U., Les foncteur dérivés de $\varprojlim$ et leurs applications en théorie des modules, Lecture Notes in Math. 254, Springer-Verlag, Berlin 1972.

[125] Jensen, C. U., On the vanishing of $\varprojlim^{(i)}$, J. Algebra 15 (1970), 151–166.

[126] Kaplansky, I., Maximal fields with valuations, Duke Math. J. 9 (1942), 303–321.

[127] Kaplansky, I., Commutative Rings, The University of Chicago Press, Chicago 1970.

[128] Kasch, F., Modules and rings, Academic Press, London–New York 1982.

[129] König, D., Theory of finite and infinite graphs, Birkhäuser, Boston 1990.

[130] Koszul, J. L., Sur un type d'algèbres différentielles en rapport avec la transgres-sion, Colloque Topologie, Bruxelles 1950, Centre Belge Rech. Math., Louvain 1951, 73–81.

[131] Krull, W. Allgemeine Bewertungstheorie, J. Reine Angew. Math. 167 (1932), 160–196.

[132] Lambek, J., A module is flat if and only if its character module is injective, Canad. Math. Bull. 7 (1964), 237–243.

[133] Lambek, J., Lectures on rings and modules, Chelsea Publishing Company, New York 1976.

[134] Lazard, D., Sur les modules plats, C. R. Acad. Sci. 258 (1964), 6313–6316.

[135] Lazard, D., Autour de la platitude, Bull. Soc. Math. France 97 (1969), 31–128.

[136] MacDonald, I. G., Sharp, R. Y., An elementary proof of the non-varnishing of certain local cohomology modules, Quart. J. Math. Oxford Ser. (2) 23 (1972), 197–204.

[137] MacLane, S., Homology, Springer-Verlag, New York 1963.

[138] Maddox, B., Absolutely pure modules, Proc. Amer. Math. Soc. 18 (1967), 155–158.

[139] Maranda, J.-M., Injective structures, Trans. Amer. Math. Soc. 110 (1964), 98–135.

[140] Martinez Hernandez J., Saorin, M., de Valle, A., Noncommutative rings whose modules have essential flat envelopes, J. Algebra 177 (1995), 434–450.

[141] Martinez Hernandez J., Relatively flat envelopes, Comm. Algebra 14 (1986), 867–884.

[142] Matlis E., Injective modules over Noetherian rings, Pacific J. Math. 8 (1958), 511–528.

[143] Matlis E., Divisible modules, Proc. Amer. Math. Soc. 11 (1960), 385–391.

[144] Matlis E., Cotorsion modules, Amer. Math. Soc. Mem. 49, Providence, R.I., 1964.

[145] Matlis E., The ring as a torsion-free cover, Israel J. Math. 37 (1980), 211–230.

[146] Matsumura, H., Commutative Algebra, Benjamin, New York 1970.

[147] Matsumura, H., Commutative ring theory, Cambridge University Press, Cam-bridge 1986.

[148] McRae, D. G., Coherent rings and homology, Ph.D. Thesis, University of Wash-ington, 1967.

[149] Megibben, C., Absolutely pure modules, Proc. Amer. Math. Soc. 26 (1970), 561–566.

[150] Nagata, M., Local rings, Tracts in Pure Appl. Math. 13, Interscience Publ., Wiley, New York 1962.

[151] Noether E., Idealtheorie in Ringbereichen, Math. Ann. 83 (1921), 24–66.

[152] Northcott, D. G., Injective envelopes and inverse polynomials, J. London Math. Soc. 68 (1974), 290–296.

[153] Nunke, R. J., Modules of extensions over Dedekind rings, Illinois J. Math. 3 (1959), 222–241.

[154] Oberst, U., and Röhrl, H., Flat and coherent functors, J. Algebra 14 (1970), 91–105.

[155] Peskine, C., Szpiro L., Dimension projective finie et cohomologie locale, Inst. Hautes Études Sci. Publ. Math. 42 (1973), 49–119.

[156] Pontryagin, L., Topological groups, L.S. Pontryagin Selected Works, Volume 2, Gordon and Breach, 1985.

[157] Raynaud, M., and Gruson, L., Critères de platitude et de projectivité, Invent. Math. 13 (1971), 1–89.

[158] Renault, G., Etudes des sous-modules compléments dans un module, Bull. Soc. Math. France, Suppl., Mem. 9, 1967.

[159] Rees, D., The grade of an ideal or module, Proc. Cambridge Philos. Soc. 53 (1957), 28–42.

[160] Rotman, J. J., An introduction to Homological Algebra, Academic Press, New York 1979.

[161] Sabbagh, G., Sur la Pureté dams les Modules, C. R. Acad. Sci. Paris Sér. A. 271 (1970), 865–867.

[162] Sah, C.-H., Cohomology of split group extensions II, J. Algebra 45 (1977), 15–68.

[163] Salce, L., Cotorsion theories for abelian groups, Sympos. Math. XXIII, Academic Press, New York 1979, 11–32.

[164] Serre, J.-P., Algèbre Locale, Multiplicités, Lecture Notes in Math. 11, Springer-Verlag, Berlin 1965.

[165] Sharp, R. Y., Local cohomology theory in commutative algebra, Quart. J. Math. Oxford Ser. (2) 21 (1970), 425–434.

[166] Sharpe, D. W., Vamos, P., Injective Modules, Cambridge University Press, Cambridge 1972.

[167] Smith, J. Local domains with Topologically T-nilpotent radical, Pacific J. Math. 30 (1969), 233–245.

[168] Stenström, B., Coherent rings and FP-injective modules, J. London Math. Soc. 2 (1970), 323–329.

[169] Strenström, B., Purity in factor categories, J. Algebra 8 (1968), 352–361.

[170] Strenström, B., Rings of quotients, Springer-Verlag, New York 1975.

[171] Strooker, J., Homological questions in local algebra, London Math. Soc. Lecture Notes Series 145, Cambridge University Press, Cambridge 1990.

[172] Switzer, R., Algebraic Topology–Homotopy and Homology, Springer-Verlag, Berlin–Heidelberg–New York 1975.

[173] Teply, M., Torsion-free injective modules, Pacific J. Math. 28 (1969), 441–453.

[174] Teply, M., Torsion-free covers II, Israel J. Math. 23 (1976), 132–135.

[175] Wakamatsu, T., Stable equivalence of self-injective algebras and a generalization of tilting modules, J. Algebra 134 (1990), 298–325.

[176] Walker, E., Richman, F., Cotorsion free, an example of relative injectivity, Math. Z. 102 (1967), 115–117.

[177] Warfield, R. B., Purity and algebraic compactness for modules, Pacific J. Math. 28 (1969), 699–719.

[178] Weibel, C. A., An introduction to homological algebra, Cambridge Stud. Adv. Math. 38, Cambridge University Press, Cambridge 1994.

[179] Weil, A., L'intégration dans les groupes topologiques et ses applications, Actualités Sci. Indust. 869, Hermann, Paris, 1940.

[180] Wisbauer, R., Foundations of module and ring theory, Gordon and Breach, 1991.

[181] Würfel, T., Über absolut reine Ringe, J. Reine Angew. Math. 262/263 (1973), 381–391.

[182] Xu, J., Flat covers of modules, Lecture Notes in Math. 1634, Springer-Verlag, Berlin 1996.

[183] Xu, J., The existence of flat covers over noetherian rings of finite Krull dimension, Proc. Amer. Math. Soc., 123 (1995), 27–32.

[184] Xu, J., Minimal injective and flat resolutions of modules over Gorenstein rings, J. Algebra 175 (1995), 451–477.

[185] Xu, J., Enochs, E. E., Gorenstein flat covers of modules over Gorenstein rings, J. Algebra, 181 (1996), 288–313.

[186] Yoshino, Y., Cohen–Macaulay modules over Cohen–Macaulay rings, London Math. Soc. Lecture Notes Series 146, Cambridge University Press, Cambridge 1990.

[187] Zariski, O., Samuel, P., Commutative algebra, Vol. 1 and 2, Princeton, N.J., Van Nostrand, 1958, 1960.

Index

$PE(M)$, 144
T-nilpotent, 119
$\mathcal{A}bs$, 166
$\mathcal{A}bs$, 177
Ann(M), 51
Ass(M), 52
Card X, 4
Ext$_n$, 179
Ext^n, 70, 179
$\mathcal{F}lat$, 177
$\mathcal{F}_{fg}$, 177
$\mathcal{F}_{fp}$, 177
$\mathcal{G}or\mathcal{F}lat$, 290
$\mathcal{G}or\mathcal{I}nj$, 271
$\mathcal{G}or\mathcal{P}roj$, 280
$\mathcal{G}or\mathcal{P}roj_{fg}$, 294
Gext$_i(M, N)$, 296
Gext$^i(M, N)$, 296
Gfext$_i(M, N)$, 297
Gtor$^i(M, N)$, 299
Gtor$_i(M, N)$, 299
$\mathcal{I}nj$, 152, 177
$\mathcal{L}$, 164
    -preenvelope, 250, 262, 283
$\mathcal{M}$, 152, 177
$Mor(\mathbf{C})$, 18
$\Omega\mathcal{G}or\mathcal{I}nj$, 316
Ω-Gorenstein
    flat, 312
    injective, 312
    injective cover, 316
    injective envelope, 317
    injective precover, 315
    injective preenvelope, 316
    projective, 312
Ord X, 3

$\mathcal{P}\mathcal{I}nj$, 177
$\mathcal{P}\mathcal{P}roj$, 177
$\mathcal{P}roj$, 177
$\mathcal{P}roj$, 152
Pext$_R^n$, 179
Soc(M), 73
Spec R, 52
Supp(M), 53
Tor^n, 180
Tor$_n$, 41, 180
$\mathcal{V}$, 307
    -cover, 307
    -dimension, 313
    -precover, 308
    -resolution, 307
$\mathcal{W}$, 307
    -dimension, 313
$\mathcal{W}_{fg}$-resolution, 308
$\mathcal{X}$-resolution, 308
depth$_I$ M, 217
depth$_{\mathrm{m}}$ R, 123
dim M, 54
gtor$_i(M, N)$, 299
$\hat{\mathbb{Z}}_p$, 36, 38
mSpec R, 52
$\mu_i(\mathfrak{p}, M)$, 215
$\pi_n(\mathfrak{p}, F)$, 195
rad(R), 15
$^{\perp}\mathcal{C}$, 152
$m(R)$, 116
$\nu_i(\mathfrak{p}, M)$, 246
$\mathbb{Z}(p^{\infty})$, 35
$\mathcal{G}_0(R)$, 258
$\mathcal{J}_0(R)$, 258

abelian category, 21

absolutely pure, 131
additive
    category, 20
    functor, 21
    map, 9
algebra
    R-, 61
    faithfully flat, 65
algebraic dual, 138
annihilator, 51
Artin–Rees lemma, 62
artinian, 46
associated prime, 52
Auslander and Buchsbaum Theorem,
    220
axiom of choice, 1

Baer's Criterion, 68, 99, 182, 186
balanced
    functor, 174, 294, 298
    map, 11
Bass invariants, 215
biadditive map, 11
bilinear map, 11
bimodule, 8
boundaries, 26

canonical
    homomorphism, 88
    module, 237
Cantor, Schröder, Bernstein Theorem,
    4
cardinality, 4
category, 17
    abelian, 21
    additive, 20
    opposite, 19
cauchy sequence, 37
chain
    complex, 25
    continuous, 160
    homotopy, 28
    map, 25
Change of rings theorem, 194

character module, 77
class, 1
    Foxby, 258
    homotopy, 28
    orthogonal, 152
    resolving, 209
closure, 37
coboundaries, 26
cocycles, 26
cofibration, 155
cofinal subset, 32
cogenerator
    injective, 77
    of a cotorsion theory, 152
coheight, 196
Cohen–Macaulay, 221
coherent, 82, 136
cohomology, 26
cokernel, 9, 21
commutative diagram, 10
Comparison Theorem, 169
complete, 37
    $\mathcal{F}$-resolution, 168
    minimal $\mathcal{F}$-resolution, 169
    module, 37
completion
    I-adic, 37, 38
    of a free $R_{\mathfrak{p}}$-module, 88
complex, 25
    chain, 25
    cochain, 26
    deleted, 42, 169
    double, 175
    Koszul, 231
    quotient, 27
composition
    in categories, 17
    series, 49
connecting homomorphism, 28
continuous chain, 160
contravariant functor, 19
convergent sequence, 37
coproduct, 21

copure
  flat, 215
  flat dimension, 215
  injective, 215
  injective dimension, 214
coreflective, 186
cosyzygy, 168
cotorsion, 117
  theory, 152
covariant functor, 18
cover
  $\mathcal{F}$-, 105, 167
  Ω-Gorenstein injective, 316
  $\mathcal{V}$-, 307
  flat, 110
  Gorenstein flat, 288
  Gorenstein injective, 270
  Gorenstein projective, 284
  injective, 121
  projective, 105, 111, 127
  torsion free, 93
covering, 106, 110, 120, 167
cycles, 26

Dedekind domain, 103, 193
depth, 217
derived functor, 170
determined by, 139
diagram
  commutative, 10
  pullback, 23
  pushout, 23
dimension, 54
  $\mathcal{F}$-, 180
  $\mathcal{G}or\mathcal{F}lat$, 290, 303
  $\mathcal{G}or\mathcal{I}nj$, 272, 303
  $\mathcal{G}or\mathcal{P}roj$, 281
  $\mathcal{G}or\mathcal{P}roj_{fg}$, 303
  λ-, 204
  μ-, 210
  copure flat, 215
  copure injective, 214
  flat, 80, 186

global, 181
  injective, 80
  Krull, 54
  projective, 80
dimension shifting, 43
direct
  limit, 31
  sum, 21
  system, 31
directed set, 31
domain
  Dedekind, 103, 193
  Prüfer, 83
double complex, 175
dual
  algebraic, 138
  Matlis, 88
duality
  local, 235
  matlis, 88
dualizing module, 235

Eakin–Nagata Theorem, 73
embedded
  finitely, 90
  prime, 56
enough
  injectives, 153
  projectives, 153
envelope
  $\mathcal{F}$-, 129, 167
  Ω-Gorenstein injective, 317
  $\mathcal{V}$-, 261
  flat, 136
  Gorenstein injective, 274
  injective, 71
  pure injective, 141, 144
  pure injective flat, 141
enveloping, 129, 167
epimorphism, 18
essential extension, 71
  Gorenstein, 277
essential submodule, 71

Gorenstein, 277
essentially zero, 238
exact
$T(-,-)$, 167
functor, 22
left, 22
makes the sequence, 168
pure, 112
right, 22
sequence of complexes, 27
sequence of modules, 9
short, 10
split, 11

faithfully flat
algebra, 65
module, 42, 65
fibration, 155
filtration, 61
I-, 61
I-good, 61
stable, 61
finitely
embedded, 90
presented, 76
flat
Ω-Gorenstein, 312
copure, 215
cotorsion theory, 163
cover, 110
dimension, 80, 186
envelope, 136
faithfully, 42, 65
module, 40
precover, 110
preenvelope, 137
resolution, 41
rigid, 189
Foxby classes, 258
free
module, 8
resolution, 10
full subcategory, 18

functor, 18
additive, 21
balanced, 174, 294, 298
contravariant, 19, 171
covariant, 18, 171
derived, 169
exact, 22
local cohomology, 229
of two variables, 20
functorial, 18

Generalized Krull Principal Ideal Theorem, 57
converse of, 57
generator of a cotorsion theory, 152
generically Gorenstein, 223
global
$\mathcal{F}$- dimension, 181
weak dimension, 186
Gorenstein
n-, 213
essential extension, 277
essential submodule, 277
extension, 274
flat, 253
flat cover, 288
flat precover, 288
flat preenvelope, 293
flat resolution, 290
generically, 223
injective, 239
injective cover, 270
injective envelope, 274
injective extension, 274
injective precover, 269
injective preenvelope, 270
injective resolution, 272
Iwanaga–, 211
projective, 246
projective cover, 284
projective precover, 279
projective preenvelope, 293
projective resolution, 280

grade, 218
graded
    module, 61
    ring, 61

Hausdorff, 36
height, 54
hereditary ring, 193
Hilbert Basis Theorem, 48
homogeneous, 61
homology module, 26
homotopic, 28
homotopy
    chain, 28
    class, 28
Horseshoe Lemma, 170

identity morphism, 17
image, 9
indecomposable module, 85
induced order, 1
inductive
    limit, 31
    system, 31
inductively ordered set, 1
injective
    FP-, 131
    Ω-Gorenstein, 312
    cogenerator, 77
    copure, 214
    cover, 121
    dimension, 80
    envelope, 71
    extension, 70
    minimal resolution, 72
    module, 68
    precover, 120
    pure, 112
    resolution, 70
    structure, 139
inverse
    limit, 33
    system, 33
irreducible module, 59

isolated associated prime, 55
isomorphic
    functors, 20
    sets, 2
isomorphism, 2, 18
Iwanaga–Gorenstein, 211

Jacobson radical, 15

kernel, 9, 21
Koszul complex, 231
Krull
    dimension, 54
    Generalized Principal Ideal The-
        orem, 57
    Intersection Theorem, 63
    Principal Ideal Theorem, 56

least element, 1
length of a module, 49
limit
    direct, 31
    inductive, 31
    inverse, 33
    of a sequence, 37
    ordinal, 3
    projective, 33
local
    cohomology functor, 229
    duality, 235
    ring, 46
localization, 44, 45
locally nilpotent, 136
long exact sequence, 28
    extended, 173

makes the sequence exact, 168
map
    additive, 9
    balanced, 11
    biadditive, 11
    bilinear, 11
    chain, 25
    functorial, 18

neat, 99
universal, 93
mapping cone, 29
Matlis
    dual, 88
    reflexive, 88
maximal
    Cohen–Macaulay, 221
    element, 1
    essential extension, 71
    Gorenstein essential extension, 278
    spectrum, 52
minimal
    $\mathcal{F}$-resolution, 169
    complete $\mathcal{F}$-resolution, 169
    Gorenstein injective extension, 278
    injective resolution, 72
mock finitely generated, 245
module, 7
    FP- injective, 131
    Ω-Gorenstein flat, 312
    Ω-Gorenstein injective, 312
    Ω-Gorenstein projective, 312
    absolutely pure, 131
    artinian, 46
    canonical, 237
    character, 77
    Cohen–Macaulay, 221
    cohomology, 26
    complete, 37
    copure flat, 215
    copure injective, 214
    cotorsion, 117
    dualizing, 235
    faithfully flat, 42, 65
    finitely embedded, 90
    finitely presented, 76
    flat, 40
    free, 8
    Gorenstein flat, 253
    Gorenstein injective, 239
    Gorenstein projective, 246
    graded, 61

homology, 26
indecomposable, 85
injective, 68
irreducible, 59
length of, 49
localization of, 44, 45
maximal Cohen–Macaulay, 221
mock finitely generated, 245
noetherian, 46
projective, 40
pure injective, 112
pure projective, 178
reduced, 241, 249
reflexive, 88
semisimple, 49
simple module, 15
strongly cotorsion, 229
torsion free, 93
unitary, 7
monomorphism, 18
morphism, 17
multiplicative subset, 44

Nakayama lemma, 15, 120
natural transformation, 20
neat map, 99
nilpotent, 15, 58
    T-, 119
    locally, 136
nilradical, 54
noetherian, 46
numbers
    cardinal, 5
    ordinal, 3

opposite category, 19
ordered set
    inductively, 1
    partially, 1
    totally, 1
ordinals, 3
orthogonal class, 152

partial resolution, 204

partially ordered set, 1
perfect, 110
Pontryagin dual, 96
Prüfer domain, 83
precover
    $\mathcal{F}$-, 105, 167
    $\mathcal{L}$-, 277
    Ω-Gorenstein injective, 315
    $\mathcal{V}$-, 308
    $\mathcal{W}$-, 259
    flat, 110
    Gorenstein flat, 288
    Gorenstein injective, 269
    Gorenstein projective, 279
    injective, 120
    projective, 111
    special $\mathcal{F}$-, 153
    torsion free, 94
precovering, 106, 120, 167
preenvelope
    $\mathcal{F}$-, 129, 167
    $\mathcal{L}$-, 250, 262, 283
    Ω-Gorenstein injective, 316
    $\mathcal{V}$-, 259
    flat, 137
    Gorenstein flat, 293
    Gorenstein injective, 270
    Gorenstein projective, 293
    pure injective, 141
    special $\mathcal{F}$-, 153
preenveloping, 129, 167
primary
    $\mathfrak{p}$-, 55, 58
    decomposition, 59
    ideal, 55
    submodule, 58
prime
    embedded, 56
    isolated associated, 55
prime divisors, 56
    minimal, 56
principle
    of transfinite construction, 6

    of transfinite induction, 6
product
    of a family of objects, 21
    of categories, 19
    tensor, 11, 231
projective
    Ω-Gorenstein, 312
    cover, 105, 110, 127
    dimension, 80
    limit, 33
    module, 40
    precover, 111
    pure, 178
    resolution, 40
pullback, 24
pure
    essential extension, 144
    exact, 112
    injective, 112
    projective, 178
    submodule, 94, 112
pure injective
    envelope, 141, 144
    preenvelope, 141
pushout, 22

quasi-Frobenius, 252
quotient
    complex, 27
    topology, 36

radical
    Jacobson, 15
    nil, 54
    of a module, 120
    of an ideal, 54
rank, 220
reduced
    module, 241, 249
    primary decomposition, 59
reflective subcategory, 189
reflexive module, 88
residue field, 46
resolution

$\mathcal{F}$-, 168
    complete, 168
    flat, 41
    free, 10
    Gorenstein flat, 290
    Gorenstein injective, 272
    Gorenstein projective, 280
    injective, 70
    left $\mathcal{G}or\mathcal{F}lat$, 290
    left $\mathcal{G}or\mathcal{I}nj$, 296
    left $\mathcal{G}or\mathcal{P}roj$, 280
    minimal injective, 72
    partial, 204
    projective, 40
    right $\mathcal{G}or\mathcal{F}lat$, 297
    right $\mathcal{G}or\mathcal{I}nj$, 272
resolving class, 209
ring
    artinian, 47
    Cohen–Macaulay, 221
    coherent, 82, 136
    Gorenstein, 211
    graded, 61
    hereditary, 193
    Iwanaga–Gorenstein, 211
    local, 46
    noetherian, 47
    of p-adic integers, 38
    perfect, 110
    quasi-Frobenius, 252
    semihereditary, 83
    semilocal, 57
    semiperfect, 112
    semisimple, 49
    Zariski, 66

Schanuel's lemma, 43, 75, 204
segment of a set, 2
semihereditary ring, 83
semilocal, 57
semiperfect, 112
semisimple, 49
sequence

M-, 217
R-, 217
cauchy, 37
convergent, 37
exact, 9, 27
extended long exact, 173
long exact, 28
short exact, 10
split exact, 11
sets
    cofinal, 32
    directed, 31
    inductively ordered, 1
    isomorphic, 2
    of morphisms, 17
    partially ordered, 1
    totally ordered, 1
    well ordered, 1
short exact sequence, 10
simple, 15
Snake lemma, 10
socle, 73
special
    $\mathcal{F}$-precover, 153
    $\mathcal{F}$-preenvelope, 153
spectrum, 52
    maximal, 52
split exact, 11
stable filtration, 61
strongly
    copure flat, 215
    copure injective, 214
    cotorsion, 229
strongly connected
    contravariantly, 173
    covariantly, 173
subcategory, 18
    coreflective, 186
    full, 18
    reflective, 189
subcomplex, 27
submodule
    essential, 71

pure, 94, 112
    superfluous, 106
subspace topology, 36, 63
superfluous submodule, 106
support, 53
system
    direct, 31
    inductive, 31
    inverse, 33
syzygy, 168

tensor product, 11
    of complexes, 231
topology, 36
    I-adic, 36
    quotient, 36
    subspace, 36, 63
torsion
    group, 35
    module, 93

torsion free
    cover, 93
    group, 35
    module, 93
totally ordered set, 1

universal
    balanced map, 11
    map, 93
    mapping problem, 11
upper bound, 1

von Neumann regular, 194

Wakamatsu lemma, 156
well ordered, 1

zero divisors, 52
Zorn's lemma, 1